Interdisciplinary Applied Mathematics

Volume 28

Problems in engineering, computational science, and the physical and biological sciences are using increasingly sophisticated mathematical techniques. Thus, the bridge between the mathematical sciences and other disciplines is heavily traveled. The correspondingly increased dialog between the disciplines has led to the establishment of the series: *Interdisciplinary Applied Mathematics.*

The purpose of this series is to meet the current and future needs for the interaction between various science and technology areas on the one hand and mathematics on the other. This is done, firstly, by encouraging the ways that mathematics may be applied in traditional areas, as well as point towards new and innovative areas of applications; and, secondly, by encouraging other scientific disciplines to engage in a dialog with mathematicians outlining their problems to both access new methods and suggest innovative developments within mathematics itself.

The series will consist of monographs and high-level texts from researchers working on the interplay between mathematics and other fields of science and technology.

Interdisciplinary Applied Mathematics

Volumes published are listed at the end of the book.

Robert E. Wyatt

Quantum Dynamics with Trajectories

Introduction to Quantum Hydrodynamics

With Contributions by Corey J. Trahan

With 139 Figures

Springer

Robert E. Wyatt
Department of Chemistry and Biochemistry
University of Texas at Austin
Austin, TX 78712
USA
wyattre@mail.utexas.edu

Mathematics Subject Classification (2000): 70-08, 76Y05, 81-02, 81-08

Library of Congress Cataloging-in-Publication Data
Wyatt, Robert E. (Robert Eugene)
Quantum dynamics with trajectories : introduction to quantum hydrodynamics / Robert E. Wyatt.
p. cm. — (Interdisciplinary applied mathematics ; 28)
Includes bibliographical references and index.
ISBN 0-387-22964-7 (acid-free paper)
1. Hydrodynamics. 2. Quantum trajectories. 3. Lagrangian functions. 4. Schrèdinger equation. 5. Quantum field theory. I. Title. II. Interdisciplinary applied mathematics ; v. 28.
QA912.W93 2005
532′.5—dc22 2004059022

ISBN-10: 0-387-22964-7 Printed on acid-free paper.
ISBN-13: 978-0387-22964-5

Printed in the United States of America. (EB)

9 8 7 6 5 4 3 2 1

springeronline.com

Preface

Remarkable progress has recently been made in the development and application of quantum trajectories as a computational tool for solving the time-dependent Schrödinger equation. Analogous methods for stationary bound states are also being developed. Each year, there have been significant extensions and improvements in the basic methodology, and applications are being made to systems with increasing complexity and dimensionality. In addition, novel quantum trajectory methods are being developed for a broad range of dynamical problems, such as mixed quantum–classical dynamics, density matrix evolution in dissipative systems, and electronic nonadiabatic dynamics.

The purpose of this book is to present recent developments and applications of quantum trajectory methods in the broader context of the hydrodynamical formulation of quantum dynamics. Although the foundations of this field were established during the first 25 years following the birth of quantum mechanics, the emphasis until recently was on interpretation rather than prediction. David Bohm's 1952 publications were especially important in establishing the foundations of this field. Methodological developments occurring during and after 1999 have made it possible to use quantum trajectories to evolve wave packets for nonstationary quantum states. The equations of motion for these trajectories are readily derived from the time-dependent Schrödinger equation and are obtained without approximation. These trajectories evolve under the influence of both classical and quantum forces, the latter bringing in all quantum effects.

Although about six chapters of this book deal with Lagrangian quantum trajectories, in which the velocity matches that of the probability fluid, other chapters deal with what will be termed post-Lagrangian quantum trajectories. In the taxonomy of quantum trajectories, Lagrangian ("Bohm-type") trajectories share part of the spectrum with non-Lagrangian trajectories, the latter following extended equations of motion with additional dynamical terms that take into account "slippage" between the probability fluid and the moving grid point. In this book, these non-Lagrangian trajectories are introduced in the broader context of adaptive moving grids, which also play a significant role in classical fluid dynamics. An important advantage of these trajectories is that they eliminate some of the problems that can arise when Lagrangian trajectories are propagated. Other extensions beyond traditional Lagrangian quantum trajectories have to do with stationary states. For these states, "pure" Lagrangian trajectories have what some view as a rather bizarre feature: they do not move at all. Extensions of the formalism to allow for quantum trajectories for which there is nonzero flux are described in the last two chapters.

There are many state-of-the-art topics covered in the following chapters that are unique to this book, including the following (not a complete list):

- The quantum trajectory equations are derived from both the Madelung–Bohmian position space and the Takabayasi phase space viewpoints.
- Detailed discussion is presented on the properties of quantum trajectories near wave function nodes.
- Evidence for Lagrangian chaos in quantum trajectory dynamics is explored, and the connection with quantum vortices is indicated.
- Function and derivative approximation on unstructured moving grids, such as those formed by Lagrangian quantum trajectories, is treated in detail.
- Adaptive moving grids for non-Lagrangian quantum trajectories are developed and applied to several examples.
- Adaptive methods for smoothing over singularities and for linking solution methods in different spatial domains are described.
- Applications to barrier scattering, decay of metastable states, and electronic energy transfer are described.
- Using fits to the density or the log derivative of the density, several methods of approximating the quantum force are described and illustrated.
- Methods are described that allow for one-at-a-time propagation of individual quantum trajectories.
- Quantum trajectory evolution in phase space is described in detail, and a novel trajectory approach for propagation of the density matrix is illustrated.
- Several new approaches to mixed quantum–classical dynamics are described, and each method is illustrated with examples.
- The quantum Navier–Stokes equation is derived, and trajectory computations of the stress tensor are illustrated.
- Methods are described for dealing with problems that can arise when propagating quantum trajectories (the derivative evaluation and node problems).
- Three trajectory approaches to stationary states are described, one of which makes close contact with semiclassical dynamics.

On the pedagogical side, a number of sections in the first half of this book (especially in Chapters 2 and 4) will be accessible to students who have had at least one course in quantum mechanics. (The reading guide in Section 1.15 lists the sections recommended for beginning students.) Some of the early chapters have been used in my graduate quantum mechanics course, and these concepts have also been presented in an undergraduate course in quantum mechanics. A simplifying aspect is that all of the trajectory equations of motion can be derived from the time-dependent Schrödinger equation by using just one elementary mathematical operation, that being differentiation. Students in elementary quantum courses who do not fear differentiation can work through the basic equations in Chapter 2 in about an hour! Also, for those wanting to go further and try it themselves, a quantum trajectory computer program is listed in Appendix 2. To provide context for the more recent developments, 11 historical comments are dispersed through the chapters.

In addition to a thorough discussion of these basic trajectory equations, there is considerable material for more advanced researchers that deals with adaptive moving grids, phase space dynamics and density matrix evolution for dissipative systems, mixed quantum–classical dynamics, and quantum trajectories for stationary states. In addition, there are about 375 references to research publications, one-third of which have appeared in the literature after 1999, and some of these results are included among the 130 figures. In order to provide background material on some topics or to give subsidiary information on specialized topics, 36 boxes containing additional information are included.

The fifteen chapters in this book cover both methodology and applications. Chapter 1 is an overview of everything that follows. It is suggested that the reader first head to Sections 1.1–1.3 for an introduction to quantum trajectories. Sections 1.4–1.13 can then be skimmed. A guide for alternative paths through the chapters is presented in figure 1.2. Chapters 2–4 cover the equations of motion and properties of quantum trajectories, and Chapters 6–9 present applications to model problems. Chapters 3 and 11 deal with phase space dynamics, but Chapter 10 should be read before Sections 11.6–11.8. Quantum trajectories for nonstationary states form the focus for Chapters 2–13, but stationary states are dealt with in Chapters 14 and 15. Problems that can arise in propagating quantum trajectories are described in Chapters 4–6, and resolutions for these problems are described in Chapters 7 and 15.

Some of the material covered in Chapters 2 and 4 overlaps material that is covered in Peter Holland's outstanding 1993 book, *Quantum Theory of Motion*. There is considerable material in Holland's book (such as spin and relativity) that is not covered in this book. However, there is a computational flavor in the present book that is lacking in Holland's book, and there are discussions here of many recent developments that, of course, could not be covered in Holland's book. In some respects, these two books complement one another.

Acknowledgments. I am very grateful for the assistance and encouragement provided by many individuals during the preparation of this book. Special thanks to Corey Trahan for writing outstanding draft versions of Chapters 6 and 7 and for revising an awkward version of Chapter 5. For a draft version of Sections 9.1–9.4, I thank Eric Bittner and Jeremy Maddox. I thank Bill Poirier for many insightful comments about the material in several of the chapters (especially Chapter 14) and for providing notes and preprints that led to some of the material in Sections 15.2 and 15.3. I thank Irene Burghardt for numerous comments that led to improved versions of Chapters 3 and 11. Edward Floyd contributed a number of comments on Chapter 14, for which I am grateful. For comments on the book, thanks to Keith Hughes and Dmytro Babyuk. For stimulating and productive collaborations over the past five years that led to several joint publications, I am indebted to Eric Bittner. In addition, for many enlightening discussions over the past five years about the material that has ended up in these chapters, I thank Gerard Parlant, Irene Burghardt, Brian Kendrick, Sonya Garashchuk, Klaus Moller, Don Kouri, and David Hoffman. For many insightful discussions and for the numerous original contributions to quantum hydrodynamics that my students and postdoctoral scientists have made since 1998, I give special thanks to Courtney Lopreore, Corey Trahan, Keith Hughes, Dmytro Babyuk, Fernando Sales Mayor, Lucas Pettey, Denise Pauler, and Brad Rowland.

For financial support that made possible our research on quantum trajectories, I thank the Robert Welch Foundation and the National Science Foundation.

For advice and assistance during the preparation of this book, I thank Achi Dosanjh, senior editor for mathematics at Springer in New York. In addition, I am grateful for the valuable assistance and expert advice provided by the production staff at Springer.

Cover Illustration. The color illustration on the cover was designed by the visualization experts at the Texas Advanced Computing Center. Reuben Reyes, David Guzman, and Karla Vega are thanked for producing this figure (using the POV-Ray ray tracing suite) and for animating the trajectory results. These quantum trajectories, shown as colored spheres at one time step, are evolving in a four-dimensional phase space, but two spatial coordinates and one momentum coordinate were used for plotting purposes. The trajectory equations of motion for this dissipative quantum system were derived from the modified Caldeira–Leggett equation, which is described in Chapter 11. The trajectories on the right side of the figure are escaping over a barrier, while those near the center are temporarily trapped in a metastable state. The value of the phase space distribution function is largest for the interior spheres and is much smaller for the boundary ones.

Austin, Texas, USA *Robert E. Wyatt*

Contents

Outline of Boxes

Historical Comments with Portraits

Section

Sources for Portraits of Physicists

Chapter 2

Bohm www.fdavidpeat.com/ideas/bohm.htm (photograph taken by Mark Edwards, professional photographer)

Lagrange jeff560.tripod.com/lagrange.jpg (photograph used after contacting Jeff Miller, email: Jeff560@aol.com)

Madelung www.th.physik.uni-frankfurt.de/~jr/gif/phys/madelung.jpg (webpage maintained by Joachim Reinhardt, email: jr@th.physik.uni-frankfurt.de)

Chapter 4

Jacobi www-gap.dcs.st-and.ac.uk/~history/PictDisplay/Jacobi.html (photograph used after contacting John O'Connor, email: joc@st-andrews.ac.uk)

Chapter 11

Kramers www.aip.org/history/newsletter/spring2000/photos.htm (photograph used by permission of the Emilio Segre Visual Archives, American Institute of Physics)

Langevin www.th.physik.uni-frankfurt.de/~jr/gif/phys/langevin.jpg (webpage maintained by Joachim Reinhardt, email: jr@th.physik.uni-frankfurt.de)

Liouville www-gap.dcs.st-and.ac.uk/~history/PictDisplay/Liouville.html (photograph used after contacting John O'Connor, email: joc@st-andrews.ac.uk)

Leggett www.physics.uiuc.edu/People/Faculty/profiles/Leggett/ (photograph used with permission of Anthony Leggett)

Wigner www.montco-pa.com/flaping/html/wigner.html (photograph used with permission of Francis Laping, photojournalist)

Chapter 12

Ehrenfest www.th.physik.uni-frankfurt.de/~jr/gif/phys/ehrenfest.jpg (webpage maintained by Joachim Reinhardt, email: jr@th.physik .uni-frankfurt.de)

Permissions for Use of Figures

Figures from the Journal of Chemical Physics, Physical Review A, Physical Review E, and Physical Review Letters have been used with the permission of the American Institute of Physics.

Figures from Advances in Chemical Physics have been used with the permission of John Wiley & Sons Inc.

Figures from Chemical Physics Letters and Physics Letters A have been used with the permission of Elsevier Ltd.

Figures from Physica Scripta have been used with the permission of The Royal Swedish Academy of Science.

Figures from the Journal of Physics: Condensed Matter have been used with the permission of the Institute of Physics.

Figures from Physica Scripta have been used with the permission of The Royal Swedish Academy of Science.

The figure from The Journal of Physical Chemistry A has been used with the permission of The American Chemical Society.

The figure from Physical Chemistry Chemical Physics has been used with the permission of the Royal Society of Chemistry.

1
Introduction to Quantum Trajectories

Quantum trajectories provide an analytical, interpretative, and computational framework for solving quantum dynamical problems. An overview is presented of methods and applications for both nonstationary and stationary quantum states.

1.1 Dynamics with Quantum Trajectories

The Schrödinger equation for both stationary and nonstationary states may be solved exactly by propagating *quantum trajectories*, at least *in principle*. The probability amplitude and the phase of the wave function are transported along these trajectories and observables may be computed directly in terms of this information. The dynamical theory governing quantum trajectories forms part of the *hydrodynamic formulation of quantum mechanics*, the logical foundations of which were established by de Broglie, Madelung, Bohm, Takabayasi, and others. Distinct from, but related to the hydrodynamic formulation is the *de Broglie–Bohm interpretation of quantum mechanics*. The foundations for these approaches were developed during the period 1926–1954, although almost nothing happened for 25 years between 1927 and 1952 [2.2, 2.3, 2.9, 3.1]. However, after another long gap, new methods for computing quantum trajectories were introduced in 1999, and this helped to create a resurgence of interest in the hydrodynamic formulation. Since then, more robust computational methods have been developed, and quantum trajectories are now being applied to a diverse range of problems, including some that were probably not anticipated by the early workers, including phase space dynamics for open quantum systems, mixed quantum–classical dynamics, and electronic nonadiabatic energy transfer.

Depending on how they are computed, investigations that employ quantum trajectories may be broadly divided into two classes. In the older and quite mature de Broglie–Bohm *analytic approach*, the time-dependent Schrödinger equation (TDSE) is first solved using conventional computational techniques (using space-fixed grids or basis set expansions). Then, individual "particles" or corpuscles are evolved along quantum trajectories $\vec{r}(t)$ with velocities generated by the "ψ-field", $d\vec{r}/dt = (\hbar/m)\,\mathrm{Im}[\vec{\nabla}\ln\psi]$. The patterns developed by these quantum trajectories as they emanate from an ensemble of "launch points" exactly define the history of the system as it evolves from the initial to the final state. The value of the synthetic route is "a means of understanding and exploring quantum behavior, that is, as a heuristic tool" [1.16]. This is the approach that was proposed and developed by de Broglie, Bohm, and others, the purpose of which is *not* to solve the TDSE per se, but to provide insight. Frequently, interpretative terms such as "pilot wave", "ontological status of the wave function", and "hidden variables" are associated with this approach. A survey of the de Broglie-Bohm interpretation is presented by Berndl et al. [1.23], and Tumulka [1.24] has presented a "dialogue" which explores many aspects of this approach. Finally, Dürr's book [1.25] *Bohmsche Mechanik als Grundlage der Quantenmechanik* provides a detailed exposition. Dürr et al. have also extended Bohmian mechanics to quantum field theory [1.34].

Since the time that Bohm's two papers were published, the analytic approach has been used to compute and interpret quantum trajectories for a diverse spectrum of physical processes, including diffraction experiments, barrier and dynamical tunneling, beam scattering from solid surfaces, and currents in molecules and electronic devices. As an example of the synthetic method, Sanz et al. studied the final angular distribution of quantum trajectories for the scattering of He atoms from corrugated Cu(110) surfaces [1.15]. In another example, Wang et al. used quantum trajectories to study the dissociative adsorption of $D_2(v = 0,1,2)$ on Cu(111) surfaces [1.14]. A detailed force analysis was carried out for the unique group of trajectories that make it into the dissociation channel. A third example concerns the analysis by Barker et al. of current distributions in quantum dots [1.22]. Many insights have been gleaned from these and related studies, some of which will be described in the following chapters (see especially Sections 4.10 and 4.11, 13.5–13.7, and 13.8 which deal with chaotic trajectories, vortex dynamics, and dynamical tunneling, respectively).

The second category of quantum trajectory methods follows the *synthetic approach*. Rather than guiding quantum trajectories with a precomputed wave function, *the trajectories and the hydrodynamic fields are computed concurrently, on the fly*. In this approach, wave packets are evolved by propagating ensembles of quantum trajectories, which become the *computational tool* for solving the quantum hydrodynamic equations of motion (QHEM). Rather than being fixed in space as in the Eulerian picture of fluid mechanics, these fluid elements move along with the probability fluid. In the special case that these elements move at the flow velocity of the fluid, this is termed the Lagrangian version of the formalism. The total force guiding each quantum trajectory includes the classical force (from the gradient of the "external" potential) plus the quantum force (from the gradient of the *quantum potential*). The latter depends on the *shape of the density surrounding each trajectory* and brings in all quantum effects. (As mentioned previously, Bohm's work was not especially concerned with this approach; rather, the focus was on "corpuscle propagation" using information gleaned from precomputed wave functions.)

The hydrodynamic formulation in the synthetic approach is considerably more than "just an interpretation"; it presents nontraditional computational techniques for solving quantum dynamical problems in addition to utilizing the descriptive terminology of fluid mechanics. An insightful analysis of trajectory approaches to quantum mechanics and their relationship to "wave pictures" has been presented by Holland [1.33].

The first implementation of the synthetic approach, the *particle method*, was developed by Weiner et al. during the period from the late 1960s through the 1970s. The computational techniques used in these studies were effectively limited to Gaussian wave packets evolving on harmonic potential surfaces [4.17, 4.18]. (Surprisingly, the first few of these papers made no mention of the much earlier work by Bohm.) Stimulated by these studies, a number of workers (including the author of this book!) attempted to use the same computational methods for wave packet evolution on anharmonic potentials, but the algorithms were too immature to permit even short time propagation. Following these early successes and failures, there was another long gap. Then, in 1999, two groups, working independently, published studies in which ensembles of quantum trajectories were evolved on anharmonic potentials. The first of these, by Lopreore and Wyatt [4.1], introduced a computational approach called the *quantum trajectory method*, QTM. The other study, by Sales Mayor, Askar, and Rabitz [4.2], developed what was called *quantum fluid dynamics*, QFD. Although different computational methods were employed, both methods solve the QHEM by evolving ensembles of quantum trajectories. (The main difference between the first QTM and QFD papers was the method used to evaluate spatial derivatives of the density and the velocity, quantities that are needed in the QHEM.)

We have already hinted that the QHEM can be solved in a number of different *pictures* (there are actually an infinite number of possibilities). In the *Eulerian picture*, fluid elements (grid points) are stationary, while in the previously mentioned *Lagrangian picture*, grid points move along trajectories with velocities matching the flow velocity of the probability fluid. In many cases, alternative pictures (referred to as the ALE, or *arbitrary Lagrangian–Eulerian picture*) that allow for *adaptive grid point motion* are more useful than the standard Eulerian or Lagrangian pictures. (Just to set the record straight, the Eulerian and Lagrangian pictures were both introduced by Euler, but it would be confusing to refer to them as "Euler I" and Euler II".) The Lagrangian and ALE pictures, in which trajectories are used to solve QHEM, provide a very different computational approach from those used in conventional formulations quantum dynamics. Perhaps it is worth emphasizing that the QHEM *can be* solved in the Eulerian picture, for example, by discretizing the equations of motion on a rigid space-time lattice. However, the full power and elegance of the hydrodynamic formulation are exposed only by *evolving quantum trajectories* in the dynamical pictures mentioned previously.

Why solve the hydrodynamic equations with *quantum* trajectories? A number of reasons are summarized in Box 1.1, but an appealing feature is that the *exact* quantum mechanical equations are being solved. *The predictive power of the*

Box 1.1. Why run quantum trajectories in the synthetic approach?

1. Exact quantum-dynamical equations of motion are solved.
2. The trajectories follow the evolving probability density.
3. A relatively small number of moving grid points (fluid elements) may be needed.
4. $x(t)$ describes where the "particle" has been, and dx/dt tells where it is going next.
5. The equations of motion bring in elementary dynamical concepts, such as forces.
6. The trajectories can be used for informative analysis, such as phase space plots.
7. New insights may arise, because the trajectories show how the process takes place.
8. New computational approaches can be developed (e.g., for density matrix evolution for mixed states and for mixed quantum–classical dynamics).
9. The computational effort scales linearly with the number of trajectories.
10. Can possibly avoid the traditional exponential scaling of computational effort with respect to the number of degrees of freedom.
11. There are no large basis sets or large fixed grids.
12. There is no need for absorbing potentials at the edges of the grids.

synthetic approach is equivalent to that of conventional quantum mechanics. In addition, these trajectories may provide a very economical way of solving the TDSE, because the trajectories follow the main features of the evolving probability density. With further improvements in the computational methods, quantum trajectory evolution may provide a robust approach for systems with more degrees of freedom than can be handled using fixed-grid methods. Another compelling reason for running quantum trajectories is that we may *gain new insights* into the dynamics. Unlike conventional computational methods, quantum trajectories provide detailed information about *how* the process takes place. These insights may lead to improved algorithms for treating systems of increasing complexity and dimensionality. This has already started to happen, as evidenced by the number of studies being reported each year. It is hoped that the prejudice displayed by some against the use of trajectories in the analytic approach will not carry over to the synthetic approach!

The appealing features of trajectories are certainly well known to those who solve dynamical problems by running classical or semiclassical trajectories. However, it is important to distinguish formally exact quantum trajectory approaches from those that "simulate" quantum dynamics in terms of classical or semiclassical trajectories, even though the latter can provide accurate and insightful results for some problems.

Unfortunately, the propagation of quantum trajectories can be problematic. In fact, using the simplest computational implementations of the basic equations of motion, there are several numerical problems that limit the propagation time of quantum trajectories, at least for some problems. The first of these is the *derivative evaluation problem*. In order to integrate the QHEM for quantum trajectories, spatial derivatives of the hydrodynamic fields are needed at the positions of the moving fluid elements. However, because these elements move at different velocities (even in the absence of an external potential), at each instant of time they form an unstructured grid, even if they were launched from a regular grid. The accurate computation of spatial derivatives, given only the function values at a set of scattered points, is a difficult problem in numerical analysis. Provided that the input fields are relatively smooth, there are several fitting algorithms, such as least squares, that provide reasonable approximate solutions to the derivative problem.

In addition to these fitting methods, there is another approach that is even more effective. Borrowing ideas from classical computational fluid dynamics, it is possible to *design moving grids* with a predetermined structure, including those with equally spaced internal points. The edge points are allowed to move, thus permitting overall translation as well as internal expansion or contraction as needed. Modifications must be made in the equations of motion to allow for these *non-Lagrangian paths*, thus leading to the *moving path transforms* of the original equations of motion. On these structured grids, standard algorithms (such as high-order finite difference or pseudospectral methods) can be used to accurately evaluate the spatial derivatives. This approach effectively solves the derivative evaluation problem.

Two additional problems concern trajectory dynamics in *nodal regions*. Near nodes associated with nonstationary states, the quantum potential undergoes rapid variations over short length scales and is singular at the exact nodal position. The first difficulty that arises in these regions is that the equations of motion for the quantum trajectories become *stiff*, meaning that that the solutions contain modes with vastly different scales in space and time. Fortunately, there are a number of excellent *implicit time integrators for stiff systems*. Although not optimal insofar as the time step is concerned, even the explicit Euler method will provide reasonable results. With use of the appropriate algorithms for stiff differential equations, the *stiff equation problem* is soluble.

The second problem that occurs in nodal regions is much more severe than those mentioned previously. In these regions, quantum trajectories may become difficult or even *impossible* to propagate accurately (using the simplest implementations of the integration algorithms). The trajectories become kinky and unstable, leading to bizarre dynamics and sudden death for the executing computer code. The use of adaptive grids helps significantly, but does not eliminate the problem. Much effort has been spent on developing methods to deal with *the node problem*. Fortunately, several ways have been found to circumvent this problem, including the use of smoothing potentials in nodal regions or the use of hybrid algorithms that combine solutions of the TDSE in the nodal regions with solutions to the hydrodynamic equations in other regions. There have been successful applications of these methods that lead to stable and accurate long-time propagation even when multiple nodes form. However, it turns out that there are other ways to "cope with the node problem", and three promising new methods developed for this purpose will be described in the final chapter of this book. Two of these methods are based on the propagation of *node-free functions.*

The emphasis in this book is on computational implementations of the synthetic route to wave packet propagation using quantum trajectories. *The focus will be on the equations of motion for quantum trajectories, the computation and properties of these trajectories, and their application to physical and chemical problems.* A number of methods are described for propagating these trajectories, and many properties of quantum trajectories are illustrated through applications to model problems. Many of these applications are for one-dimensional model problems, but a few apply to problems in much higher dimensionality. The literature on applications of quantum trajectory methods to stationary quantum states is relatively sparse. However, three approaches devised for these states will be described in Chapters 14 and 15. In the following chapters, quantum trajectory approaches developed during and after 1999 will be applied to a number of processes, including barrier tunneling, electronic transitions, decoherence, and mixed quantum–classical dynamics. In addition, new quantum trajectory approaches will be described for evolving the density matrix and the phase space distribution function for dissipative quantum systems.

Regarding item number 10 in Box 1.1, it is frequently mentioned that conventional computational methods for solving the Schrödinger equation *scale exponentially* with respect to the number of degrees of freedom. As pointed out by Poirier, this is not a fundamental limitation; rather, it is an artifact associated with the use of direct product basis sets [1.18]. Recently, several promising new methods may spell an end to exponential scaling [1.19–1.21]. One of these methods, which combines wavelet basis sets with phase space truncation schemes, has extended the number of degrees of freedom that may be considered from about 6 to about 15 [1.18]. Further extensions are expected in the number of degrees of freedom that can be handled on current computers.

It is not currently known in detail how methods using ensembles of quantum trajectories scale with respect to the number of degrees of freedom. Not many multidimensional studies have been performed (however, there are some examples; see [7.19, 8.2, 8.5]). A more significant feature is that the trajectories follow along with the evolving density. Because of this, it is not necessary to cover large regions where there is little activity, just in case something happens there in the future. In addition, very few trajectories are required in regions where the hydrodynamic fields are relatively smooth. However, when the dynamics become more complex (for example, when nodes and ripples form in the density), additional trajectories are needed to resolve the details. As the dynamics develops in time, trajectories can be added or deleted, depending on the smoothness of the hydrodynamic fields. How many trajectories are required to do the job is highly dependent on the nature of the system under study and on control parameters, such as the average energy of the wave packet. As additional multidimensional examples are explored, it is expected that these scaling issues will become resolved.

The following sections of this chapter provide an overview of the main topics covered in this book, but the more technical parts can be skipped on a first reading.

An overview of two of the topics covered in this book (the use of adaptive grids and the use of fitting techniques to develop approximations to the quantum force) appeared in *Computing in Science and Engineering* [1.1].

1.2 Routes to Quantum Trajectories

As shown on the left side of figure 1.1, for nonstationary quantum states, there are two independent routes to the equations of motion for quantum trajectories. The first of these begins with the wave function expressed in position space and leads to the equations developed by Madelung and Bohm. The second route follows the work of Takabayasi and begins with the quantum distribution function expressed in phase space. We will return to the phase space approach later in this section, after the QHEM are previewed. (Although the goals are different, some of the same equations appear in both the analytic and synthetic approaches. However, there are important equations used in the synthetic approach to trajectory propagation, such as equations 1.13, 1.14, 1.18, and 1.20, which are not needed in the de Broglie–Bohm interpretation.)

There are also several routes to quantum trajectories for stationary states, all of which emphasize the importance of the quantum stationary Hamilton–Jacobi equation. The oldest of these, pioneered by Edward Floyd, leads to *Floydian trajectories*, the second one (developed by Faraggi and Matone) begins with a broad unifying concept, *the quantum equivalence principle*, and the most recent one

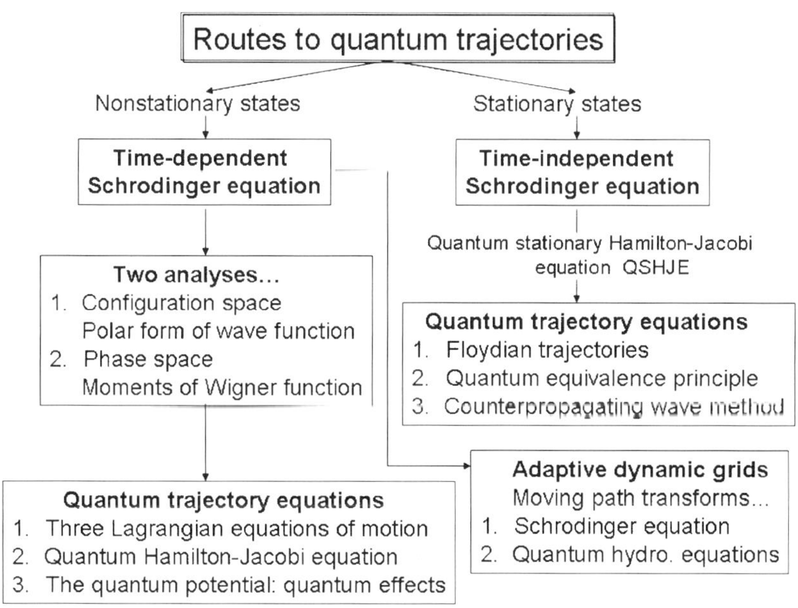

FIGURE 1.1. Routes to quantum trajectories for both stationary and nonstationary states.

(developed by Poirier) is based on expansion of the wave function in terms of *counterpropagating waves*. The latter approach provides a significant reconciliation of the hydrodynamic formulation with semiclassical mechanics. We will delay further mention of these approaches until Sections 1.12 and 1.13.

Starting from the TDSE, Madelung carried out the first derivation of the Eulerian version of the hydrodynamic equations in 1926 [2.9]. This work was greatly extended by Bohm's 1952 papers [2.2, 2.3], which showed how quantum effects, originating from the *quantum potential*, influence the motions of microscopic particles. These topics form the subject of Chapter 2, especially Section 2.2, a brief summary of which will be presented in this section.

Madelung began by writing the complex-valued time dependent wave function in polar form, given by (assuming a one-dimensional example)

$$\Psi(x,t) = R(x,t)e^{iS(x,t)/\hbar}, \tag{1.1}$$

in which both $R(x,t)$, the amplitude, and $S(x,t)$, the *action function,* are real-valued. In addition, and this is quite important, it is assumed that $R(x,t) \geq 0$ at all points. The probability density associated with this wave function is $\rho(x,t) = R(x,t)^2$. In order to develop the Eulerian version of the hydrodynamic equations, we begin by substituting equation 1.1 into the TDSE. After some manipulations, we obtain a system of two *coupled partial differential equations*. The *first equation* is the *continuity equation*

$$\frac{\partial \rho(x,t)}{\partial t} = -\frac{\partial j(x,t)}{\partial x} = -\frac{\partial}{\partial x}\left[\rho(x,t)v(x,t)\right]. \tag{1.2}$$

The probability flux is $j(x,t) = \rho(x,t)v(x,t)$, in which $v(x,t)$, the *flow velocity* of the probability fluid, is determined by the spatial derivative of the action function,

$$v(x,t) = \frac{1}{m}\frac{\partial S(x,t)}{\partial x}. \tag{1.3}$$

The *second equation* that results from this substitution is the *quantum Hamilton–Jacobi equation*, given by

$$-\frac{\partial S(x,t)}{\partial t} = \frac{1}{2m}\left(\frac{\partial S(x,t)}{\partial x}\right)^2 + V(x) + Q(x,t). \tag{1.4}$$

Except for the final term on the right, it has the same form as the classical Hamilton-Jacobi equation for the classical action function. The first term on the right side is the *flow kinetic energy*, the second term is the "classical" potential energy, and the final term is the Bohm *quantum potential.*

Because of its explicit dependence on $\hbar$, *the quantum potential Q brings all quantum effects into the hydrodynamic formulation*. The quantum potential can be expressed in several ways, from the R-amplitude or the probability density:

$$Q(x,t) = -\frac{\hbar^2}{2m}\frac{1}{R}\frac{\partial^2 R}{\partial x^2} = -\frac{\hbar^2}{2m}\rho^{-1/2}\frac{\partial^2 \rho^{1/2}}{\partial x^2}. \tag{1.5}$$

We note that Q depends on the curvature of the R-amplitude, as measured by the second derivative. A significant feature is the presence of R in the denominator.

Near a node in the wave function, the quantum potential becomes very large, unless it happens that the curvature of R is zero at this point. (Such inflection points occur at nodes in wave functions for bound stationary states.) In general, Q becomes very large near a node, and the sign can be either positive or negative.

It will be shown in Chapter 2 that the local kinetic energy associated with a wave packet can be decomposed into *flow kinetic energy* and *shape kinetic energy* [4.3], the latter being determined by the curvature of the amplitude. The shape kinetic energy is the same function that has been identified as the quantum potential. The quantum potential is a measure of the shape induced *internal stress*; this quantity is relieved when the packet "flattens out" as much as possible, subject to constraints imposed by the potential energy.

Prior to Bohm's work, it was appreciated that *quantum mechanics is nonlocal*: every part of a quantum system depends on every other part and is subject to organization by the whole. A significant feature of Bohm's work is that *the quantum potential was identified as the origin of nonlocality*. In addition to the classical force, the quantum potential leads to an additional force, the *quantum force*, acting to guide the trajectory. As a result, in the hydrodynamic formulation *the trajectories lose their independence* and are organized (correlated) by the quantum potential. If Q is neglected in the quantum Hamilton–Jacobi equation, then the resulting classical equations describe a *local theory*.

The preceding equations of motion for $\rho(x, t)$ and $S(x, t)$ have been expressed in the *Eulerian picture*: an observer at a *point fixed in space* watches the wave packet move by. In order to recast these equations into a form appropriate for propagating trajectories, we will take a different viewpoint. If we follow along a (yet to be determined) trajectory $x(t)$ moving with speed v, the rate of change in the function $f(x, t)$ is given by the *total time derivative*

$$\frac{df}{dt} = \frac{\partial f}{\partial t} + \frac{dx}{dt}\frac{\partial f}{\partial x} = \frac{\partial f}{\partial t} + v\frac{\partial f}{\partial x}, \tag{1.6}$$

in which the first term is the rate measured by the stationary observer, and the second term converts us to the moving frame. The observer following along this trajectory is in the *Lagrangian frame*, and equation 1.6 is used to convert time derivatives from the Eulerian frame into the Lagrangian frame.

We will now transform the Eulerian version of the quantum Hamilton–Jacobi equation into an equation of motion for the quantum trajectories. The first step is to transform this into an equation of motion for the spatial derivative of the action, $\partial S/\partial x$. (This procedure, developing *equations of motion for spatial derivatives*, will be significantly extended in Chapter 10.) If we operate on both sides of equation 1.4 with $\partial/\partial x$, and then use equation 1.6 to convert to the Lagrangian frame, we obtain the Newtonian-type equation of motion

$$m\frac{dv}{dt} = -\frac{\partial}{\partial x}(V + Q) = f_c + f_q. \tag{1.7}$$

The right side of this equation shows that the total force guiding the trajectory is the sum of the classical force, $f_c = -\partial V/\partial x$, and the quantum force, $f_q = -\partial Q/\partial x$.

Box 1.2. Summary of equations of motion for quantum trajectories

$$\frac{d\rho(\vec{r},t)}{dt} = -\rho(\vec{r},t)\vec{\nabla}\cdot\vec{v}(\vec{r},t), \tag{1}$$

$$m\frac{d\vec{v}}{dt} = -\vec{\nabla}(V+Q) = \vec{f}_c + \vec{f}_q, \tag{2}$$

$$\frac{dS}{dt} = \frac{1}{2m}\vec{\nabla}S\cdot\vec{\nabla}S - (V(\vec{r}) + Q(\vec{r},t)) = L, \tag{3}$$

$$Q(\vec{r},t) = -\frac{\hbar^2}{2m}\frac{1}{R(\vec{r},t)}\nabla^2 R(\vec{r},t), \tag{4}$$

$$\rho(\vec{r},t) = R(\vec{r},t)^2, \tag{5}$$

$$\vec{v}(\vec{r},t) = (1/m)\vec{\nabla}S(\vec{r},t). \tag{6}$$

This *quantum Newtonian equation*, which describes *how an object moves*, plays a major role in the hydrodynamic formulation.

The *Lagrangian version* of the quantum Hamilton–Jacobi equation, a rate equation for the action function, is determined by the *quantum Lagrangian*, which is evaluated along the trajectory

$$\frac{dS}{dt} = \frac{1}{2m}\left(\frac{\partial S}{\partial x}\right)^2 - (V(x) + Q(x,t)) = L. \tag{1.7}$$

The quantum Lagrangian is the excess of the flow kinetic energy over the total potential energy, the latter now including the quantum potential in addition to the classical potential.

We are now in a position to summarize the three quantum trajectory equations of motion in Box 1.2. The equations of motion in the first half of this box are expressed in the Lagrangian frame, and the three subsidiary equations in the lower half relate the quantum potential and the density to the R-amplitude. In addition, the flow velocity is related to the gradient of the action.

Only two years after Bohm's papers appeared, Takabayasi presented a very different derivation of the hydrodynamic equations [3.1]. Rather than substituting the polar form of the wave function into the TDSE, he began with a quantum-mechanical distribution function, the Wigner function, in (x, p) phase space. This derivation will be presented in Chapter 3. (It was in 1932 that Wigner introduced a quantum phase space distribution function, $W(x, p, t)$, that has *some* features in common with the classical probability distribution [3.12].) Takabayasi's seminal contribution was to develop equations of motion for the *momentum moments* of the Wigner function. These moments are defined by the expression

$$\bar{p}_n(x,t) = \int_{-\infty}^{\infty} p^n W(x,p,t)\,dp. \tag{1.9}$$

The equations of motion for these moments form an infinite hierarchy, with the rate of change of one moment dependent on both lower and higher moments. For pure states, in which the state of the system is described by a wave function, the lowest two moments form a closed set, and the hierarchy terminates. Remarkably, the resulting equations for these moments are identical to the Bohm equations of motion for the probability density and the flow momentum.

The coupled rate equations for the momentum moments are applicable to both *pure states and statistical mixtures*. For *pure states*, when these equations are compared with the Lagrangian equations of motion in Box 1.2, it is revealed that the momentum appearing in the latter equations is the *average momentum* at each position x of the underlying phase space distribution function. Furthermore, for *mixed states*, a generalization of the quantum force has been derived [3.9–3.11] and this depends on the momentum spread of the phase space distribution function. During the course of developing these concepts are developed in Chapter 3, we will develop a deeper understanding of the trajectory equations, and the possibility of significant extensions to more general systems will be revealed (including dissipative systems coupled to a thermal environment). These studies will be described in Chapter 3, related methods for dissipative systems will be presented in Chapter 11, and in Chapter 12 we will use the momentum moments to develop trajectory representations for mixed quantum–classical systems.

1.3 The Quantum Trajectory Method

In the preceding section, we indicated that the equations of motion for quantum trajectories can be derived from two quite different viewpoints. In this section, it will be shown that these equations can be integrated *on the fly* to generate the density, action, and complex-valued wave function for an ensemble of quantum trajectories. In this *synthetic approach* to quantum hydrodynamics, the wave function is not precomputed in advance of the trajectory analysis; rather, the hydrodynamic fields and the quantum trajectories are computed concurrently. The *quantum trajectory method* (QTM), introduced by Lopreore and Wyatt in 1999 [4.1], is a computational implementation of the synthetic approach. The QTM will be described in this section, and many additional details will be presented in Chapter 4.

In the QTM, wave packet evolution is described in terms of a relatively small number of *correlated fluid elements* evolving along quantum trajectories. The QTM is initiated by discretizing the initial wave packet in terms of N *fluid elements*, small chunks of the probability amplitude. The equations of motion for the fluid elements are integrated in lockstep fashion, from one time step to the next. Along each quantum trajectory, the amplitude and action are computed by integrating two coupled equations of motion, and from these, the wave function is readily synthesized. These fluid elements are correlated with one another through the *quantum potential*, from which is derived the quantum force. *Through the quantum potential, each fluid element is influenced by the motions of the other elements, and the ensuing correlation brings all quantum effects into the dynamics.*

There are several ways in which the equations presented in Box 1.2 may be combined to form computational schemes. The *first approach, the force version*, uses the continuity equation, the quantum Newtonian equation, and the quantum Hamilton–Jacobi equation:

$$\frac{d\rho}{dt} = -\rho\vec{\nabla}\cdot\vec{v}, \tag{1.10}$$

$$m\frac{d\vec{v}}{dt} = -\vec{\nabla}V - \vec{\nabla}Q, \tag{1.11}$$

$$\frac{dS}{dt} = L(t) = \frac{1}{2}mv^2 - (V + Q). \tag{1.12}$$

In the early papers on the QTM [4.3–4.7], these equations were used to evolve ensembles of quantum trajectories on several anharmonic potentials. A disadvantage is that spatial derivatives of the quantum potential are needed in equation 1.11, and this brings additional sources of error into the evaluation of the quantum trajectories.

The *second approach, the potential energy version*, again includes the continuity equation and the quantum Hamilton–Jacobi equation. However, the Newtonian-type equation, equation 1.11, is not used at all. As a result, the equations in the second set are:

$$\frac{d\rho}{dt} = -\rho\vec{\nabla}\cdot\vec{v}, \tag{1.13}$$

$$\frac{dS}{dt} = L(t) = \frac{1}{2}mv^2 - (V + Q), \tag{1.14}$$

$$\frac{d\vec{r}}{dt} = \vec{v} = \frac{1}{m}\vec{\nabla}S. \tag{1.15}$$

This set has the advantage that the quantum force is not explicitly evaluated; however, the gradient of the action must be evaluated in order to integrate equation 1.15 to find the quantum trajectories.

If it were not for the spatial derivatives appearing in the three functions $\vec{\nabla}\cdot\vec{v}$, Q, and $\vec{\nabla}S$, the equations of motion would be straightforward to integrate. However, the presence of these terms makes propagation of quantum trajectories a difficult problem in numerical analysis. The reason is that the hydrodynamic fields are available only at the positions of the fluid elements, and these are at locations dictated by the equations of motion. Even if the fluid elements are launched from a regular grid, after a short time interval they will generally form an unstructured grid. Later, in Chapter 5, we will consider in detail several algorithms, including *moving least squares*, which can be used to evaluate these derivatives. Another approach, described in Chapter 7, uses adaptive grids to deal with the derivative evaluation problem.

Along the quantum trajectory $\vec{r}(t)$ leading from an initial point $(\vec{r}_0, t_0)$ to the final point $(\vec{r}, t)$, the rate of change in the density is given by equation 1.10. When the probability density is expressed in terms of the R-amplitude, this equation becomes

$dR/dt = (-1/(2R))\vec{\nabla} \cdot \vec{v}$. Integration of this equation gives the R-amplitude in terms of the value at the initial point

$$R(\vec{r}, t) = \exp\left[-\frac{1}{2}\int_{t_0}^{t}(\vec{\nabla} \cdot \vec{v})\, d\tau\right] R(\vec{r}_0, t_0). \tag{1.16}$$

In addition, from equation 1.12, the exponential of the action at the final point is

$$e^{iS(\vec{r},t)/\hbar} = \exp\left[\frac{i}{\hbar}\int_{t_0}^{t} L(\tau)d\tau\right] e^{iS(\vec{r}_0,t_0)/\hbar}. \tag{1.17}$$

After multiplying these two equations, we obtain an equation for computing the wave function along the trajectory [4.3]

$$\psi(\vec{r}, t) = \exp\left[-\frac{1}{2}\int_{t_0}^{t}(\vec{\nabla} \cdot v)\, d\tau\right] \exp\left[\frac{i}{\hbar}\int_{t_0}^{t} L(\tau)d\tau\right] \psi(\vec{r}_0, t_0). \tag{1.18}$$

The product of the two exponentials is the *hydrodynamic propagator* (as a reminder, both integrals are evaluated along the quantum trajectory). This propagator, along with other aspects of the hydrodynamic theory, has been discussed by Holland [1.33].

Jacobi studied the relationship between transported volume elements and expressed the ratio of the new to the old volume elements in terms of what is now called the *Jacobian*, $dV(t_0 + dt) = J(t_0 + dt, t_0)dV(t_0)$. A slight extension of the material in Box 4.5 shows that the Jacobian evaluated along the quantum trajectory is given by

$$J(t, t_0) = \exp\left[\int_{t_0}^{t} \vec{\nabla} \cdot \vec{v}\, d\tau\right]. \tag{1.19}$$

This equation states that if the velocity field has positive divergence (the velocity vectors "point away from each other"), then the Jacobian (and the volume element) will increase along the flow. Using equation 1.19, the Jacobian may be incorporated into the equation giving the time dependent wave function along the quantum trajectory,

$$\psi(\vec{r}, t) = J(t)^{-1/2} \exp\left[\frac{i}{h}\int_{t_0}^{t} L(\tau)d\tau\right] \psi(\vec{r}_0, t_0) \tag{1.20}$$

This equation is now routinely used to compute the wave function along quantum trajectories.

There are two important "no trespassing" rules that govern the conduct of quantum trajectories [4.16]. *Rule 1:* Quantum trajectories cannot cross through the same point in space-time. Or, put another way, only one quantum trajectory can arrive from the past at a given space-time point. *Rule 2:* Quantum trajectories cannot pass through nodes or nodal surfaces. It is the quantum potential that keeps quantum trajectories away from nodal surfaces.

In semiclassical quantum mechanics, the *initial value representation* (IVR) plays a significant role [4.36]. As the name implies, integrations are performed over "initial values" of the coordinates. A quantum trajectory version of the IVR is very useful for calculating *time correlation functions* [4.37–4.39], one of the simplest being the overlap between a stationary monitor function and a time evolving wave packet:

$$C(t) = \langle\psi(t)|\phi(0)\rangle = \int dx(t)\psi(x(t))^*\phi(x(0)). \tag{1.21}$$

It will be shown in Chapter 4 that the quantum trajectory-IVR version of this *cross-correlation function* is readily obtained by making two substitutions in this equation. The resulting IVR expression is then

$$C(t) = \int dx(0)\psi(x(0))^*\phi(x(0))\, J(x(t), x(0))^{1/2} e^{-iS(x(t),x(0))/\hbar}. \tag{1.22}$$

In this form, *integration is performed only at* $t = 0$ over the positions from which the quantum trajectories were launched. The integrand brings in the overlap distribution between the two functions *only at the initial time,* and this is multiplied by a time-dependent factor, the latter evaluated for the unique trajectory that makes the trip from $x(0)$ to $x(t)$. An important feature is that *Monte Carlo sampling* can be used to select the initial coordinates for the quantum trajectories [4.39].

Nodes or quasi-nodes frequently develop in wave packet scattering, for example, on the "reactant" side of a potential energy barrier. (The term *quasi-node* refers to a region where the amplitude reaches a local minimum and attains a small value but does not become exactly zero.) The quantum potential forces fluid elements away from nodal regions as the amplitude $R(x, t)$ heads toward zero. This increased separation of fluid elements in nodal regions is termed *inflation.* In other regions, typically where the density reaches a local maximum, *compression* occurs when the fluid elements are forced together [4.6]. The net result is that the separations between the fluid elements become very nonuniform. This in turn leads to a *sampling problem*: there is an oversupply of information in regions of compression and an undersupply in inflationary regions near nodes. It is for this reason that adaptive grid methods are very useful, if not essential, in nodal regions; these will be described later, in Section 1.6, and in more detail in Chapter 7.

1.4 Derivative Evaluation on Unstructured Grids

In this section, several approximate methods will be described for computing the spatial derivatives that are required to integrate the equations of motion for quantum trajectories. Many additional details will be provided in Chapter 5. Because part of this book is dedicated to Lagrangian quantum hydrodynamics, we will focus on methods capable of handling multidimensional approximation on unstructured grids. The goal is to find an approximate function denoted by $g(\vec{r})$ that accurately represents the data set $\{\vec{r}_i, f_i\}_{i=1}^{N}$. Once $g(\vec{r})$ is obtained, we will analytically calculate and use its spatial derivatives. There are two general ways for finding $g(\vec{r})$ from a given data set: interpolation and fitting. In *interpolation*, the input function values are *exactly reproduced* by the interpolant. Although the interpolant matches the data at each grid point, its derivatives may not be very accurate, because the approximate function may wildly fluctuate between the input points [5.2]. In contrast, a *functional fit* does not exactly reproduce the data, and it may "smooth" over the input function values. The fit is usually obtained by minimizing an error functional that depends on both the input data and the parameters embedded in $g(\vec{r})$.

The various methods for derivative computation on unstructured grids that will be described later in Chapter 5 are surveyed below.

(a) *Least squares methods.* A powerful approach that is frequently used for fitting multidimensional data on unstructured grids is the least squares technique [5.2]. In the moving least squares (MLS) variant of this algorithm, a *local* fit is evaluated around each of the N data points. In order to develop this fit, we will form a *stencil* involving n_p points surrounding the reference point denoted by $\vec{r}_k$. When various reference points are considered, the patches representing the domains of influence overlap one another, thus providing for very important *grid correlation*.

In order to find the least squares fit, we assume that the input function can be locally expanded in a *basis set* having the dimension $(n_b + 1)$. Each basis function, denoted by $P_j(\vec{r} - \vec{r}_k)$, where $j = 0,1,2,\ldots,n_b$, depends on the locations of both the observation point $\vec{r}$ and the reference point $\vec{r}_k$. The approximate function expanded in terms of these functions is then given by

$$g(\vec{r}) = \sum_{i=0}^{n_b} a_i P_i(\vec{r} - \vec{r}_k). \tag{1.23}$$

Because the local expansion coefficients are evaluated separately around each grid point, this is termed the *moving least squares* method. In order to derive equations for the expansion coefficients, a *variational criterion* is used. For points within the stencil, the *lack of fit* is measured by the sum of the squares of the errors,

$$\Omega_k = \sum_{j=1}^{n_p} \left(f_j - \sum\nolimits_{i=0}^{n_b} a_i P_i(\vec{r}_j - \vec{r}_k) \right)^2. \tag{1.24}$$

In order to make the fit even more local to the reference point, the *weighted errors* can be used,

$$\Omega_k = \sum_{j=1}^{n_p} w_j [f_j - \sum_{i=0}^{n_b} a_i P_i(\vec{r}_j - \vec{r}_k)]^2, \tag{1.25}$$

in which the nonnegative weight function is peaked at the reference point.

The *variational criterion* is then as follows: as we search the $(n_b + 1)$–dimensional coefficient space, we seek the "best" vector that minimizes the variance. To find the vector a_{best}, we set the partial derivatives of the variance to zero and this leads to a matrix equation for the expansion coefficients. Before presenting these equations, we will define two matrices. The P-matrix has rows labeled by grid points and columns labeled by basis functions (the i, j element $P_{i,j} = P_j(\vec{r}_i - \vec{r}_k)$ stores the j-th function evaluated at grid point $\vec{r}_i$). In addition, the diagonal *weight matrix* has nonzero elements only on the diagonal. In terms of these two matrices and the data vector f, we obtain the *normal equations* of least squares analysis, given by $P^t WPa = P^t Wf$. Formally, the solution vector is given by $a = (P^t WP)^{-1} P^t Wf$. This method has been used extensively in applications of the quantum trajectory method, as Chapters 6 and 8 will testify.

Although each algorithm described in this section is capable of providing approximate values for partial derivatives on unstructured grids, only the MLS method has been utilized in quantum trajectory calculations for systems with multiple degrees of freedom (see Chapter 8). However, the computational requirements of the MLS can be demanding, since it requires refitting around each grid point. Because of this, in high-dimensionality, expansions beyond local quadratic forms may not be practical for the matrix-oriented version. Dynamic MLS, described next, may provide enough of a reduction in computational effort to permit the use of larger basis sets.

(b) *Dynamic least squares.* In most applications of least squares, the fitting function $g(\vec{r})$ is expanded in a basis set, and the expansion coefficients are calculated at each time step by solving the corresponding matrix equations. From the *dynamic least squares* (DLS) [5.20, 5.19] viewpoint, the best coefficients are found at the *minimum of an effective potential energy surface.* To find the coefficient values at this minimum, equations of motion are derived for the coefficients, and the coefficients are evolved until relaxation occurs on the effective potential surface. Because the effective potential energy should have a minimum at the "best fit" in equation 1.23, this suggests using the variance given in equation 1.25. As time moves along, equations of motion guide the coefficient vector to seek out the minimum of this time dependent effective potential energy.

Only one study has used the DLS method to obtain solutions for the quantum trajectory equations of motion. In this study [5.20], it was found that the DLS method gave a four to five fold improvement in CPU time compared with the usual matrix least squares method.

(c) *Fitting with distributed approximating functionals (DAFs).* These functionals have been used extensively for fitting, interpolation, extrapolation, solving differential equations, and signal processing. For example, at the reference point x_k, the DAF approximation on a uniform one-dimensional grid is obtained from the expansion [5.26, 5.27]

$$g(x_k) = \sum_{i=k-m}^{k+m} S(x_i - x_k) f_i, \tag{1.26}$$

in which the DAF function $S(x_i - x_k)$ is centered at the i-th point. There is only example for which this formalism was used in the quantum trajectory method [5.29].

(d) *Tessellation-fitting method.* Another derivative approximation scheme, based on tessellation and function fitting, has been developed by Nerukh and Frederick [5.30]. A *tessellation* is a tiling of a region by *simplex figures* (lines, triangles, and tetrahedra in one, two, and three dimensions, respectively) without gaps or overlaps [5.31]. Following tessellation, an approximation to the function is obtained around a reference grid point, from which the spatial derivatives are evaluated. This method was applied to the scattering of an initial Gaussian wave packet from an Eckart barrier augmented by harmonic potentials in one or two additional degrees of freedom [5.30].

(e) *Finite element method.* Finite element techniques [5.32] have been used for the evaluation of spatial derivatives in order to propagate quantum trajectories in two degrees of freedom [5.33]. We will again consider a reference point and a set of neighboring grid points that lie on the *physical grid*. The technique then involves two main steps. (1) *The Mapping Step.* First, this small set of distributed points on the physical grid is mapped to a *computational grid* with the coordinates (ξ, η). This mapping is designed so that the points on the computational grid form a simple geometry (such as a square in two dimensions). (2) *The Interpolation Step.* Next, local interpolation on the finite element in the computational space is performed, and the derivatives at the grid points are evaluated by differentiating this interpolant. In the final step, these derivatives are mapped back into the physical space. This method was used to propagate quantum trajectories in studies of photodissociation dynamics [5.33].

1.5 Applications of the Quantum Trajectory Method

The quantum trajectory method has been used to obtain numerical solutions to the QHEM for a number of wave packet scattering problems (see [6.1–6.15] and the related references [6.16, 6.17]). In the examples presented in Chapters 6 and 8, we will see that the QTM is a very reliable computational method when applied to certain problems, especially when the hydrodynamic fields possess sufficient "smoothness".

In Chapter 6, we will examine quantum trajectory solutions for four relatively simple but illustrative examples. In the first three models, excellent results are obtained using fewer grid points and larger time steps than needed for the direct integration of the TDSE on a fixed grid. The first model problem is certainly the simplest case, the evolution of a free Gaussian wave packet, where each quantum trajectory is governed only by its initial conditions and the quantum force. In the next three examples, the propagation of quantum trajectories under the influence of an external potential is considered. In one example, wave packet scattering from an Eckart barrier, computational difficulties are encountered in regions where the amplitude is reflected from the barrier, and nodes or quasi-nodes occur. These problems arise due to rapid variations in the quantum potential and quantum force near nodal regions. In addition, the fluid elements form an unstructured mesh, and from the viewpoint of derivative evaluation, we would prefer to have the trajectories more equally distributed. In the following section, and in Chapters 7 and 15, new algorithms that alleviate the node and derivative evaluation problems will be described.

In Chapter 8, applications of the quantum trajectory method will be made to three wave packet scattering problems that progressively involve increasing complexity in the dynamics and the number of degrees of freedom. The first example deals with wave packet evolution in a system in which a quantum superposition is prepared [8.1]. We will follow the time dependent hydrodynamic fields for two cases, dependent on whether the system-environment coupling is "on" or "off". Later, in Chapter 13, we will return to this decoherence model and describe the classical and quantum components of the stress tensor, which enter the quantum Navier–Stokes equation describing the rate of change in the momentum density (probability density times momentum).

The second example moves to a system with more degrees of freedom: the time dependent *decay of a metastable state* on a potential surface involving a reaction coordinate coupled to a ten mode harmonic reservoir [8.2]. As in the decoherence example mentioned previously, swarms of quantum trajectories are evolved, and the wave function is synthesized along the individual quantum trajectories. A set of quantum trajectories will be illustrated, and the wave function will be displayed.

The third study is an extension of previous quantum trajectory computations on electronic energy transfer [8.3, 8.4]. In the earlier studies, an ensemble of quantum trajectories was launched on the lower of two coupled potential curves. Transfer of amplitude and phase to the upper curve occurs smoothly and continuously within a localized coupling region. For the example described in Chapter 8, the dimensionality of the problem will be significantly increased [8.5] to allow for the interaction of two eleven-dimensional potential surfaces. On each of these potential surfaces, the QHEM are integrated to find the probability density and action function along each of the trajectories. A series of plots will be presented to illustrate wave packet evolution on each of these coupled potential surfaces.

1.6 Beyond Bohm Trajectories: Adaptive Methods

Derivative evaluation on unstructured grids and trajectory instability near nodal regions create computational difficulties for integration of the QHEM. Fortunately, adaptive methods can be used to solve the first problem and to circumvent the second problem. These methods facilitate the computation of spatial derivatives, remove "kinkiness" in the quantum trajectories, improve stability of the integration scheme in nodal regions, and as a result, permit accurate propagation of trajectory ensembles for longer times. In this section, and in greater detail in Chapter 7, it will be shown how to construct *designer grids* to evolve *non-Lagrangian quantum trajectories*. Using transforms of the hydrodynamic equations to allow for these non-Lagrangian trajectories, it is possible to obtain solutions on *dynamic adaptive grids,* wherein the points move according to user designed algorithms. Another adaptive technique involves *modification of the force* acting on the trajectory to ease passage through nodal regions. With the addition of *artificial viscosity* to the equations of motion, kinkiness near nodes is alleviated, thus permitting propagation to longer times. In addition, it is also possible, during certain time intervals, to solve *different dynamical equations in different spatial regions*. In this hybrid technique, the solution to the TDSE within a nodal region is linked to the solution of the hydrodynamic equations outside of the "ψ-patch". These techniques will be briefly described in the remainder of this section, and a number of examples using these methods for one-dimensional wave packet scattering will be described in Chapter 7.

Dynamical equations and adaptive grids. It is possible to have *absolute control* over the path followed by each fluid element when using the arbitrary Lagrangian–Eulerian (ALE) method. In these *designer grids, the particle velocities are not equal to the flow velocity of the probability fluid.* Similar ALE methods have proven extremely successful in solving classical fluid and solid dynamical problems [7.1–7.2].

In order to allow for arbitrary grid point velocities, the *moving path transforms* of the Lagrangian version of the equations of motion must be derived. This is done by relating the total time derivative along the defined path to the space-fixed partial derivative. Using the *grid velocity* term $\dot{\vec{x}}$, the relationship between these two time derivatives is

$$\frac{d}{dt} = \frac{\partial}{\partial t} + \dot{\vec{x}} \cdot \vec{\nabla}. \tag{1.27}$$

On substitution of this total time derivative into the Eulerian equations of motion, the *moving path transforms of the hydrodynamic equations* become

$$\frac{dS(\vec{r},t)}{dt} = (\dot{\vec{x}} - \vec{v}) \cdot (m\vec{v}) + \frac{1}{2} m\vec{v} \cdot \vec{v} - Q - V, \tag{1.28}$$

$$\frac{d\rho(\vec{r},t)}{dt} = (\dot{\vec{x}} - \vec{v}) \cdot \vec{\nabla}\rho - \rho\vec{\nabla} \cdot \vec{v}. \tag{1.29}$$

These equations have been manipulated to isolate the new term $\vec{w} = \dot{\vec{x}} - \vec{v}$, called the *slip velocity*. Depending on the slip velocity, three cases can arise:

$$\begin{array}{llll} 1. & \dot{\vec{x}} = 0, \quad \vec{w} = -\vec{v}, & & \text{Eulerian grid,} \\ 2. & \dot{\vec{x}} = \vec{v}, \quad \vec{w} = 0, & & \text{Lagrangian grid,} \\ 3. & \dot{\vec{x}} \neq \vec{v} \text{ and } \dot{\vec{x}} \neq 0, & \vec{w} \neq -\vec{v} \text{ and } \vec{w} \neq 0, & \text{ALE grid.} \end{array} \tag{1.30}$$

For case 1, the grid points are fixed in space, and the original Eulerian equations of motion are recovered. For case 2, the grid points are locked in concerted motion with the fluid and move along with the fluid's flow velocity. Fewer grid points are needed for this Lagrangian grid because the quantum trajectories follow the flow. For case 3, the grid points are not fixed, nor do they follow the fluid's flow. In fact, this condition for an ALE grid allows for grid speeds to be assigned arbitrarily. There are many ways to do this, including application of the *equidistribution principle*, wherein the grid points dynamically adapt according to specific features of the solution (i.e., the solution gradient or the curvature). A simpler alternative is to force the grid points to sweep out a regular grid with uniform spacings. It was mentioned earlier that this greatly facilitates the evaluation of spatial derivatives. In Chapter 7, we will describe several studies that utilize the ALE method to obtain solutions to the hydrodynamic equations.

Adaptive grids with the equidistribution principle. Using the equidistribution principle, it is possible to choose the grid speeds so as to *adapt the trajectory locations* to the underlying fields as they evolve in time. The purpose is to guide fluid elements to regions of high activity, including those where the solution gradients and curvatures are large. A number of studies dealing with the numerical solution of evolutionary partial differential equations [7.8–7.10] have distributed the grid points so that a *monitor* function is equally distributed over the field,

$$\int_{x_i}^{x_{i+1}} M(x)dx = \text{constant}, \tag{1.31}$$

or in the discrete form, $M_i(x_{i+1} - x_i) = \text{constant}$. Most monitors are designed to sense specific information about the evolving solutions, and subsequently, use this information to redistribute the positions of the grid points. Unfortunately, however, instantaneous movement of the grid points does not ensure smooth dynamic adaptation of the grid, and the grid paths may oscillate and become numerically unstable in time. This problem can be overcome by using a modification [7.10] that provides for smooth spatial and temporal adaptation.

Using a monitor function based on the wave function curvature, this method was used to solve the *moving-path transform of the TDSE* [7.14], given by

$$i\hbar\frac{d\psi(x,t)}{dt} = -\frac{\hbar^2}{2m}\frac{\partial^2\psi}{\partial x^2} + V(x,t)\psi + i\hbar\dot{x}\frac{\partial\psi}{\partial x}. \tag{1.32}$$

Using this scheme, it was possible to obtain accurate solutions at long propagation times while significantly reducing the number of grid points compared to the standard Eulerian scheme.

Adaptive smoothing in the equations of motion. Although the ALE method helps to capture the solution in nodal regions, kinks in the trajectories and hydrodynamic fields in these regions can be severe. Similar features in classical hydrodynamic fields, such as shock fronts, create numerical instabilities similar to those encountered in the quantum hydrodynamic equations. A standard approach is to introduce a new term into the equations of motion to smooth over these problems, this being the *artificial viscosity*. The purpose of this term is to moderate the effective potential as the particle encounters the disturbance. When added to the trajectory equations of motion, the artificial viscosity softens the quantum force and prevents nodes from fully forming. Using a combination of the ALE method for grid design and artificial viscosity to smooth the kinks, it has been possible to obtain accurate long-time transmission probabilities for several wave packet barrier scattering problems [7.12, 7.19].

Adaptive dynamics in hybrid algorithms. To directly tackle problems near nodes, hybrid algorithms have been introduced [7.15]. They are adaptive in the sense of solving different dynamical equations in different spatial regions. For example, during a certain time interval, the solution of the TDSE in a nodal region can be linked with the solution of the hydrodynamic equations in the node-free regions. In Chapter 7, several successful applications of hybrid methods will be presented.

1.7 Approximations to the Quantum Force

Several methods have been developed for approximating the quantum force. In order to develop these models, a parameterized function is fit to either the density or the log derivative of the density. Because they are formulated in terms of simple sums over information carried by the quantum trajectories, these methods can be readily extended to higher dimensionality. However, the lowest-level fitting models cannot capture variegated features in the density, such as multiple ripples and nodes. These methods and results are described in greater detail in Chapter 9.

Statistical analysis was used Maddox and Bittner [9.1] to obtain a density that is *most likely* to represent data carried by the trajectories. In this scheme, it is assumed that the density can be approximated by summing contributions from M Gaussians,

$$\rho(r) = \sum_{m=1}^{M} p(r, c_m), \tag{1.33}$$

in which $p(r, c_m)$ is the *joint probability* that a randomly chosen fluid element that is located at position r also "belongs to" the m-th Gaussian labeled by c_m. Each Gaussian is parameterized by a *weight* $p(c_m)$, a *mean position vector* μ_m, and a vector of *variances* σ_m (or a full *covariance matrix* C_m if necessary). An iterative procedure, *expectation-maximization* (EM) [9.7], was used to find the parameters that best approximate the input data. This fitted density was then used to compute an approximate quantum force that drives the ensemble of trajectories.

This procedure was used to determine ground state densities, which are generally much simpler than their excited state and nonstationary counterparts. The ground state can be realized from an initial non-stationary state by adding a small *damping term* (to remove the excess kinetic energy) $-\gamma m v$ to the equation of motion,

$$m\dot{v} = -\nabla(V + Q) - \gamma m v. \tag{1.34}$$

As the distribution becomes increasingly narrow, the quantum force becomes very strong and requires the ensemble to maintain a finite width. When equilibrium is reached at longer times, the quantum force exactly counterbalances the classical force, the ensemble no longer evolves in time, and the resulting distribution can be used to compute expectation values. As an example, this method will be used in Chapter 9 to compute the ground state for a two mode model for the stretch-bend states of methyl iodide.

A global method, *least squares fitting to a sum of Gaussians*, was used by Garashchuk and Rassalov [9.2, 9.3] to develop approximations to the quantum force. The global approximation for the density was expressed in terms of a sum of Gaussians, and the parameters were determined by minimization of the least squares error functional

$$J(s) = \int (\rho(x) - f(x;s))^2 \, dx, \tag{1.35}$$

in which s denotes the set of time-dependent parameters. This method was applied to wave packet scattering from an Eckart barrier, where different numbers of Gaussians were used to compute the energy-resolved transmission probability [9.2, 9.3]. However, in these global fitting methods, the total energy is not strictly conserved.

An alternative way of approximating the quantum force is to fit the *log derivative of the density* $\mu(x) = \partial \ln(\rho)/\partial x$, with a parameterized function $g(x;s)$ [9.4, 9.5]. The fitting procedure for $\mu(x)$ can be either global or local, and the resulting fit may be more sensitive to local features of the density. Several examples of the global fitting procedure will be provided later, in Chapter 9. In the *local fitting approach* [9.6], the first step is to divide the coordinate space into a series of *domains.* Associated with each domain is a space fixed *domain function* $\Omega_l(x)$, which may be chosen to reach a large value in the specified domain, but which extends into other domains. The parameters in the fitting function are then evaluated *separately* for each domain. In practice, linear fitting functions with different parameters can be used in each of a small number of domains, although this will not provide high accuracy near nodes. This method has been applied to the collinear hydrogen exchange reaction and to several one-dimensional problems involving anharmonic potentials.

1.8 Propagation of Derivatives Along Quantum Trajectories

The applications of quantum trajectories presented in the preceding sections have employed time-consuming fitting techniques, such as least squares, to evaluate spatial derivatives of the fields around each fluid element in the evolving ensemble. In this section, two related but distinct methods [10.1, 10.12] are described that *circumvent the use of fitting techniques to evaluate spatial derivatives.* In the method of Trahan, Hughes, and Wyatt [10.1], a system of *exact analytic equations of motion* is derived for the spatial derivatives of C and S that appear in the wave function $\psi(r,t) = \exp(C(r,t) + iS(r,t)/\hbar)$. This method is referred to as the DPM, which stands for *derivative propagation method.* In the second method that is based on the propagation of spatial derivatives, that of Liu and Makri [10.12], a system of coupled equations for spatial derivatives of the density are derived from the conservation relation for the weight along a quantum trajectory (the weight is the product of the probability density and the volume element). These derivatives are evaluated in terms of density derivatives at the starting point for the trajectory x_0 along with the *coordinate sensitivities* $\partial^m x(t)/\partial x_0{}^m$, which in turn are evaluated from the *trajectory stability matrix* and its spatial derivatives. This method is referred to as the TSM, the *trajectory stability method.*

In both of these methods, various orders of spatial derivatives are coupled together in an infinite hierarchy, and low-order truncations lead to approximate quantum trajectories. Because function fitting is eliminated, use of these methods leads to orders of magnitude reduction in the computational effort compared with the propagation of an ensemble of linked trajectories. Significantly, *quantum effects can be included at various orders of approximation, there are no basis set expansions, and single quantum trajectories (rather than correlated ensembles) can be propagated, one at a time.* For some problems, including barrier transmission, excellent results have been obtained. However, these approximate quantum trajectories run into problems in capturing interference effects, including those that occur on the "back side" of barrier scattering problems.

As a prelude to the DPM, it is useful to review the one-dimensional version of the Eulerian quantum hydrodynamic equations,

$$\begin{aligned} \frac{\partial C}{\partial t} &= -\frac{1}{2m}\left[S_2 + 2C_1 S_1\right], \\ \frac{\partial S}{\partial t} &= -\frac{1}{2m}(S_1)^2 + \frac{\hbar^2}{2m}\left[C_2 + (C_1)^2\right] - V, \end{aligned} \tag{1.36}$$

in which the order of the spatial derivative is denoted by a subscript, for example $S_n = \partial^n S/\partial x^n$. An alternative to propagation of an ensemble of correlated fluid elements is obtained by *propagation of the required spatial derivatives* along *individual trajectories*. In order to provide a simple illustration, we will directly differentiate equations 1.36 two times with respect to x, giving

$$\begin{aligned} \frac{\partial C_1}{\partial t} &= -\frac{1}{2m}\left[S_3 + 2C_1 S_2 + 2C_2 S_1\right], \\ \frac{\partial C_2}{\partial t} &= -\frac{1}{2m}\left[S_4 + 2C_1 S_3 + 4C_2 S_2 + 2C_3 S_1\right], \\ \frac{\partial S_1}{\partial t} &= -\frac{1}{2m}\left[2S_1 S_2\right] + \frac{\hbar^2}{2m}\left[C_3 + 2C_1 C_2\right] - V_1, \\ \frac{\partial S_2}{\partial t} &= -\frac{1}{2m}\left[2S_1 S_3 + 2(S_2)^2\right] + \frac{\hbar^2}{2m}\left[C_4 + 2C_1 C_3 + 2(C_2)^2\right] - V_2. \end{aligned} \tag{1.37}$$

In these equations, not only is there back-coupling from higher derivatives to lower ones, but there is additional up-coupling; for example, $\partial C_2/\partial t$ is coupled to the higher derivatives S_4, S_3, and C_3. As a result, unless these higher derivatives vanish, the four equations form the beginning of an infinite hierarchy of coupled equations. In order to make progress, we assume that the hydrodynamic fields are smooth enough to be well approximated by low-order polynomial expansions in the displacements around a trajectory. *Truncation of the derivative hierarchy is equivalent to assuming polynomial smoothness at some order.* Conversion of the time derivatives in equations 1.37 from the Eulerian frame to those appropriate for motion along an arbitrary path $r(t)$ is again made through the relation $(d/dt) = (\partial/\partial t) + \dot{r}(t)\cdot\nabla$, where $\dot{r}$ is the *path velocity*. In order to demonstrate some features of the DPM, computational results will be presented in Chapter 10 for two model problems.

A distinct though related derivative propagation scheme has been implemented by Liu and Makri [10.12]. In this scheme, the *probability density is propagated along individual quantum trajectories*. The starting point is the *conservation relation for the weight evaluated along the trajectory*

$$\rho(x,t)dx(t) = \rho(x_0,0)dx(0) = \rho_0(x_0)dx(0). \tag{1.38}$$

Along the trajectory $x(t)$, launched from the starting point x_0, the increment $dx(t)$ may increase or decrease. From this equation, we obtain the density in terms of the initial value and the inverse of the Jacobian evaluated along the trajectory

$$\rho(x,t) = \rho(x_0,0)\frac{\partial x(0)}{\partial x(t)} = \rho(x_0,0)J(x(t),x(0))^{-1}. \tag{1.39}$$

Since the quantum force may be expressed in terms of spatial derivatives of the density, the first task is to evaluate the derivatives of the density. This is done by repeatedly applying $\partial/\partial x$ to both sides of equation 1.39. In the resulting equations, spatial derivatives of the density at time t are expressed in terms of derivatives of this quantity *at the starting point* (these are known from the specified form of the initial wave packet) along with derivatives of the Jacobian. The derivatives of the initial position with respect to the final position are evaluated in terms of the inverse of the Jacobian and its derivatives. Evaluation of these quantities involves calculation of derivatives of the final position reached by the trajectory with respect to the initial position. This brings us to the *trajectory stability matrix* (TSM), which is well known in classical dynamics. There are four sensitivity coefficients, and these are arranged to form the *stability matrix*

$$M(t) = \begin{bmatrix} \dfrac{\partial x(t)}{\partial x(0)} & \dfrac{\partial x(t)}{\partial p(0)} \\ \dfrac{\partial p(t)}{\partial x(0)} & \dfrac{\partial p(t)}{\partial p(0)} \end{bmatrix}. \tag{1.40}$$

An important feature is that the (1,1) element of this matrix is the Jacobian that we are seeking. The stability matrix evolves in time according to the first-order matrix differential equation

$$\frac{dM(t)}{dt} = T(t)M(t) \tag{1.41}$$

(the matrix $T(t)$ will be defined later, in Chapter 10). Differentiation of equation 1.41 leads to the order m sensitivities $\partial^m x(t)/\partial {x_0}^m$. This scheme, in common with the DPM, leads to an infinite hierarchy of coupled equations, which are then truncated at low-order. Quantum effects are approximately captured and quantum trajectories may be propagated one at a time. Using a low-order version of this scheme, energy-resolved transmission probabilities have been computed for the Eckart barrier scattering problem, and these results are in reasonable agreement with the exact results.

The two methods introduced in this section both involve the solution of truncated systems of coupled equations of motion for spatial derivatives of the hydrodynamic fields along individual quantum trajectories. These spatial derivatives bring in *regional nonlocality*, so that the DPM and TSM are not sensitive to *long-range correlations,* which would, for example, allow the trajectory to respond to events (such as node formation) occurring some distance away. Because of this fundamental limitation, the propagation of ensembles of correlated trajectories, as described in the preceding sections, still plays a significant role.

1.9 Trajectories in Phase Space

For *open quantum systems*, such as those coupled to a thermal reservoir, phase space formulations play a key role. In one degree of freedom, for example, the goal is to describe the evolution of the quantum distribution function $W(x, p, t)$ in the two-dimensional phase space characterized by the coordinates $\{x, p\}$. In addition to providing informative pictures of the dynamics, average values can be calculated directly from this distribution function. We will see in Chapter 3 that for an *isolated quantum system*, the time evolution of the Wigner phase space distribution function is governed by the Wigner–Moyal equation [11.15, 11.31]. When the quantum subsystem is *coupled to a thermal environment*, the equations of motion become more complicated. The Caldeira-Leggett equation is a well-known evolutionary equation for the quantum distribution function for an open system [11.30]. This equation and a variant that has two additional smoothing terms, the Diosi equation [11.27, 11.28], are described in Chapter 11.

It has been only within the past few years that quantum trajectories have been used to evolve quantum phase space distribution functions. Donoso and Martens [11.2–11.4] developed a computational method for propagating ensembles of "entangled trajectories" and they used this approach to obtain solutions to the classical Kramers equation and the Wigner equation. In addition, Trahan and Wyatt used a version of the derivative propagation method to develop solutions for several classical and quantum phase space equations of motion [11.1, 10.14].

Classical and quantum-mechanical phase space equations of motion will be described in greater detail in Chapter 11, but a brief overview will be presented in this section. For an *isolated classical system*, the flow of probability density in phase space is addressed by the *Liouville equation* [11.5], which gives the rate of change in the density at a fixed point $\{x, p\}$

$$\frac{\partial W}{\partial t} = -\frac{p}{m}\frac{\partial W}{\partial x} + \frac{\partial V}{\partial x}\frac{\partial W}{\partial p}. \tag{1.42}$$

For *open classical systems*, those that can exchange energy with the surroundings, Kramers derived an important equation [11.7] that governs the evolution of the density for a subsystem in contact with a heat bath maintained at an equilibrium temperature T. The subsystem and the environment are linked through the friction coefficient γ. The Kramers equation is given by

$$\frac{\partial W}{\partial t} = -\frac{p}{m}\frac{\partial W}{\partial x} + \frac{\partial V}{\partial x}\frac{\partial W}{\partial p} + \gamma\frac{\partial}{\partial p}\left[p + mk_B T\frac{\partial}{\partial p}\right]W. \tag{1.43}$$

The first two terms on the right side are formally the same as the Liouville equation. In addition, the first of the two terms involving γ is the *dissipative term* and the second one leads to *momentum diffusion.*

In quantum mechanics, one of several equations that govern the evolution of a phase space distribution function is the Wigner–Moyal (WM) equation [11.15, 11.31, 11.16], which will be introduced in Chapter 3,

$$\frac{\partial W}{\partial t} = -\frac{p}{m}\frac{\partial W}{\partial x} + \frac{\partial V}{\partial x}\frac{\partial W}{\partial p} - \frac{\hbar^2}{24}\frac{\partial^3 V}{\partial x^3}\frac{\partial^3 W}{\partial p^3} + \frac{\hbar^4}{1920}\frac{\partial^5 V}{\partial x^5}\frac{\partial^5 W}{\partial p^5} + O(\hbar^6). \tag{1.44}$$

Unlike the nonnegative classical distribution function, the Wigner function may become negative in some regions of phase space.

The inclusion of dissipative effects in the dynamics of *open quantum systems* is a topic with a rich history [11.21]. The well-known Caldeira–Leggett (CL) equation [11.30] applies in the small friction coefficient, high temperature limit ($k_B T \gg E_0$, where E_0 is the zero-point energy of the uncoupled system). The phase space version of the CL equation has the same form as the Wigner equation, with the addition of the same two friction terms that appear in the Kramers equation. Including only the lowest-order quantum term, this equation has the form

$$\frac{\partial W}{\partial t} = -\frac{p}{m}\frac{\partial W}{\partial x} + \frac{\partial V}{\partial x}\frac{\partial W}{\partial p} - \frac{\hbar^2}{24}\frac{\partial^3 V}{\partial x^3}\frac{\partial^3 W}{\partial p^3} + \gamma\frac{\partial (pW)}{\partial p} + \gamma m k_B T\frac{\partial^2 W}{\partial p^2}. \tag{1.45}$$

In a careful rederivation of the CL equation, Diosi [11.27, 11.28] found two additional smoothing terms that add to the right side of the CL equation. The Wigner, CL, and Diosi equations will be described further in Chapter 11.

In order to develop trajectory equations from the phase space equations of motion [11.2–11.4], we begin by writing each of these equations in the form of a *continuity equation* involving the divergence of the flux vector $\vec{J}$,

$$\frac{\partial W}{\partial t} = -\vec{\nabla}\cdot\vec{J}. \tag{1.46}$$

The phase space gradient operator is $\vec{\nabla} = (\partial/\partial x, \partial/\partial p)$ and the flux is defined by the product of the density and the velocity, $\vec{J} = W\vec{V}$, where $\vec{V} = (V_x, V_p)$ is the phase space velocity vector. The velocity components are defined when the right side of each phase space equation of motion is written in the divergence form given by equation 1.46. A significant feature is that the velocity component in the p direction (which is the force acting on the trajectory) can be decomposed into a classical (local) component along with a nonlocal density dependent term $V_p = dp/dt = F_{local} + F_{nonlocal}(W)$ where F_{local} includes both the classical force $-\partial V/\partial x$ and the dissipative term $-\gamma x$. *The nonlocal force component brings quantum, frictional, and thermal effects into the trajectory dynamics.*

To replace step-by-step propagation of an ensemble of linked trajectories, a different method for propagating classical or quantum phase space trajectories will be described in Chapter 11. Analytic equations of motion for the partial derivatives of the distribution function with respect to x and p will be derived and these quantities will be propagated along the trajectories concurrently with the density itself. This method is an application of the derivative propagation method, which was described in Section 1.8. An enormous benefit is that single trajectories may be propagated, one at a time, and fitting is no longer required to compute the spatial derivatives that are required in the equations of motion.

A topic closely related to evolution of the phase space distribution function is the time dependence of the density matrix $\rho(x, x', t)$. (These two functions are related through a Fourier transform.) In order to develop a trajectory approach, Maddox and Bittner [11.32, 11.33] substituted the polar form of the density matrix into the Caldeira–Leggett equation and derived equations of motion for the amplitude and phase. The evolution of the density matrix was studied by propagating ensembles of correlated trajectories. A novel feature of this approach is that decoherence is associated with the flow of trajectories away from the diagonal axis. As the density matrix becomes more concentrated about the diagonal, the quantum potential increases, which forces trajectories away from the diagonal axis. At equilibrium, squeezing in the off diagonal direction is exactly counterbalanced by the outward force arising from the quantum potential. In Chapter 11, trajectory flow associated with density matrix evolution will be illustrated for several one-dimensional potentials.

1.10 Mixed Quantum–Classical Dynamics

Because of the large number of degrees of freedom in many systems of physical interest, it is not possible to study the wave packet dynamics using conventional quantum-mechanical techniques. This has led to the development of mixed quantum–classical methods that employ a quantum mechanical description of "light" particles such as electrons and protons, and classical trajectory evolution for "heavy" particles such as nuclei. As will be seen in this section and in more detail in Chapter 12, there are consistent and accurate ways of handling this problem that make use of quantum trajectories for the quantum subsystem.

Two different trajectory formalisms for mixed quantum-classical dynamics will be briefly described for a system in which the coordinates for the quantum and classical subsystems are denoted by q and Q, while p and P denote the corresponding momenta. The approach developed by Burghardt and Parlant [12.3] starts with the Wigner distribution in the four-dimensional phase space and leads to equations of motion in a three-dimensional partial hydrodynamic phase space. Because of the starting point in this derivation, this will be referred to as the *phase space method.* In the second approach, developed independently by Prezhdo and Brooksby [12.5] and by Gindensperger, Meier, and Beswick [12.4], we start with the wave function in the two-dimensional *configuration space* and make approximations in the equations of motion for the quantum trajectories. Because of the starting point for this derivation, this approach is referred to as the *configuration space method.*

In the *phase space method* [12.3], we begin with the Wigner distribution function $W(q,p,Q,P,t)$ in the *full phase space*. This four-dimensional phase space is then *contracted* to the three-dimensional *partial hydrodynamic phase space* in which the three independent coordinates are given by (q,Q,P). In this reduction, the quantum momentum, now an average value, becomes a function of the three remaining independent variables. The equations of motion are derived, analogous to the moment methods introduced in Chapter 3, by computing *partial moments* $\bar{p}_n = \langle p^n W \rangle$ of the distribution function $W(q,p,Q,P,t)$ with respect to *only* the quantum momentum p. Assuming that the system is in a *pure state*, there are only two independent momentum moments, the 0-moment, which is the same as the reduced density, and the 1-moment, which is related to the *average value* of the quantum momentum. Although p denotes an *average value*, it is important to emphasize that P is not averaged. The resulting equations of motion in this approach are given by

$$\boxed{Quantum:} \quad \frac{dq}{dt} = \frac{p}{m},$$

$$\frac{dp}{dt} = -\frac{\partial V(q,Q)}{\partial q} + f_q(q,Q,P), \tag{1.47}$$

$$\boxed{Classical:} \quad \frac{dQ}{dt} = \frac{P}{M},$$

$$\frac{dP}{dt} = -\frac{\partial V(q,Q)}{\partial Q}. \tag{1.48}$$

These equations are augmented by the continuity equation

$$\frac{d\rho(q,Q,P)}{dt} = -\rho(q,Q,P)\frac{\partial v(q,Q,P)}{\partial q}, \tag{1.49}$$

in which the flow velocity is $v(q,Q,P) = p(q,\ Q,\ P)/m$. In addition, the hydrodynamic force in equation 1.47 is given by the gradient of the momentum variance,

$$f_q(q,Q,P) = -\frac{1}{m\rho(q,Q,P)}\frac{\partial\sigma(q,Q,P)}{\partial q}. \tag{1.50}$$

in which $\sigma(q,Q,P)$ measures the "thickness" of the momentum distribution with respect to the quantum momentum p. In equations 1.47 and 1.48, it is only through the quantum force that $\hbar$ appears explicitly. For the quantum subsystem, the quantum force on the right side of equation 1.47 is an expected consequence of the reduction from the four-dimensional phase space to the partial hydrodynamic phase space. However, it is important to note that the quantum force and momentum p now carry two additional labels Q and P, which define the projection of the trajectory in the classical subspace.

A relatively simple model involving two coupled harmonic oscillators will be used in Chapter 12 to illustrate this mixed quantum–classical method [12.3]. Unlike other methods for mixed quantum–classical dynamics, this approach provides the *exact* quantum-mechanical solution for this system.

In addition to the phase space method just described, two closely related methods that employ quantum trajectories for mixed quantum–classical dynamics have been developed. The mixed quantum–classical Bohmian (MQCB) method of Gindensperger, Meier, and Beswick [12.4] and the method of Prezhdo and Brooksby [12.5] display a number of similarities, but were derived in different ways and differ in details of implementation. In order to set the stage for these two methods, we begin by applying Bohmian mechanics to a two-degree-of-freedom system. The system is a pure state, and quantum trajectory equations are derived for *both* degrees of freedom. The time-dependent wave function for the system is written in the usual polar form

$$\psi(q,Q,t) = R(q,Q,t)e^{iS(q,Q,t)/\hbar}, \tag{1.51}$$

and this is substituted into the TDSE. After some algebra, we end up with the expected continuity equation and equations of motion for the quantum trajectories. Integration of these equations provides the density and momenta $(p(t), P(t))$ along the quantum trajectory given by $(q(t), Q(t))$. It is important to note that *both* of the momenta, $p(t)$ and $P(t)$, are hydrodynamic (average) values. So far, *no approximations* have been made.

We will now make *several approximations* that lead to equations of motion for a quantum subsystem interacting with a classical subsystem [12.4]. For this purpose, assume that the reduced mass for the classical subsystem is much larger than that for the quantum subsystem and that the second derivatives with respect to the "classical" coordinate Q of the action $S(q,Q,t)$ and the amplitude $R(q,Q,t)$ are *very small* and can be neglected. As a result, the quantum and classical subsystems, with coordinates and momenta now denoted by (q,p) and (Q,P), evolve according to the approximate equations of motion

$$\boxed{Quantum:} \quad \frac{dq}{dt} = \frac{p(q,Q,t)}{m},$$
$$\frac{dp(q,Q,t)}{dt} = -\frac{\partial V(q,Q)}{\partial q} - \frac{\partial U(q,Q,t)}{\partial q}, \tag{1.52}$$

$$\boxed{Classical:} \quad \frac{dQ}{dt} = \frac{P(q,Q,t)}{M},$$
$$\frac{dP(q,Q,t)}{dt} = -\frac{\partial V(q,Q)}{\partial Q} - \frac{\partial U(q,Q,t)}{\partial Q}. \tag{1.53}$$

In addition, the quantum potential (denoted by U to avoid confusion with the classical coordinate) is approximated by

$$U(q,Q,t) = -\frac{\hbar^2}{2m}\rho(q,Q,t)^{-1/2}\frac{\partial^2\rho(q,Q,t)^{1/2}}{\partial q^2}. \tag{1.54}$$

In equations 1.52 and 1.54, the time dependence of the classical coordinate enters as a parameter, and equations 1.53 depend parametrically on the time-dependent coordinate for the quantum trajectory. Note that the gradient with respect to Q of the quantum potential enters the equation for dP/dt, but a further approximation involves neglect of this term. Rather than integrating the continuity equation along the quantum trajectories, as in the quantum trajectory method, Gindensperger et al. elect to directly solve the TDSE for the quantum subsystem, the Hamiltonian of which depends on the time-dependent potential $V(q, Q(t))$. In Chapter 12, several applications of the MQCB method will be described.

1.11 Additional Topics in Quantum Hydrodynamics

Both the stress tensor and vortex dynamics play important roles in both classical [13.1] and quantum fluid dynamics. In the quantum domain, the stress tensor has terms that are formally equivalent to those appearing in the classical dynamical equations along with additional $\hbar$ – dependent quantum contributions. Also in the quantum realm, vortices that form around nodes in the wave function appear in calculations on electron transport in wave guides and molecular wires, rearrangement collisions, and beam scattering from the surfaces of solids. Unlike those in classical fluid dynamics, a remarkable feature of quantum vortices is that they carry quantized *circulation* and angular momentum. In this section, the origin and interpretation of the stress tensor and of quantized vortices will be briefly described. In Chapter 13, a number of examples will be given, especially for quantum vortex dynamics.

The Navier–Stokes equations in classical fluid dynamics express the rate of change in the momentum density (linear momentum times the fluid density) as the sum of two contributions: the force density arising from external fields along with a contribution from "internal" forces, the latter expressed in terms of the stress tensor. The stress tensor appearing in the quantum version of the Navier–Stokes equation includes both classical (not explicitly $\hbar$ – dependent) and quantum components. The classical component depends only on the flow velocity, while the quantum component depends on spatial variations of the density.

The role of the stress tensor in the one dimensional version of the quantum mechanical equations of motion can be easily described. The analysis for this case leads to an equation for the rate of change in the momentum density $(\rho m v)$ expressed in terms of the gradient of the *stress* (a scalar for this case) plus the classical force density

$$\frac{\partial(\rho m v)}{\partial t} = -\frac{\partial \Pi}{\partial x} - \rho \frac{\partial V}{\partial x}. \tag{1.55}$$

This equation is *formally the same* as one of the *Navier–Stokes equations* in classical fluid dynamics. The stress term is given by

$$\Pi = \rho m v^2 + \rho m \left\{ \frac{\hbar^2}{4m^2} \frac{1}{\rho^2} \left(\frac{\partial \rho}{\partial x} \right)^2 \right\} - \frac{\hbar^2}{4m} \frac{\partial^2 \rho}{\partial x^2}. \tag{1.56}$$

The first term on the right, lacking explicit dependence on $\hbar$, is the *classical stress*, the second term is the *quantum stress* and the final term, the *quantum pressure*, is denoted by P. The quantum component of the stress tensor can be compactly expressed in terms of the *osmotic velocity*, $u = -(D/\rho)(\partial \rho/\partial x)$, where the *quantum diffusion coefficient* is $D = \hbar/(2m)$. The stress can then be expressed in terms of the pressure and the two velocities u and v, $\Pi = P + \rho m(v^2 + u^2)$, in which the first and third quantum terms depend explicitly on $\hbar$.

Some ingredients of the quantum Navier–Stokes equation for the momentum density were exposed in the one-dimensional example, but the multidimensional version in Chapter 13 will bring additional terms into the equations. The diagonal elements of the stress tensor produce compressive deformation, and the off-diagonal shear terms lead to squashing deformation of a small fluid element. The stress tensor again has both *classical and quantum components*. These elements are given by

$$\Pi_{ij} = P\delta_{ij} + \rho m\left(v_i v_j + u_i u_j\right) = \rho m v_i v_j + \Pi_{ij}^{quantum}. \tag{1.57}$$

The (quantum) diagonal pressure term is $P = -(\hbar^2/(2m))\nabla^2\rho$ and the flow and diffusive velocity components are given by

$$v_i = \frac{1}{m}\nabla_i S, \quad \text{and} \quad u_i = -\frac{D}{\rho}\nabla_i \rho. \tag{1.58}$$

In common with the one-dimensional example, the quantum components of the stress tensor depend on derivatives of the density, rather than on the components of the flow velocity. In contrast, the "classical" component only depends on components of the flow velocity.

There are very few quantum systems for which the stress tensor has been computed and only one case for which the computation and analysis were carried out using quantum trajectories. In Chapter 8, the quantum trajectory method will be used to analyze the mechanism for suppression of an interference feature in a model two-mode system. In Chapter 13, we will demonstrate the manner in which components of the stress tensor contribute to the total stress for this model [13.18].

We will now turn to the second topic presented in Chapter 13, quantum vortices. In quantum mechanics, *vortices form around nodes in the wave function* (except in one degree of freedom). Quantum trajectories surrounding the vortex core form approximately circular loops, the wave function phase undergoes a 2π winding (or integer multiples thereof), and the circulation integral and angular momentum are quantized. In vortex dynamics, a fundamental role is played by the *circulation integral*, a measure the *strength* of the vortex. For a two degree of freedom problem, let C denote a small closed curve encircling the point (x_0, y_0). The tangent vector to a point on the curve is $\vec{l}(x, y)$. The line integral of the tangential component of the flow velocity is the *circulation integral*

$$\Gamma = \oint_C \vec{v}\cdot d\vec{l} = \frac{1}{m}\oint_C \vec{p}\cdot d\vec{l} = \frac{1}{m}\oint_C \vec{\nabla}S\cdot d\vec{l} = \frac{1}{m}\Delta S, \tag{1.59}$$

where $\Delta S = S_2 - S_1$ is the change in action for one transit around the loop. If the loop *does not* pass through a wave function node, the action is continuous, and the net change around the loop is zero, $\Delta S = 0$. Now assume that the tiny loop C *encircles a node* in the wave function at the point (x_0, y_0). The phase of the wave function is no longer a continuous function around the loop, the net phase change is $\Delta S/\hbar = n \cdot 2\pi$, and from equation 1.59, *the circulation integral is quantized*: $\Gamma = \Delta S/m = n(2\pi)\hbar/m$. The integer n can be positive or negative, the sign determines the *vortex chirality,* and the magnitude measures the *state of excitation* of the vortex.

Examples of quantum vortices, most of them for problems with two degrees of freedom, are described in Chapter 13. In all of these examples, the *analytic approach to quantum hydrodynamics* was followed: the wave function was analyzed after first solving the Schrödinger equation using conventional computational techniques.

1.12 Quantum Trajectories for Stationary States

For a stationary bound state at energy E, since the R-amplitude and the density are independent of time, the continuity equation takes the simplified form

$$\frac{\partial}{\partial x}\left(R(x)^2 \frac{\partial S(x)}{\partial x}\right) = 0. \tag{1.60}$$

It is customary in Bohmian mechanics to take the action to be constant, $S =$ constant (the constant might as well be zero), which does lead to a solution of the continuity equation. However, because this leads to vanishing of the momentum, $p = \partial S/\partial x = 0$, the Bohm trajectories *just sit at one place,* because the classical and quantum forces exactly counterbalance one another. In Chapter 14, we will introduce two related but distinct approaches for stationary states, developed independently by Floyd [14.1] and by Faraggi and Matone (FM) [14.15], in which the quantum trajectories have the appealing feature that they evolve with time. A recent approach for stationary states, which has the appealing feature of combining concepts and techniques from semiclassical mechanics with those of the hydrodynamic formulation, will be described in Section 1.13, and in more detail in Chapter 15 [15.1, 15.2].

Both Floyd and FM emphasize the fundamental role played by the stationary state version of the quantum Hamilton–Jacobi equation. For these states, the action function can be separated into space and time parts, $S(x,t) = W(x) - Et$, where $W(x)$ is the *reduced action,* or Hamilton's *characteristic function*. The quantum Hamilton–Jacobi equation then becomes

$$\frac{1}{2m}\left(\frac{\partial W}{\partial x}\right)^2 + V + Q = E, \tag{1.61}$$

which is referred to as the *quantum stationary Hamilton–Jacobi equation*, QSHJE.

Over 20 years ago, Floyd began developing a novel trajectory approach for *bound stationary states*. This approach is quite different from the Bohmian-hydrodynamic formulation that has been surveyed in the preceding sections. Floyd defined the *energy dependent modified potential* $U = V + Q$, in which Q is the quantum potential. Floyd pointed out that for a given eigenvalue, *there are an infinite number of modified potentials* $U_1, U_2, \ldots$, and associated with each of these is a trajectory $x_1(t), x_2(t), \ldots$. Each *Floydian trajectory* $x_j(t)$ evolves on the corresponding potential U_j and the pair $\{U_j(x), x_j(t)\}$ determines a *microstate* of the Schrödinger equation. Each of these *different microstates specifies the same wave function* and the same eigenvalue. Because the microstates do not arise directly from the Schrödinger wave equation, both Floyd and FM regard the QSHJE as *more fundamental than the* Schrödinger *equation,* in the sense of providing additional information that is "lost" in going to wave function solutions of the Schrödinger equation.

Beginning in the period 1998–1999, Faraggi and Matone proposed and then began exploring the consequences of a *quantum equivalence principle* [14.13]. This deep and far-reaching postulate provides a novel route for developing quantum mechanics. Using *the equivalence of physical systems under coordinate transformations*, they were able to *derive* the form of the quantum potential for stationary states. It is important to appreciate that they did not start from the Schrödinger equation, assume a form for the wave function, and end up with an expression for the quantum potential. Furthermore, they propose that *quantum mechanics itself may arise from this principle.* A key equation in their work, as well as that of Floyd, is the *quantum stationary Hamilton–Jacobi equation*, which Faraggi and Matone *deduce* from the equivalence principle. This formulation also leads to trajectory solutions, and these are analogous to those introduced years earlier by Floyd. The aim of Chapter 14 is to present an introduction to Floydian trajectories and to the equivalence principle of Faraggi and Matone.

1.13 Coping with Problems

In the preceding sections, we have seen that there are two main impediments to the straightforward implementation of quantum trajectory methods: the *derivative evaluation problem* and the *node problem.* An effective way of dealing with the derivative problem will be reviewed in this section. In addition to the approaches introduced in Chapter 7 for dealing with the node problem (hybrid methods and use of the artificial viscosity), Chapter 15 introduces two new methods for handling this issue. These methods, both of which make use of the *superposition principle*, are briefly described later in this section.

Coping with the derivative evaluation problem. In order to propagate quantum trajectories, spatial derivatives of the hydrodynamic fields are needed at the positions of the moving fluid elements. Difficulties with derivative evaluation are exacerbated when there are regions of inflation and compression, which lead to undersampling and oversampling problems. To make matters worse, inflation and compression occur concurrently with the formation of ripples and nodes, regions where the quantum potential undergoes rapid changes in both space and time.

In Chapter 7, the ALE method for developing adaptive moving grids will be described. Rather than using Lagrangian grids with points that move at the flow velocity of the probability fluid, the grid point velocities may be defined to produce *structured grids with uniform grid point spacings* (these are "non-Lagrangian grids"). In order to do this, the trajectories and hydrodynamic fields are propagated concurrently by integrating the *moving-path transforms of the hydrodynamic equations of motion.* In order to permit overall expansion or contraction of the translating grid, it is still convenient to use Lagrangian trajectories for the edge points. However, at each time step, the internal grid points are *forced to be equally spaced.* Spatial derivatives may then be accurately evaluated using any of the standard methods that have been developed for this type of grid.

Coping with the node problem. Surprisingly, there are methods that can be used to *completely circumvent the node problem.* In order to introduce the first two of these methods, let us assume that the time-dependent hydrodynamic fields $R(x, t)$ and $C(x, t)$ show evidence of node formation in some region. If we were to advance further in time, some of these incipient nodes (prenodes) would likely develop into full-fledged nodes. *In order to avoid the nodes that form at later times, we will take advantage of the superposition principle. The wave function at time t is decomposed into components that are node-free.* Each component may be propagated in the hydrodynamic formulation. Two methods for doing this will be previewed in this section, and examples of these decompositions will be demonstrated in Chapter 15.

The first approach, which we will term the *counterpropagating wave method* (CPWM), is based on the analysis of Poirier [15.1, 15.2]. The wave function for the nonstationary state is decomposed into a *carrier wave* multiplied by a real-valued prefactor that, unlike the Bohm preexponential factor $R(x, t)$, *can take on negative values.* The Poirier decomposition is then

$$\psi(x, t) = P(x, t)e^{iS_0(x,t)/\hbar}, \tag{1.62}$$

in which the *carrier momentum* is $p_0 = \partial S_0/\partial x$. It is required that the carrier action function be smooth and *not jump* by $\pm\pi\hbar$ when passing through a node. The real-valued prefactor, which is allowed to become negative, is decomposed into counterpropagating node-free components

$$P(x, t) = q(x, t)\cos[\delta(x, t)/\hbar] = \frac{1}{2}\left[q(x, t)e^{i\delta(x,t)/\hbar} + q(x, t)e^{-i\delta(x,t)/\hbar}\right], \tag{1.63}$$

in which the q-amplitude is smooth, node-free, and nonnegative, $q > 0$. It is clear that if any nodes develop in $P(x, t)$, they must do so because of the oscillating cosine factor. Using this decomposition, the overall wave function may be written

$$\psi(x, t) = \frac{1}{2}\left[q(x, t)e^{iS_+(x,t)/\hbar} + q(x, t)e^{iS_-(x,t)/\hbar}\right] = \psi^{(+)}(x, t) + \psi^{(-)}(x, t). \tag{1.64}$$

The momenta associated with the components $\psi^{(\pm)}$ are given by $p_\pm(x, t) = p_0(x, t) \pm p_\delta(x, t)$, in which $p_\delta = \partial\delta/\partial x$. Relative to an observer riding the carrier wave $\exp(iS_0/\hbar)$, the counterpropagating components move at speeds given by $v_+ = p_\delta/m$ and $v_- = -p_\delta/m$. Each of the two components, $\psi^{(+)}$ and $\psi^{(-)}$ is node-free. However, nodes can develop in ψ due to the superposition of these two components. An example of this decomposition for a nonstationary wave function will be described in Section 15.4. For stationary states, additional detail is provided in Box 15.1.

The second approach for coping with the node problem, termed the *covering function method*, was developed by Babyuk and Wyatt [15.3]. The superposition principle is still involved, but instead of decomposing the wave function $\psi(x,t)$, a new function (called the "total" function) is built by *augmenting* the "actual" function with a "covering" function

$$\psi_T(x,t) = \psi(x,t) + \psi_C(x,t). \tag{1.65}$$

The purpose of the covering function is to "cover up" the nodes in ψ. The total function may have ripples in the region where the actual function had nodes, but, by design, it is guaranteed to be free of nodes.

When prenodes develop and the monitor function signals impending problems, the superposition in equation 1.65 is formed. Then, the node-free functions ψ_T and ψ_C are propagated (using the hydrodynamic equations, not the TDSE) for the time interval T, and the actual function is recovered at the later time

$$\psi(x,t+T) = \psi_T(x,t+T) - \psi_C(x,t+T). \tag{1.66}$$

The function $\psi(x,t+T)$ can be examined for nodes and prenodes by testing the quantum potential to see if it is above a specified threshold value. If there are no indications of problems, then the hydrodynamic fields are propagated by integrating the QHEM. However, if there are still problems with $\psi(x,t+T)$, then this function is covered again, "re-covered", using a new covering function. This sequence is then repeated as many times as necessary to advance the wave function to the target time. In Chapter 15, the CFM will be applied to wave packet scattering problems in one and two degrees of freedom.

The third way to deal with the node problem, termed here the *complex amplitude method* (CAM), was suggested by Garashchuk and Rassolov (GR) [2.32]. The following non-Madelung form is used for the wave function

$$\psi(x,t) = r(x,t)\chi(x,t)e^{is(x,t)/\hbar}, \tag{1.67}$$

in which the *complex-valued function* $\chi(x,t)$ is chosen so that the real-valued amplitude $r(x,t)$ and the action $s(x,t)$ are smooth functions. The amplitude $r(x,t)$ describes the smooth envelope, and $\chi(x,t)$ builds in local features, including nodes. When this form for the wave function is substituted into the TDSE, coupled equations of motion can be derived for the functions $r(x,t)$ and $s(x,t)$. In addition, a method was suggested for finding the amplitude function $\chi(x,t)$. In applications of this method to the propagation of various initial states for the harmonic oscillator and for a Gaussian initial wave packet scattering from an Eckart barrier, a linear approximation was used for $\chi(x,t)$, $\chi(x,t) = a_0(t) + a_1(t)x$. The three methods previewed in this section for handling the node problem are quite promising and may greatly extend the applicability and robustness of the hydrodynamic formulation.

It has been only since 1999 that quantum trajectories have been used as a computational technique for directly solving the time-dependent Schrödinger equation. Since that time, a number of promising methods have been introduced to greatly enhance and extend the basic methodology. Over the coming years, it appears that these and yet unforeseen hydrodynamic approaches may be extended to systems of increasing complexity and dimensionality.

1.14 Topics Not Covered

• *Surreal Bohm trajectories?*
Are Bohm trajectories "real", in the sense of being measurable? In an interesting series of papers, some claim that they are, while others claim otherwise [1.2–1.6]. In the synthetic approach, we will not concerned with whether the individual quantum trajectories followed by fluid elements can somehow be detected.
• *Quantum trajectories and the geometric phase.*
Several references deal with this issue [1.7–1.9]. It would be interesting to integrate quantum trajectories on the fly in systems where the geometrical phase makes its appearance.
• *Intrinsic spin.*
Chapters 9 and 10 in Holland's book provide a thorough introduction to this topic [2.23]. As an example, the scattering of beams of particles with $s = 1/2$ is illustrated.
• *Fermions and bosons.*
Although a few papers deal with formal aspects of identical particles in Bohmian mechanics, there is an absence of computational studies [2.23, 1.10]. Identical particle scattering would be a fruitful area for investigation.
• *Bohmian mechanics and measurements.*
The so-called measurement problem [1.11, 2.23] was of keen interest to David Bohm, and this continues to be a topic of interest.
• *Ginsburg–Landau equation and quantum trajectories.*
The equations of motion for quantum trajectories in superconductors have been derived [1.12], but there have been no on the fly calculations.
• *Semiclassical mechanics.*
This vast topic is beyond the scope of this book, but is well documented in review articles and books [4.36, 1.13].
• *Quantum trajectories and the uncertainty relations.*
McLafferty derived an uncertainty relation for quantum trajectories [1.17] and earlier work on uncertainty and measurement theory is in Holland's book [2.23].
• *Quantum hydrodynamics and density-functional theory.*
The connections between density-functional theory and the hydrodynamic formulation of quantum mechanics have been explored in a number of studies, especially those by Ghosh and Deb [1.26–1.29]. Applications to atomic collision and field ionization processes have also been made [1.30–1.32]. These are fruitful topics for further study.

1.15 Reading Guide

Figure 1.2 shows various paths through Chapters 2–15, depending on the interests and goals of the reader. After skimming Chapter 1, it is recommended that the reader head for Chapters 2 and 4, where the Bohmian route to quantum trajectories is presented and where many properties of these trajectories are described. From Chapter 4, the reader can then proceed to Chapters 6 and 8. Chapter 15 deals with "coping with problems" that arise in computational implementations of the hydrodynamic formulation. This should be the termination of all routes thorough the intervening chapters.

Following Chapters 1 and 2, an alternative path would then lead to Chapter 3 on the phase space route to the hydrodynamic equations. After completing Chapter 10 on derivative propagation methods, the reader would be prepared for phase space dynamics and mixed quantum–classical dynamics in Chapters 11 and 12.

Another path would be to cover Chapters 2, 3, and 4, and then branch to any of the following chapters. However, Chapters 10, 11, and 12 should be read in sequence. Chapters 5 through 9, 13, 14, and 15 can be read in any order after Chapter 4, although Chapter 15 makes some reference to the adaptive methods covered in Chapter 7.

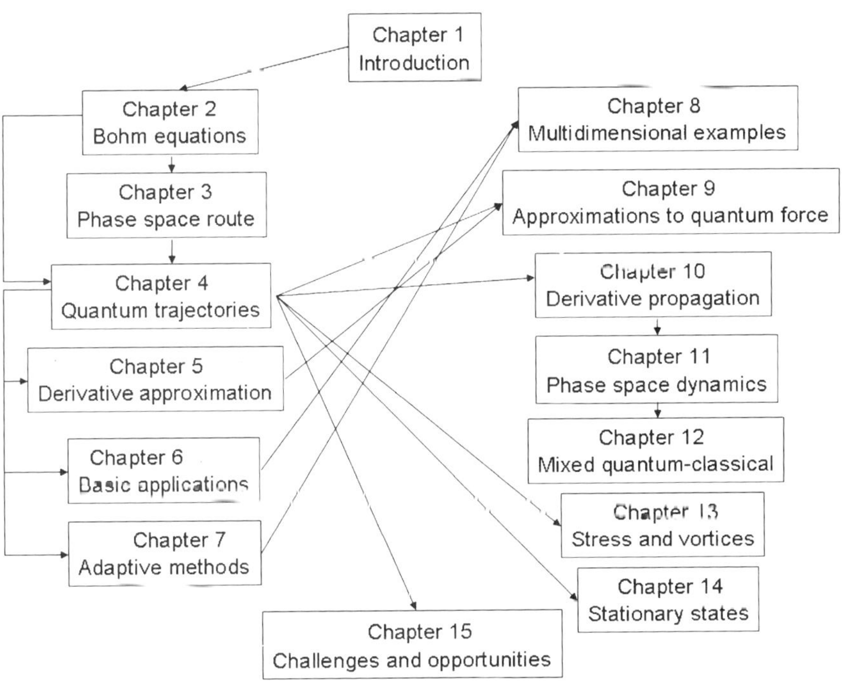

FIGURE 1.2. Alternative paths through the 15 chapters on *Quantum Dynamics with Trajectories*.

Reading guide for students

The following sections provide an introduction to the methods and applications:
Overview: 1.1–1.3, 1.5, 1.6.
Hydrodynamic equations: 2.1–2.3, 2.5, 2.6.
Quantum trajectories: 4.1–4.3, 4.5, 4.7
Elementary applications: 6.1–6.4.
Adaptive dynamic grids: 7.1–7.3

References

1.1. R.E. Wyatt and E.R. Bittner, *Using quantum trajectories and adaptive grids to solve quantum dynamical problems*, Computing in Science and Engineering **5**, 22 (2003).
1.2. B.-G. Englert, M.O. Scully, G. Sussman, and H. Walther, *Surrealistic Bohm trajectories*, Z. Naturforsch. **47** a, 1175 (1992).
1.3. C. Dewdney, L. Hardy, and E.J. Squires, *How late measurements of quantum trajectories can fool a detector*, Phys. Lett. A **184**, 6 (1993).
1.4. M.O. Scully, *Do Bohm trajectories always provide a trustworthy physical picture of particle motion*? Physica Scripta, T **76**, 41 (1998).
1.5. M.O. Terra Cunha, *What is surrealistic about Bohm trajectories?* arXiv:quant-ph/9809006 (3 Sept. 1998).
1.6. Y. Aharonov, B.-G. Englert and M.O. Scully, *Protective measurements and Bohm trajectories*, Phys. Lett. A **263**, 137 (1999).
1.7. C. Philippidis, D. Bohm, and R.D. Kaye, *The Aharonov–Bohm effect and the quantum potential*, Il Nuovo Cimento, **71** B, 75 (1982).
1.8. R.E. Kastner, *Geometrical phase effect and Bohm's quantum potential*, Am. J. Phys. **61**, 852 (1993).
1.9. A. Mostafazadeh, *Quantum adiabatic approximation, quantum action, and Berry's phase*, arXiv:quant-ph/9606021 (19 June 1996).
1.10. H.R. Brown, E. Sjoqvist, and G. Bacciagaluppi, *Remarks on identical particles in de Broglie-Bohm theory*, Phys. Lett. A **251**, 229 (1999).
1.11. J.I. Usera, *An approach to measurement by quantum-stochastic-parameter averaged Bohmian mechanics*, arXiv:quant-ph/0001054 (18 Jan 2000).
1.12. J. Berger, *Extension of the de Broglie–Bohm theory to the Ginsburg-Landau equation*, arXiv:quant-ph/0309143 (19 Sep 2003).
1.13. M. Brack and R.J. Bhaduri, *Semiclassical Physics* (Addison-Wesley, Reading, MA, 1997).
1.14. Z.S. Wang, G.R. Darling, and S. Holloway, *Dissociation dynamics from a de Broglie-Bohm perspective*, J. Chem. Phys. **115** , 10373 (2001).
1.15. A.S. Sanz, F. Borondo, and S. Miret-Artes, *Causal trajectories description of atom diffraction by surfaces*, Phys. Rev. B **61**, 7743 (2000).
1.16. G.E. Bowman, *Bohmian mechanics as a heuristic device: Wave packets in the harmonic oscillator*, Am. J. Phys. **70**, 313 (2002).
1.17. F. McLafferty, *On quantum trajectories and an uncertainty relation*, J. Chem. Phys. **117**, 10474 (2002).
1.18. B. Poirier, *Using wavelets to extend quantum dynamics calculations to ten or more degrees of freedom*, J. Theoret. Comp. Chem. **2**, 65 (2003).
1.19. B. Poirier and J.C. Light, *Efficient distributed Gaussian basis for rovibrational spectroscopy calculation*, J. Chem. Phys. **113**, 211 (2000).

1.20. H.-G. Yu, *Two-layer Lanczos iteration approach to molecular spectroscopic calculation*, J. Chem. Phys. **117**, 8190 (2002).
1.21. X.-G. Wang and T. Carrington, Jr., *A contracted basis-Lanczos calculation of vibrational levels of methane: Solving the Schrödinger equation in nine dimensions*, J. Chem. Phys. **119**, 101 (2003).
1.22. J.R. Barker, R. Akis, and D.K. Ferry, *On the use of Bohm trajectories for interpreting quantum flows in quantum dot structures*, Superlattices and Microstructures **27**, 319 (2000).
1.23. K. Berndl, M. Daumer, D. Dürr, S. Goldstein, and N. Zanghi, *A survey of Bohmian mechanics,* Il Nuovo Cimento **110B**, 735 (1995).
1.24. R. Tumulka, *Understanding Bohmian mechanics: A dialogue,* Am. J. Phys. **79**, 1220 (2004).
1.25. D. Dürr, *Bohmsche Mechanik als Grundlage der Quantenmechanik* (Springer, Berlin, 2001).
1.26. S.K. Ghosh and B.M. Deb, *Densities, density-functionals, and electron fluids*, Phys. Repts. **92**, 1 (1982).
1.27. B.M. Deb and S.K. Ghosh, *Schrödinger fluid dynamics of many-electron systems in a time-dependent density-functional framework*, J. Chem. Phys. **77**, 342 (1982).
1.28. S.K. Ghosh and M. Berkowitz, *A classical fluid-like approach to the density-functional formalism of many-electron systems*, J. Chem. Phys. **83**, 2976 (1985).
1.29. S. Kümmel and M. Brack, *Quantum fluid dynamics from density-functional theory*, Phys. Rev. A **64**, 022506 (2001).
1.30. Vikas and B.M. Deb, *Ground-state electronic energies and densities of atomic systems in strong magnetic fields through a time-dependent hydrodynamical equation*, Int. J. Quantum Chem. **97** , 701 (2004).
1.31. B.K. Dey and B.M. Deb, *Stripped ion–helium atom collision dynamics within a time-dependent quantum fluid density functional theory*, Int. J. Quantum Chem. **67**, 251 (1998).
1.32. A.K. Roy and S.-I Chu, *Quantum fluid dynamics approach for strong-field processes: Application to the study of multiphoton ionization of high-order harmonic generation of He and Ne atoms in intense laser fields*, Phys. Rev. A **65**, 043402 (2002).
1.33. P. Holland, *Computing the wave function from trajectories: particle and wave pictures in quantum mechanics and their relation*, Annals of Physics (NY), to be published.
1.34. D. Dürr, S. Goldstein, R. Tumulka, and N. Zanghi, *Bohmian mechanics and quantum field theory*, Phys. Rev. Lett. **93**, 090402 (2004).

2
The Bohmian Route to the Hydrodynamic Equations

The Madelung–Bohm route to the hydrodynamic equations is developed. Coupled partial differential equations for the amplitude and phase of the wave function are derived from the Schrodinger equation.

2.1 Introduction

For several years during the late 1940s and early 1950s, David Bohm set out to investigate some features that he found troubling in what is now known as the Copenhagen interpretation of quantum mechanics. Bohm's efforts led to his insightful text *Quantum Mechanics*, published in 1951 [2.1]. It is said that as he wrote this text, he became even more uncomfortable with some aspects of the Copenhagen interpretation. These troubling features concerned the completeness of quantum mechanical descriptions of physical systems, the closely related Einstein–Podolsky–Rosen (EPR) gedanken experiment, and quantum measurement theory, especially the issue concerning the so-called collapse of the wave function, a topic that still leaves most beginners (and some experts!) in quantum mechanics perplexed. In any case, during this same period Bohm began working on his new interpretation of quantum theory, and this culminated in two papers that appeared in *Physical Review* in 1952, just after he arrived in São Paulo, Brazil [2.2, 2.3]. Bohm was puzzled by the reaction, actually, the absence thereof, that his papers created in the theoretical physics community. It was the insightful work of John Bell, especially his work that led to what are now referred to as Bell's theorems, that helped to put Bohm's work in a favorable light [2.4]. John Bell, whom some regard as the most influential scholar of quantum mechanics during the last half of the twentieth century, claimed that he was strongly influenced by Bohm's work. This fascinating history is covered in David Peat's biography of David Bohm [2.5], *Infinite Potential*, and in Jeremy Bernstein's biography of John Bell that appears in his book [2.6] *Quantum Profiles*.

There is considerable material about Bohmian mechanics on Sheldon Goldstein's website [2.7]. (Especially interesting is the email correspondence between Sheldon Goldstein and Nobel prize winner Steven Weinberg). The links on this site entitled, Bohmian Mechanics, Collaboration Bohmian Mechanics, and Bohmian Mechanics at Innsbruck are especially recommended. The latter link to the homepage of Gebhard Grubl at Innsbruck leads to nice pictures of Bohm trajectories. The "Collaboration" links several groups, including those of Detlef Dürr and Nino Zanghi. The Dürr website also contains links to material about Bohmian mechanics. An overview of some aspects of the de Broglie–Bohm interpretation appears in Goldstein's two-part series, *Quantum Theory without Observers* [2.30], which appeared in *Physics Today*. Finally, there is even a novel [2.8], *Properties of Light*, whose central character, an aging theoretical physicist, is loosely based on the life of David Bohm. Unfortunately, the (highly improbable) theme revolving around the physicist's beautiful young daughter, also a theoretical physicist, is pure fiction.

It was aptly stated by Bowman that "Bohmian mechanics is at once conservative and radical. It is conservative in its determinism, and in its retention of the fundamental classical concepts of trajectory and particle. Yet it is radical in its rejection of precisely those highly non-classical ideas that are fundamental to the long-dominant Copenhagen interpretation of quantum mechanics" [2.27].

Just four months after Bohm's two papers appeared in *Physical Review*, Takehiko Takabayasi, at Nagoya University, submitted the first of a series of remarkable papers (which are less well known than they deserve to be) in which he developed many of the ideas and equations that now appear in the hydrodynamical formulation of quantum mechanics [2.28]. Readers will appreciate the penetrating insights expressed in his papers. The first of these provides an interpretation of the quantum potential and makes the first connections with phase space ensembles, a topic to which he returned in later papers (this topic will be explored in Chapter 3).

Historical comment. David J. Bohm was born in 1917 in Wilkes-Barre, Pennsylvania. After graduating from Penn State in 1939, he began graduate studies at Cal Tech, but transferred to the University of California, Berkeley, where he received his Ph.D. in 1943 under J. Robert Oppenheimer. For the remainder of the war, he worked on problems in plasma physics at the Radiation Laboratory at Berkeley. After the war, he became an assistant professor at Princeton University. His influential text *Quantum Theory* was published in 1951. In 1950, he was charged with refusing to answer questions before the House Un-American Activities Committee. He was acquitted in 1951, but his days at Princeton were over: the university had already suspended him. Even with Einstein's intervention on his behalf, Princeton refused to renew his contract. While at Princeton, Bohm began to develop a deterministic nonlocal hidden variable theory, which was published in two papers in 1952. After leaving Princeton, Bohm then took a position in physics at the University of São Paulo, Brazil. In 1955, he moved to Israel and spent two years at the Technion in Haifa. He then moved to Britain, occupying a position at the University of Bristol until 1961. In 1959, with his student Yakir Aharonov, he discovered the predecessor of what came to be known as the geometric phase. He then became professor of theoretical physics at Birkbeck College of London University until he retired in 1987. David Bohm remained active in physics until he died in London in 1992.

In Section 2.2, the Madelung–Bohm derivation of the Eulerian version of the quantum hydrodynamic equations of motion (QHEM) is presented. These equations are common to both the de Broglie–Bohm interpretation and the hydrodynamic formulation of quantum mechanics. The difference is the purpose for which they are used (see Section 1.1). One of the equations derived in this early work is related to the classical Hamilton–Jacobi equation, which is described in Section 2.3. The field equations of classical dynamics are described in Section 2.4. These equations are obtained when one term is neglected in the quantum hydrodynamic equations. The Bohm quantum potential and the quantum version of the Hamilton–Jacobi equation are described in Sections 2.5 and 2.6. In Section 2.7, short comments about de Broglie's pilot wave interpretation and Bohm's hidden variable theory conclude this chapter.

2.2 The Madelung–Bohm Derivation of the Hydrodynamic Equations

In his 1926 paper, published only a few months after Schrodinger's celebrated series of six papers, Madelung carried out the first derivation of the hydrodynamic equations [2.9]. He began by writing the complex-valued time-dependent wave function in polar form. In order to obtain the polar form, we graphically represent the complex number $c = a + ib$ (the Cartesian form) as a vector in two-dimensional space with axes "real part" and "imaginary part" (an Argand diagram). The distance from the point c to the origin is $r = (a^2 + b^2)^{1/2}$, and the angle between this vector and the x-axis is $\phi = \tan^{-1}(b/a)$. Thus, the Cartesian and polar forms for c are expressed as

$$c = a + ib = re^{i\phi}. \tag{2.1}$$

For the time-dependent wave function (one space coordinate will be assumed for now) $\Psi(x, t)$, which is a solution to the time-dependent Schrödinger equation (TDSE), the Cartesian and polar forms are related by

$$\Psi(x, t) = A(x, t) + iB(x, t) = R(x, t)e^{iS(x,t)/\hbar}, \tag{2.2}$$

in which both $R(x, t)$, the amplitude, and $S(x, t)$, the *action function*, are real-valued functions. It is important that the amplitude is assumed to be nonnegative at every point: $R \geq 0$. The phase of the wave function is $S(x, t)/\hbar$, and the action function, by design, carries the same units as $\hbar$, namely, energy $\times$ time. Note that the wave function is invariant to a "parallel shift" in the action: $\Psi(x, t)$ does not change under $S \to S + 2\pi n\hbar$, where $n = \pm 1, \pm 2, \ldots$. The reason for using the term "action function" will be described later in this chapter. The probability density associated with this wave function is $\rho(x, t) = \Psi(x, t)^* \Psi(x, t) = R(x, t)^2$. It is frequently convenient to represent the amplitude in exponential form, $R(x, t) = \exp C(x, t)$, where $C(x, t)$ is designated the "C-amplitude" (also see Box 6.1). The wave function then becomes $\Psi(x, t) = \exp[C(x, t) + iS(x, t)/\hbar]$, where $\mathbb{S} = C + iS/\hbar$ is the *complex-phase*. The polar form for the wave function used by Madelung and Bohm is compared with several other forms in Box 2.1 (which can be skipped on a first reading).

Historical comment. Erwin Madelung was born in 1881 in Bonn, Germany. He obtained his Dr. Phil. at the University of Göttingen in 1905. After several years at the Universities of Kiel and Münster, he joined the faculty of the University of Frankfurt, where he served from 1921 until 1950. He is especially known for contributions in solid state physics.

Box 2.1. Other forms for the wave function

The Madelung–Bohm polar form for the complex-valued wave function requires that the amplitude and action function be real-valued. In addition, at every point, $R(x, t) \geq 0$, and near a node in the amplitude, $S(x, t)$ must change by $\pm\pi\hbar$. There are other ways to express $\psi(x, t)$ for nonstationary problems and $\psi(x)$ for stationary states, some of which are listed below.

Nonstationary states

• In Chapter 15, we will use a related form (suggested by Poirier), $\psi(x, t) = P(x, t)\exp[iS_0(x, t)/\hbar]$, where both the amplitude and phase are real-valued [15.1]. However, the P-amplitude will be allowed to change sign, and the continuous function S_0 will not jump by $\pm\pi\hbar$ at nodes. Furthermore, the P-amplitude will be written in trigonometric form, $P(x, t) = q(x, t)\cos\delta(x, t)$, where the amplitude is nonnegative, $q(x, t) > 0$. The latter form may also be used for stationary states.

• In Box 8.4, another form for the wave function, $\psi(x, t) = A(x, t)$ $\exp[iS(x, t)/\hbar]$, will be used, where both the amplitude and the phase may be *complex-valued*. Burant and Tully derived equations of motion for these functions [8.22] and then made approximations before obtaining solutions for model electronic nonadiabatic problems.

• A form for the wave function related to the previous one has been used by Garashchuk and Rassolov [2.32]. They expressed the preexponential complex amplitude in factored form, $A(x, t) = r(x, t)\chi(x, t)$, in which $r(x, t)$ and the action function $s(x, t)$ are real-valued smooth functions and where the complex-valued function$\chi(x, t)$ describes local features, such as nodes. The use of this function will be described further at the end of Section 15.3.

• In Box 2.6, the wave function will be expressed in terms of the complex action, $\psi(x, t) = \exp[iF(x, t)/\hbar]$, and an equation of motion will be presented for $F(x, t)$. Time-dependent WKB approximations [2.10] can be derived by expanding $F(x, t)$ in ascending powers of $\hbar/i$, $F = S_0 + (\hbar/i)S_1 + (\hbar/i)^2$ $S_2 + \ldots$ (The WKB method in quantum mechanics, named after independent work by Wentzel, Kramers, and Brillouin, is a combination of two much older techniques, the phase integral method of Carlini (1783–1862) and the connection formulas derived by Jeffreys (1915).) If we retain only the first two terms of this expansion, the equations of motion are

$$-\partial S_0/\partial t = (1/2m)(\partial S_0/\partial x)^2 + V = H(\partial S_0/\partial x, x), \quad (1)$$

$$-\partial S_1/\partial t = (1/m)(\partial S_0/\partial x)(\partial S_1/\partial x) + (1/2m)\partial S_0^2/\partial x^2. \quad (2)$$

The first equation is the Hamilton–Jacobi equation of classical mechanics. The solution is real-valued in the region covered by classical trajectories. In addition, the second equation is a form of the continuity equation. This equation can be converted to the standard form when it is recognized that the probability density is given by $\rho = \exp[2S_1]$. There are a number of studies dealing with time-dependent WKB approximations, but these will not be our principal concern [2.11–2.14].

Stationary states

(Quantum trajectories for stationary states are the subject of Chapters 14 and 15.)

• For a stationary state, $-\partial S_0/\partial t = E$, and equation 1 becomes the *eikonal equation*,

$$(\partial S_0/\partial x)^2 = 2m(E - V). \tag{3}$$

Micha has developed and applied various eikonal methods, especially for electronic nonadiabatic processes [2.15–2.17]. (In geometrical optics, the eikonal (from the Greek, meaning "image") equation specifies the phase of the wave front, and light rays form orthogonal trajectories to these fronts. The eikonal equation for the phase is $(\vec{\nabla} S)^2 = n(\vec{r})^2$, where the right side is the square of the refractive index of the medium [2.18]. Comparison with equation 3 shows that $[2m(E - V)]^{1/2}$ plays the role of the refractive index.)

• The time-independent Schrödinger equation can be expressed in the form

$$\psi''(x) + G(x)\psi(x) = 0, \tag{4}$$

in which the "prime notation" denotes the second derivative, and where $G(x)$ is energy dependent. If $u_1(x)$ and $u_2(x)$ are linearly independent solutions of this equation satisfying the conditions $u_1(x_0) = 1, u_2(x_0) = 0, u_1'(x_0) = 0, u_2'(x_0) = a$, then we will define the function $w(x) = [u_1^2(x) + u_2^2(x)]^{1/2}$. This function satisfies the *Ermakov–Pinney differential equation* [2.19, 2.20]

$$w''(x) + G(x)w(x) = a^2/w(x)^3. \tag{5}$$

In terms of this function, the general solution to the original equation can then be expressed as

$$\psi(x) = Cw(x)\sin[a\int_{x_0}^{x} w(x)^{-2}dx - D], \tag{6}$$

in which C and D are arbitrary constants and where the argument of the sine function contains a type of *phase integral* [2.21]. The procedure outlined here is *Milne's method* [2.22].

It is important to note that there is an important situation in which the polar form of the wave function is *not useful*. At a node, a point at which $\Psi = 0$, we have $R = 0, C \to -\infty$, and S is undefined. However, for points that are not too close to the node, the polar form can be very useful. The behavior of the time evolving wave packet as a node passes over a *fixed point* is illustrated in figure 2.1. At time t_1, before the node reaches this point, both the real and imaginary parts of the wave function take on negative values, as shown in the top panel. The phase angle $(S/\hbar)$ at this time is about $(7/8) \cdot (2\pi)$, and the R-amplitude is "large" (bottom panel).

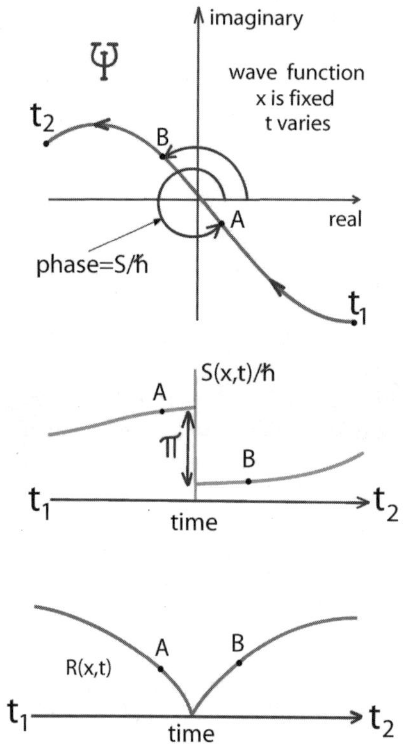

FIGURE 2.1. The complex-valued wave function (top), the phase of the wave function (middle), and the R-amplitude (bottom) versus time as a node passes over a fixed point. Points A and B denote times just before and just after the node passes over this fixed point. Note that the phase suddenly decreases by π as the node passes over this point.

As time increases, the node approaches this fixed point. At point A in the figure, the node is about to arrive and the R-amplitude has decreased from its value at the starting time t_1. As the node passes over this point, R goes to zero, and the phase (middle panel) is undefined. Then, at point B, a short time after the node has passed over the fixed point, R again increases (bottom panel). As time advances and we pass from point A to point B, the phase drops by π, as shown in the middle panel. The phase is well defined for times just before and just after the node passes over the point, but it is not defined at the instant that the node crosses the point. At later times, near the time t_2, the phase is about $(3/8) \cdot 2\pi$ and R is slowly increasing. Also note that for times close to that at which the node passes the fixed point, the R-amplitude forms a typical V-shaped cusp.

We are now ready to continue with the derivation of the hydrodynamic equations. We begin by substituting the polar form for the wave function into the TDSE,

$$\left[-\frac{\hbar^2}{2m}\frac{\partial^2}{\partial x^2} + V(x)\right]\Psi(x,t) = i\hbar\frac{\partial\Psi(x,t)}{\partial t}. \tag{2.3}$$

The resulting equation is then split into two equations for the real and imaginary parts. After some rearrangement, we obtain a system of two coupled partial differential equations. These equations will be described separately.

The first equation of the pair is

$$\frac{\partial R(x,t)^2}{\partial t} = -\frac{\partial}{\partial x}\left[R(x,t)^2 \cdot \frac{1}{m}\frac{\partial S(x,t)}{\partial x}\right]. \tag{2.4}$$

With different notation for the various terms, this equation may be recast into the form of a *continuity equation*. First, we identify $R(x,t)^2$ with the probability density $\rho(x,t)$. In addition, associated with $\Psi(x,t)$ is the *probability flux* (or *current*), given by

$$j(x,t) = \frac{\hbar}{2mi}\left[\Psi(x,t)^* \frac{\partial \Psi(x,t)}{\partial x} - \Psi(x,t)\frac{\partial \Psi(x,t)^*}{\partial x}\right]. \tag{2.5}$$

This quantity gives the rate at which probability flows past a fixed point; the "units" are probability/sec. If we insert the polar form of the wave function into equation 2.5, we obtain the flux in terms of the density and the derivative of the action

$$j(x,t) = \rho(x,t) \cdot \frac{1}{m}\frac{\partial S(x,t)}{\partial x}. \tag{2.6}$$

In classical fluid flow, the flux is given by $j(x,t) = \rho(x,t)v(x,t)$, where $v(x,t)$ is the *flow velocity* of the fluid. In equation 2.6, we will make this association and refer to the *flow velocity of the probability fluid* as the function

$$v(x,t) = \frac{1}{m}\frac{\partial S(x,t)}{\partial x}. \tag{2.7}$$

Now, returning to equation 2.4, the term in brackets on the right side is just the probability flux, so that we finally obtain the standard form of the continuity equation

$$\frac{\partial \rho(x,t)}{\partial t} = -\frac{\partial j(x,t)}{\partial x} = -\frac{\partial}{\partial x}\left[\rho(x,t)v(x,t)\right]. \tag{2.8}$$

This equation asserts that the rate of change in the probability density at a fixed point is proportional to the imbalance (the derivative with respect to x) in the flux. Furthermore, if the flux is decreasing with respect to x, then the density will increase at this point.

The one-dimensional version of the continuity equation was given in equation 2.8. Before leaving the continuity equation, we note that the three-dimensional form of this equation in the Eulerian picture is given by

$$\frac{\partial \rho(\vec{r},t)}{\partial t} = -\vec{\nabla} \cdot \vec{j}(\vec{r},t) = -\vec{\nabla} \cdot \left[\rho(\vec{r},t)v(\vec{r},t)\right]. \tag{2.9}$$

The right side involves the divergence of the flux, which is a vector field. A positive value of the divergence near a point is evident graphically when the flux vectors point away from each other. In terms of the polar form of the wave function, the flux is given by $j(\vec{r},t) = \rho(\vec{r},t)(1/m)\vec{\nabla} S(\vec{r},t)$.

The second equation that results from substituting the polar form of the wave function into the TDSE is (we will continue with a one-dimensional example)

$$-\frac{\partial S(x,t)}{\partial t} = \frac{1}{2m}\left(\frac{\partial S(x,t)}{\partial x}\right)^2 + V(x) + Q(x,t). \tag{2.10}$$

This equation is the Eulerian version of the *quantum Hamilton–Jacobi equation*. The rationale for this designation will be considered in the next section. The first term on the right side, the *flow kinetic energy*, may also be expressed in the form $T_{flow}(x,t) = p(x,t)^2/2m$, where the *flow momentum* is $p(x,t) = mv(x,t) = \partial S(x,t)/\partial x$. The second term on the right side of equation 2.10 is the "classical" potential, and the last term is the Bohm *quantum potential*. We will give an explicit equation for Q in Section 2.5, but for now we will point out that Q is the only term that involves explicit dependence on $\hbar$. Both R (or C) and S also depend implicitly on $\hbar$. It is important to appreciate that equations 2.9 and 2.10 are exact and that they are equivalent to the TDSE.

The preceding equations (expressed in the Eulerian picture) could be solved numerically on a rigid space-time lattice. So far, there is no hint that trajectories could be used to evolve the coupled partial differential equations for $S(x,t)$ and $\rho(x,t)$. The trajectory viewpoint will begin to make its appearance just after equation 2.12 in the next section.

2.3 The Classical Hamilton–Jacobi Equation

If the quantum potential is dropped from the right side of equation 2.10, we obtain the *classical Hamilton–Jacobi equation*

$$-\frac{\partial S_c(x,t)}{\partial t} = \frac{1}{2m}\left(\frac{\partial S_c(x,t)}{\partial x}\right)^2 + V(x). \tag{2.11}$$

In classical Hamiltonian dynamics, the aim is to solve this partial differential equation for $S_c(x,t)$, which is known as the classical action, or *Hamilton's principal function*. In Hamiltonian mechanics, the momentum associated with the moving particle is $p = mv = \partial S_c/\partial x$, so that $S_c(x,t)$ is the *generator* of the classical trajectory. Using this momentum relation in equation 2.11, we associate the right side with the total energy of the trajectory,

$$-\frac{\partial S_c(x,t)}{\partial t} = \frac{p^2}{2m} + V(x) = E. \tag{2.12}$$

We will continue with the classical Hamilton–Jacobi equation after a short detour.

The equations of motion presented so far in this chapter have been expressed in the *Eulerian picture*. An observer watches the unfolding events while standing at a fixed point x. We can always tell that we are in this frame when we see the partial derivative of a function with respect to time, $\partial f(x,t)/\partial t$, which means that this rate is computed at fixed x ($\partial/\partial t$ is the *local rate operator*). There is another viewpoint that is essential for developing trajectory equations of motion. Imagine that an observer follows along a *trajectory* $x(t)$ moving with speed $v = p/m =$

$(1/m)\,\partial S/\partial x$. During the small time interval dt, the change in this function, *as seen by this moving observer*, as she moves from $x(t)$ to $x(t+dt)$, is given by

$$df = \frac{\partial f}{\partial t}dt + \frac{\partial f}{\partial x}dx. \tag{2.13}$$

Dividing by dt then gives the rate of change in the function as seen by the moving observer:

$$\frac{df}{dt} = \frac{\partial f}{\partial t} + \frac{dx}{dt}\frac{\partial f}{\partial x} = \frac{\partial f}{\partial t} + v\frac{\partial f}{\partial x}. \tag{2.14}$$

This observer is moving along in the *Lagrangian picture*, and equation 2.14 is used to convert time derivatives from the Eulerian picture into the Lagrangian picture. In this equation, the last term, arising from a small change in position, is referred to as the *convective term*. This equation plays a key role in the trajectory formulation of quantum mechanics.

In figure 2.2, an imaginary observer follows along the curve $x(t)$ as the clock advances from t to $t+dt$. This observer carefully measures the change in the function, df, and then completes the exercise by dividing this quantity by dt: *rate* $= df/dt$.

We will now return to the classical Hamilton–Jacobi equation in equation 2.11. If we differentiate this equation with respect to x, and use the relations $\partial S/\partial x = mv$ and $\partial^2 S/\partial x^2 = m\,(\partial v/\partial x)$, we obtain the Eulerian version of the *equation of motion*

$$m\left[\frac{\partial}{\partial t} + v\frac{\partial}{\partial x}\right]v = -\frac{\partial V}{\partial x}. \tag{2.15}$$

However, from equation 2.14, we recognize the term in brackets as the Lagrangian time derivative operator $d/dt = \partial/\partial t + v\,\partial/\partial x$. In addition, the right side is

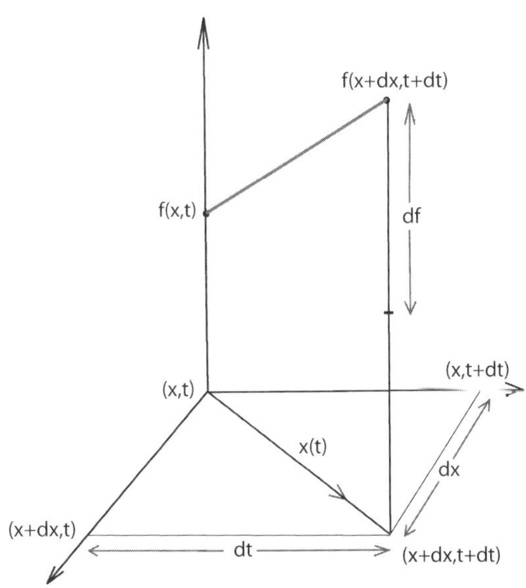

FIGURE 2.2. The observer moves along the curve $x(t)$ and computes df/dt, the rate of change in the function. The observer is moving along at the speed dx/dt and *is* the moving frame.

recognized as the classical force f_c. We have thus completed the link between the classical Hamilton–Jacobi equation and the Newtonian equation of motion for the classical trajectory

$$m\frac{dv}{dt} = f_c. \tag{2.16}$$

There is even more that can be done with the classical Hamilton–Jacobi equation. Again using equation 2.11, we can convert the time derivative of S_c into the Lagrangian frame $\partial S_c/\partial t = dS_c/dt - v\,(\partial S_c/\partial x) = dS_c/dt - mv^2$. After substituting this relation into equation 2.12, we obtain

$$\frac{dS_c}{dt} = \frac{1}{2}mv^2 - V = L_c. \tag{2.17}$$

This equation states that as we move along the classical trajectory $x(t)$, the rate of change in the action is determined by the *classical Lagrangian*, $L_c = T - V$, which is the excess of the kinetic energy over the potential energy. When evaluated numerically, this quantity can be positive, negative, or zero.

Historical comment. Although frequently considered to be a "French mathematician", Joseph-Louis Lagrange was actually born in Turin, Italy, in 1736. However, his scientific work was done in France, where he was widely viewed as "the greatest mathematician of the eighteenth century". His work *Mechanique Analytique* was a masterpiece, although it has the dubious distinction of being the first treatise on mechanics that does not contain a single diagram. Lagrange is also known for his work on the theory of differential equations and variational calculus. He died in Paris in 1813.

If we follow a classical trajectory between a starting point (x_1, t_1) and an ending point (x_2, t_2), the change in action is given by integrating equation 2.17 along the trajectory,

$$S(t_2) = S(t_1) + \int_{t_1}^{t_2} \left(\frac{1}{2} m v^2 - V(x) \right) dt, \tag{2.18}$$

in which it is understood that the integrand is evaluated as a function of time along the trajectory.

There is another way to develop the classical trajectory that is closely related to equation 2.18. Imagine that we hold the starting and ending space-time points fixed. The classical trajectory linking these two points is denoted by $x_c(t)$, and the action accumulated along this trajectory is S_c (this is the integral on the right side of equation 2.18). Now consider a *path* $x(t)$ *not* the same as the classical trajectory, linking the same two points. At each time, the deviation from the classical trajectory is $y(t) = x(t) - x_c(t)$. These quantities are shown in figure 2.3. The action evaluated along this path is the functional $S[x(t)]$. An important theorem, the *principle of extreme action* (Hamilton's principle) states that of all the "rubber band" paths linking these fixed endpoints, the classical trajectory is the one of *extreme action*. This variational principle for the classical trajectory is

$$extremize \left[\int_{t_1}^{t_2} \left(\frac{1}{2} m \left(\frac{dx(t)}{dt} \right)^2 - V(x(t)) \right) dt \right]. \tag{2.19}$$

Extremization of the integral with respect to variation of the path leads to the classical trajectory.

Before completing this section, we will convert the Eulerian version of the continuity equation into the Lagrangian frame. The Eulerian time derivative of

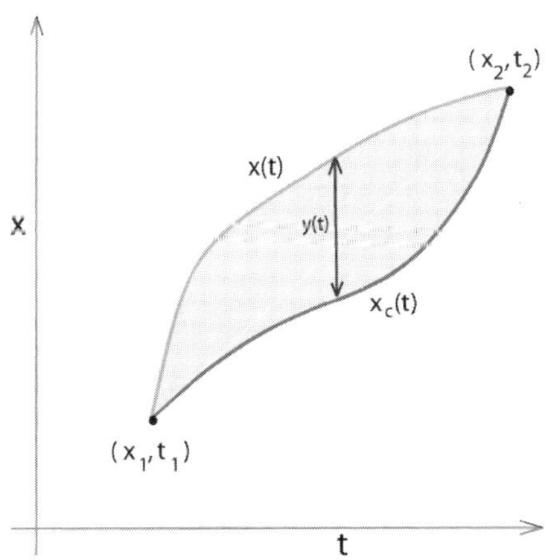

FIGURE 2.3. The classical trajectory $x_c(t)$ and a "distorted path" $x(t)$ linking initial point (x_1, t_1) and final point (x_2, t_2). At each time, the deviation between the path and the trajectory is $y(t)$.

Box 2.2. The classical wave function

On occasion, it is useful to construct "*classical*" *wave functions*. (These wave functions, also called "semiclassical" wave functions, play a major role in semiclassical mechanics.) For example, Haug and Metiu constructed such a function for use in modeling photoabsorption spectra in an alkali atom-He dimmer [2.29]. The classical wave function, in polar form, is written

$$\psi_{cl}(x,t) = R_{cl}(x,t)e^{iS_{cl}(x,t)/\hbar}, \tag{1}$$

in which R_{cl} and S_{cl} are the classical amplitude and action functions. From a swarm of classical trajectories, the density around one of the trajectories is computed, and the amplitude is then given by $R_{cl}(x,t) = \sqrt{\rho_{cl}(x)}$. The action along a trajectory is given by

$$S_{cl}(x,t) = S_{cl}(x_0,t_0) + \int_{t=t_0}^{t} L_{cl}(t)dt, \tag{2}$$

in which $L_{cl}(t)$ is the *classical Lagrangian*, again evaluated along the trajectory. The action can also be computed as an average over the values carried by neighboring trajectories. Although *the dynamics are classical*, the method leads to an approximate quantum probability amplitude.

The classical wave function satisfies the "*classical time-dependent Schrödinger equation*"

$$i\hbar\frac{\partial\psi_{cl}}{\partial t} = -\frac{\hbar^2}{2m}\frac{\partial^2\psi_{cl}}{\partial x^2} + V\psi + \frac{\hbar^2}{2m}\frac{1}{|\psi_{cl}|}\frac{\partial^2|\psi_{cl}|}{\partial x^2}. \tag{3}$$

The final term in this equation is $-Q$ and this has the effect of erasing quantum effects from R_{cl} and S_{cl} when this equation is converted into the *classical hydrodynamic equations of motion*. The classical TDSE is discussed by Holland [2.23].

a function is given by $\partial f/\partial t = df/dt - \vec{v}\cdot\vec{\nabla}f$. When this is used for the left side of equation 2.9 and the right side is expanded, it is seen that the term $\vec{v}\cdot\vec{\nabla}\rho$ cancels from both sides. We are then left with the *Lagrangian form of the continuity equation*, $d\rho/dt = -\rho\vec{\nabla}\cdot\vec{v}$, where the right side brings in the divergence of the velocity field.

Box 2.2 is concerned with a use for the classical action, namely, to build the "classical wave function".

2.4 The Field Equations of Classical Dynamics

The equations developed in the preceding section allow us to formalize the *hydrodynamic field version of classical dynamics* (also see pages 55–58 in [2.23]). Rather than focusing on a single trajectory launched from position $\vec{r}_0$ with velocity

Box 2.3. Field equations of classical dynamics

Lagrangian Frame

$$\frac{d\rho(\vec{r},t)}{dt} = -\rho(\vec{r},t)\vec{\nabla}\cdot v(\vec{r},t), \quad (1)$$

$$m\frac{d\vec{v}}{dt} = -\vec{\nabla}V = \vec{f}_c, \quad (2)$$

$$\frac{dS}{dt} = \frac{1}{2m}\vec{\nabla}S\cdot\vec{\nabla}S - V(\vec{r}) = L_c, \quad (3)$$

$$\vec{v}(\vec{r},t) = \vec{p}(\vec{r},t)/m = (1/m)\vec{\nabla}S(\vec{r},t). \quad (4)$$

$\vec{v}_0$ at time t_0, we launch a large number of trajectories from the set of initial conditions $\{\vec{r}_0, \vec{v}_0\}$. At this initial time, the probability density of trajectories near point $\vec{r}$ is $\rho(\vec{r}, t_0)$. This density is the fraction of the total number of trajectories (N) that are found in a small element of volume dV, divided by this volume: $\rho(\vec{r}, t_0) = dN(\vec{r}, t_0)/(N \cdot dV)$. As time proceeds, individual trajectories evolve according to the equations of motion, and the density of trajectories near this point changes to $\rho(\vec{r}, t)$. Associated with the flow of this probability fluid near point $\vec{r}$ are two additional fields, the flow velocity $v(\vec{r}, t)$ and the action function $S(\vec{r}, t)$. The two coupled fields $\rho(\vec{r}, t)$ and $S(\vec{r}, t)$ provide a complete description of this probability fluid. If we now follow the flow in the Lagrangian picture, these two fields evolve according to the equations that are summarized in Box 2.3. Although it may not seem so important at this point, it is very significant that the Newtonian equation for the evolution of each trajectory (the second equation in the box) does *not involve the density of trajectories*. The cloud consists of a *collection on noninteracting points* moving along through space-time under the influence of (only) the external potential $V(\vec{r})$. Each of these trajectories responds to a *local force that is independent of the probability density*.

2.5 The Quantum Potential

There are a number of comments to make about the quantum Hamilton–Jacobi equation given in equation 2.10. As mentioned earlier, because of its explicit dependence on $\hbar$, the quantum potential Q brings all quantum effects into the hydrodynamic formulation. The quantum potential can be computed in three ways: from the R-amplitude, the probability density, and the C-amplitude,

$$Q(x,t) = -\frac{\hbar^2}{2m}\frac{1}{R}\frac{\partial^2 R}{\partial x^2} = -\frac{\hbar^2}{2m}\rho^{-1/2}\frac{\partial^2 \rho^{1/2}}{\partial x^2} = -\frac{\hbar^2}{2m}\left[\frac{\partial^2 C}{\partial x^2} + \left(\frac{\partial C}{\partial x}\right)^2\right]. \quad (2.20)$$

Focusing on computation of the quantum potential from the R-amplitude, we note that Q depends on the curvature of R, as measured by the second derivative.

Although R is always nonnegative, the curvature in the numerator can be positive, negative, or zero, and as a result, Q can also be positive, negative, or zero. A significant feature is the presence of R in the denominator. It is also important to realize, because the density and amplitude evaluated along a trajectory are time-dependent, that the quantum potential is also time-dependent. As a consequence, *the total energy evaluated along a quantum trajectory is not constant*, although the average energy for the ensemble of trajectories is conserved (for time-independent external fields).

Near a node in the wave function, the quantum potential *may* blow up, but this does not always happen. Poirier has classified nodes into types [2.24], depending on what happens to the *numerator* of the expression for Q. For *Type I nodes*, the point where $R = 0$ is also an inflection point, so that $\nabla^2 R = 0$. In this case, which includes stationary bound states, Q is *formally* well behaved. For *Type II nodes*, $\nabla^2 R \neq 0$ at the nodal point and Q is singular. Nodes in nonstationary wave packets generally fall into this category. Near this type of node, Q becomes very large and the sign can be either positive or negative.

Sometimes Q is referred to as "the mysterious quantum potential", even though it arises quite simply in the derivation of the quantum Hamilton–Jacobi equation. In this derivation, Q arises through the action of the kinetic energy operator on the polar form of the wave function, from the term

$$\frac{\partial^2}{\partial x^2}\left[\mathrm{Re}^{iS/\hbar}\right]. \tag{2.21}$$

As a consequence, the quantum potential must be a measure of the kinetic energy associated with this wave function. To carry this idea further, we will consider the *local kinetic energy*, defined by

$$T(x,t) = \mathrm{Re}\left[\psi^*\left[-\frac{\hbar^2}{2m}\frac{\partial^2}{\partial x^2}\right]\psi/\left(\psi^*\psi\right)\right], \tag{2.22}$$

where the term in the inner brackets is the kinetic energy operator. If the polar form of the wave function is substituted into this expression and we then take the real part of the result, we obtain a *decomposition of the local kinetic energy in terms of a flow component and a shape component* [2.31]:

$$T(x,t) = T_{flow} + T_{shape} = \frac{1}{2m}\left(\frac{\partial S}{\partial x}\right)^2 + \left(-\frac{\hbar^2}{2m}\frac{1}{R}\frac{\partial^2 R}{\partial x^2}\right). \tag{2.23}$$

In this decomposition, the *flow kinetic energy* depends only on the derivative of S, while the *shape kinetic energy* depends on the second derivative of R. The shape kinetic energy is the same quantity that we identified with the quantum potential. (Although page 93 in [2.23] does show the partitioning of the local kinetic energy, the terms *flow* and *shape* were not used.)

Given a wave function, it is possible for all of the kinetic energy to be flow energy, or it could be entirely shape energy. For example, the plane wave given by

$\psi(x) = A\exp(ip_0x/\hbar)$, where A is a constant, has $T_{flow} = p_0^2/2m$ and $T_{shape} = 0$. On the other hand, the Gaussian wave packet given by $\psi(x) = A\exp\left(-\beta x^2\right)$ has $T_{flow} = 0$ and $T_{shape} = \left(-\hbar^2/(2m)\right)\left(4\beta^2x^2 - 2\beta\right)$. Clearly, a real-valued wave function (one where $S = 0$ at all points) always has the property $T_{flow} = 0$ and will have $T_{shape} = 0$ only if the R-amplitude is free of curvature.

The quantum potential can also be regarded as a measure of the shape-induced *internal stress* in the wave packet; this quantity is relieved when the packet "flattens out". The quantum potential is closely related to the stress tensor for the probability fluid, a relationship that will be explored in Chapter 13. A related feature is that the quantum potential is *invariant to scale changes* in the R-amplitude. This means that under the transformation $R \to \lambda R$, where λ is the scale factor, the quantum potential is invariant. The quantum potential is thus independent of the "height" of R and depends only on its shape. One more important feature of the quantum potential is that it introduces *contextuality* into the dynamics. This means that even though Q changes with time, it always "remembers" the initial conditions placed on the wave function at the initial time.

Prior to Bohm's work, it was appreciated that *quantum mechanics is nonlocal*: every part of a quantum system depends on every other part and is *subject to organization by the whole*. This is true even in the absence of an external potential $V(\vec{r})$. However, when we examine the Schrödinger equation, the source of this nonlocality is not at all obvious. *One of the most significant features of Bohm's work is that the quantum potential was explicitly identified as the source of this nonlocality*. If Q is neglected in the quantum Hamilton–Jacobi equation, then the resulting classical equations describe a *local theory*. Earlier in this chapter, it was pointed out that the evolution of a classical dynamical system can be viewed in terms of the evolution of a cloud of noninteracting points. As we shall see, the quantum potential brings quantum trajectories under the influence of nonlocal effects and the resulting trajectories lose their independence: they are "all in it together". This results in additional changes in the trajectory, beyond what would be expected from the classical force acting alone. Features of the quantum potential that have been described in this section are summarized in Box 2.4.

Box 2.4. Features of the quantum potential

$$Q = -(\hbar^2/(2m))R^{-1}\nabla^2 R$$

- Introduces all quantum effects into the hydrodynamic equations
- Gives the shape contribution to the total kinetic energy
- Measures the curvature-dependent internal stress
- Influences quantum trajectories through the quantum force $\vec{f}_q = -\vec{\nabla} Q$
- Introduces contextuality (dependence on initial wave function)
- Is the source of nonlocality in the dynamics
- For nonstationary states, diverges on a nodal surface where $R = 0$

2.6 The Quantum Hamilton–Jacobi Equation

Now that all components of the one-dimensional version of the quantum Hamilton–Jacobi equation are in place, we will turn to the corresponding equation in three-dimensions. The Eulerian version of equation 2.10 in multiple dimensions becomes

$$-\frac{\partial S(x,t)}{\partial t} = \frac{1}{2m}\vec{\nabla}S \cdot \vec{\nabla}S + V(x) + Q(x,t), \tag{2.24}$$

where the quantum potential depends on the Laplacian of the R-amplitude, $Q = -\left(\hbar^2/(2m)\right) R^{-1}\nabla^2 R$. The first term on the right side of this equation, the flow kinetic energy, depends on the flow momentum, which in turn depends on the gradient of the action function, $\vec{p} = \vec{\nabla}S$. In addition, the flow velocity is $\vec{v} = (1/m)\vec{\nabla}S$. It is a property of the gradient that if we construct a surface on which S has a constant value, then at a point on this surface, the vector representing the flow momentum is orthogonal to this surface. Figure 2.4 shows several momentum vectors drawn perpendicular to a constant surface S. (Although $\vec{p}$ is generally orthogonal to a surface of constant S, this is *not* always the case. For example, in the presence of a magnetic field, $\vec{p} = \vec{\nabla}S - (e/c)\vec{A}$, where $\vec{B} = \vec{\nabla} \times \vec{A}$ is the magnetic field (see page 44 in [2.23]). Another example is presented in Section 11.10.)

If we take the gradient of equation 2.24 and then convert the Eulerian time derivative into the Lagrangian frame, the result is that the quantum Hamilton–Jacobi equation is converted into a Newtonian equation of motion

$$m\frac{d\vec{v}}{dt} = -\vec{\nabla}(V + Q) = \vec{f}_c + \vec{f}_q. \tag{2.25}$$

The right side of this equation is the total force acting on the trajectory. This force is the sum of the classical force $\vec{f}_c = -\vec{\nabla}V$ and the quantum force $\vec{f}_q = -\vec{\nabla}Q$. Even when the classical force vanishes, the so-called free particle, the point will generally be under the influence of a nonvanishing quantum force. This quantum Newtonian

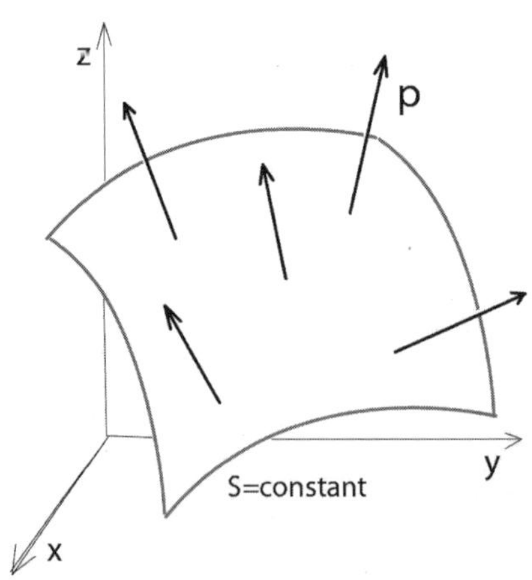

FIGURE 2.4. Several momentum vectors orthogonal to a surface of constant S.

equation, which tells how an object moves, plays a major role in Bohm's version of quantum mechanics. A very important feature is that the nonlocal quantum potential enters the equation determining the trajectory. To emphasize this feature, the total force may be written

$$\vec{f}_{total} = \vec{f}_c + \vec{f}_q = \vec{f}_{local} + \vec{f}_{nonlocal}(\rho), \tag{2.26}$$

in which the last term brings in all quantum effects. Just as the density affects the trajectory, the trajectory affects the density through the continuity equation. The trajectory and the density riding along it are fundamentally linked through the coupled hydrodynamic equations of motion.

Analogous to equation 2.17, the *Lagrangian version* of the quantum Hamilton–Jacobi equation is determined by the *quantum Lagrangian* evaluated along the trajectory

$$\frac{dS}{dt} = \frac{1}{2m}\vec{\nabla}S \cdot \vec{\nabla}S - (V(x) + Q(x,t)) = L. \tag{2.27}$$

In common with the classical case, the quantum Lagrangian is the excess of the flow kinetic energy over the potential energy, the potential now including the quantum potential in addition to the classical potential. Along the *quantum trajectory* linking an initial point $(\vec{r}_1, t_1)$ to a final point $(\vec{r}_2, t_2)$, the change in action is computed by integrating the quantum Lagrangian along the trajectory

$$\Delta S = \int_1^2 L(t)dt. \tag{2.28}$$

Again, as in the classical case, if we consider 'rubber band' deformed paths linking these two fixed points, the action is extremized along the quantum trajectory. This is the *quantum version of Hamilton's principle*, which was mentioned earlier with respect to equation 2.19.

We are now in a position to summarize the three equations of motion that form the foundation of quantum hydrodynamics. These are the continuity equation, the Newtonian equation for the flow acceleration, and the Hamilton–Jacobi equation relating the rate of change in the action to the quantum Lagrangian. These three equations, along with the subsidiary equations that relate the quantum potential and the density to the R-amplitude, and the flow velocity to the gradient of the action, are summarized in Box 2.5. It is important to appreciate that the first two equations in this box are *coupled* partial differential equations for ρ and $\vec{v}$; these two hydrodynamic fields *codetermine* each other.

Although $\vec{r}(t)$ specifies the location of a fluid element at each time, in the Lagrangian picture a more complete notation would also indicate the starting location for the trajectory. Notation such as $\vec{r}(\vec{r}_0, t)$ might be used for this purpose, but for brevity we will usually not specify the starting location. Given a set of starting conditions, the mapping $\{\vec{r}_0 \rightarrow \vec{r}(t)\}$ specifies the *flow*.

In Box 2.1, it was mentioned that another form for the wave function is $\psi(x,t) = \exp[iF(x,t)/\hbar]$, where $F(x,t)$ is the complex-valued action. In Box 2.6,

Box 2.5. Equations of quantum hydrodynamics

Lagrangian Frame

$$\frac{d\rho(\vec{r},t)}{dt} = -\rho(\vec{r},t)\vec{\nabla}\cdot\vec{v}(\vec{r},t), \tag{1}$$

$$m\frac{d\vec{v}}{dt} = -\vec{\nabla}(V+Q) = \vec{f}_c + \vec{f}_q, \tag{2}$$

$$\frac{dS}{dt} = \frac{1}{2m}\vec{\nabla}S\cdot\vec{\nabla}S - (V(\vec{r}) + Q(\vec{r},t)) = L, \tag{3}$$

$$Q(\vec{r},t) = -\frac{\hbar^2}{2m}\frac{1}{R(\vec{r},t)}\nabla^2 R(\vec{r},t), \tag{4}$$

$$\rho(\vec{r},t) = R(\vec{r},t)^2, \quad v(\vec{r},t) = \vec{p}(\vec{r},t)/m = (1/m)\vec{\nabla}S(\vec{r},t), \tag{5}$$

$$\psi(\vec{r},t) = \mathrm{R}(\vec{r},t)\mathrm{e}^{iS(\vec{r},t)/\hbar}. \tag{6}$$

Box 2.6. Wave function in terms of the complex-valued action

Rather than the Madelung–Bohm polar form for the wave function, where R and S are real-valued, another form is given by

$$\psi(x,t) = \exp[iF(x,t)/\hbar], \tag{1}$$

in which F is the *complex action*. This action is related to the C-amplitude and the action function S introduced in Section 2.2 through the relation $F = -i\hbar C + S$, and the probability density associated with this wave function is $\rho(x,t) = \exp[-(2/\hbar)\mathrm{Im}(F)]$. If this form is substituted into the TDSE, we obtain a modified version of the Hamilton–Jacobi equation given by

$$-\frac{\partial F}{\partial t} = \frac{1}{2m}\left(\frac{\partial F}{\partial x}\right)^2 + V - \frac{i\hbar}{2m}\frac{\partial^2 F}{\partial x^2}. \tag{2}$$

(Of course, if F is decomposed into real and imaginary parts, equation 2 then can be related to the hydrodynamic equations that have been presented in Box 2.5.

John [2.25] defined the complex-valued momentum by the guidance relation

$$m\dot{x} = \frac{\partial F}{\partial x} = \frac{\hbar}{i}\frac{1}{\psi}\frac{\partial\psi}{\partial x}, \tag{3}$$

so that equation 2 can also be written

$$-\frac{\partial F}{\partial t} = \frac{1}{2m}\left(\frac{\partial F}{\partial x}\right)^2 + V - \frac{i\hbar}{2}\frac{\partial\dot{x}}{\partial x}. \tag{4}$$

Since equation 3 generally requires that the velocity be complex-valued, John suggested that the coordinate is also complex-valued, $x = x_r + ix_i$, where x_r is the "physical coordinate".

Complex paths $x_i(t)$ against $x_r(t)$ have been plotted for several one-dimensional systems, including the harmonic oscillator, the potential step, and the spreading Gaussian wave packet. For "static" bound stationary states that are described in Chapter 14, these trajectories do not "sit in one place", as do Bohmian trajectories.

the equation of motion for the complex action is described, and a trajectory approach based on this equation is indicated.

2.7 Pilot Waves, Hidden Variables, and Bohr

For a short time following the birth of quantum mechanics, de Broglie worked on the *pilot wave interpretation of quantum mechanics*. The principal equation of this viewpoint appears near the bottom of Box 2.5. The wave influences the motion of the particle through the *guidance equation*, $v(\vec{r}, t) = \vec{p}(\vec{r}, t)/m = (1/m)\vec{\nabla} S(\vec{r}, t)$. In this interpretation, the wave function, through the gradient of its phase, guides the particle in such a way that the particle's speed depends on how rapidly the phase changes in space. De Broglie started on this interpretation in 1926 and continued until the famous Solvay Congress in Brussels in late 1927. Following some critical comments at this meeting, he dropped this approach, although he did return to similar work after Bohm's papers appeared in 1952. Holland provided the following concise description of de Broglie's contribution [2.23]:

> He recognized a dual role for the ψ−function; not only does it determine the likely location of a particle, it also influences the location by exerting a force on the orbit. It thus acts as a "pilot-wave" that guides the particle.

Holland also comments that in this interpretation, the wave function is "an agent that causes paths to curve". This idea is symbolized in figure 2.5.

In the titles to his two 1952 papers, Bohm used the term *hidden variable*, a phrase that seems to imply "mysterious, obscure physical quantities" [2.27]. In so-called Bohmian mechanics, the particle is guided by the forces that appear in the second equation in Box 2.5,

$$m\frac{d\vec{v}}{dt} = -\vec{\nabla}(V + Q) = \vec{f}_c + \vec{f}_q. \tag{2.29}$$

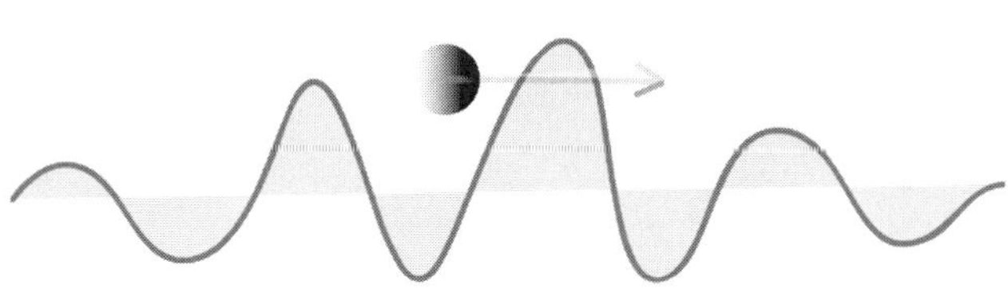

FIGURE 2.5. The pilot wave guiding a particle. The momentum of the particle is determined by the gradient of the phase of the underlying wave function, $\vec{p} = \vec{\nabla} S$.

The acceleration of the particle has both classical and quantum components, the latter depending on the gradient of the quantum potential. The so-called "*hidden variables*" *are the position and momentum (or velocity) of the particle*; they are "hidden" because the Schrödinger equation and its wave function solution do not seem to provide information about particle trajectories. Bohm's major contribution was to demonstrate that quantum mechanics can be reformulated (not approximated!) so that these dynamical variables emerge to play the dominant role. In addition, the de Broglie–Bohm interpretation is referred to as an *ontological theory* with the emphasis focused on *what is* rather than *what is measured.* This view is quite different from the extreme view expressed by Bohr, as quoted in Jammer's book [2.26]:

> There is no physical world, there is only an abstract physical description. *It's wrong to think the task of physics is to find out how nature is.* (Italics added for emphasis.)

(Mermin pointed out [2.33] that this frequently quoted statement was attributed to Bohr by his associate Aage Petersen. The quoted statement does not occur in Bohr's writings.) It is no wonder that past (and possibly some current) adherents to the Copenhagen dogma have had such a hard time with Bohm's ideas.

References

2.1. *D. Bohm, Quantum Theory* (Prentice-Hall, New York, 1951); reprinted as a Dover volume in 1989.
2.2. D. Bohm, *A suggested interpretation of the quantum theory in terms of 'hidden variables' I*, Phys. Rev. **85**, 166 (1952).
2.3. D. Bohm, *A suggested interpretation of the quantum theory in terms of 'hidden variables' II*, Phys. Rev. **85**, 180 (1952).
2.4. J.S. Bell, *Speakable and unspeakable in quantum mechanics* (Cambridge, New York, 1993).
2.5. F.D. Peat, *Infinite Potential, The Life and Times of David Bohm* (Addison-Wesley, Reading, Mass., 1997).
2.6. J. Bernstein, *Quantum Profiles* (Princeton University Press, Princeton, NJ, 1991).
2.7. www.math.rutgers.edu/~oldstein/
2.8. R. Goldstein, *Properties of Light* (Houghton Mifflin, Boston, 2000).
2.9. V.E. Madelung, *Quantentheorie in hydrodynamischer form*, Z. Physik, **40**, 322 (1926).
2.10. W. Pauli, *General Principles of Quantum Mechanics* (Springer-Verlag, New York, 1980).
2.11. M.P.A. Fisher, *Resonantly enhanced quantum decay: A time-dependent Wentzel–Kramers–Brillouin approach*, Phys. Rev. B **37**, 75 (1988).
2.12. L. Raifeartaigh and A. Wipf, *WKB properties of time-dependent Schrödinger system*, Found. Phys. Lett. **18**, 307 (1987).
2.13. H.J. Korsch and R. Mohlenkamp, *A note on multidimensional WKB wavefunctions: Local and global semiclassical approximation*, Phys. Lett. A **67**, 110 (1978).
2.14. C. Sparber, P.A. Markowich, and N.J. Mauser, *Wigner functions versus WKB-methods in multivalued geometrical optics*, arXiv:math-ph/0109029 (20 Mar. 2002).
2.15. D.A. Micha, *A self-consistent eikonal treatment of electronic transitions in molecular collisions*, J. Chem. Phys. **78**, 7138 (1983).

2.16. J.A. Olson and D.A. Micha, *A self-consistent eikonal treatment of diabatic rearrangement: Model $H^+ + H_2$ calculations*, J. Chem. Phys. **80**, 2602 (1984).
2.17. D.A. Micha, *Time-dependent many-electron treatment of electronic energy and charge exchange in atomic collisions*, J. Phys. Chem. A **103**, 7562 (1999).
2.18. M. Born and E. Wolf, *Principles of Optics* (Cambridge University Press, Cambridge, 1999), Ch. III.
2.19. A.B. Nasser, *Scattering via invariants of quantum hydrodynamics*, Phys. Lett. A **146**, **89** (1990).
2.20. E. Pinney, *The nonlinear differential equation, $y'' + p(x)y + cy^{-3} = 0$*, Proc. Am. Math. Soc. **1**, 681 (1950).
2.21. N. Froman and P.O. Froman, *Physical problems solved by the phase integral method* (Cambridge University Press, Cambridge, 2001).
2.22. W.E. Milne, *The numerical determination of characteristic numbers*, Phys. Rev. **35**, 863 (1930).
2.23. P.R. Holland, *The Quantum Theory of Motion* (Cambridge University Press, Cambridge, 1993).
2.24. B. Poirier, *Reconciling semiclassical and Bohmian mechanics: I. Stationary states*, J. Chem. Phys. **121**, 4501 (2004).
2.25. M.V. John, *Modified de Broglie–Bohm approach to quantum mechanics*, Found. Phys. Lett. **15**, 329 (2002).
2.26. M. Jammer, *The philosophy of quantum mechanics* (Wiley, New York, 1974).
2.27. G.E. Bowman, *Bohmian mechanics as a heuristic device: Wave packets in the harmonic oscillator*, Am. J. Phys. **70**, 313 (2002).
2.28. T. Takabayasi, *On the formulation of quantum mechanics associated with classical pictures*, Prog. Theor. Phys. **8**, 143 (1952).
2.29. K. Haug and H. Metiu, *A test of the possibility of calculating absorption spectra by mixed quantum–classical methods*, J. Chem. Phys. **97**, 4781 (1992).
2.30. S. Goldstein, *Quantum theory without observers, Parts one and two*, Physics Today, March–April, 1998.
2.31. R.E. Wyatt, *Quantum wave packet dynamics with trajectories: wave function synthesis along quantum paths*, Chem. Phys. Lett. **313**, 189 (1999).
2.32. S. Garashchuk and V. Rassolov, *Modified quantum trajectory dynamics using a mixed wavefunction representation*, J. Chem. Phys. **121**, 8711 (2004).
2.33. N. David Mermin, *What's wrong with this quantum world?* Physics Today, Feb. 2004; also, *Letters to the Editor*, Physics Today, Oct. 2004.

3
The Phase Space Route to the Hydrodynamic Equations

Classical and quantum equations of motion for momentum moments of phase space distribution functions provide another route to the position space hydrodynamic equations

3.1 Introduction

In 1954, less than two years after Bohm's two papers appeared, Takabayasi submitted a manuscript containing a very different derivation of the hydrodynamic equations [3.1]. Rather than starting in position space and substituting the polar form of the wave function into the TDSE, he began with the time-dependent Wigner distribution function in phase space. The seminal contribution was to develop equations of motion for the momentum moments of the Wigner function. These equations form an infinite hierarchy, with the rate of change of one moment coupled to both lower and higher momentum moments. For pure states, in which the system can be represented in terms of a wave function, the lowest two moments form a closed set, and the hierarchy terminates. The resulting equations for the 0-moment and the 1-moment are identical to the Bohm equations of motion for the probability density and the momentum. However, the phase space route provides additional insight into the terms in the Bohm equations, especially regarding the interpretation of the momentum and the quantum force. As pointed out by Takabayasi, "*the quantum force may be looked upon as an apparent force appearing as a result of projecting the phase space ensemble onto the configuration space*". In addition, he pointed out that the phase space route is also applicable to mixed states, such as those involving a thermal distribution over eigenstates.

Over the intervening years, the phase space route has been frequently used to develop quantum hydrodynamic equations for the current in electronic devices, such as resonant tunneling diodes (RTD), semiconductor surface layers, and molecular wires. These studies include the following. In 1981, Iafrate et al. [3.2] developed rate equations for the first few moments of the Wigner–Boltzmann equation, which includes a collision term. In 1990, Frensley [3.3] presented a thorough analysis of rate equations for moments of the Wigner function and used this formalism for extensive computational studies of RTDs. In 1994, Gardner [3.4] derived rate equations for both the Boltzmann equation and the Wigner–Boltzmann equation and used these to compute current–voltage curves for RTDs. In 1997, Gasser and Markowich [3.5] studied semiclassical and classical limits of quantum transport equations that were derived from momentum moments of the Wigner function. In 2000, Gardner and Ringhofer [3.6] summarized the moment equations and presented computational results for RTDs. In 2003, Degond and Ringhofer described an entropy minimization method for closing the moment system [3.22].

Wigner functions are frequently used in quantum optics, for analysis of states of the electomagnetic field and for developing equations of motion for fields interacting with matter [3.7]. Wigner functions and other quantum phase space distributions have been used to analyze for possible chaotic behavior in systems for which the underlying classical system is chaotic. This and related topics, including the semiclassical limit of the Wigner function, are described in the excellent book by Tabor [3.19]. In 1985, McLafferty also explored the relationship between the Wigner function and the Bohmian equations [3.23].

In addition to these studies, in 1993 Muga, Sala, and Snider published an insightful paper that explored the connection between moments of classical and quantum phase space distributions [3.8]. Using the Liouville equation in the classical case or the Wigner–Moyal equation for the quantum case, they developed rate equations for the momentum moments of the corresponding phase space distribution function. They pointed out that "*an internal potential is defined as a common object to both classical statistical mechanics and to quantum mechanics*". In addition, they compared plots of the momentum moments for classical and quantum phase space distributions obtained for scattering from an Eckart barrier. This interesting example will be described later, in Section 3.7.

In spite of the powerful ideas in these studies, prior to 2000, the phase space route to the hydrodynamic equations did not have a very large impact on research activities involving quantum dynamics, including dynamical studies in chemical physics. However, this situation may very well change due to the investigations by Burghardt and coworkers [3.9–3.11] and by Bittner et al. [3.20, 3.21]. Some of these studies will be described in detail later in this chapter. In addition, Chapter 11 contains additional developments on dissipative quantum systems and also presents trajectory approaches to the time evolution of the density matrix, and Chapter 12 describes the development of trajectory representations for mixed quantum–classical systems. The studies by Burghardt, Cederbaum, and Moller [3.9–3.11] make explicit connections in the moment hierarchy between systems described as *pure states and as mixtures* (these terms are described in Box 3.1). They also relate the moment approach for pure states with the Bohmian equations that were presented in Chapter 2. Surprisingly, these connections have been largely absent from the literature. Furthermore, for mixed states, a generalization of the quantum force was developed, and this also depends on the momentum spread, measured by the variance, of the underlying phase space distribution. The moment hierarchy was applied by these authors to situations involving dissipation, and

Box 3.1. Density matrix and the density operator

The probability density for a *pure state*, one described by a wave function, is given by $\rho(x) = \psi^*(x)\psi(x)$, which in Dirac notation is written $\rho(x) = \langle x|\psi\rangle\langle\psi|x\rangle$. (Recall that in Dirac notation, the position space probability amplitude and its conjugate are written $\psi(x) = \langle x|\psi\rangle$ and $\psi^*(x) = \langle\psi|x\rangle$.) The ket-bra operator in the middle of this expression is the *density operator*, $\hat{\rho} = |\psi\rangle\langle\psi|$ (defined as an *outer product* of $|\psi\rangle$ and $\langle\psi|$), and the position space "diagonal matrix element" of this operator is $\rho(x) = \langle x|\hat{\rho}|x\rangle$, where $|x\rangle$ is an eigenket of the position operator. A generalization of the latter expression allows for "off-diagonal matrix elements" of the density operator, $\rho(x, x') = \langle x|\hat{\rho}|x'\rangle$, where x and x' can be thought of as (continuous) labels on the rows and columns of the density matrix. For example, for an electron in the ground state of a box of length L, the x, x' matrix element of the density operator is given by

$$\rho(x, x') = \sqrt{\frac{2}{L}}\sin(\pi x/L)\sqrt{\frac{2}{L}}\sin(\pi x'/L). \tag{1}$$

For a *mixed state*, a statistical mixture, the density operator can be written in terms of a linear combination of density operators for a set of eigenstates,

$$\hat{\rho} = \sum_{j=1}^{N} w_j|\psi_j\rangle\langle\psi_j| = \sum_{j=1}^{N} w_j\hat{\rho}_j, \tag{2}$$

where the weights are nonnegative, $w_j \geq 0$, and sum to unity (the normalization condition)

$$\sum_{j=1}^{N} w_j = 1. \tag{3}$$

For this mixed state, the x, x' matrix element of the density operator is given by

$$\begin{aligned}\rho(x, x') = \langle x|\hat{\rho}|x'\rangle &= \sum_{j=1}^{N} w_j\langle x|\psi_j\rangle\langle\psi_j|x'\rangle \\ &= \sum_{j=1}^{N} w_j\psi_j^*(x)\psi_j(x').\end{aligned} \tag{4}$$

An example would be the density matrix for a Boltzmann (thermal) distribution at temperature T over the particle-in-a-box energy eigenstates

$$\rho(x, x') = \sum_{j=1}^{N} \frac{e^{-E_j/k_BT}}{Z(T)}\sqrt{\frac{2}{L}}\sin(j\pi x/L)\sqrt{\frac{2}{L}}\sin(j\pi x'/L), \tag{5}$$

in which $Z(T)$ is the partition function at absolute temperature T and k_B is Boltzmann's constant. For this mixed state, the probability density would be the diagonal element of this matrix:

$$\rho(x) = \sum_{j=1}^{N} \frac{e^{-E_j/k_B T}}{Z(T)} \frac{2}{L} \sin^2(j\pi x/L). \tag{6}$$

The mixed state density matrix plays a major role in the quantum dynamics of open systems, those coupled to an environment. Trajectory representations for open systems will be treated in detail in Chapter 11.

to coupled electronic states. As these and related concepts are developed in this chapter, not only will we develop a deeper understanding of the quantum trajectory equations, but significant extensions to more general types of systems will be described.

In order to introduce phase space concepts in a familiar context, Section 3.2 describes classical trajectories evolving in the two-dimensional (x, p) phase space. Section 3.3 introduces the Wigner quantum-mechanical phase space distribution function and its equation of motion, the Wigner–Moyal equation. Following the work of Burghardt and Cederbaum [3.9, 3.10], momentum moments of the Wigner function are introduced in Section 3.4. Equations of motion for these moments are developed in Section 3.5. For pure state systems, an important connection is made with the Bohmian trajectory equations that were introduced in Chapter 2. A similar moment analysis is made for classical phase space probability densities in Section 3.6. In Section 3.7, classical and quantum moments are compared for scattering from an Eckart barrier [3.8]. Comparisons between Liouville phase space, the one introduced in Section 3.2 (wherein x and p are the independent variables), with the *hydrodynamic phase space* (where momenta are now dependent variables) are described in Section 3.8 [3.11]. Concluding remarks appear in Section 3.9.

3.2 Classical Trajectories and Distribution Functions in Phase Space

For a classical trajectory in a system with one degree of freedom, separate plots may be made to show the time dependence of the coordinate and momentum. Another way to display the motion is to combine the separate pictures and plot the momentum versus the coordinate. At each time, a dot is made to locate $p(t)$ at the corresponding value of the position $x(t)$. The dots are then connected in this *phase space* to show the time dependence of the trajectory. Multiple trajectories launched from different initial points are frequently plotted together on the same diagram. The concept is readily extended to higher dimensionality: for a system with N coordinates and momenta $\{x_i, p_i\}$, the phase space is spanned by the $2N$ axes labeled by these independent variables.

One of the simplest examples for describing phase space orbits in one-dimension is the harmonic oscillator. For such an oscillator with frequency ω (rad/sec) and mass m, the Hamiltonian is $H(p, x) = p^2/(2m) + (m\omega^2/2)x^2$. For a trajectory having total energy E, the momentum and coordinate for the trajectory are constrained to satisfy the equation

$$\left(\frac{p}{\sqrt{2mE}}\right)^2 + \left(\frac{x}{\sqrt{2E/(m\omega^2)}}\right)^2 = 1. \tag{3.1}$$

This quadratic form is the equation for an ellipse having semimajor and semiminor axes $\sqrt{2mE}$ and $\sqrt{2E/(m\omega^2)}$. The latter equations show that trajectories with increasing energies sweep out larger elliptic orbits.

A different and possibly more interesting example is provided by the metastable oscillator, for which a cubic term is added to the harmonic potential, $V(x) = (1/2)m\omega^2x^2 - (1/3)bx^3$, where b is the (positive) anharmonicity parameter. This potential has a near-harmonic well for small values of x, but as x increases, there is a barrier of height $V^* = (m\omega^2)^3/(6b^2)$ at the position $x^* = (m\omega^2)/b$. For $x > x^*$, the potential decreases and becomes negative when $x > (3/2)x^*$. A plot of this metastable potential is shown in figure 10.1. Quantum trajectories for this metastable system will be described later, in Chapters 10 and 11.

Classical trajectories may easily be computed for this potential by numerically integrating Hamilton's equations of motion. A phase space diagram showing 40 trajectories is displayed in figure 3.1. There are a number of comments to make about this figure. Trajectories labeled 1 through 9 cycle around the origin in bound orbits having energies below the barrier height. As the energy approaches V^* from below, the orbits show increasing deviations from the elliptic shape that characterizes the low energy, nearly harmonic orbits. Orbit number 10 has an energy just above V^* and is the first of the unbound sequence of scattering states labeled 10 through 24. These orbits have one inner turning point on the hard wall of the potential that occurs for $x < 0$. For the lower energy members of this set, the orbits labeled 10 through 14 for example, it is clearly noted that the momentum decreases near the barrier maximum at the position x^*. There is a special orbit (not shown in this figure) with energy $E = V^*$, called the *separatrix*, that divides the bound and unbound sets. The third group of orbits, numbered 25 through 40, all have energies below V^*, and are also unbound states. They have inner turning points to the right of the barrier maximum at x^*. Some of the orbits in the bound region have energies equal to those in this outer unbound region, but these pairs of orbits are disconnected in phase space by a region where the potential energy exceeds the total energy, $E < V$. Quantum trajectories can make the connection between these two regions by tunneling between the classical turning points located just to the left and right of the barrier maximum. We will return to this example later in this chapter.

In addition to plotting the trajectory for each member of an ensemble, there is an alternative way to describe the phase space dynamics of a classical system. In classical statistical mechanics, the probability of finding trajectories in a small phase space cell is frequently defined in terms of a continuous time-dependent probability density. In terms of this density, denoted by $W_{cl}(x, p, t)$, the probability of finding the system (i.e., of finding dN trajectories out of the total number N) having coordi-

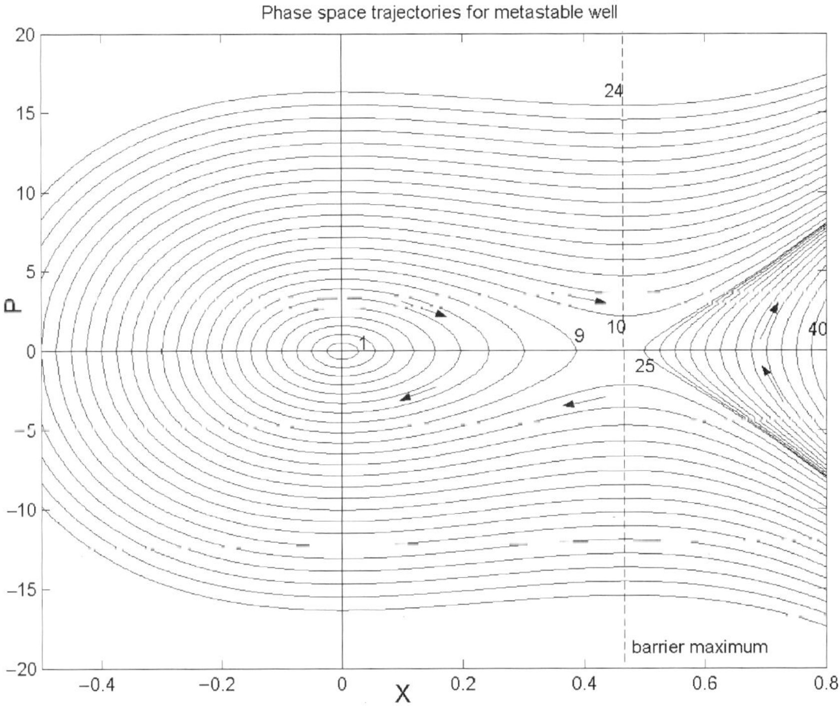

FIGURE 3.1 Phase space plot for 40 *classical trajectories* evolving on the metastable well potential. The parameter values used in this calculation are (in atomic units): $m = 2000$, $\omega = 0.01$, $b = 0.4276$. The barrier maximum is located at $x^* = 0.47$, and the barrier height is $V^* = 1600$ cm^{-1}. Trajectories 1–9 circulate around the bound region near $x = 0$. Trajectories 10–24 have energies above the barrier maximum and are unbound. Trajectories 25–40 are also unbound and have energies *below* the barrier maximum. They approach the barrier from the asymptotic region at large values of x.

nates and momenta in the small box of area $dA = dxdp$ centered at the point (x, p) is given by $dP(x, p, t) = W_{cl}(x, p, t)dxdp$ (where $dP(x, p, t) = dN(x, p, t)/N$). (See the first part of Section 11.2, which also deals with classical phase space distributions.) This density is nonnegative, $W_{cl}(x, p, t) \geq 0$, and satisfies the usual normalization condition

$$\iint W_{cl}(x, p, t)dxdp = 1. \tag{3.3}$$

Given the initial condition at $t = 0$, this probability fluid can be visualized as *flowing through phase space*, a picture of the dynamics that evokes a hydrodynamic description of the flow. The equation of motion for the classical phase space probability, the Liouville equation, will be given in equation 3.11 and in Section 11.2. Extension of the concept of phase space distributions to quantum systems is described in the following section.

3.3 The Wigner Function

In 1932, Wigner introduced a quantum mechanical phase space distribution function [3.12] that has some, but not all, of the features of a classical probability distribution. An excellent introduction to the Wigner function appears in the textbook by Bialynicki-Birula et al. [3.13], and Wigner functions for both energy eigenstates and time-dependent wave packets for the square well potential are described by Belloni et al. [3.24]. Considerable detail about these functions is given in the 1984 review article *Distribution Functions in Physics*, by Hillery et al. [3.14]. In addition, the 1983 review article by Carruthers and Zachariasen focuses on *quantum collision theory with phase space distributions* [3.15].

Wigner formulated the time-dependent quantum distribution function $W(x, p, t)$ in terms of the Fourier transform of a certain bilinear form involving the wave function $\psi(x, t)$ and its conjugate $\psi^*(x, t)$. Before defining the Wigner transform, consider the *overlap function* between the wave function at the "advanced" point $x + r/2$ with its conjugate at the "retarded" point $x - r/2$. This overlap, a type of spatial *correlation function*, is

$$O(x, r) = \psi^*(x + r/2, t)\psi(x - r/2, t), \tag{3.4}$$

where r, the separation, will also be called the *hopping distance* (Schleich refers to this as the *jump distance* [3.7]). When $r = 0$, the overlap is just the familiar probability density, and as r increases, the overlap will decay to zero, at least for wave functions that are localized in position space. As an example, for the first excited state of a harmonic oscillator, the wave function is given by

$$\psi(x) = Nxe^{-x^2/(2\alpha^2)}, \tag{3.5}$$

where N is the normalization factor and α scales the width of the function. If we choose $\alpha = 2$ and let $x = 0$, the overlap as a function of the hopping distance r is shown in figure 3.2. When $r = 0$, the overlap vanishes for this example, but it drops into negative valleys before approaching zero at large positive or negative values of the hopping distance.

The Wigner phase space distribution function is obtained at each value of x by computing the *Fourier components of the overlap function*. This function thus measures the distribution of plane waves having different de Broglie wavelengths $\lambda = h/p$ in the overlap function. The Wigner integral transform of the overlap kernel is given by

$$W(x, p, t) = \frac{1}{2\pi\hbar} \int \psi^*(x + r/2, t)\psi(x - r/2, t)e^{ipr/\hbar} dr. \tag{3.6}$$

(When viewing this equation, it is worth pointing out that x and p are dynamical variables, *not* quantum-mechanical operators.) For the example shown in figure 3.2, the overlap function is real-valued and symmetric in r, so the integral becomes the cosine transform of the overlap function.

Before giving several examples of Wigner functions, we will list some additional general features possessed by this function.

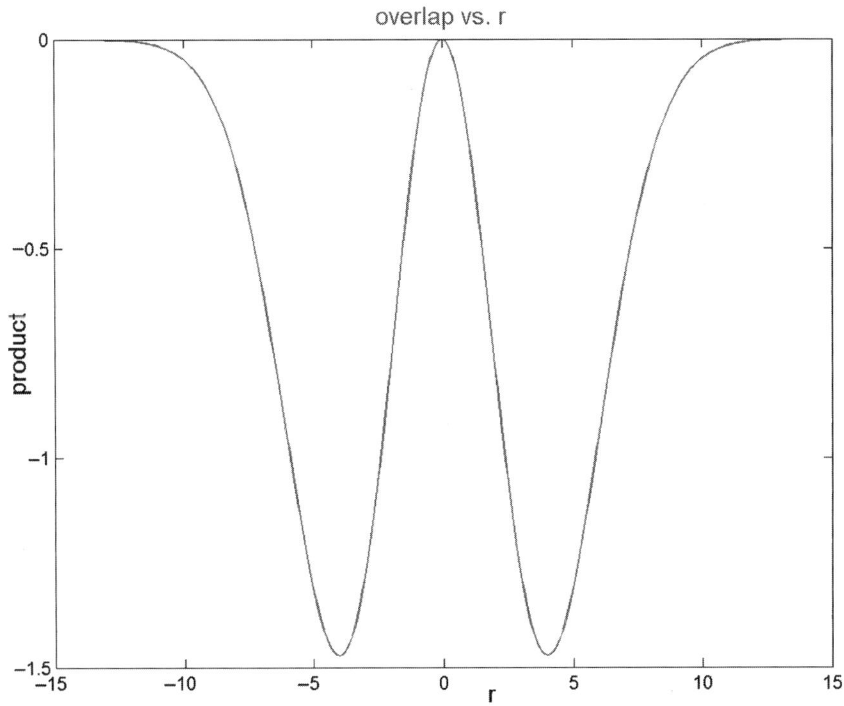

FIGURE 3.2. The overlap kernel for the first excited state of the harmonic oscillator, plotted versus the hopping distance r. (The normalization factor has been set to unity.)

(a) The Wigner function is real-valued and normalized to unity.

(b) The integrals over p or x give the x-space or p-space probability densities (marginal probability distributions), respectively:

$$\rho(x,t) = |\psi(x,t)|^2 = \int W(x,p,t)\,dp,$$
$$\rho(p,t) = |\phi(p,t)|^2 = \int W(x,p,t)\,dx, \tag{3.7}$$

where $\phi(p,t)$ is the *momentum space wave function* obtained from $\psi(x,t)$ through Fourier transformation,

$$\phi(p,t) = \frac{1}{\sqrt{2\pi\hbar}} \int_{-\infty}^{\infty} \psi(x,t) e^{ipx/\hbar}\,dx.$$

(c) There is an upper limit on the magnitude of the Wigner function [3.7]

$$|W(x,p,t)| \le \frac{1}{\pi\hbar} \approx 0.318 \quad \text{(in a.u.)}. \tag{3.8}$$

In contrast, for a classical phase space distribution function, there is no bound on the height of the probability function. In classical mechanics, a spiked distribution is allowed; an example is a *single trajectory* having the coordinates $\{x(t), p(t)\}$, for which the corresponding time-dependent distribution function is given by $W_{cl}(x, p, t) = \delta(x - x(t))\delta(p - p(t))$.

(d) The Wigner function usually has *negative basins,* and for this reason it cannot be a true probability density; rather, it is referred to as a *quasi-probability distribution function.* In many cases, $W(x, p, t)$ shows *diffraction fringes* and multiple small-amplitude ripples, some of which go negative. These basins are caused by the interference of amplitudes, such as occurs in barrier scattering [3.7]. Only a Gaussian wave packet moving on a potential of at most quadratic dependence on x yields a Wigner function that is nonnegative at all times. Clearly, most potential surfaces and wave packets do not satisfy these criteria, and at some time or other they will develop negative basins. An example will be given later in Box 11.1, and the two examples given later in this section also illustrate this feature.

(e) The mapping from $\psi(x, t)$ to a phase space distribution function that satisfies property (b) listed previously is *not unique.* However, among various possible transforms, the Wigner mapping is one of the simplest that possesses these properties. Another phase space distribution that satisfies property (b) (in a trivial way) is given by $\rho(x, p, t) = |\psi(x, t)|^2|\phi(p, t)|^2$. This distribution function, unlike the Wigner function, has the nice feature that at all phase space points, $\rho(x, p, t) \geq 0$.

The time evolution of the Wigner function is determined by the evolution of the wave function that appears in the overlap kernel on the right side of equation 3.6. In order to derive a *rate equation for the Wigner function*, we will operate on both sides of equation 3.6 with $\partial/\partial t$. When this operator acts on the wave function (or its conjugate) in the integrand, we use the Schrödinger time-dependent wave equation to replace $\partial\psi/\partial t$ with $(-i/\hbar)\hat{H}\psi$. After integrating by parts, the terms involving $\nabla^2\psi$ lead to the term $(-p/m)\partial W/\partial x$. In addition, the potential energy in the term $[V(x + r/2) - V(x - r/2)]$ will be expanded around the point x in a Taylor series in the hopping distance r. In this way, we finally obtain the *equation of motion for the Wigner function* [3.12]:

$$\begin{aligned}\frac{\partial W}{\partial t} &= -\frac{p}{m}\frac{\partial W}{\partial x} + \frac{\partial V}{\partial x}\frac{\partial W}{\partial p} + \sum_{k=1}^{\infty}\frac{\left(-\hbar^2/4\right)^k}{(2k+1)!}\frac{\partial^{2k+1}V}{\partial x^{2k+1}}\frac{\partial^{2k+1}W}{\partial p^{2k+1}},\\ &= -\frac{p}{m}\frac{\partial W}{\partial x} + \frac{\partial V}{\partial x}\frac{\partial W}{\partial p} - \frac{\hbar^2}{24}\frac{\partial^3 V}{\partial x^3}\frac{\partial^3 W}{\partial p^3} + \frac{\hbar^4}{1920}\frac{\partial^5 V}{\partial x^5}\frac{\partial^5 W}{\partial p^5} + \ldots,\\ &= -\frac{p}{m}\frac{\partial W}{\partial x} + \frac{2}{h}\sin\left(\frac{\hbar}{2}\frac{\partial}{\partial x}\frac{\partial}{\partial p}\right)V(x)W(x, p, t).\end{aligned} \tag{3.10}$$

In this last version, it is important to recognize that $\partial/\partial x$ operates only on the potential energy and that $\partial/\partial p$ acts only on the Wigner function. In the infinite series on the right side of lines one and two, each term involving an even power of $\hbar$ brings in an odd spatial derivative of the potential multiplying the same odd power derivative of the Wigner function with respect to momentum. In the last line of equation 3.10, the infinite series in the preceding versions involving odd derivatives of the potential energy (including the second term on the right) has been summed to give an analytic expression. Equation 3.10 is formally known as the *Wigner–Moyal equation* [3.16].

If the "quantum terms" involving explicit $\hbar$ dependence are set to zero, the resulting equation governs the time evolution of a classical probability distribution. *Liouville's equation of motion for the classical probability function* is

$$\frac{\partial W_{cl}}{\partial t} = -\frac{p}{m}\frac{\partial W_{cl}}{\partial x} + \frac{\partial V}{\partial x}\frac{\partial W_{cl}}{\partial p}. \tag{3.11}$$

A number of additional comments about this equation and its solutions appear in the first part of Section 11.2. Unlike the solutions of the Wigner–Moyal equation, the classical distribution function is nonnegative at every phase space point. Because it formally reduces to the Liouville equation in the classical limit $\hbar \to 0$, the Wigner–Moyal equation of motion is also called the *quantum Liouville equation.*

If the potential energy can be expressed as an n-th degree polynomial, the infinite series on the right side of equation 3.10 terminates. An important case is the quadratic potential function $V(x) = a + bx + cx^2$, for which there are no terms on the right side of equation 3.10 that bring in $\hbar$. A uniform electric field and the harmonic oscillator potential provide examples of this type of potential. In this case, *the Liouville equation generates the exact quantum dynamics in phase space, and all quantum effects are subsumed into the initial conditions.* As an alternative to numerical integration of the Liouville equation, classical trajectories in an ensemble may be integrated, one at a time. From Chapter 2, recall that the position space evolution of quantum trajectories evolving on a quadratic potential (or no potential at all) must include the influence of the quantum potential. This example brings out an important distinction between Bohmian trajectories evolving in position space and Wigner dynamics in phase space.

We will now consider several examples of Wigner functions. First, for the ground and first excited states of the harmonic oscillator, the Wigner transforms are given by

$$W(x,p) = (1/(\pi\hbar))\exp\left[-\left(\left(\frac{x}{\alpha}\right)^2 + \left(\frac{p\alpha}{\hbar}\right)^2\right)\right], \tag{3.12}$$

$$W(x,p) = N'\left[2\left(\left(\frac{x}{\alpha}\right)^2 + \left(\frac{p\alpha}{\hbar}\right)^2\right) - 1\right]\exp\left[\left(\left(\frac{x}{\alpha}\right)^2 + \left(\frac{p\alpha}{\hbar}\right)^2\right)\right], \tag{3.13}$$

where $\alpha = \sqrt{\hbar/m\omega}$ (which has units of length) [3.13]. For the first excited state, the preexponential factor determines the sign of the Wigner function. In this case, for phase space points within the elliptic boundary defined by the equation $(x/\alpha)^2 + (p\alpha/\hbar)^2 = 1/2$, this function takes on negative values. For the parameter value $\alpha = 1$, these two Wigner functions are shown in figure 3.3. The Wigner function for the excited state has a volcano shape, with a negative basin edged by a circular

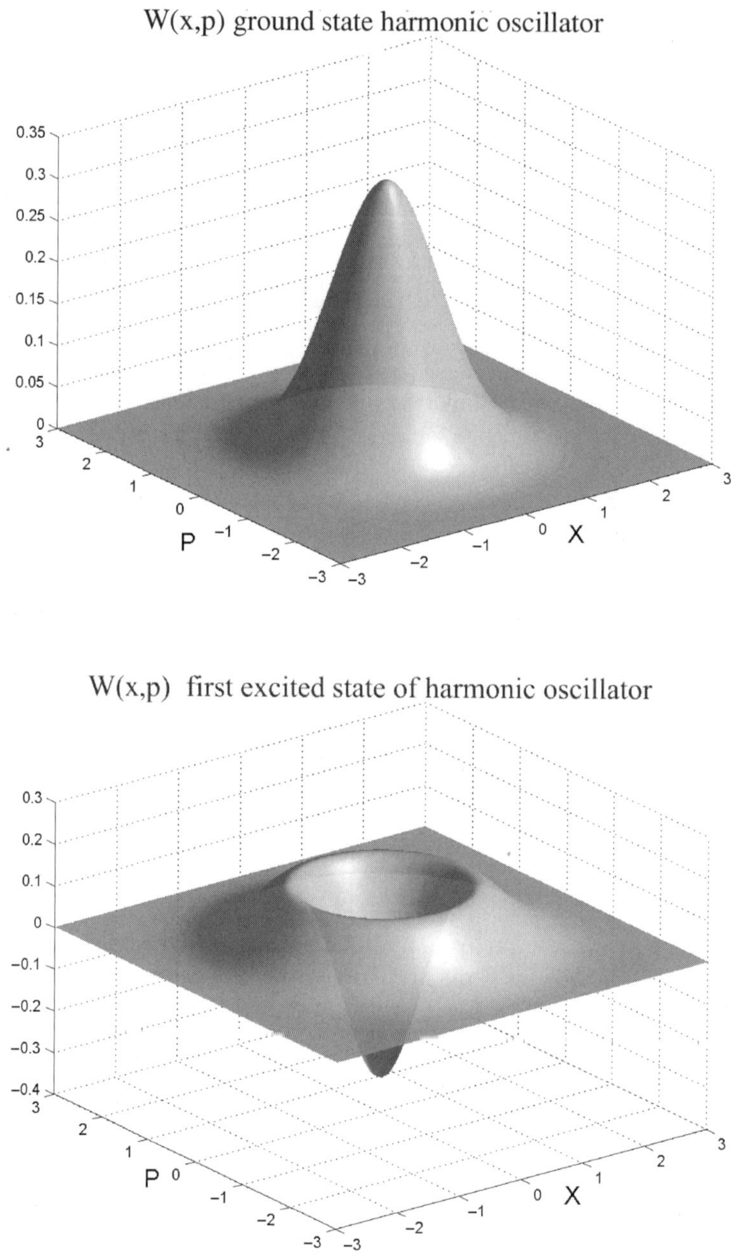

FIGURE 3.3. Wigner functions for the ground and first excited harmonic oscillator states. In equations 3.12 and 3.13, the width parameter has the value $\alpha = 1$, and the coordinates are in atomic units. The Wigner function for the excited state has a negative basin surrounding the origin.

rim. Wigner functions for the higher harmonic oscillator excited states always have negative basins in some regions of phase space.

As a second example, we will return to the *metastable well potential*, for which the classical phase space was described in Section 3.2. If the system is prepared in the harmonic ground state of the potential, the Wigner transform of the initial Gaussian wave function is concentrated within an elliptic region approximately defined by $|x| < 0.4$ and $|p| < 8$. In order to develop the time evolution of this Wigner function in the metastable well potential, the Wigner–Moyal equation was integrated numerically on a two-dimensional grid in phase space. (The fixed-grid finite difference algorithm used for these calculations is described in Box 11.8.) Before presenting these computational results, we will comment on some qualitative features of the time evolution. At early times, we expect that positive momentum components of the initial distribution will begin to twist to the right toward the barrier region near $x^* = 0.47$. As time proceeds, part of this distribution will move across the barrier to form a tube within which the density will gain momentum as the potential decreases beyond the barrier region.

Turning to the computational results, figures 3.4 and 3.5 show the Wigner distribution at two times, $t = 6$ fs and $t = 14$ fs, respectively. Because of the anharmonicity of the potential, the peak of distribution gradually shifts toward the barrier region while simultaneously the main density tube begins to form near the momentum value $p = 5$. The maximum density within the central region gradually decreases, dropping from about 0.33 at 6 fs to about 0.29 at 14 fs. Possibly the

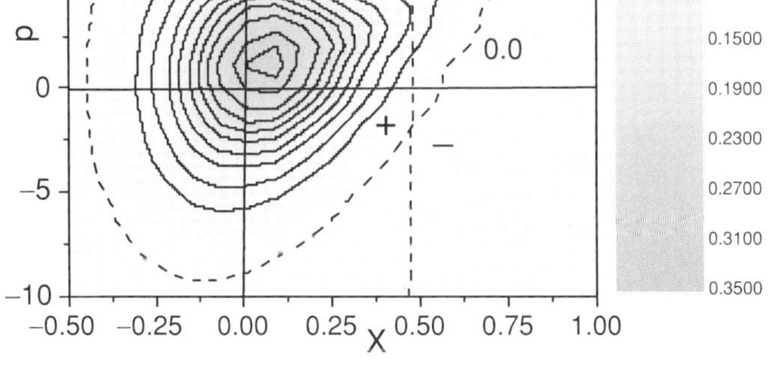

FIGURE 3.4. The Wigner function for the metastable well potential at $t = 6$ fs. The border between negative and positive regions is marked by the dashed contour. The position of the barrier maximum is shown by the vertical dashed line. One contour is shown within the negative basin that forms near $p = 7.5$.

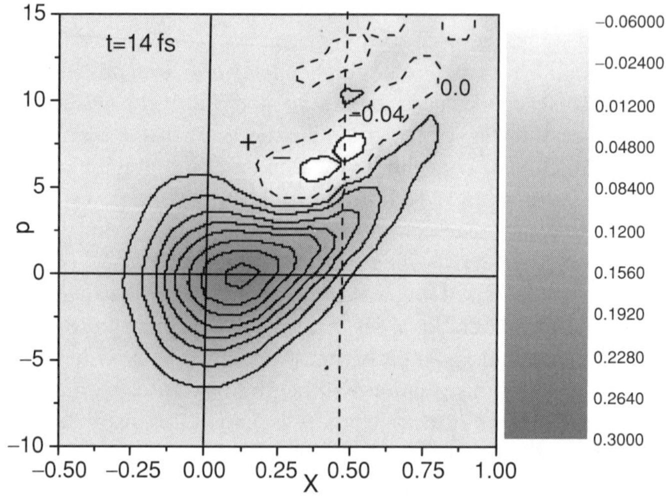

FIGURE 3.5. The Wigner function for the metastable well potential at $t = 14$ fs. The border between negative and positive regions is marked by the dashed contour. The position of the barrier maximum is shown by the vertical dashed line. One contour line is shown within the negative basin that forms near $p = 7$.

most interesting feature shown in these two figures is the shallow negative basin that forms just above the main density tube (i.e., on the high momentum side); within this basin the distribution function has the value -0.03 at 6 fs and then decreases slightly more to -0.05 at 14 fs. Presumably, this basin forms due to interference between forward-moving and reflected components in the vicinity of the barrier maximum. We will return to further analysis of this example near the end of the next section.

3.4 Moments of the Wigner Function

We are now ready to introduce the momentum moments of the Wigner function [3.1, 3.16, 3.9–3.11]. These moments are obtained by multiplying the Wigner function by p^n, $n = 0, 1, 2, \ldots$, and then integrating over p:

$$\bar{p}_n(x, t) = \int_{-\infty}^{\infty} p^n W(x, p, t) dp = \langle p^n W \rangle. \tag{3.14}$$

Takabayasi aptly refers to this as "*projection . . . onto the coordinate space*" [3.1]. Each moment (actually a *moment density*, because it is still a "density" as far as the x variable is concerned) is a function of x and t and is thus a *local quantity* (because it depends on the value of x). The first few moments are

$$\rho(x,t) = \bar{p}_0(x) = \int_{-\infty}^{\infty} W(x,p,t)\,dp = \langle W \rangle,$$

$$\bar{p}_1(x,t) = \int_{-\infty}^{\infty} p\,W(x,p,t)\,dp = \langle pW \rangle,$$

$$\bar{p}_2(x,t) = \int_{-\infty}^{\infty} p^2 W(x,p,t)\,dp = \langle p^2 W \rangle. \tag{3.15}$$

in which the bracket notation $\langle .. \rangle$ refers to integration over p (the resulting expression is still a function of x). The 0-moment is the same as the probability density. The 1-moment is related to the flux (current density) $j(x,t) = \bar{p}_1(x,t)/m$, and the 2-moment is related to the kinetic energy density $T(x,t) = \bar{p}_2(x,t)/(2m)$.

By substituting the explicit form for the Wigner function *for pure states* into equation 3.14, Iafrate et al. [3.2] obtained the following expressions for the first few moments:

$$\begin{aligned}
\bar{p}_1(x,t) &= p(x,t)\rho(x,t),\\
\bar{p}_2(x,t) &= p^2(x,t)\rho(x,t) - \frac{\hbar^2}{4}\rho(x,t)\frac{\partial^2 \ln \rho(x,t)}{\partial x^2},\\
\bar{p}_3(x,t) &= p^3(x,t)\rho(x,t) - \frac{\hbar^2}{4}\rho(x,t)\left[3p(x,t)\frac{\partial^2 \ln \rho(x,t)}{\partial x^2} + \frac{\partial^2 p(x,t)}{\partial x^2}\right].
\end{aligned} \tag{3.16}$$

Each moment (except for $n = 1$) has the general form

$$\bar{p}_n(x,t) = p^n(x,t)\rho(x,t) + \text{quantum terms}, \tag{3.17}$$

where the first part, $p(x,t)^n \rho(x,t)$, is the "classical flow" term, and (for pure states) the *average momentum* is given by the gradient of the action, $p(x,t) = \partial S(x,t)/\partial x$. (Equation 3.17 is also valid for mixed states, but different quantum terms come in beyond the classical terms; see equation 18 in Burghardt et al. [3.9].) In addition, for pure states, the moments for $n > 1$ are completely determined by the 0-moment and the 1 moment. This simplicity does not carry over to the case of mixed states, although for some approximations (such as a Gaussian density), the hierarchy breaks at the 3-moment [3.9–3.11].

Another quantity that will be useful in the next section is the *variance* in p, the 2-moment relative to the mean value of the momentum:

$$\sigma(x,t) = \int_{-\infty}^{\infty} (p - p(x,t))^2 W(x,p,t)\,dp. \tag{3.18}$$

This quantity can also be written

$$\sigma(x,t) = \bar{p}_2(x,t) - p(x,t)^2\rho(x,t). \tag{3.19}$$

This variance is a measure of the "fatness" of the distribution and also measures *momentum fluctuations* about the mean value of the momentum. Takabayasi referred to $\sigma(x,t)$ as an element (a diagonal element) of the *dispersion tensor* [3.1]. In Section 3.7, an example will be given that demonstrates the surprising feature that this (quantum) variance can become negative. This cannot happen for classical probability distributions, because $W_{cl}(x,p,t) \geq 0$ at every phase space point.

For a pure state, substituting $\bar{p}_2(x,t)$ from equation 3.16 into equation 3.19 gives the momentum variance in terms of derivatives of the probability density

$$\sigma(x,t) = -\frac{\hbar^2}{4}\rho(x,t)\frac{\partial^2 \ln \rho(x,t)}{\partial x^2}, \tag{3.20}$$

in which we note that the classical flow term $p(x,t)^2\rho(x,t)$, has canceled. Expressing the density for a pure state in terms of the R-amplitude $\rho = R^2$, the variance becomes

$$\sigma(x,t) = -\frac{\hbar^2}{2}\left[R\frac{\partial^2 R}{\partial x^2} - \left(\frac{\partial R}{\partial x}\right)^2\right]. \tag{3.21}$$

In the following section, the quantum potential and the quantum force will be expressed in terms of $\sigma(x,t)$ and its x-derivative.

The variance in equation 3.21 is closely related to a quantity that will be introduced in Chapter 13. The quantum version of the Navier–Stokes equation, which determines the rate of change in the momentum density, contains a term $-\partial\Pi/\partial x$ (see equation 13.5) involving the derivative of the stress tensor (a scalar for the one-dimensional case). The stress, given in equation 13.6, contains both classical and quantum components. The classical component depends on the flow velocity (related to the gradient of the action), and the quantum component depends on derivatives of the density and thus measures the *shape stress* in the probability fluid. The *quantum shape stress is related to the momentum variance* in equation 3.21 through the equation $\Pi_{qu}(x,t) = \sigma(x,t)/m$.

In order to provide one example of momentum moments computed from the Wigner function, we will return to the now familiar metastable well potential. Using the Wigner function at $t = 6$ fs (shown previously in figure 3.4) the low-order moments were computed. Figure 3.6 shows the 0-moment (position space probability density), 1-moment, 2-moment, momentum variance, and average momentum plotted as functions of x. The probability density has shifted slightly toward the soft part of the potential at positive values of the x-coordinate. Due to the tube of density extending to the upper right corner of figure 3.4, the average value of the momentum increases significantly when x is greater than about 0.2. The 2-moment follows the shape of the probability density, except near the barrier maximum ($x^* = 0.47$), where it is approximately constant. The momentum variance follows the shape of the 2-moment, but with a smaller value, especially near the barrier maximum.

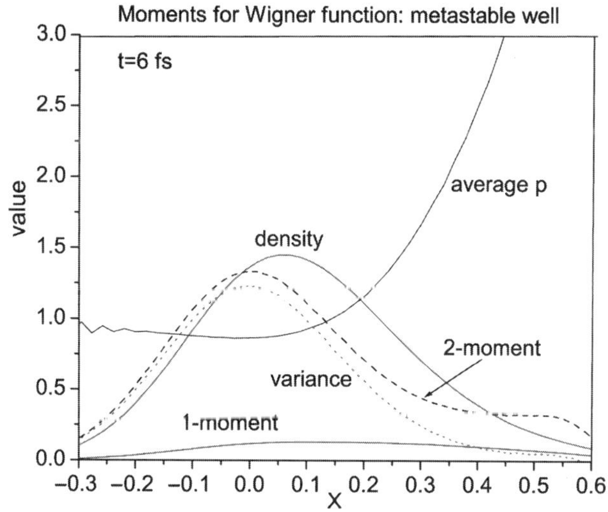

FIGURE 3.6. Momentum moments computed from the Wigner function shown earlier in figure 3.4. The 0-moment (position space probability density) multiplied by a factor of 100, 1-moment, 2-moment, variance, and average momentum (in atomic units) are plotted as functions of x.

3.5 Equations of Motion for the Moments

The equations of motion for the momentum moments of the Wigner function are obtained by multiplying the Wigner–Moyal equation, equation 3.10, by p^n and integrating over the momentum. Before giving several examples, the following assumptions are made concerning boundary conditions on the Wigner function and its momentum derivatives:

$$W(x, p, t) \xrightarrow{p\to\pm\infty} 0,$$
$$\frac{\partial^n W(x, p, t)}{\partial p^n} \xrightarrow{p\to\pm\infty} 0. \tag{3.22}$$

To give an example, in order to evaluate the rate of change in the 0-moment, we simply integrate both sides of the Wigner–Moyal equation with respect to p and obtain

$$\frac{\partial\langle W\rangle}{\partial t} = -\frac{1}{m}\left\langle p\,\frac{\partial W}{\partial x}\right\rangle + \frac{\partial V}{\partial x}\left\langle\frac{\partial W}{\partial p}\right\rangle - \frac{\hbar^2}{24}\frac{\partial^3 V}{\partial x^3}\left\langle\frac{\partial^3 W}{\partial p^3}\right\rangle + \dots. \tag{3.23}$$

Now, using the expressions

$$\left\langle\frac{\partial W}{\partial p}\right\rangle = W(x, \infty, t) - W(x, -\infty, t) = 0,$$
$$\left\langle\frac{\partial^3 W}{\partial p^3}\right\rangle = \frac{\partial^2 W}{\partial p^2}\Big|_{p\to-\infty}^{p\to+\infty} = 0, \quad \left\langle\frac{\partial^n W}{\partial p^n}\right\rangle = \frac{\partial^{n-1} W}{\partial p^{n-1}}\Big|_{p\to-\infty}^{p\to+\infty} = 0, \tag{3.24}$$

we then obtain from equation 3.23

$$\frac{\partial \langle W \rangle}{\partial t} = -\frac{1}{m}\frac{\partial \langle pW \rangle}{\partial x}, \qquad \text{or} \qquad \frac{\partial \bar{p}_0}{\partial t} = -\frac{1}{m}\frac{\partial \bar{p}_1}{\partial x}. \tag{3.25}$$

Since $\rho(x,t) = \bar{p}_0(x) = \langle W \rangle$ and $\bar{p}_1(x,t) = p(x,t)\rho(x,t) = \langle pW \rangle$, we finally obtain

$$\frac{\partial \rho(x,t)}{\partial t} = -\frac{1}{m}\frac{\partial p(x,t)\rho(x,t)}{\partial x}. \tag{3.26}$$

This equation is recognized as the *Eulerian version of the continuity equation.*

A second example will be given. To obtain the equation of motion for the 1-moment, we multiply the Wigner–Moyal equation by p and integrate over the momentum to obtain

$$\frac{\partial \langle pW \rangle}{\partial t} = -\frac{1}{m}\left\langle p^2 \frac{\partial W}{\partial x}\right\rangle + \frac{\partial V}{\partial x}\left\langle p\frac{\partial W}{\partial p}\right\rangle - \frac{\hbar^2}{24}\frac{\partial^3 V}{\partial x^3}\left\langle p\frac{\partial^3 W}{\partial p^3}\right\rangle + \cdots. \tag{3.27}$$

The second integral on the right is evaluated by integrating by parts:

$$\left\langle p\frac{\partial W}{\partial p}\right\rangle = -\langle W \rangle = -\bar{p}_0. \tag{3.28}$$

Evaluation of the remaining integrals and applying boundary conditions gives

$$\frac{\partial \bar{p}_1}{\partial t} = -\bar{p}_0\frac{\partial V}{\partial x} - \frac{1}{m}\frac{\partial \bar{p}_2}{\partial x}. \tag{3.29}$$

Note that the rate of change in the 1-moment is coupled on the right side to both the 0-moment and the 2-moment.

If we now use the relations $\bar{p}_1 = p\rho$ and $\sigma = \bar{p}_2 - p^2\rho$ (where $p(x,t)$ is again the average momentum), we obtain, after some algebra,

$$m\left[\frac{\partial}{\partial t} + v\frac{\partial}{\partial x}\right]v = -\frac{\partial V}{\partial x} - \frac{1}{m}\frac{1}{\rho}\frac{\partial \sigma}{\partial x}. \tag{3.30}$$

This is a Newtonian-type equation of motion, expressed in the Eulerian frame, for the flow velocity $v(x,t) = p(x,t)/m$. However, after transforming to the Lagrangian frame, the left side can be written $m(dv/dt) = dp/dt$. Burghardt et al. [3.9] identify the second term on the right side as the *hydrodynamic force*

$$F_{hydro} = -\frac{1}{m}\frac{1}{\rho}\frac{\partial \sigma}{\partial x}. \tag{3.31}$$

This force may take on both positive and negative values.

There are several very important interelated comments to make about equations 3.30 and 3.31.

(a) The *flow velocity is obtained, for each value of x, from the momentum average of the phase space Wigner function,*

$$v(x,t) = p(x,t)/m = \bar{p}_1(x,t)/(m\rho(x,t)). \tag{3.32}$$

(b) Even though the notation does an excellent job of hiding this feature, the momentum $p(x,t)$ that appears in the Bohmian equations of motion is the *momentum average* of an underlying phase space distribution function.
(c) The hydrodynamic force is determined by the spatial imbalance (measured by the derivative) of the momentum variance of the underlying phase space distribution. In addition, as mentioned in Chapter 2, it may also be viewed as an internal, density-dependent stress.
(d) The form for the hydrodynamic force given in equation 3.31 is valid for both pure states and mixtures and is a generalization of the Bohm quantum force (which was derived for pure states).
(e) Both the phase space description (evolution of the Wigner function generated by the Wigner–Moyal equation) and the hydrodynamic description (evolution of momentum moments) are *exact formulations* of quantum dynamics.

The equations of motion for the next two higher moments follow in the same way: we multiply the Wigner–Moyal equation by p^2or p^3 and then integrate over p. After evaluating some of the integrals by parts, the following equations are obtained:

$$\frac{\partial \bar{p}_2}{\partial t} = -\frac{1}{m}\frac{\partial \bar{p}_3}{\partial x} - 2\frac{\partial V}{\partial x}\bar{p}_1,$$

$$\frac{\partial \bar{p}_3}{\partial t} = -\frac{1}{m}\frac{\partial \bar{p}_4}{\partial x} - 3\frac{\partial V}{\partial x}\bar{p}_2 - \frac{\hbar^2}{4}\frac{\partial^3 V}{\partial x^3}\bar{p}_0. \tag{3.33}$$

Note that when the left side involves the derivative of an odd moment, then the right side brings in only even moments, with the lowest moment being $\bar{p}_0$. Likewise, an equation for the time derivative of an even moment involves only odd moments on the right side, with the lowest moment being $\bar{p}_1$. Finally, the above equation for $\partial \bar{p}_3/\partial t$ is the first one to bring in terms from the Wigner–Moyal equation that have *explicit $\hbar$ dependence*.

In each equation of motion for one of the momentum moments, there is only one term on the right side involving a higher moment, this being $-(1/m)\partial \bar{p}_{n+1}/\partial x$. This type of term has a classical origin, coming from the *convective term* (also called the drift or streaming term) $-(p/m)\partial/\partial x$ in the Liouville equation. The remaining moments on the right side are of lower order than the moment on the left side whose time derivative is being evaluated.

There is considerable simplification in these equations of motion when we are dealing with pure states. In this case, only the lowest two moments, $\bar{p}_0$ and $\bar{p}_1$, are independent; all higher moments may be expressed in terms of the probability density and the flux, ρ and j. Equation 3.30 has a form reminiscent of the Bohm equation of motion in Box 2.5. However, the hydrodynamic force in equation 3.30 is not a recognizable version of the quantum force that appeared earlier, in Chapter 2. In order to make the connection between equation 3.31 and Box 2.5, we will first substitute equation 3.21 for the momentum variance into equation 3.31

for the hydrodynamic force:

$$F_{hydro} = -\frac{1}{m\rho}\frac{\partial}{\partial x}\left[-\frac{\hbar^2}{4}\rho\frac{\partial^2 \ln \rho}{\partial x^2}\right]. \tag{3.34}$$

Next, after expressing the density in terms of the R-amplitude, $\rho = R^2$, we obtain

$$F_{hydro} = \frac{\hbar^2}{2m}\frac{1}{R^2}\left[R\frac{\partial^3 R}{\partial x^3} - \frac{\partial R}{\partial x}\frac{\partial^2 R}{\partial x^2}\right]. \tag{3.35}$$

However, if we start with the Bohm quantum potential expressed in terms of the curvature of the R-amplitude and then evaluate the quantum force

$$F_{qu} = -\frac{\partial Q}{\partial x} = -\frac{\partial}{\partial x}\left[-\frac{\hbar^2}{2m}\frac{1}{R}\frac{\partial^2 R}{\partial x^2}\right], \tag{3.36}$$

we obtain the *same expression* for F_{hydro} that was given in equation 3.35. For pure states, these two forces are identical. Thus the equation of motion given in equation 3.30 is identical with the Bohm equation that was derived in Chapter 2.

To give one example, for a freely evolving Gaussian wave packet, $R(x,t) = N\exp\left[-\beta(t)(x - v(x,t)t)^2\right]$, the hydrodynamic (quantum) force is given by $F_{hydro}(x,t) = (2\hbar^2\beta(t)^2/m)(x - v(x,t)t)$, where the term $v(x,t)t$ locates the moving center of the spreading wave packet. Thus, elements of the wave packet furthest from the center feel the largest quantum force, while the quantum force vanishes at the center of the packet.

In the case of mixed states, the rate equation hierarchy for the momentum moments does not terminate, but in some cases well-defined truncation schemes may be applied. Lill, Haftel, and Herling [3.17] and Burghardt and Cederbaum [3.9–3.10] have independently described truncation schemes based on Gaussian approximations for the density matrix; Bittner et al. [3.21] have discussed truncation schemes based on properties of cumulant expansions; and Ploszajczak and Rhodes-Brown [3.18] described methods for obtaining a closed self-consistent set of equations. Finally, in Section 3.1, we mentioned that Degond and Ringhofer described an entropy minimization method for truncating the moment system [3.22].

3.6 Moment Analysis for Classical Phase Space Distributions

In the preceding sections, we have developed equations of motion for the momentum moments associated with a quantum mechanical phase space distribution function. A similar approach may be extended to the (nonequilibrium) normalized classical phase space density $W_{cl}(x, p, t)$, which is (unlike the Wigner function, in general) positive definite, $W_{cl}(x, p, t) \geq 0$, at all points in phase space. Momentum moments for both classical and quantum phase space distributions were described by Muga et al. [3.8] (see the discussion and plots in Section 3.7 of this study) and more recently in the detailed analyses by Burghardt and coworkers [3.9–3.11].

The *classical momentum moments*, defined analogously to equations 3.14–3.15, form a family of functions at each value of x:

$$\bar{p}_n^{cl}(x,t) = \int_{-\infty}^{\infty} p^n W_{cl}(x,p,t)\,dp = \langle p^n W_{cl}\rangle. \tag{3.37}$$

Likewise, the local density in position space and the average momentum are given by

$$\rho_{cl}(x,t) = \bar{p}_0^{cl}(x,t) = \int_{\infty}^{\infty} W_{cl}(x,p,t)\,dp = \langle W_{cl}\rangle,$$

$$p_{cl}(x,t) = \langle p W_{cl}\rangle \,/\, \langle W_{cl}\rangle = \bar{p}_1^{cl}(x,t)/\bar{p}_0^{cl}(x,t). \tag{3.38}$$

and the variance in the momentum distribution, by analogy to equation 3.18, is given by

$$\sigma_{cl}(x,t) = \int_{-\infty}^{\infty} (p - p_{cl}(x,t))^2 W_{cl}(x,p,t)\,dp. \tag{3.39}$$

Equations of motion for the classical momentum moments are derived by multiplying the Liouville equation by p^n and integrating over p. For reference in this section, the Liouville equation is (see equation 3.11)

$$\frac{\partial W_{cl}(x,p,t)}{\partial t} = -\frac{p}{m}\frac{\partial W_{cl}(x,p,t)}{\partial x} + \frac{\partial V}{\partial x}\frac{\partial W_{cl}(x,p,t)}{\partial p}. \tag{3.40}$$

After multiplying by p^n and integrating over the momentum, we obtain

$$\frac{\partial\langle p^n W_{cl}\rangle}{\partial t} = -\frac{1}{m}\left\langle p^{n+1}\frac{\partial W_{cl}}{\partial x}\right\rangle + \frac{\partial V}{\partial x}\left\langle p^n \frac{\partial W_{cl}}{\partial p}\right\rangle, \tag{3.41}$$

and to be more explicit, the first two equations in the hierarchy are

$$\begin{aligned}\frac{\partial \bar{p}_0^{cl}}{\partial t} &= -\frac{1}{m}\frac{\partial \bar{p}_1^{cl}}{\partial x},\\ \frac{\partial \bar{p}_1^{cl}}{\partial t} &= -\frac{1}{m}\frac{\partial \bar{p}_2^{cl}}{\partial x} - \frac{\partial V}{\partial x}\bar{p}_0^{cl},\end{aligned} \tag{3.42}$$

while the general equation of motion for the n-moment becomes

$$\frac{\partial \bar{p}_n^{cl}}{\partial t} = -\frac{1}{m}\frac{\partial \bar{p}_{n+1}^{cl}}{\partial x} - n\frac{\partial V}{\partial x}\bar{p}_{n-1}^{cl}. \tag{3.43}$$

As in the quantum-mechanical equations of motion, there is up-coupling to the next higher moment in the first term on the right and down-coupling to the previous lower moment in the second term. The up-coupling is always a "convective" flow term.

In terms of the *local density* and the *average momentum*, the first two moment equations become

$$\frac{\partial \rho_{cl}(x,t)}{\partial t} = -\frac{1}{m}\frac{\partial p_{cl}(x,t)\rho_{cl}(x,t)}{\partial x},$$

$$m\left[\frac{\partial}{\partial t} + v_{cl}\frac{\partial}{\partial x}\right] v_{cl} = -\frac{\partial V}{\partial x} + F^{cl}_{hydro}. \tag{3.44}$$

where the *classical hydrodynamic force* is determined by the spatial derivative of the variance

$$F^{cl}_{hydro}(x,t) = -\frac{1}{m}\frac{1}{\rho_{cl}(x,t)}\frac{\partial \sigma_{cl}(x,t)}{\partial x}. \tag{3.45}$$

Equations 3.44 and 3.45 are analogous, term by term, to the equations of motion for quantum trajectories. However, in the quantum case, the momentum variance depends on $\hbar$, whereas of course, $\sigma_{cl}(x,t)$ does not. The equations of motion for the first two moments usually *do not form a closed set*, because the 2-moment appearing in σ_{cl} is not decoupled from the higher momentum moments. These two equations were also derived by Walkup in 1991 [3.25] (the last two terms in his equation 7 are equivalent to equation 3.45).

The significant point is that *the classical hydrodynamic force has exactly the same form and interpretation as the quantum (hydrodynamic) force (for both pure states and mixtures).* In both cases, *this internal force arises through the reduction of equations of motion from phase space to position space.* In addition, *the momentum that appears in the reduced equations of motion in position space is the local (at each value of x) average momentum associated with the underlying classical phase space distribution.*

The impression that one would get from Chapter 2 is that the quantum force has no classical analogue. However, in this chapter, we have seen that *the Bohm quantum force for pure states is a special case of a more general force, the hydrodynamic force, that makes its appearance in reduced equations of motion in both classical and quantum mechanics.* The unifying feature is the continuum mechanics of continuous distributions in phase space; classical *or* quantum distributions give rise to a hydrodynamic force when the dynamics are reduced to position space by developing equations of motion for the momentum moments.

There is one important way in which the quantum and classical phase space distributions functions can differ [3.11]. The quantum distribution must always be "spread out" (as implied by equation 3.8), but the classical function can become arbitrarily sharp. In the latter case, this needle-shaped δ−function has a vanishing momentum variance $\sigma \to 0$, and all of the momentum moments have the form $\bar{p}^{cl}_n(x,t) = p(x,t)^n\delta(x - x_{cl}(t))$. As a result, the hydrodynamic force vanishes, and the equations of motion generate the dynamics of a *single classical trajectory*. However, as pointed out previously, if the classical distribution is spread out, analogous to a quantum distribution, then the momentum variance will be nonzero, and the hydrodynamic force will be nonvanishing.

3.7 Time Evolution of Classical and Quantum Moments

An informative comparison of time-dependent classical and quantum momentum moments for the scattering of an initial Gaussian phase space distribution from a repulsive barrier was presented in the 1993 study by Muga et al. [3.8]. The Eckart barrier is given by $V(x) = V_0\,[\cosh(ax)]^{-2}$, and the initial Gaussian distribution (used for both the classical and quantum cases) is given by

$$W(x, p, 0) = \frac{1}{\pi\hbar} \exp\left[-(x - x_0)^2/(2\delta^2) - 2\delta^2(p - p_0)^2/\hbar^2\right], \tag{3.46}$$

where (x_0, p_0) locates the average initial position and momentum. The parameter values used in this study (in atomic units) are $m = 1$, $E = 5$ (initial average energy), $p_{th} = \sqrt{10} = 3.16$ (classical momentum threshold for transmission), $x_0 = -10$, and $p_0 = 3$. Figure 3.7 shows the classical and quantum position space densities at four times, t =1, 3, 5, and 7 a.u. Although the transmitted densities (for $x > 0$)

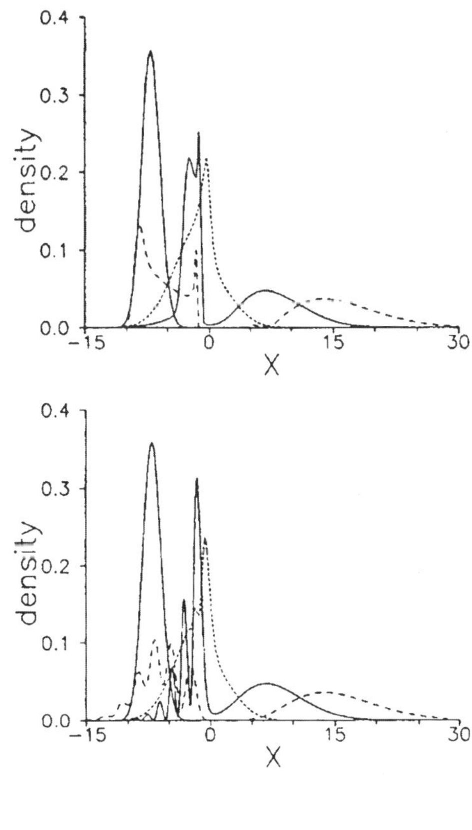

FIGURE 3.7. Probability densities vs. position at four times [3.8]: $t = 1$ (solid), $t = 3$ (dotted), $t = 5$ (solid), $t = 7$ (dashed). The top panel shows the classical densities, and the lower panel shows the quantum densities.

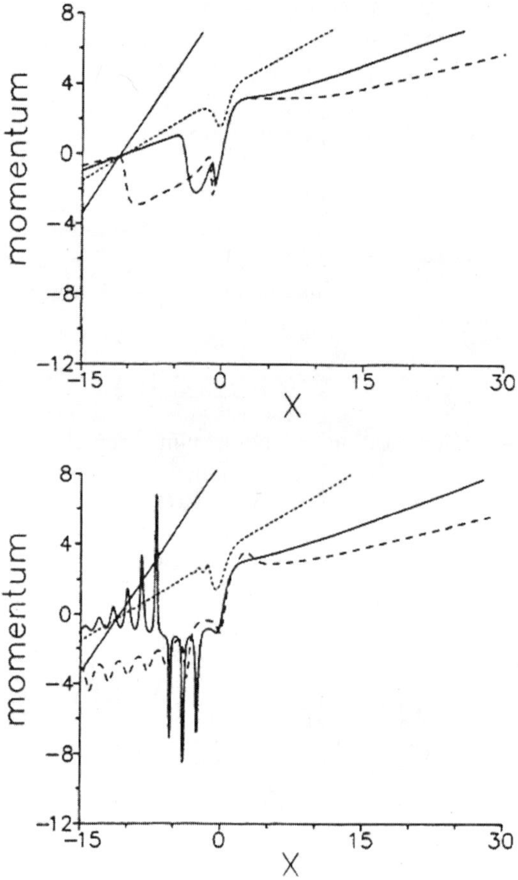

FIGURE 3.8. Average momentum versus position at four times [3.8], for the same conditions as figure 3.7. The top panel shows the classical momenta, and the lower panel shows the quantum momenta.

are quite similar, the reflected densities for the quantum case exhibit multiple interference ripples that are not present in the classical case.

The average momenta are plotted against x in figure 3.8, and the momentum variance is shown in figure. 3.9. There are several comments to be made about the results shown in these figures. In the transmitted region ($x > 0$) in figure 3.8, the classical and quantum average momenta are in excellent agreement. However, in the reflected region ($x < 0$) at $t = 5$ (solid curve), the quantum momentum shows large positive or negative spikes at the positions of the minima in the quantum density in figure 3.7. These interference effects are absent in the classical case and as a result, the spatial variation of the classical momentum is much smoother in this region. The momentum variance shown in figure 3.9 for both the classical and quantum cases is essentially zero in the transmitted region at all times. As a result, the hydrodynamic force has almost no effect in this region. However, in the reflected region, there is a large momentum variance for both the classical and

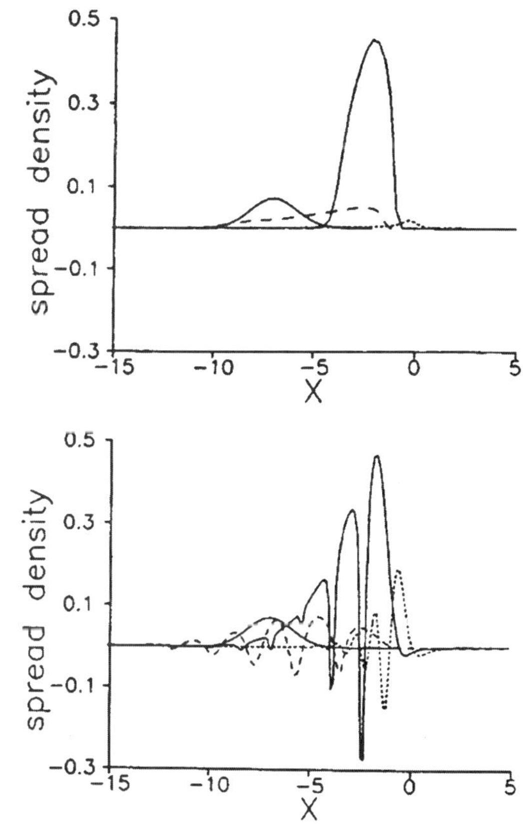

FIGURE 3.9. Momentum variance versus position at four times [3.8], for the same conditions as figure 3.7. The top panel shows the classical variance, and the lower panel shows the quantum variance.

quantum cases, especially at the time $t = 5$. In addition, and this may be surprising, *the quantum momentum variance can become negative*, due to negative basins that develop in the Wigner function just to the left of the barrier maximum. This feature is closely related to the negative spikes in the average momentum that occur in this region during the same time interval.

3.8 Comparison Between Liouville and Hydrodynamic Phase Spaces

In the *Liouville phase space* that was introduced in Section 3.2, x and p are *independent variables*, and a trajectory is located by the pair of coordinates $\{x(t), p(t)\}$. A contour map of the Wigner function, when viewed "vertically" at a fixed value of x, displays the distribution of momenta at this position. Clearly, there is not a unique value of the momentum at each position (except for a single trajectory). Both Takabayasi [3.1] and Burghardt and Moller [3.11] exploit advantages of the *hydrodynamic phase space*: the *average momentum for a hydrodynamic trajectory* $p(x, t)$ is plotted versus x. *For both pure states and statistical mixtures, at each value of the position, there is a unique value for the momentum, the average value.* This average momentum, which is the momentum that appears in the Bohmian equations of motion, is a *dependent variable*. For a pure state, this momentum is given by the gradient of the action function, $p(x, t) = \partial S(x, t)/\partial t$. In hydrodynamic phase space, the density riding along a trajectory is given by $\rho_{hydro}(x, t) = \rho(x, t)\delta(p - p(x, t))$, where the $\delta-$function provides the momentum constraint. (This relation was pointed out by Takabayasi; see equation 5.5 in [3.1].) As emphasized in the preceding sections, in hydrodynamic phase space, classical and quantum trajectories evolve under the influence of both the classical force (given by the gradient of the potential) and the hydrodynamic force. The latter results from the reduction from the full Liouville phase space to the "restricted" hydrodynamic phase space, in which only the $x-$coordinate retains the role of an independent variable.

3.9 Discussion

Rather than taking the Bohmian approach to the quantum trajectory equations that were developed in Chapter 2, in which we started in position space and substituted the polar form of the wave function into the TDSE, in this chapter the analysis began with the time-dependent Wigner distribution in phase space. The seminal idea, introduced by Takabayasi [3.1] and expanded recently [3.9–3.11, 3.20, 3.21], is to develop a system of equations of motion for the momentum moments of the Wigner function in position space. In general, these equations form an infinite hierarchy, with the rate of change of one moment coupled to the next higher moment and to a series of lower moments. For the pure states that were emphasized in this chapter (the system can be represented in terms of a wave function), the hierarchy terminates, and the rate equations for the lowest two moments form a closed set.

In this chapter, for both pure states and mixtures, the coupled rate equations for position space momentum moments of the phase space distribution function were derived. These dynamical equations can be viewed as the projection of phase space dynamical equations (such as the Wigner–Moyal equation) into position. For pure states, when these equations were compared with the Bohmian trajectory equations developed previously in Chapter 2, it was revealed that the momentum appearing in the latter equations is the average value at each position of the underlying phase space distribution function. For mixed states, the hydrodynamic force (a generalization of the quantum force) was described, as was its dependence on the momentum variance of the phase space distribution. In addition, rate equations for the momentum moments were also developed for *classical* phase space probability distributions. Even in this case, trajectories in an ensemble move in position space under the influence of a hydrodynamic force. Finally, we described distinctions between the Liouville phase space that was introduced in Section 3.2

and the hydrodynamic phase space, in which the momentum plays the role of a dependent variable. As these concepts were developed in this chapter, not only did we develop a deeper understanding of the various terms appearing in the Bohmian trajectory equations, but the possibility of significant extensions to more general types of systems was revealed. In addition, and this will be very important for the developments to be described later, in Chapter 11, the phase space route is also applicable to open quantum systems coupled to thermal environments.

References

3.1. T. Takabayasi, *The formulation of quantum mechanics in terms of an ensemble in phase space*, Prog. Theor. Phys. **11**, 341 (1954).

3.2. G.J. Iafrate, H.L. Grubin, and D.K. Ferry, *Utilization of quantum distribution functions for ultra-submicron device transport*, J. de Physique Colloq. **42**, 307 (1981).

3.3. W.R. Frensley, *Boundary conditions for open quantum systems driven far from equilibrium*, Rev. Mod. Phys. **62**, 745 (1990).

3.4. C.L. Gardner, *The quantum hydrodynamic model for semiconductor devices*, SIAM J. Appl. Math. **54**, 409 (1994)

3.5. I. Gasser and P.A. Markowich, *Quantum hydrodynamics, Wigner transforms and the classical limit*, Asym. Anal. **14**, 97 (1997).

3.6 C.L. Gardner and C. Ringhofer, *Numerical simulation of the smooth quantum hydrodynamic model for semiconductor devices*, Comp. Methods Appl. Mech. Eng. **181**, 393 (2000).

3.7. W.P. Schleich, *Quantum Optics in Phase Space* (Wiley-VCH, Berlin, 2001).

3.8. J.G. Muga, R. Sala, and R.F. Snider, *Comparison of classical and quantal evolution of phase space distribution functions*, Physica Scripta **47**, 732 (1993).

3.9. I. Burghardt and L.S. Cederbaum, *Hydrodynamic equations for mixed quantum states. I. General formulation*, J. Chem. Phys. **115**, 10303 (2001).

3.10. I. Burghardt and L.S. Cederbaum, *Hydrodynamic equations for mixed quantum states. II. Coupled electronic states*, J. Chem. Phys. **115**, 10312 (2001).

3.11. I. Burghardt and K.B. Moller, *Quantum dynamics for dissipative systems*, J. Chem. Phys. **117**, 7409 (2002).

3.12. E.P. Wigner, *On the quantum correction for thermodynamic equilibrium*, Phys. Rev. **40**, 749 (1932).

3.13. I. Bialynicki-Birula, M. Cieplak, and J. Kaminski, *Theory of Quanta* (Oxford University Press, New York, 1992), Ch. 13.

3.14. M. Hillery, R.F. O'Connell, M.O. Scully, and E.P. Wigner, *Distribution functions in physics: Fundamentals*, Phys. Rep. **106**, 121 (1984).

3.15. P. Carruthers and F. Zachariasen, *Quantum collision theory with phase-space distributions*, Rev. Mod. Phys. **55**, 245 (1983).

3.16. J.E. Moyal, *Quantum mechanics as a statistical theory*, Proc. Camb. Phil. Soc. **45**, 99 (1949).

3.17. J.V. Lill, M.I. Haftel, and G.H. Herling, *Semiclassical limits in quantum-transport theory*, Phys. Rev. A **39**, 5832 (1989).

3.18. M. Ploszajczak and M.J. Rhodes-Brown, *Approximation scheme for the quantum Liouville equation using phase-space distribution functions*, Phys. Rev. Lett, **55**, 147 (1985).

3.19. M. Tabor, *Chaos and integrability in nonlinear dynamics* (Wiley, New York, 1989).
3.20. J.B. Maddox and E.R. Bittner, *Quantum dissipation in the hydrodynamic moment hierarchy: A semiclassical truncation strategy*, J. Phys. Chem. B **106**, 7981 (2002).
3.21. E.R. Bittner, J.B. Maddox, and I. Burghardt, *Relaxation of quantum hydrodynamic modes*, Int. J. Quantum Chem. **89**, 313 (2002).
3.22. P. Degond and C. Ringhofer, *Quantum moment hydrodynamics and the entropy principle*, J. Stat. Phys. **112**, 587 (2003).
3.23. F. McLafferty, *On quantum trajectories and an approximation to the Wigner path integral*, J. Chem. Phys. **83**, 5043 (1985).
3.24. M. Belloni, M.A. Doncheski, and R.W. Robinett, *Wigner quasi-probability distribution for the infinite square well: Energy eigenstates and time-dependent wave packets*, Am. J. Phys. **72**, 1183 (2004).
3.25. R.E. Walkup, *A local-Gaussian approximation for the propagation of a classical distribution*, J. Chem. Phys. **95**, 6440 (1991).

4
The Dynamics and Properties of Quantum Trajectories

In the quantum trajectory method, a wave packet is evolved by propagating an ensemble of correlated trajectories. Details of the quantum trajectory method are presented, and a number of properties of quantum trajectories are described.

4.1 Introduction

In the preceding two chapters, the equations of motion in the hydrodynamical formulation of quantum mechanics were developed from two quite different viewpoints. In the first part of this chapter, it will be shown that the *hydrodynamical equations can be integrated on the fly to generate the density, action, and complex-valued wave function along a set of quantum trajectories.* The wave function is not precomputed in advance of the hydrodynamic analysis; rather, trajectory propagation and wave function generation occur concurrently at each time step. The quantum trajectory method (QTM), reported by Lopreore and Wyatt in 1999 [4.1], is a computational implementation of the synthetic approach to the hydrodynamic formulation of quantum mechanics. Rabitz and coworkers developed a related approach, *quantum fluid dynamics* (QFD), which was reported later the same year [4.2]. In both QTM and QFD, essentially the same hydrodynamic equations are solved, but there are differences in the computational approaches used to generate the quantum trajectories. In this chapter, the formulation of the QTM will be described in some detail, and in later chapters additional interpretative and computational features will make their appearance.

In Section 4.2, the equations of motion for quantum trajectories will be described in detail. In Section 4.3, it will be shown how the complex-valued time dependent wave function can be synthesized along each quantum trajectory. Synthesis of the wave function along quantum trajectories will be contrasted with the Feynman path integral approach for wave function propagation in Section 4.4. A Jacobian enters the equation for propagation of the wave function along quantum trajectories, and this is described in Section 4.5. Time-dependent matrix elements, such as correlation functions, may be expressed as integrals over the initial coordinates, thus leading to the initial value representations that are described in Section 4.6. Quantum trajectories obey two important noncrossing rules; these are described in Section 4.7. The dynamics of quantum trajectories near nodes in the wave function are described in detail in Section 4.8. In the following three sections, we switch from the synthetic to the analytic approach for the computation of quantum trajectories. Using a number of illustrations from the literature, the properties of chaotic quantum trajectories are described in Sections 4.9–4.11. Finally, Section 4.12 deals with the question, Why weren't quantum trajectories computed by the synthetic route just after 1952, the year that Bohm's papers were published?

4.2 Equations of Motion for the Quantum Trajectories

The flow of quantum-mechanical probability density through configuration space is that of a *compressible fluid*. The evolution of this fluid will be described in terms of a relatively small number of *correlated fluid elements* evolving along quantum trajectories. In the quantum trajectory method, the initial wave packet (which is assumed to be known) is discretized in terms of *N fluid elements*, small chunks of the probability fluid. The equations of motion for this set of fluid elements are integrated in lockstep fashion, from one time step to the next. Along each trajectory, the probability density and action function are computed by integrating two coupled equations of motion. From the density and action, the complex-valued wave function is easily synthesized. These fluid elements are correlated with one another through the *quantum potential*. This forces each evolving fluid element to be influenced by the motions of the other elements, even when the external (classical) potential vanishes. The resulting correlation brings all quantum effects into the dynamics. However, if the quantum potential is neglected, the evolution is that of a classical ensemble of N independent mass points. In figure 4.1, two evolving ensembles are shown: in the upper panel, dotted lines represent some of the correlations between fluid elements in the quantum ensemble, while in the lower panel, the mass points move independently along classical trajectories. An analogy to quantum and classical motion is described in Box 4.1.

The equations of motion used in the QTM have already been presented in Box 2.5. There are several ways in which these equations may be combined to give slightly different computational approaches. The *first approach* uses the continuity equation, the Newtonian-type equation, and the quantum Hamilton–Jacobi equation (to find the action along the trajectory). These equations are referred to as the *force version of the hydrodynamic equations*. These equations are

$$\frac{d\rho}{dt} = -\rho\vec{\nabla}\cdot\vec{v}, \tag{4.1}$$

$$m\frac{d\vec{v}}{dt} = -\vec{\nabla}V - \vec{\nabla}Q, \tag{4.2}$$

$$\frac{dS}{dt} = L(t) = \frac{1}{2}mv^2 - (V+Q). \tag{4.3}$$

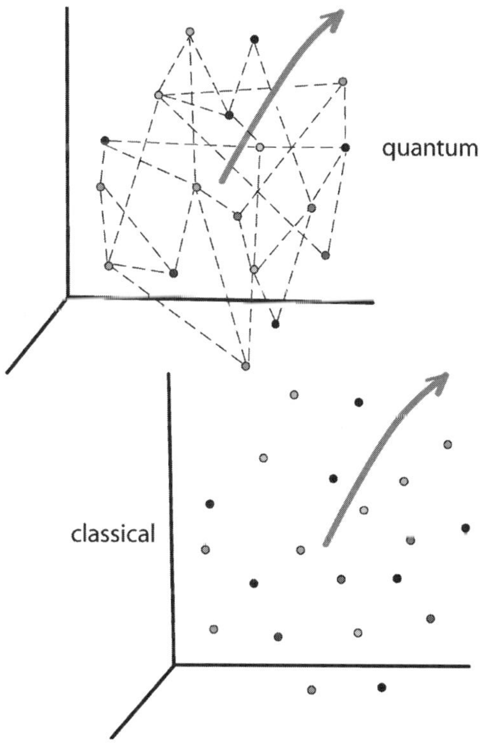

FIGURE 4.1. Quantum and classical ensembles: motions of the fluid elements in the upper panel are correlated through the quantum potential (dotted lines), while for the classical ensemble in the lower plot, the mass points are not aware of each other and follow independent trajectories.

Box 4.1. Classical versus quantum class exams

(Based on an analogy by Craig Martens, University of California at Irvine)

The classical exam: The professor walks into the classroom at 9 AM carrying the exam papers. He announces to the class, "I will now pass out the exams. The exam is over at 9:50. Remember, no talking and of course do not look at anyone else's paper. OK, you may begin." (In this case, the students "propagate" independently.)

The quantum exam: The professor walks into the classroom at 9 AM carrying the exam papers. He announces to the class, "I will now pass out the exams. The exam is over at 9:50. Remember, anytime that you have a thought or whenever you write down something on the exam paper, say it out loud. I want each of you to know what everyone else is doing and thinking. OK, you may begin." (The students propagate as a correlated ensemble! Question: How should the grades be assigned?)

These equations were used in this form in several of the early papers on the QTM [4.3–4.7]. A disadvantage is that spatial derivatives of the quantum potential are needed in equation 4.2, and this brings an additional source of error into the evaluation of the trajectory. In addition to equations 4.1–4.3, the *trajectory equation* is added to find the quantum trajectory from the velocity:

$$\frac{d\vec{r}}{dt} = \vec{v} \quad \Rightarrow \quad \vec{r}(t + \Delta t) = \vec{r}(t) + \vec{v} \cdot \Delta t. \tag{4.4}$$

(A *trajectory* is the solution to the initial value problem $d\vec{r}/dt = v(\vec{r}, t)$, given $\vec{r}(0)$. Physically, this is a long exposure time photograph of an illuminated fluid element [4.49]. In contrast, a *streamline* is a solution to the equation $d\vec{r}/ds = v(\vec{r}, t)$, where t is a constant and s is a parameter. In the latter case, we picture the velocity field at a fixed time.)

The *second set of equations* used to develop quantum trajectories includes the continuity equation, the Hamilton–Jacobi equation, and the trajectory equation. However, the Newtonian equation for the flow acceleration, equation 4.2, is not used at all. As a result, the three equations in the *potential energy version of the hydrodynamic equations* (so named because of the potential energy terms in the Hamilton–Jacobi equation) are as follows:

$$\frac{d\rho}{dt} = -\rho\vec{\nabla} \cdot \vec{v}, \tag{4.5}$$

$$\frac{dS}{dt} = L(t) = \frac{1}{2}mv^2 - (V + Q), \tag{4.6}$$

$$\frac{d\vec{r}}{dt} = \vec{v} = \frac{1}{m}\vec{\nabla}S. \tag{4.7}$$

This set has the advantage that the quantum force is not explicitly evaluated; however, the gradient of the action must be evaluated in order integrate equation 4.7 to find the trajectory. It should be noted that the quantum potential influences the action through integration of equation 4.6 and when the gradient of the action is evaluated in equation 4.7. The time integration of these equations is described in terms of "pseudocode" in Box 4.2 and a FORTRAN program for integrating the QTM equations is provided in Appendix 2.

At each instant of time, the *state of the system* is specified by the *descriptor*

$$D(t) = \{\vec{r}_i(t), R_i(t), S_i(t)\}_{i=1}^{N},$$

which lists the location of each fluid element along with the amplitude and action function at the position of each element. From the latter information, the wave function can be constructed. In the descriptor, the velocity of each fluid element $\vec{v}_i(t)$ need not be specified, because this can be obtained from the gradient of the action function.

If it were not for the spatial derivatives that appear in the three functions $\vec{\nabla} \cdot \vec{v}$, Q, and $\vec{\nabla}S$, equations 4.5–4.7 would be straightforward to integrate. However, the presence of these terms makes integration of these equations a difficult problem in numerical analysis. The reason in that information (i.e., hydrodynamic fields)

Box 4.2. The basic QTM computer program

An example QTM FORTRAN program is presented in Appendix 2. In this box, "pseudocode" is presented to illustrate the main features of the QTM computer program for scattering of a wave packet from an Eckart barrier in one degree of freedom.

1. *Parameter values*
   time step, dt; number of time steps, $ntime$; number of fluid elements, N.
   Least squares: number of basis functions, n_b; number of points in stencil, n_p.
   Initial grid point spacing, δx.
2. *Parameters for initial wave packet*
   center of initial wave packet, x_0; width parameter, β; initial trans. energy, E_0.
   particle mass, m.
3. *Eckart barrier parameters*
   barrier height, V_0; center of barrier, x_b; width parameter, α
4. *Function initialization*
   grid points x_i; speed v_i; density ρ_i; action S_i
5. *Time loop*: *do* $k = 1, ntime$
   Eckart potential and classical force at grid points V_i, fc_i
   C-amplitude at each grid point $C_i = (1/2)\ln(\rho_i)$
   Quantum potential
      Call MLS: input C_i; output the derivatives $d1C_i$, $d2C_i$
      Compute Q_i from $d1C_i$ and $d2C_i$
   Quantum force
      Call MLS: input Q_i; output the quantum force fq_i
   Update the action
      Compute Lagrangian $L_i = (1/2)mv_i^2 - (V_i + Q_i)$
      New action: $S_i \to S_i + L_i \cdot dt$
   Update positions and velocities
      New position $x_i \to x_i + v_i \cdot dt$
      New velocity $v_i \to v_i + (fc_i + fq_i) \cdot dt/m$
   Compute $\partial v/\partial x$ for later use in updating density
      Call MLS: input v_i; output $dvdx_i$
   Update density
      New density $\rho_i \to \rho_i \cdot \exp[-dvdx_i \cdot dt]$
6. *End time loop*

is available only at the positions of the fluid elements, and the locations of these elements are dictated by the equations of motion. Even if the fluid elements start at $t = 0$ on a regular grid, after a short time, due to the influence of the total force acting on each element, the fluid elements will form an unstructured grid (unstructured in the sense that the fluid elements do not form a Cartesian mesh).

Box 4.3. Features of the quantum trajectory method

- Trajectories follow the main features of the density
- Evolve an ensemble of N correlated trajectories
- Integrate coupled equations of motion for density and action
- Trajectory dynamics correlated through the quantum potential
- The quantum potential introduces all quantum effects
- Can synthesize wave function along each evolving trajectory
- No large basis sets or space fixed grids
- No absorbing potentials at edges of region to absorb amplitude
- Only the "classical" potential is needed, not the force
- Could compute classical $V(\vec{r})$ on the fly: a quantum MD algorithm
- Computational effort scales as N, the number of fluid elements
- Computer code can be parallelized

The spatial derivatives involved in the equations of motion bring *nonlocal effects* into the dynamics; it is through the spatial derivatives that each fluid element is influenced by the surrounding hydrodynamic fields.

For many years, the evaluation of accurate derivatives on unstructured grids has been considered to be one of the most challenging and important problems in classical numerical analysis. There still is not a general method of "industrial strength" that is accurate and dependable for multidimensional problems, but a number of good approximate techniques have been developed for specific applications. Later, in Chapter 5, we will consider several algorithms in detail that can be used to approximately evaluate spatial derivatives on unstructured grids. In addition, adaptive moving grids, discussed in Chapter 7, allow for accurate and efficient derivative evaluation.

A number of features of the quantum trajectory method are summarized in Box. 4.3. Parallelization of the QTM computer code is mentioned at the bottom of the list in this box. Based on the two papers by Weatherford, Banicescu, and coworkers [4.8, 4.9], Box 4.4 goes into parallelization in more detail.

4.3 Wave Function Synthesis Along a Quantum Trajectory

Along a quantum trajectory $\vec{r}(t)$ leading from an initial point $(\vec{r}_0, t_0)$ to the final point $(\vec{r}, t)$, the rate of change in the density is given by equation 4.1. When the density is expressed in terms of the R-amplitude, the continuity equation becomes $dR/dt = (-1/(2R))\vec{\nabla} \cdot \vec{v}$. This equation is easily integrated to give the new R-amplitude in terms of the value at the initial time:

$$R(\vec{r}, t) = \exp\left[-\frac{1}{2}\int_{t_0}^{t} (\vec{\nabla} \cdot \vec{v})_{\vec{r}(\tau)} d\tau\right] R(\vec{r}_0, t_0). \tag{4.8}$$

Box 4.4. PQTM: the parallelized version of the QTM computer program

(This material is an "extra" topic that can be skipped on a first reading of this chapter.)

Before describing the strategy that Weatherford et al. [4.8, 4.9] used for parallelization of the quantum trajectory program, we will first outline some general features of parallel computational architectures. Based on the organization of memory and the address space of the memory, the three major classes of parallel computers are listed below.

(a) *Shared memory.* There is a single large address space that can be accessed relatively uniformly by all of the processors. These systems usually have a large physical memory and a small number of processors.
(b) *Distributed memory.* There are many individual address spaces, each of which is local to a single processor. There is a high penalty in latency and bandwidth to address the memory associated with another processor.
(c) *Mixed memory model (clustered memory).* A small number of processors (maybe 2 to 10) in one node share local memory. Each processor has fast access to the local memory on this node, but there is a stiff penalty to access the memory attached to a different node.

By inserting compiler directives into the code so as to parallelize certain control loops, Weatherford et al. developed a parallel version of the QTM program for a shared memory environment. The system that they used had 8 processors and 4 GB of shared memory. In the QTM program, there are two main nested control loops, the outer one over time and the inner one over the individual fluid elements. Then for each fluid element, there are three control loops that have computational dependence on other fluid elements. These three sections of code use the moving least squares algorithm to evaluate the following derivatives: (1) given the R-amplitudes from the previous time step, the quantum potential is evaluated; (2) given the quantum potential, the quantum force is evaluated; (3) after updating the positions and velocities of all fluid elements, the divergence of the velocity field is evaluated at the position of each fluid element. At the conclusion of each of these derivative evaluation steps, the parallel processes are synchronized.

Using this PQTM code, Weatherford et al. studied the parallel speedup as a function of the number of fluid elements used to discretize the wave packet and with respect to the number of processors. Using 111 fluid elements for a one-dimensional scattering problem, the speedup was a factor of 5.2 with 8 processors. With a smaller number of fluid elements (21), the speedup was smaller (2.9 with 8 processors). A number of other results are reported in the two papers cited previously. The parallelization of quantum hydrodynamic programs will be especially important for studies in higher dimensionality.

Box 4.5. $\vec{\nabla} \cdot \vec{v}$ is the rate of what?

We will first consider two trajectories in one dimension. At time t, the trajectories, at positions x and $x + d(t)$, are separated by the small distance $d(t)$. The two trajectories then move distances Δx_1 and Δx_2 as the time advances a small amount to $t + \Delta t$. The new separation between the trajectories is then

$$\begin{aligned} d(t + \Delta t) &= d(t) + \Delta x_2 - \Delta x_1, \\ &= d(t) + v(x + d(t)) \cdot \Delta t - v(x) \cdot \Delta t, \\ &= d(t) + \frac{\partial v}{\partial x} d(t) \cdot \Delta t. \end{aligned} \tag{1}$$

As a result, the change in separation between the two trajectories is

$$\Delta d = \frac{\partial v}{\partial x} d(t) \cdot \Delta t, \tag{2}$$

and the change in separation divided by the initial separation becomes

$$\frac{\Delta d}{d} = \frac{\partial v}{\partial x} \cdot \Delta t. \tag{3}$$

The left side of this equation is the *fractional change in the separation* between the two trajectories during the time interval Δt. The rate of change in this quantity is the one-dimensional analogue of the divergence

$$\frac{(\Delta d/d)}{\Delta t} = \frac{\partial v}{\partial x}. \tag{4}$$

In multiple dimensions, we obtain

$$\begin{matrix}\text{Rate of change in fractional} \\ \text{change of volume element} \\ \text{defined by trajectories at corners}\end{matrix} = \vec{\nabla} \cdot \vec{v}. \tag{5}$$

Thus the divergence of the velocity locally measures the rate of change in a geometric quantity. If the velocity field has positive divergence, then the local volume element is expanding. Arguments identical to those made above can also be made for classical trajectories; for example, see the study by Stodden and Micha [4.10].

The divergence of the velocity field that appears in this equation is integrated along the trajectory $\vec{r}(t)$. We note that $\vec{\nabla} \cdot \vec{v}$ has units of a rate, 1/time. In Box 4.5, we answer the question "rate of what?" In addition, from the equation for dS/dt given in equation 4.6, the exponential of the action at the final point is given by

$$e^{iS(\vec{r},t)/\hbar} = \exp\left[\frac{i}{\hbar} \int_{t_0}^{t} L(\tau) d\tau\right] e^{iS(\vec{r}_0,t_0)/\hbar}. \tag{4.9}$$

We will now multiply the two preceding equations to obtain an equation for updating the wave function along the trajectory. The two exponential factors in the following expression update the R-amplitude and the phase of the initial wave function

$$\psi(\vec{r}, t) = \exp\left[-\frac{1}{2}\int_{t_0}^{t} (\vec{\nabla}\cdot v)_{\vec{r}(\tau)} d\tau\right] \exp\left[\frac{i}{\hbar}\int_{t_0}^{t} L(\tau) d\tau\right] \psi(\vec{r}_0, t_0). \tag{4.10}$$

This equation builds in nonlocality through the dependence of the velocity on the quantum force and the dependence of the quantum Lagrangian on the quantum potential [4.3]. The product of the two exponentials in equation 4.10 is the *hydrodynamical (wave function) propagator* (both integrals are evaluated along the specified quantum trajectory). Holland has elaborated further on expressions such as equation 4.10, which link wave and "particle" concepts [1.33].

There is another way to derive the wave function propagator in equation 4.10. If we evolve along a quantum trajectory from time t to time $t + dt$, the new wave function is given by

$$\psi(t + dt) = \psi(t) + \frac{\partial \psi}{\partial t} dt + (\vec{v}\cdot\vec{\nabla}\psi)\, dt. \tag{4.11}$$

If we now use the TDSE to evaluate $\partial\psi/\partial t$, use $\vec{v} = (1/m)\vec{\nabla}S$ in the last term, and use the polar form of the wave function throughout, then after some algebra we obtain

$$\psi(t + dt) = \left\{1 - \frac{1}{2}(\vec{\nabla}\cdot\vec{v})dt + \frac{i}{\hbar}\left[\frac{1}{2}mv^2 - (V + Q)\right] dt\right\} \psi(t). \tag{4.12}$$

The term within the braces $\{\ldots\}$ is the *wave function propagator* for the time increment dt. After a number of these small time steps, the composite propagator is identical to equation 4.10.

For those familiar with Feynman path integrals, equation 4.10 is quite remarkable. We will comment further on this equation after a brief diversion to review the elementary features of Feynman's approach to quantum mechanics.

4.4 Bohm Trajectory Integral Versus Feynman Path Integral

In Feynman's path integral formulation of quantum mechanics [4.11, 4.12], the wave function at point $(\vec{r}, t)$ is constructed by integrating over contributions from different starting points $\vec{r}_0$ at the initial time t_0:

$$\psi(\vec{r}, t) = \int K(\vec{r}, t; \vec{r}_0, t_0)\psi(\vec{r}_0, t_0) dr_0. \tag{4.13}$$

In this expression, reading from the right, we multiply the amplitude to be at the starting point $(\vec{r}_0, t_0)$, namely, the initial wave function evaluated at that point, times the transition amplitude for making the hop from the starting point to the viewing point $(\vec{r}, t)$. We then sum over the contributions from all of the initial points. This synthesis of the wave function at *one point* by gathering contributions from *many points* at an earlier time is the quantum-mechanical expression of *Huygens' principle* in wave optics.

The transition amplitude, or Feynman propagator, in equation 4.13 is obtained by summing the exponential phase factor $\exp(iS_c[\vec{r}(t)]/\hbar)$ over all paths $\vec{r}(t)$ linking the initial and final points:

$$K(\vec{r}, t; \vec{r}_0, t_0) = N \int_{\vec{r}_0, t_0}^{\vec{r}, t} e^{iS_c[\vec{r}(t)]/\hbar} D\vec{r}(t). \tag{4.14}$$

The various paths linking these points all carry the common normalizing factor (N), or magnitude, but they contribute different phase factors. The expression on the right side of equation 4.14 (rather than the one in equation 4.13) is a "path integral". The standard notation

$$\int_{initial}^{final} (\ldots) D\vec{r}(t) \approx \sum_{paths} (\ldots), \tag{4.15}$$

indicates that we are to sum contributions to the integrand over all paths going between the initial and final points. The *classical action* evaluated for one of these paths is given by

$$S_c[\vec{r}(t)] = \int_{t_0}^{t} \left[\frac{1}{2} m(d\vec{r}(t)/dt) \cdot (d\vec{r}(t)/dt) - V(\vec{r}(t))\right] dt, \tag{4.16}$$

where the quantity in brackets is recognized as the *classical Lagrangian*. The notation $S_c[\ldots]$ is used to emphasize that the action is a *functional* of the path. The Feynman propagator is a kind of wave function; if the initial wave function in equation 4.13 is spiked at point $\vec{r}_0 = \vec{a}$, $\psi(\vec{r}_0, t_0) = \delta(\vec{r}_0 - \vec{a})$, then the resulting wave function at point $\vec{r}$ *is* the propagator $\psi(\vec{r}, t) = K(\vec{r}, t; \vec{a}, t_0)$. The classical trajectory linking the two endpoints, assuming that there is one, gives the stationary phase contribution to the integrand in equation 4.14.

The contrast between the Feynman and Bohm constructions of the wave function at the point $(\vec{r}, t)$ is striking. Feynman says that we must sum over an *infinite number of independent but interfering paths*, the phase for each path determined by a Lagrangian of "classical" form. In the hydrodynamic formulation, however, we need to propagate only along *one quantum trajectory, which is influenced by the underlying hydrodynamic fields.* However, this quantum trajectory is "aware" of the wave function in the surrounding region through the influence of the quantum potential and the curvature of the action ($\vec{\nabla} \cdot \vec{v} = (1/m)\nabla^2 S$). Nonlocality thus enters the Feynman and Bohmian propagation schemes in very different ways. Additional relationships between Bohm trajectories and Feynman paths are described by Abolhasani and Golshani [4.45].

Historical comment. Feynman and Bohm crossed paths many times. Feynman enjoyed visiting Brazil during carnival season, and some of these visits overlapped the period that Bohm was teaching in São Paulo (1951–1955). In 1952, Feynman took time from bongo drumming to spend several days with Bohm at Copacabana Beach. During these meetings, it is not known whether Bohm ever attempted the bongo drums. Many interesting comments about these quite different personalities are found in David Peat's biography of Bohm [4.13].

4.5 Wave Function Propagation and the Jacobian

Geometry plays an important role in the interpretation of equation 4.10. At time t, imagine a volume element $dV(t)$, possibly having the shape of a square or rectangle for the two-dimensional case, such as shown in figure 4.2. The corners of this element are defined by the positions of trajectories. For example, in one dimension, two trajectories located at $x_0 - dx/2$ and $x_0 - dx/2$ border the element of length given by $dl = dx$. In figure 4.2, four trajectories at the corners $\{a, b, c, d\}$ define an element whose area is dA. As time advances by the increment dt, the trajectories move to new positions according to the equations of motion. Again referring to figure 4.2, the four trajectories at the corners of the starting box move to the new locations $\{a', b', c', d'\}$. By virtue of the movement of each of these corners, the volume element changes into $dV(t + dt)$. What is the relationship between the final and initial volumes?

Jacobi studied the relationship between transported volume elements and expressed the ratio of the new to the old volume elements in terms of what is now called the *Jacobian:*

$$dV(t_0 + dt) = J(t_0 + dt, t_0)dV(t_0). \tag{4.17}$$

A slight extension of the analysis given in Box 4.5 can be used to show that when time advances from t_0 to t, the Jacobian is given by

$$J(t, t_0) = \exp\left[\int_{t_0}^{t} \vec{\nabla} \cdot \vec{v}\, dt\right]. \tag{4.18}$$

This equation states that if the velocity field has positive divergence (the velocity vectors "point away from each other"), then the Jacobian (and the volume element)

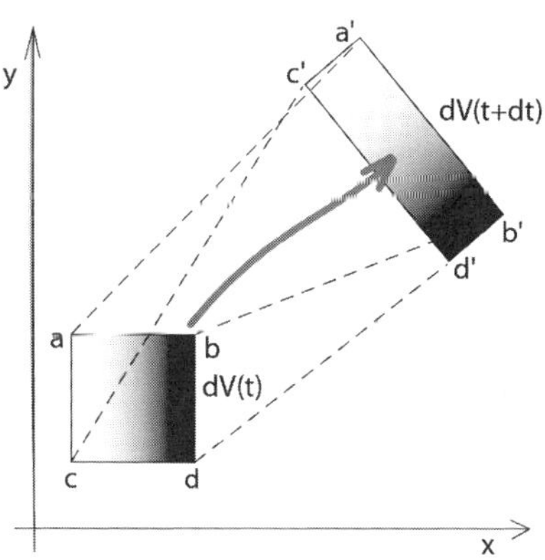

FIGURE 4.2. Area elements at times t and $t + dt$. The ratio of the new area to the old area is given by the Jacobian of the transformation.

will increase along the flow. In this expression, the divergence is evaluated along the trajectory. When the flow is incompressible, $\vec{\nabla} \cdot \vec{v} = 0$, the Jacobian is invariant, but this is *not* usually the case with regard to the flow of probability density in quantum mechanics. (In Chapter 11, we will see that *classical* flow in phase space is incompressible; this is Liouville's theorem.)

The Jacobian in equation 4.18 is actually the solution to the differential equation

$$\frac{dJ}{dt} = +(\vec{\nabla} \cdot \vec{v})J, \tag{4.19}$$

and it might not be surprising to note that this equation has *almost* the same form (note the + sign on the right) as the Lagrangian version of the continuity equation for the probability density. The initial condition on the solution is $J(t_0, t_0) = 1$, and this has already been used in equation 4.18. (Equation 4.19 is referred to as Euler's equation, even though there are many different equations bearing this designation.) The quantity $\kappa = (\vec{\nabla} \cdot \vec{v})$, sometimes referred to as the *compressibility* of the system [4.40], measures the degree of expansion or contraction of a bundle of trajectories.

Historical comment. Carl Gustav Jacob Jacobi was born in Potsdam in 1804 and died in Berlin in 1851. He made important contributions to elliptic functions, mechanics, partial differential equations, and variational calculus.

The Jacobian introduced in this section is connected to equation 4.10 in the following way: the first exponential on the right side of this equation is closely related to the Jacobian of the transformation of volume elements that occurs along the flow. Using equation 4.18, we obtain equations for the time-dependent wave function and the probability density along the quantum trajectory:

$$\psi(\vec{r},t) = J(t)^{-1/2} \exp\left[\frac{i}{\hbar}\int_{t_0}^{t} L(\tau)d\tau\right]\psi(\vec{r}_0,t_0),$$

$$\rho(\vec{r},t)J(t,t_0) = \rho(\vec{r}_0,t_0). \tag{4.20}$$

The latter equation expresses *conservation of the product* ρJ along the quantum trajectory. This result is one way to express the continuity equation in the Lagrangian frame. As the volume element changes along the flow, the density adjusts accordingly such that this product always retains the value specified by the initial condition.

It is straightforward to prove that the quantity ρJ is conserved along the trajectory. We have

$$\frac{d(\rho J)}{dt} = \rho\frac{dJ}{dt} + J\frac{d\rho}{dt} = \rho\left[J\vec{\nabla}\cdot\vec{v}\right] + J\left[-\rho\vec{\nabla}\cdot\vec{v}\right] = 0. \tag{4.21}$$

This equation, giving the result $\rho(t)J(t) =$ constant along a quantum trajectory, was used effectively by Garashchuk and Rassolov in studies dealing with the one-dimensional scattering of wave packets on Eckart potential barriers [4.14, 4.15]. They defined $w = \rho\, dV$ as the conserved *weight* along the trajectory. For barrier transmission, the probability of being on the product side of the barrier at time t is given by

$$P(t) = \int_{product} \rho\, dV, \tag{4.22}$$

in which the integral is taken over the product region. The trajectory version of this equation is

$$P(t) = \sum_j w_j, \tag{4.23}$$

in which the sum over the weights is taken over all quantum trajectories that have made it to the product side of the barrier. It is perhaps surprising to note that evaluation of the density along each trajectory is not required.

4.6 The Initial Value Representation for Quantum Trajectories

In semiclassical approximations to quantum mechanics, the *initial value representation* (IVR) plays a significant role [4.36]. As the name implies, integrations are performed over "initial values" of the coordinates in expressions involving sums over classical paths. A quantum trajectory version of the IVR has been developed by Bittner [4.37] for use in calculating various time correlation functions. Zhao and Makri [4.38] have compared semiclassical and Bohmian versions of the IVR, and Liu and Makri [4.39] have implemented the quantum trajectory version of the IVR. Several examples will be given below to illustrate how the IVR can be used in quantum trajectory calculations of time correlation functions. All of these relations are *formally exact.*

A familiar example of a correlation function involves the time-dependent overlap between a stationary monitor function and a time-evolving wave packet:

$$C(t) = \langle \psi(t)|\phi(0)\rangle = \int dx(t)\psi(x(t))^{*}\phi(x(0)). \tag{4.24}$$

The quantum trajectory version of the IVR for this *cross-correlation function* is readily obtained by making the following two substitutions in this equation. First, the Jacobian evaluated along the quantum trajectory is used to relate the volume element at time t to the initial value at time $t = 0$ (this is the one-dimensional version of equation 4.17):

$$dx(t) = J(x(t), x(0)) \cdot dx(0). \tag{4.25}$$

The Jacobian evaluated along the *unique* quantum trajectory launched from the source point $x(0)$ is given by (see equation 4.18)

$$J(x(t), x(0)) = \exp\left[\int_0^t \vec{\nabla} \cdot \vec{v}(x(\tau))\, d\tau\right], \tag{4.26}$$

in which the integrand, $\vec{\nabla} \cdot \vec{v} = (1/m)\partial^2 S/\partial x^2$, is evaluated along this trajectory. The numerical value of the Jacobian may increase or decrease along the trajectory depending on whether neighboring trajectories are diverging or converging, respectively. The Jacobian depends on the spatial curvature of the action, which in turn is a function of the quantum potential.

The second substitution that we will use is the relationship between the wave function at the trajectory position $x(t)$ and the value at the starting position (see equation 4.20).

$$\psi(x(t)) = J(x(t), x(0))^{-1/2} e^{iS(x(t),x(0))/\hbar}\psi(x(0)). \tag{4.27}$$

In this expression, the two terms multiplying the wave function at the initial point perform the following functions: the Jacobian factor updates the R-amplitude along the trajectory, and the exponential term updates the phase. Substitution of the two preceding relations into the correlation function, equation 4.24, then yields

$$C(t) = \int dx(0) \cdot \psi(x(0))^{*}\phi(x(0)) \cdot J(x(t), x(0))^{1/2} e^{-iS(x(t),x(0))/\hbar}. \tag{4.28}$$

In this form, *the integration is performed at t* $= 0$ *over the positions from which the quantum trajectories are launched.* The integrand brings in the overlap distribution between the two functions *only at the initial time*, and this is multiplied by a time-dependent factor, the latter evaluated for the unique trajectory that makes the trip from $x(0)$ to $x(t)$.

Another application of the quantum trajectory version of the IVR involves computation of the *average value of a coordinate-dependent operator* $A(t)$ [4.37, 4.39]. The expectation value is defined by the usual expression

$$\langle A(t)\rangle = \langle\psi(t)|\hat{A}|\psi(t)\rangle = \int dx(t)|\psi(x(t))|^2 A(x(t)). \tag{4.29}$$

In the IVR form appropriate for quantum trajectories, we have

$$\langle A(t)\rangle = \int dx(0)|\psi(x(0))|^2 A(x(t)). \tag{4.30}$$

In this version, the *initial probability density* is multiplied by the time-varying value for the dynamical variable evaluated along the trajectory. This integrand is then summed over all starting positions for the trajectories.

A special case of the preceding average concerns calculation of the time-dependent *barrier transmission probability* [4.39]. Assuming that the barrier maximum is located at $x_b = 0$, the transmission probability is the total density that makes it to the product side of the barrier:

$$P(t) = \int_{-\infty}^{\infty} dx(t)|\psi(x(t))|^2 H(x(t)), \tag{4.31}$$

in which $H(x(t))$ is the Heaviside step function (which takes on the value of unity if the argument is nonzero). In the quantum trajectory version of the IVR, this quantity is written

$$P(t) = \int_{-\infty}^{\infty} dx(0)|\psi(x(0))|^2 H(x(t)). \tag{4.32}$$

Trajectories on the front end of the wave packet (between the trailing trajectory located at $x_{\min}$ and the leading trajectory at the position $x_{\max}$) make it over the barrier, so the transmission probability may be written

$$P(t) = \int_{x_{\min}}^{x_{\max}} dx(0)|\psi(x(0))|^2 H(x(t)). \tag{4.33}$$

This integral is over the initial positions for those trajectories that make it to the product side of the barrier.

Bittner used the quantum trajectory version of the IVR in conjunction with the derivative propagation method (DPM, described in Chapter 10) to compute time correlation functions for several model one-dimensional potentials: the harmonic oscillator, a harmonic oscillator with a quartic perturbation, and an inverted Gaussian potential [4.37]. Liu and Makri have used this formalism to compute transmission probabilities for the Eckart barrier [4.39]. These authors mentioned an important feature for implementations of this method to compute correlation functions for multidimensional problems: *Monte Carlo sampling* can be used to select the initial coordinates for the quantum trajectories. This method was tested on the calculation of the time-dependent width of a squeezed state in a harmonic potential.

4.7 The Trajectory Noncrossing Rules

There are two important "no trespassing" rules that govern the conduct of quantum trajectories. These rules are well described in Holland's book [4.16].

Rule 1. Quantum trajectories cannot cross through the same point in space-time. Or, put another way, only one quantum trajectory can arrive from the past at a given space-time point. Of course, a given point in space may have multiple trajectories pass through it, as long as they do so at different times. Again, this points out the significant difference between quantum trajectories and Feynman paths. What might happen if two quantum trajectories arrived at the same space-time point? There would be two wave function values; the wave function would no longer be single valued. This is not good.

The difference between the dynamics of classical and quantum trajectories can be illustrated for N "particles" on a line striking an immovable wall. Each classical trajectory bounces off the wall and crosses the ones that are still incoming. For the quantum particles, as they approach the wall, *they all turn around in unison*, never crossing each other. Like well-mannered guests at a cocktail party, they always keep a polite distance apart.

This noncrossing rule is illustrated later, in Chapter 6, where the time dependence of an ensemble of quantum trajectories evolving in a harmonic potential is compared with the dynamics of a like number of classical trajectories (which periodically cross paths).

Rule 2. Quantum trajectories cannot pass through nodes or nodal surfaces. The quantum potential keeps quantum trajectories away from nodal surfaces. What might happen if a quantum trajectory could make it through an exact node? Equation 4.10 would then require that the density vanish along the trajectory after it passed through the node. These zero-density quantum trajectories ("ghost trajectories") do not need to be considered.

In a region where a node is forming, trajectories can hop across the developing node, but for the instant that the exact node is formed, this hopping must cease. However, as the postnode heals and the density begins to fill in (assuming that this is the case), trajectories can again resume passage through this region. The complex dynamics of quantum trajectories near nodes are described in more detail in the next section.

4.8 Dynamics of Quantum Trajectories Near Wave Function Nodes

Near nodes in the wave function, quantum trajectories are subjected to rapidly changing quantum forces, which can cause them to become kinky. The trajectories certainly remain continuous, but experience sudden changes in position and velocity. Zhao and Makri [4.38] have carefully investigated the positions, momenta, and forces acting on quantum trajectories in the bound potential given by $V(x) = p^2/(2m) + (1/2)m\omega^2x^2 + 0.1x^4$. Near the multiple nodes that form in a wave packet evolving in this potential well, quantum trajectories develop high-frequency components and experience sudden shifts in direction. The quantum force acting on a trajectory exhibits sudden variations, and these can be large in magnitude. As a result, it is difficult to integrate the quantum hydrodynamic equations of motion by directly propagating Lagrangian trajectories. Some of the techniques discussed later, in Chapters 7, 14, and 15, may be very useful for this anharmonic potential.

In the remainder of this section, the dynamics of quantum trajectories near one particular node that forms in a scattering wave function will be illustrated [4.6]. We usually start at $t = 0$ with all of the fluid elements launched from equally spaced grid positions. (It is certainly possible to sample the initial positions according to the density, so that the grid points are small where the density is large.) At each time step during the ensemble evolution, the trajectory locations $\{x_i(t)\}$ define a *moving unstructured grid*, with nonuniform nearest neighbor spacings. The nonuniformity that develops includes local features termed *inflation and compression* [4.6]. (The term inflation was used by Pinto-Neto and Santini [4.35] in reference to the influence of the quantum force on Bohm trajectories near the singularity in the Wheeler–DeWitt equation, which was used in the study of a model quantum cosmology.)

The example that will be described in this section concerns the one-dimensional scattering of an initial Gaussian wave packet from an Eckart barrier. The Eckart potential is given by $V(x) = V_0 \sec h^2 [a(x - x_b)]$, where $V_0 = 8,000\,\text{cm}^{-1}$ is the barrier height, $a = 0.4$ determines the barrier width, and $x_b = 7.0$ locates the barrier maximum. The mass used in these studies was $m = 2000$, and the integration time step was $\Delta t = 4$ a.u. or 0.097 fs. (Atomic units are used throughout, except where indicated otherwise.) The number of fluid elements was 120, the center of the wave packet was to the left of the barrier at $x_0 = 2$, and the initial translational energy was $E_0 = 4000\,\text{cm}^{-1}$.

Examples of compression and inflation that occur during the scattering of this wave packet are shown in figure 4.3. For the reflected region to the left of the barrier, each dot shows the R-amplitude at the location of 79 fluid elements at time step 470. At this time, the main part of the wave packet is reflecting back to the left in the direction from which the initial packet was launched toward the barrier. The transmitted part of the wave packet (involving an additional 41 particles) for $x > x_b$ is not shown. For this particular time step, compression of the fluid elements occurs near $x = 2.6$ and 3.6 a.u.

In the reflected region to the left of the barrier, nodes or quasinodes develop, and fluid elements are forced away from these regions when $R(x)$ heads toward zero. (The term *quasi-node* refers to a region where the amplitude reaches a local minimum and attains a small value but does not become exactly zero. Quasi-nodes include *prenodes* and *postnodes*: during the course of time, a prenode usually evolves into a node and then into a postnode. However, in some cases, the node doesn't completely form, so the sequence is prenode→postnode.) The increased separation of fluid elements as they move away from a nodal region toward regions of higher density is a manifestation of inflation. Examples are shown in figure 4.3: the density of fluid elements is lower near the quasi-nodes that form at the three positions $x = 1.5$, 2.1, and 2.8 a.u. The net result of compression and inflation

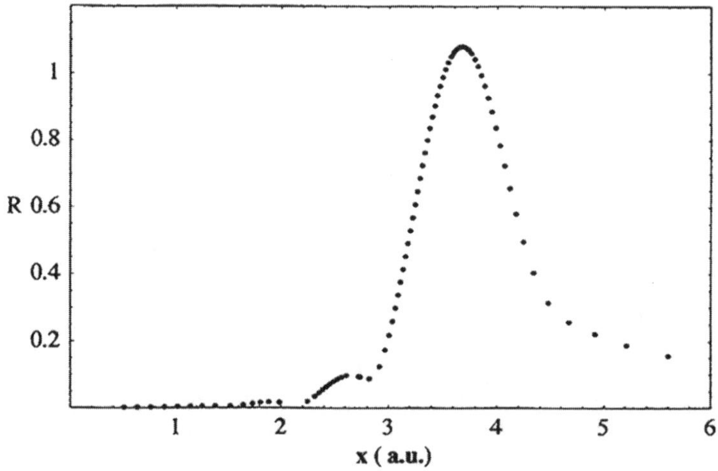

FIGURE 4.3. Amplitude function $R(x)$ for wave packet scattering from an Eckart barrier at time step 470 ($t = 45.5$ fs) [4.6]. The dots show the positions of the fluid elements. The transmitted region to the right of the barrier maximum ($x > 7$ a.u.) is not shown.

is that the interparticle spacings become very nonuniform. This in turn leads to a *sampling problem*: there is an oversupply of information in regions of compression and an undersupply in inflationary regions near nodes.

Several features of trajectory dynamics near nodes will now be examined. Near time step 457, a node develops at the position $x = 3.1$ a.u. in the reflected component of the wave packet. This node is on the leading edge of the main part of the reflected density. Starting around time step 420, a prenode begins to form and as time proceeds, this prenode gradually develops into the "full" node at $x = 3.1$ a.u. After time step 457, this node begins the healing process and the resulting postnode gradually disappears by time step 520. The whole process of node formation and decay thus takes place within about 100 time steps. The R-amplitude and the quantum potential near this position are shown in figure 4.4. Near the position of the node, $R(x)$ has a "V" shape, and the quantum potential takes on large values.

Although quantum trajectories cannot cross nodes, *they can cross over prenodes and postnodes*. As a prenode gradually forms, trajectories can hop from one side to the other. *Since the "exact" node forms only at one instant of time, the nodal noncrossing rule is of little consequence*. However, what is significant is that inflation occurs during the whole history: prenode→node→postnode. Trajectories are forced away from the "central region" of the node and the quantum force imparts a high velocity to them. As a result, when trajectories cross from one side of a quasi-node to the other, they must do so very quickly.

The features just described are illustrated in figure 4.5, which shows the quantum potential and the trajectory positions at three time steps for the reflected portion of the wave packet. Recall that a node develops at $x = 3.1$ a.u. at time step 457. On time step 440 (top plot), trajectories 1–26 are to the left of the center of the prenode, and trajectories 27–120 are to the right. Inflation near the center of the prenode

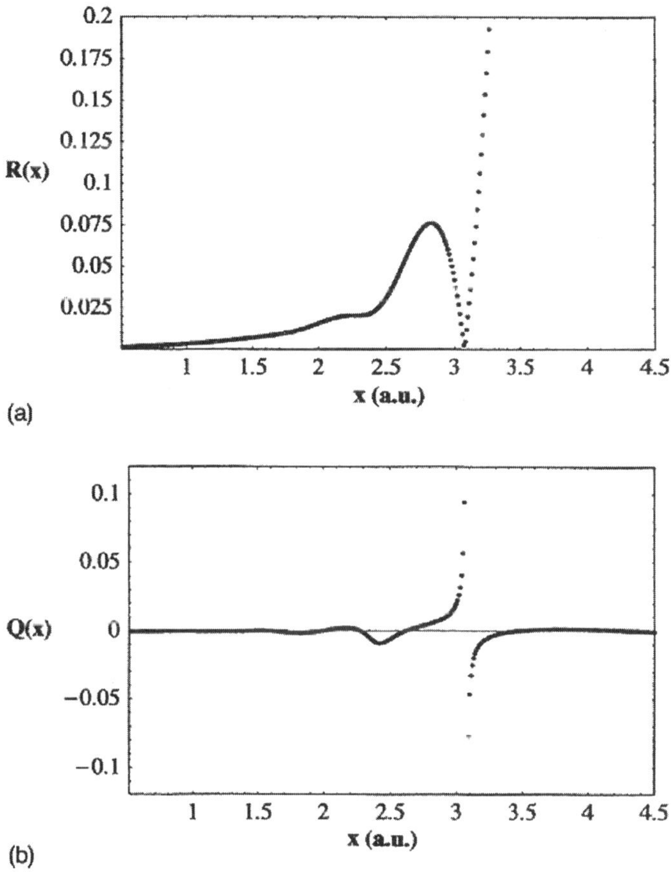

FIGURE 4.4. R-amplitude (a) and quantum potential (b) at time step 457 [4.6]. Only the reflected region near the node at $x = 3.1$ a.u. is shown in this figure.

and compression to the right of this position are evident. Just before this time step, trajectory 27 crossed from the left to the right of the center of the nodal region. Advancing to time step 450 (middle plot), trajectory 26 has just joined the trajectories to the right of the center, and trajectory 25, now located to the left of the center, is ready to join those on the right. On time step 460 (lower plot), just as the node starts to decay, trajectories 25–120 are located to the right of the center of this postnode. As time proceeds, inflation near the center of the postnode diminishes as healing continues. It is clear from this analysis that during the birth–death history of this node, a few nearby trajectories can hop from one side of the quasi-node to the other.

The transport of a few trajectories from one side of a quasi-node to the other presents computational problems because of the interconnected issues of inflation and sampling. When particles inflate away from the center of the quasi-node, accurate computation of the quantum potential and quantum force becomes difficult. In this region, both Q and f_q vary rapidly, and because trajectories on opposite

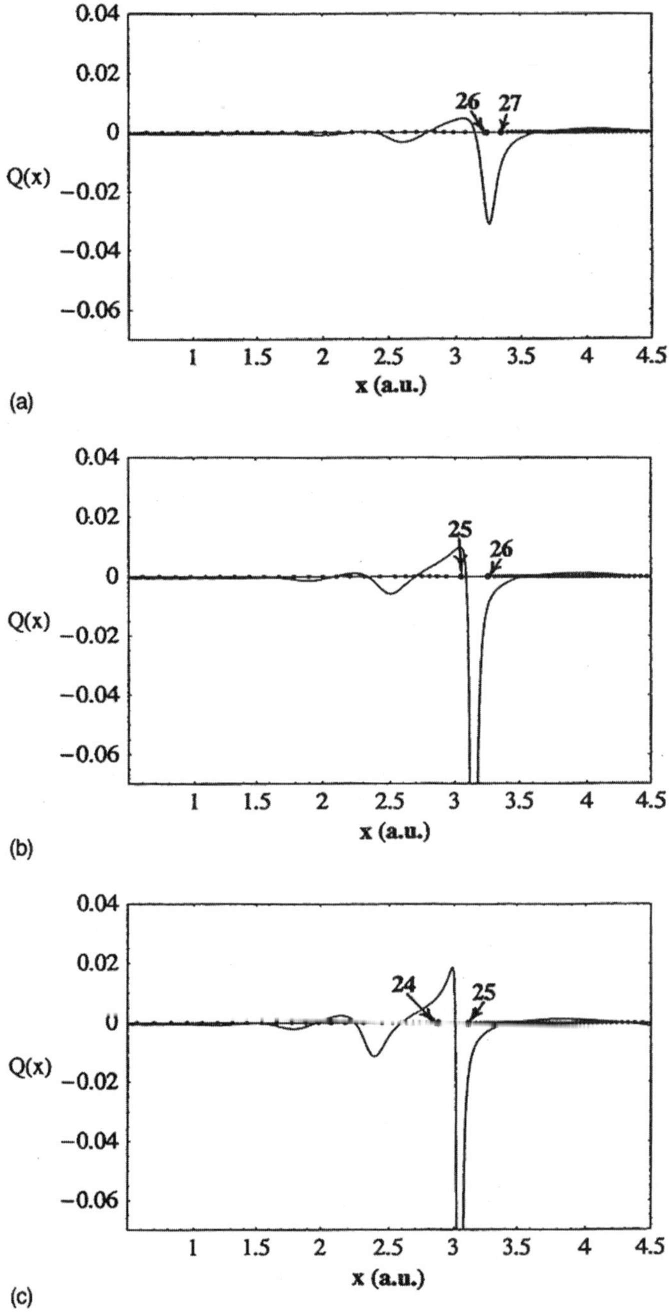

FIGURE 4.5. Quantum potential (continuous curve) and locations of quantum trajectories (dots on horizontal axis) at three time steps: (a) 440, (b) 450, (c) 460 [4.6]. The labels on the two trajectories on either side of the quasi-node are indicated.

sides of the quasi-node are relatively far apart, great care is required to compute accurate spatial derivatives. It is for this reason that the adaptive grid methods described later, in Chapter 7, are very useful, if not essential.

4.9 Chaotic Quantum Trajectories

Chaos in nonlinear deterministic classical systems is ubiquitous. Due to the efforts of many workers, this topic is considered to be relatively well understood. There are accepted diagnostic tools for identifying chaotic classical trajectories and abundant examples that clearly demonstrate the complex properties of classical trajectories in chaotic systems. The sensitive dependence of trajectories on initial conditions is one of the prime indicators of chaotic behavior. In a chaotic classical system, initially close trajectories may exponentially separate, at least over some time intervals. Among the quantities used to quantify the degree of classical chaos are *finite time Lyapunov exponents* and the *power spectra* of trajectories; these are described in more detail in Boxes 4.6 and 4.7, respectively.

Box 4.6. The Lyapunov exponent

The *Lyapunov exponent* is a widely used geometric measure of the rate of exponential separation of a pair of evolving trajectories. The more positive the exponent, the faster the trajectories move away from each other. In order to define this quantity, consider a reference trajectory $\vec{r}_0(t)$ and a neighboring trajectory $\vec{r}(t)$. At each time, the distance between the trajectories is denoted by $d(t)$, and the initial value $d(0)$ is assumed to be very small. This distance is usually computed in configuration space, but it is sometimes computed in phase space, where momentum differences are included in the calculation. The formal definition of the exponent is then

$$\lambda = \lim_{t\to\infty}\left[\frac{1}{t}\ln\frac{d(t)}{d(0)}\right]. \tag{1}$$

In practice, relatively short time runs are used to estimate the exponent. In many applications to trajectory data, λ is plotted as a function of time and changes in the value frequently occur, especially in bound systems.

The value obtained for the exponent is used to categorize the trajectory:

a. $\lambda > 0$: these divergent trajectories appear to repel each other, and the trajectories are classified as chaotic.
b. $\lambda = 0$: these trajectories display neutral, stable dynamics.
c. $\lambda < 0$: these convergent trajectories appear to attract each other, and they are frequently observed in dissipative systems.

The following website provides additional examples:
http:// hypertextbook.com/chaos/43.shtml

Box 4.7. Power spectrum of a trajectory

The *power spectrum* $F(\omega)$ of a time-dependent signal $x(t)$, with input data provided in the time interval from $t = 0$ to $t = T$, gives the distribution of frequencies present in the signal. The frequency distribution is determined by the finite-limit transform, which is similar to the familiar Fourier transform

$$F(\omega) = N \left| \int_{t=0}^{t=T} x(t) e^{i\omega t} dt \right|^2 \tag{1}$$

where N is a normalization factor, which we can choose to have the value $N = 1$.

For the simple case of a monochromatic input signal, $x(t) = x_0 \sin(\omega_0 t)$, the power spectrum for small values of the detuning $\Delta = \omega - \omega_0$ is given by

$$F(\omega) = (x_0 T/2)^2 \sin c^2(\Delta T/2), \tag{2}$$

where the *cardinal sine function* is defined by $\sin c(z) = \sin(z)/z$. At the origin, $z = 0$, the function $\sin c$ reaches its maximum value of unity. There are nodes at $z = n\pi$, for $n = \pm 1, \pm 2, \ldots$. As a result, for a fixed value of T, the power spectrum has a maximum at $\Delta = 0$ and exhibits damped secondary maxima between the zeros at $\Delta = \pm n \cdot 2\pi/T$. As T increases, the peaks look more like delta function spikes, with peak values of $(x_0 T/2)^2$.

If the signal has more components, for example

$$x(t) = \sum_{j=1}^{N} x_j \sin(\omega_j t), \tag{3}$$

then a plot of the power spectrum for long signal times exhibits a series of sharp spikes located at the frequencies $\{\omega_j\}$, with heights determined by the coefficients $\{x_j\}$. The spectrum can be rather complicated, but there are noticeable gaps between the sharp lines. An example is shown in figure 4.6 (c), and this will be described in the following section. For a complicated input signal with a large number of frequency components, such as the function

$$x(t) = \sum_{j=1}^{\infty} x_j \exp(i\omega_j t), \tag{4}$$

a plot of the power spectrum looks continuous and "grassy", with many peaks of various heights in each frequency range $\Delta\omega$. The input signal in this case is said to be noisy or "chaotic". An example is shown in figure 4.7 (c), and this will also be described in the next section.

The situation regarding chaos in quantum mechanics is not so clear. (Several definitions of quantum chaos are given in the review by Chirikov [4.33].) Arguments are sometimes made that dismiss the role of chaos in quantum mechanics (see the discussion in the first part of [4.27]), including the linearity of the

Schrödinger equation, recurrence of wave functions in bounded systems, and the dynamical stability of wave functions. However, criteria have been proposed and applied to model systems in order to identify quantum chaos (see [4.28] for a more complete discussion). These criteria frequently seek to identify *Eulerian chaos* through analysis of the time series of a dynamical variable at *one location* in space. In addition, the statistical distribution of the nearest-neighbor spacings between energy eigenvalues has frequently been used as a "static" monitor for chaos. When the energy level spacing statistics are analyzed, the Wigner distribution is usually found when the underlying classical system is chaotic, while Poisson statistics have been found when the corresponding classical system is regular (nonchaotic). (These distributions and their relationship to possible chaos are well described in the book by Blumel and Reinhardt [4.34].) However, it has been found that some systems that are chaotic in the classical limit (the hydrogen atom in a magnetic field and a two-dimensional quartic oscillator, for example) have eigenvalue spacings that deviate significantly from the anticipated Wigner statistical distribution. The result of these and many other investigations indicates that the existence and identification of Eulerian quantum chaos is still an active research topic.

In conventional, "Copenhagen style", approaches to quantum mechanics, deterministic trajectories for single particles or for fluid elements in a wave packet are not part of the formalism. However, as we have described in the preceding chapters, it is possible in both the de Broglie–Bohm interpretation and in the hydrodynamic formulation of quantum mechanics to make detailed analyses of the properties of quantum trajectories, so that we may attempt to answer the following series of questions:

- Using the same diagnostic tools that were developed for classical systems, can some quantum trajectories be classified as chaotic?
- If so, do chaotic quantum trajectories develop only in systems wherein the underlying classical system is chaotic?
- Or can chaotic quantum trajectories develop in systems where the underlying classical system is regular?
- Do the statistical properties of the energy level spacing distribution have any relationship to the presence or absence of chaotic quantum trajectories?

In the course of analyzing trajectory results on model systems, we will provide answers to some of these questions. Trajectory analysis for both classical and quantum systems can identify *Lagrangian chaos* associated with evolving fluid elements [4.47].

An important feature that Bohmian mechanics brings to the study of possible quantum chaos is that *no reference to the corresponding classical system need be made.* The analysis of quantum trajectories is completely independent of the properties of any reference system obtained by changing the size of $\hbar$. However, it may be very informative to make such a comparison.

The idea of studying quantum chaos by analyzing the properties of Bohmian trajectories was suggested by Dürr et al. in 1992 [4.44]. Beginning around 1993, the search for possible chaotic trajectories in quantum mechanics has been addressed computationally in a number of systems of different dimensionalities that employ various forms for the potential energy. In addition, several diagnostic tools have been used to detect and quantify possible chaotic behavior in quantum trajectories. An overview of some of these studies is presented below.

Dimensionality of system and form of the potential energy:

1-D systems: parabolic barrier [4.27], double square well [4.28]
2-D systems: rectangular well [4.1, 4.43], square well [4.2], [4.9], circular well [4.2], [4.29], right triangular billiard [4.41], Henon-Heiles potential [4.23, 4.26], kicked rotor (standard quantum map) [4.24], quartic oscillator [4.27], stadium billiard [4.30], anisotropic harmonic oscillator [4.31], quantum cat map (charged particle in unit square driven by periodic field) [4.32]
3-D systems: cubic box [4.22], hydrogen atom in external field [4.25]

Diagnostic tools used to identify possible chaotic quantum trajectories:

Positive Lyapunov exponent: [4.21], [4.24], [4.25], [4.27], [4.31], [4.32], [4.43]
Plots of trajectories or surface of section plots: [4.21], [4.22], [4.25], [4.28], [4.29], [4.30]
Power spectrum of trajectory: [4.28], [4.29]
Other: [4.23], [4.26], [4.31]

For one-dimensional motion confined to a finite interval, it has been shown that quantum trajectories cannot be chaotic [4.42, 4.43]. This is in spite of computational studies that claim evidence of chaos in such systems.

In the following section, we will address some of the questions that were brought up in the preceding paragraphs by considering several examples selected from this list of references. In each example, the analytical route to quantum hydrodynamics was followed: quantum trajectories were calculated from precomputed time-dependent wave functions.

4.10 Examples of Chaotic Quantum Trajectories

The first example [4.29] concerns the dynamics of a particle in a two dimensional box of sides L, the *square billiard problem*. The Hamiltonian, of course, is separable, and classical trajectories are *not chaotic*. The easily drawn periodic orbits involve trajectories that bounce from the walls to form simple geometric patterns. However, in this study, the quantum trajectories may, *depending on the initial conditions* (this is a very important qualifier) execute very complex patterns. At $t = 0$, a wave packet was prepared by taking a linear combination of a few of the lowest eigenfunctions for the particle in the two-dimensional box. These eigenfunctions are denoted by $u_{mn}(x, y)$, where (m, n) are the integer quantum numbers for the (x, y) modes, respectively. In this study, quantum trajectories associated with two initial linear combinations were studied:

$$\begin{aligned}\psi_1(x, y, t = 0) &= u_{11}(x, y) + u_{12}(x, y) + iu_{21}(x, y),\\ \psi_2(x, y, t = 0) &= u_{12}(x, y) + iu_{21}(x, y) + u_{23}(x, y).\end{aligned} \tag{4.34}$$

The starting positions (x_0, y_0) for the trajectories were (0.8, 0.5) for the first case and (0.5, 0.25) for the second case. The system of units is such that $\hbar = 2M = 1$, and the unit of length is the side of the box L.

For the first linear combination, a quantum trajectory is shown in figure 4.6 (a). The trajectory passes through the center of the box on every cycle, and it avoids the walls due to the strong repulsive quantum potential near the four edges. As the trajectory passes through the phase space plane defined by the condition $x = 0$, the value of the velocity component v_y can be plotted against y. The simple pattern formed by the dots shown in figure 4.6 (b) confirms that the motion is quasiperiodic. The power spectrum $F(\omega)$ generated by the coordinate $x(t)$ is shown in figure 4.6 (c). The sparse distribution of "sticks" indicates that a small number of frequencies are associated with this regular motion.

For the second linear combination, a quantum trajectory is shown in figure 4.7 (a). The trajectory associated with this simple wave function is very complex. The irregular scattering of points in the phase space surface of section, see figure 4.7 (b), indicates that the motion is chaotic. The power spectrum $F(\omega)$, generated by the coordinate $x(t)$, is shown in figure 4.7 (c). The grassy broadband structure confirms that many frequencies with different strengths are associated with this very irregular motion. Each of the three plots in figure 4.7 is indicative of chaotic motion. The components in the linear combinations in equation 4.34 clearly have a profound influence on the nature of the quantum trajectories developed from these initial wave functions.

How complicated must the wave function be in order to generate Lagrangian chaos? This issue was addressed by Makowski et al. [4.48], who showed that a linear combination of two (properly chosen) stationary state wave functions will suffice.

The second example of chaotic dynamics deals with classical and quantum trajectory studies of a hydrogen atom under the influence of an external oscillating electromagnetic field [4.25]. In the dipole approximation, the interaction potential between the atom and the oscillating field is given by $V(\vec{r}, t) = -\vec{\mu}(r) \cdot \hat{n} E_0 \sin(\omega t)$, where the dipole operator is $\vec{\mu} = (er)\hat{r}$, in which $\hat{r}$ is a unit vector along the electron–proton axis and $\hat{n}$ specifies the field polarization direction. The electric field, whose strength is adjusted by the parameter $\varepsilon = eE_0$, is chosen to lie in the plane of the initial unperturbed classical orbit. (The value $\varepsilon = 1$ corresponds to the force 1.14×10^3 eV/cm.) The H atom is initially prepared in the eigenstate with quantum numbers $n = 15, l = 14, m = 0$, and it interacts with the electric field for the time interval $10 \cdot 2\pi/\omega$. When the interaction is turned off, the wave function for the H atom is a superposition of unperturbed states.

Before turning to results for the quantum trajectories, some aspects of the *classical* motion will be described. The diagnostic test for regular or chaotic character

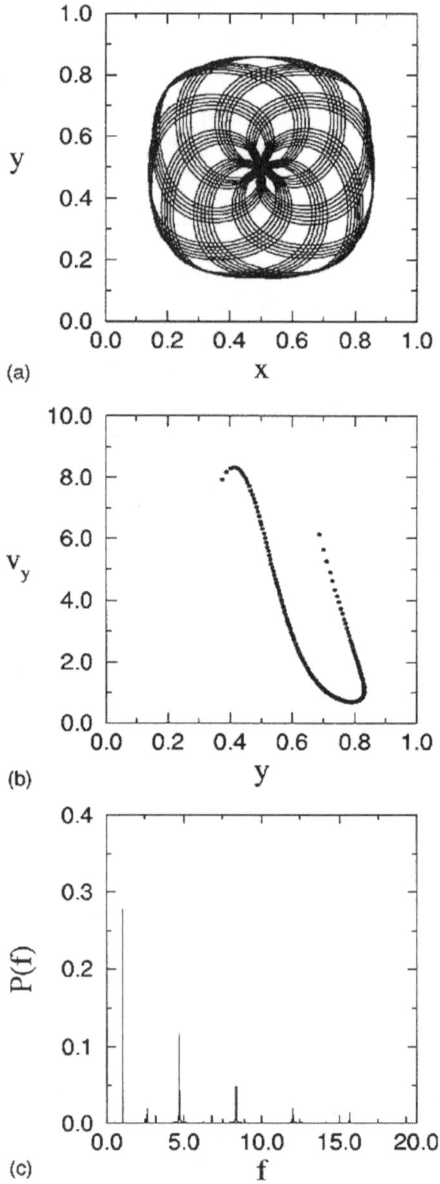

FIGURE 4.6. Quantum trajectory [4.29] associated with the initial wave function ψ_1: (a) trajectory in the (x, y) plane; (b) phase space plot in the (v_y, y) plane; (c) power spectrum (arbitrary units) for the time series of $x(t)$.

will be stroboscopic surfaces of section for the radial motion. In this technique, for discrete times given by $t_n = n(2\pi/\omega)$, the radial momentum for the trajectory is plotted against the value of the electron–proton distance. For low field strengths, ε less than about 2, the classical trajectories are regular and spatially localized, as shown in the left plot of figure 4.8 for the value $\varepsilon = 1.9$. However, as ε is gradually increased above 2, the trajectories display chaotic character, the electron moves

FIGURE 4.7. Quantum trajectory [4.29] associated with the initial wave function ψ_2: (a) trajectory in the (x, y) plane; (b) phase space plot in the (v_y, y) plane; (c) power spectrum (arbitrary units) for the time series of $x(t)$.

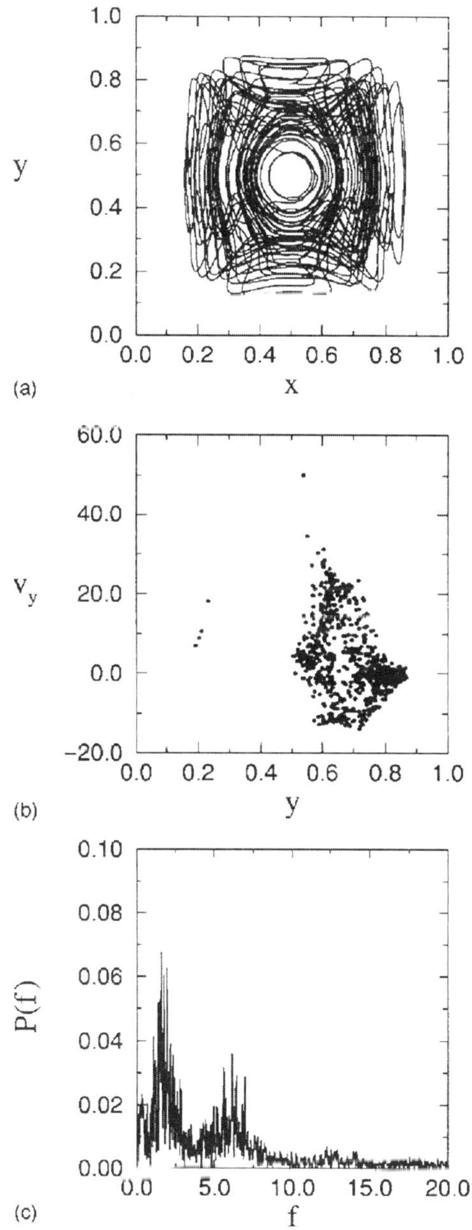

far from the nucleus, and eventually the atom ionizes, as shown in the right side of this figure for $\varepsilon = 2.1$. In addition, when $\varepsilon < 2$, the classical Lyapunov exponent is close to zero, and it gradually increases when ε is increased above this value (see figure 1 in [4.25]). The field strength $\varepsilon = 2$ thus corresponds to the classical chaotic threshold.

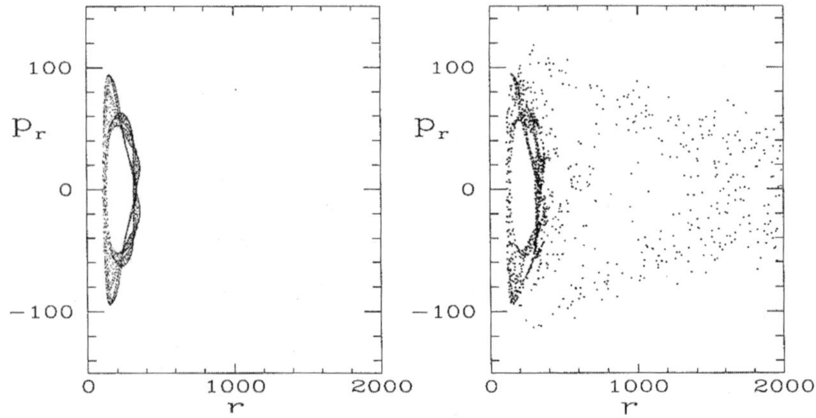

FIGURE 4.8. Poincaré surfaces of section for the (classical) radial momentum and the radial coordinate for two values of the field strength near the chaotic threshold [4.25]: the left plot displays regular motion for $\varepsilon = 1.9$; the right plot displays chaotic motion, with embedded remnants of regular motion, for $\varepsilon = 2.1$.

For the quantum trajectories, a modified definition of the Lyapunov exponent was used as the monitor for possible irregular behavior. The quantum Lyapunov exponent in this study was defined as the weighted time average of the local rate of exponential separation of nearby trajectories. Along a sample trajectory, the following sum over time steps was performed,

$$\lambda_q = \frac{1}{N \cdot \Delta t} \sum_{n=1}^{N} w_n \ln\left(\frac{|d(t_n)|}{|d(t_{n-1})|}\right), \tag{4.35}$$

where N is the number of integration steps and $t_n = n \cdot \Delta t$. The quantity $d(t)$ is the distance between the sample trajectory and an initially close neighboring trajectory. The weights are chosen to be proportional to the probability density

$$w_n = \frac{|\psi_n|^2}{\sum_{m=1}^{N} |\psi_m|^2}, \tag{4.36}$$

in which ψ_n is the wave function along the sample trajectory at time t_n. In figure 4.9, the quantum Lyapunov exponents, calculated according the preceding definition, are plotted against the field strength parameter. For comparison, the classical exponent λ_c is also shown in the same figure. It is evident that there is a clear increase in the quantum Lyapunov exponent near $\varepsilon = 2$, at approximately the same value for which the classical exponent increases. In the interval given by $2 < \varepsilon < 13$, the quantum exponent hovers around 0.05, while for $\varepsilon > 13$ there is a significant increase in the exponent. This figure also provides evidence for the quantum attenuation of classical chaos. In this study, the quantum ionization threshold was not located, and plots were not displayed of the quantum trajectories below and above the threshold that occurs near $\varepsilon = 2$. Further analysis of the quantum trajectories in this system would be very interesting.

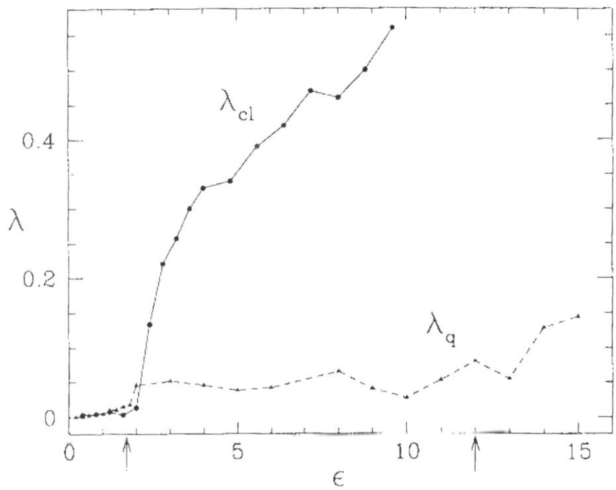

FIGURE 4.9. Quantum and classical Lyapunov exponents as a function of the field strength [4.25]. The two arrows, at $\varepsilon = 1.7$ and $\varepsilon = 12$, are analytical estimates (for a one-dimensional case) of the classical chaotic threshold and for the quantum delocalization threshold.

4.11 Chaos and the Role of Nodes in the Wave Function

In the studies by Frisk [4.22] and Koncl and Makowski [4.21], and for the first example mentioned in the preceding section [4.25], chaotic quantum trajectories can arise in systems where the potential energy vanishes, where the energy level spectrum is regular, and where there are no *chaotic classical trajectories*. In these systems, it is the *quantum potential that causes irregular motion to develop* in the quantum trajectories. In addition, in systems where multiple nodes develop, the quantum potential is very complex, and some of the quantum trajectories, responding to the strong quantum force near these regions, develop high velocities and are forced to execute very complex aperiodic motions. Node formation and chaotic quantum trajectories are thus intimately connected.

In detailed studies of quantum trajectories in a rectangular billiard, Wu and Sprung [4.43] found that a bundle of trajectories will experience large linear and angular velocity gradients near vortices. These effects, along with sudden changes in the size and shape of volume elements near vortices, are the most important factors responsible for chaotic trajectories. Because of the significant role played by nodes, it is not true that quantum mechanics always tends to *suppress* classically chaotic behavior. In fact, the opposite may sometimes be the case, as in a square billiard problem where the potential energy vanishes.

In Chapter 13, we will describe an important feature of the probability flux that occurs near nodes in wave functions. Vortices carrying quantized circulation form around nodal points in two dimensions and around thread or ring vortices in higher dimensionality. To give an example, the interconnection between quantized vortices and the regular or chaotic features of quantum trajectories was studied

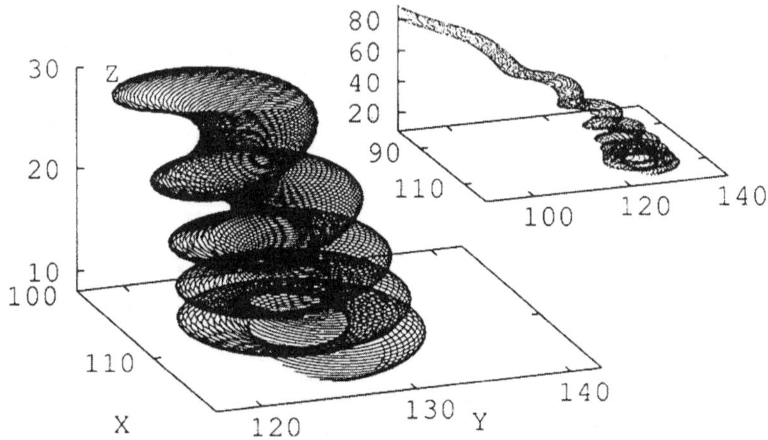

FIGURE 4.10. A quantum trajectory trapped around a moving nodal line in the hydrogen atom [4.46]. The top part of the figure shows the entire trajectory over the time interval 0 to 50,000 a.u., and the lower figure shows a magnification of a portion of this trajectory.

for hydrogen atom wave packets. Falsaperla and Fonte [4.46] formed a wave packet at the initial time $\psi_b(r, \theta, \phi, t = 0)$, by taking a superposition of nine eigenstates for $n = 10$ and one from $n = 9$. Near nodal lines in this nonstationary state, plots were made to show the time dependence of quantum trajectories and Lyapunov exponents. Two quantum trajectories associated with this wave packet are plotted in figures 4.10 and 4.11. In figure 4.10, the trajectory is trapped around a moving nodal line. The trajectory generates a spiral curve as it moves along the nodal line. The Lyapunov exponent for this trajectory, shown as the lower

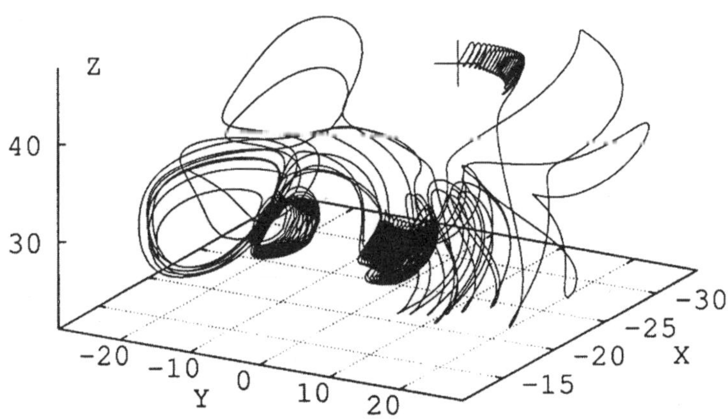

FIGURE 4.11. A quantum trajectory that leaves one nodal line and is trapped by another nodal line [4.46]. The point marked with a cross at the upper right is the starting point. The time interval is the same as that in figure 4.10.

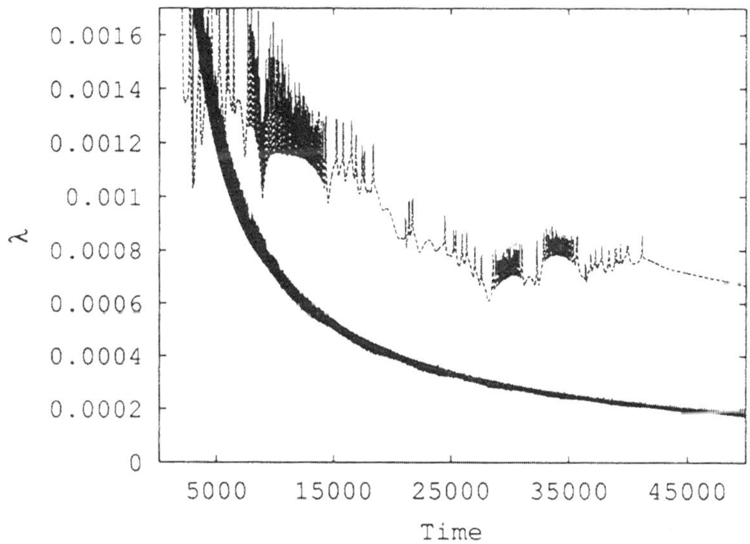

FIGURE 4.12. Largest Lyapunov exponents as a function of time [4.46]. The lower solid curve is for the regular trajectory shown in figure 4.10 and the upper dashed curve is for the trajectory shown in figure 4.11, which displays intermittent chaos.

solid curve in figure 4.12, exhibits an approximately exponential decay as time increases. However, for the trajectory shown in figure 4.11, the dynamics are completely different. This trajectory shows regular motion near one nodal line and then develops intermittent chaos as it moves over to another nodal line, where it again spirals in regular motion. The Lyapunov exponent for this trajectory, shown as the upper dashed curve in figure 4.12, displays short bursts of rapidly fluctuating values. An important conclusion follows from these studies: quantum trajectories display regular spiral-type dynamics close to nodal lines, but chaotic motion may arise during the time interval after the trajectory leaves one nodal line and is later captured by another one.

4.12 Why Weren't Quantum Trajectories Computed 50 Years Ago?

Almost 50 years passed between the publication of Bohm's papers and the 1999 development of the QTM and QFD methods. Why was there such a long gap before quantum trajectories and the wave function were computed on the fly, by directly integrating the quantum hydrodynamic equations of motion? What developments transpired between 1952 and 1999 that permitted this approach to begin to bear fruit? The equations of motion for quantum trajectories were known by 1952, but their use as a computational tool for solving the time-dependent Schrödinger equation was not fully appreciated, or at least demonstrated, until 1999. One possibly is that there was lingering prejudice against the use of trajectories in quantum mechanics. Another reason that the early researchers avoided the numerical solution of these equations stems from their nonlinearity; coupled systems of nonlinear differential equations containing possible singularities are notoriously difficult to solve. Issues of stability and accuracy still play a significant role in constructing computational algorithms for systems of this type.

In addition to the profound advances in computer hardware that occurred after 1952, developments in applied mathematics and computational engineering played a significant role in the recent development of computational approaches to the quantum hydrodynamic equations. In particular, advances in the structural engineering and computational fluid mechanics communities concerning function fitting algorithms (and their use for computing function derivatives on unstructured grids) and adaptive dynamic grids played a major role in the QTM work by Wyatt and coworkers. In addition, finite element techniques are also useful for this purpose, and these techniques played a prominent role in the work by Rabitz and coworkers. Before these developments in 1999, the earlier computational studies by Weiner and Askar [4.17, 4.18] around 1970 (Gaussian wave packets evolving on harmonic potentials) and the studies by Scheid and coworkers [4.19, 4.20] in the 1980s (PIC or particle in a cell calculations) were important for taking the first steps in this new direction in quantum dynamics. These studies were momentarily noticed in the literature, but did not "catch on", as evidenced by the lack of follow-on studies in the dynamics community. Part of the reason was that the computational techniques used to evaluate the quantum potential and the quantum force lacked the accuracy to propagate quantum trajectories accurately on anharmonic potential surfaces even for short time scales.

References

4.1. C.L. Lopreore and R.E. Wyatt, *Quantum wave packet dynamics with trajectories*, Phys. Rev. Lett. **82**, 5190 (1999).

4.2. F. Sales Mayor, A. Askar, and H.A. Rabitz, *Quantum fluid dynamics in the Lagrangian representation and applications to photodissociation problems*, J. Chem. Phys. **111**, 2423 (1999).

4.3. R.E. Wyatt, *Quantum wave packet dynamics with trajectories: wave function synthesis along quantum paths*, Chem. Phys. Lett. **313**, 189 (1999).

4.4. C.L. Lopreore and R.E. Wyatt, *Quantum wave packet dynamics with trajectories: Reflections on a downhill ramp potential*, Chem. Phys. Lett. **325**, 73 (2001).

4.5. E.R. Bittner and R.E. Wyatt, *Integrating the quantum Hamilton–Jacobi equations by wave front expansion and phase space analysis*, J. Chem. Phys. **113**, 8888 (2001).

4.6. R.E. Wyatt and E.R. Bittner, *Quantum wave packet dynamics with trajectories: Implementation with adaptive Lagrangian grids*, J. Chem. Phys. **113**, 8898 (2001).

4.7. C.L. Lopreore, R.E. Wyatt, and G. Parlant, *Electronic transitions with quantum trajectories*, J. Chem. Phys. **114**, 5113 (2001).

4.8. R.G. Brook, P.E. Oppenheimer, C.A. Weatherford, I. Banicescu, and J. Zhu, *Solving the hydrodynamic formulation of quantum mechanics: A parallel MLS method*, Int. J. Quantum Chem. **85**, 263 (2001).

4.9. R.K. Vadapalli, C.A. Weatherford, I. Banicescu, R.L. Carino, and J. Zhu, *Transient effect of a free particle wave packet in the hydrodynamic formulation of the time-dependent Schrödinger equation*, Int. J. Quantum Chem. **94**, 1 (2003).
4.10. C.D. Stodden and D.A. Micha, *Generating wave functions from classical trajectory calculations: The divergence of streamlines*, Int. J. Quantum Chem.: Symposium **21**, 239 (1987).
4.11. R.P. Feynman, *Space-time approach to non-relativistic quantum mechanics*, Rev. Mod. Phys. **30**, 24 (1948).
4.12. R.P. Feynman and A.R. Hibbs, *Quantum mechanics and path integrals* (Addison-Wesley, Reading MA, 1965).
4.13. F.D. Peat, *Infinite Potential, The Life and Times of David Bohm* (Addison-Wesley, Reading, Mass., 1997).
4.14. S. Garashchuk and V.A. Rassolov, *Semiclassical dynamics based on quantum trajectories*, Chem. Phys. Lett. **364**, 562 (2002).
4.15. S. Garashchuk and V.A. Rassolov, *Semiclassical dynamics with quantum trajectories: formulation and comparison with the semiclassical initial value representation propagator*, J. Chem. Phys. **118**, 2482 (2003).
4.16. P.R. Holland, *The Quantum Theory of Motion* (Cambridge Press, New York, 1993).
4.17. J.H. Weiner and Y. Partom, *Quantum rate theory for solids. II. One-dimensional tunneling effects*, Phys. Rev. **187**, 187 (1969).
4.18. A. Askar and J.H. Weiner, *Wave packet dynamics on two-dimensional quadratic potential surfaces*, Am. J. Phys. **39**, 1230 (1971).
4.19. G. Terlecki, N. Grun, and W. Scheid, *Solution of the time-dependent Schrödinger equation with a trajectory method and application to $H^+ + H$ scattering*, Phys. Lett. A **88**, 33 (1982).
4.20. P. Zimmerer, M. Zimmermann, N. Grun, and W. Scheid, *Trajectory method for the time-dependent Schrödinger and Thomas-Fermi equations*, Comp. Phys. Comm. **63**, 21 (1991).
4.21. S. Konkel and A.J. Makowski, *Regular and chaotic causal trajectories for the Bohm potential in a restricted space*, Phys. Lett. A **238**, 95 (1998).
4.22. H. Frisk, *Properties of the trajectories in Bohmian mechanics*, Phys. Lett. A **227**, 139 (1997).
4.23. P.K. Chattaraj and S. Sengupta, *Quantum fluid dynamics of a classically chaotic oscillator*, Phys. Lett. A **181**, 225 (1993).
4.24. U. Schwengelbeck and F.H.M. Faisel, *Definition of Lyapunov exponents and KS entropy in quantum dynamics*, Phys. Lett. A **199**, 281 (1995).
4.25. G. Iacomelli and M. Pettini, *Regular and chaotic quantum motions*, Phys. Lett. A **212**, 29 (1996).
4.26. S. Sengupta and P.K. Chattaraj, *The quantum theory of motion and signatures of chaos in the quantum behavior of a classically chaotic system*, Phys. Lett. A **215**, 119 (1996).
4.27. G.G. de Polavieja, *Exponential divergence of neighboring quantal trajectories*, Phys. Rev. A **53**, 2059 (1996).
4.28. O.F. de Alcantara Bonfim, J. Florencio, and F.C. Sa Barreto, *Quantum chaos in a double square well: an approach based on Bohm's view of quantum mechanics*, Phys. Rev. E **58**, 6851 (1998).
4.29. O. F. de Alcantara Bonfim, J. Florencio, and F.C. Sa Barreto, *Chaotic dynamics in billiards using Bohm's quantum mechanics*, Phys. Rev. E **58**, R2693 (1998).

4.30. D.A. Wisniacki, F. Borondo, and R.M. Benito, *Dynamics of trajectories in chaotic systems*, Europhys. Lett. **64**, 441 (2003).
4.31. R.H. Parmenter and R.W. Valentine, *Deterministic chaos and the causal interpretation of quantum mechanics*, Phys. Lett. A **201**, 1 (1995).
4.32. F.H.M. Faisal and U. Schwengelbeck, *Unified theory of Lyapunov exponents and a positive example of deterministic quantum chaos*, Phys. Lett. A **207**, 31 (1995).
4.33. B.V. Chirikov in W.D. Heiss (ed.), *Chaos and Quantum Chaos* (Springer, New York, 1992).
4.34. A. Blumel and W.P. Reinhardt, *Chaos in Atomic Physics* (Cambridge University Press, Cambridge, 1997).
4.35. N. Pinto-Neto and E. Santini, *Must quantum spacetimes be Euclidean?* Phys. Rev. D **59**, 123517 (1999).
4.36. W.H. Miller, *The semiclassical initial value representation: A potentially practical way of adding quantum effects to classical molecular dynamics*, J. Phys. Chem. A **105**, 2942 (2001).
4.37. E.R. Bittner, *Quantum initial value representations using approximate Bohmian trajectories*, J. Chem. Phys. **119**, 1358 (2003).
4.38. Y. Zhao and N. Makri, *Bohmian versus semiclassical description of interference phenomena*, J. Chem. Phys. **119**, 60 (2003).
4.39. J. Liu and N. Makri, *Monte Carlo Bohmian dynamics from trajectory stability properties*, J. Phys. Chem. A **108**, 5408 (2004).
4.40. M.E. Tuckerman, Y. Liu, G. Ciccotti, and G.J. Martyna, *Non-Hamiltonian molecular dynamics: Generalizing Hamiltonian phase space principles to non-Hamiltonian systems*, J. Chem. Phys. **115**, 1678 (2001).
4.41. J.A. de Sales and J. Florencio, *Quantum chaotic trajectories in integrable right triangular billiards*, Phys. Rev. E **67**, 016216 (2003).
4.42. S. Goldstein, *Absence of chaos in Bohmian mechanics*, Phys. Rev. E **60**, 7578 (1999).
4.43. H. Wu and D.W.L. Sprung, *Quantum chaos in terms of Bohm trajectories*, Phys. Lett. A **261**, 150 (1999).
4.44. D. Dürr, S. Goldstein, and N. Zanghi, *Quantum chaos, classical randomness, and Bohmian mechanics*, J. Stat. Phys. **68**, 259 (1992).
4.45. M. Abolhasani and M. Golshani, *The path integral approach in the frame work of causal interpretation*, Annal. Found. L. de Broglie, **28**, 1 (2003).
4.46. P. Falsaperla and G. Fonte, *On the motion of a single particle near a nodal line in the de Broglie–Bohm interpretation of quantum mechanics*, Phys. Lett. A **316**, 382 (2003).
4.47. J.M. Finn and D. del-Castillo-Negrete, *Lagrangian chaos and Eulerian chaos in shear flow dynamics*, Chaos **11**, 816 (2001).
4.48. A.J. Makowski, P. Peploswski, and S.T. Dembinski, *Chaotic causal trajectories: the role of the phase of stationary states*, Phys. Lett. A **266**, 241 (2000).
4.49. J.M. Ottino, *The Kinematics of Mixing: Stretching, Chaos, and Transport* (Cambridge University Press, New York, 1989).

5
Function and Derivative Approximation on Unstructured Grids

The accurate computation of spatial derivatives on unstructured grids is a challenging problem. Accurate spatial derivatives are required for integration of the quantum-hydrodynamic equations of motion. Several methods for evaluating these derivatives will be described.

5.1 Introduction

In this chapter, we will discuss several methods for continuous function approximation. These methods will be used to approximate spatial derivatives needed at each time step in order to integrate the hydrodynamic equations of motion. Because a large portion of this book is dedicated to Lagrangian quantum dynamics, we will focus on those methods capable of handling multidimensional approximation on unstructured grids. This is a very challenging problem in numerical analysis and computational engineering and science [5.1].

The problem is as follows: given the data set $\{\vec{r}_i, f_i\}_{i=1}^{N}$, the goal is to find an approximate function $g(\vec{r})$ that accurately represents the true function $f(\vec{r})$ within this data region. (Sometimes, it is not necessary to assume a form for $g(\vec{r})$; rather, the algorithm is smart enough to *adaptively* seek it out internally [5.3].) Once $g(\vec{r})$ is obtained, we can analytically calculate and use its spatial derivatives as approximations to those of $f(\vec{r})$. The functional form of $g(\vec{r})$ will usually have a number of linear parameters (and possibly a few nonlinear parameters) that need to be either optimized or adapted to the input data. Of course, we would like to keep the total number of parameters to a minimum, since this will greatly decrease the computational effort. Unfortunately, however, by doing this we impose a *complexity restriction* on the form of $g(\vec{r})$. In other words, if only a small number of parameters are used, we are assuming that the true solution has some degree of *smoothness* (meaning that the spatial derivatives beyond some order vanish). This may or may not be valid for the function $f(\vec{r})$, but we hope for the best.

There are two general ways for finding the approximate $g(\vec{r})$ from a given data set, these being interpolation and fitting. In *interpolation*, the input function values are *exactly reproduced* by the interpolant. Although the interpolant is exact at each grid point, its derivatives may not be, and sometimes they may wildly oscillate in the gaps between these points [5.2]. In contrast, a functional *fit* may not exactly represent the original data set; rather, it may "smooth" over the original data. If the data set is not fully trusted (maybe significant errors crop up from previous numerical calculations or from experimental noise), this smoothing may be extremely beneficial. In addition, we will also see that fitting can sometimes be used to smooth over singularities or "kinks" obtained from ill-behaved solutions. The fit is obtained by minimizing an error functional that depends on the input data and the parameters embedded in $g(\vec{r})$. In statistics, this minimization procedure is the basis for *regression analysis* [5.3]. In addition, in the emerging field of *data mining*, similar techniques are used to explore patterns and regularities in huge data sets in spaces with many independent variables [5.3].

Especially important is the broad category of *meshless methods*, which do not require prespecified spatial distributions or connectivity patterns among the input data points. The unstructured mesh can be readily generated by a relatively simple algorithm. The review article by Belytschko et al. surveys this field up to about 1996 [5.1]. A more recent comprehensive review is provided in the book *Mesh Free Methods*, by G. R. Liu [5.36]. The starting point for the development of meshless methods in fluid dynamics seems to be *smoothed particle hydrodynamics* (SPH), developed especially by Monaghan and coworkers [5.4]. (A thorough exposition of SPH is presented in the book by Liu and Liu [5.37].) This method has been used for modeling astrophysical phenomena without boundaries, such as dust clouds and exploding stars, and for other applications in fluid dynamics. Meshless methods currently being developed are part of the trend toward *local representations* of complex data sets. At least in part because of their use in computational mechanics and classical fluid dynamics, meshless methods are becoming increasingly robust, and the literature in this area is rapidly expanding, as evidenced by the number of journal articles appearing each year.

Before dealing with specific methods later in this chapter, it is important to build a list of some desirable (and overlapping) requirements for general function approximation. This is nothing short of a "wish list", and no one algorithm has yet to satisfy all the listed elements.

- The algorithm should be as automatic as possible, so that minimal user intervention is required when different data sets are analyzed.
- The approximating function$g(\vec{r})$should have as few adjustable parameters as possible.
- Smoothness is highly desirable: the approximating function should avoid serpentine behavior between the grid points.
- The algorithm and approximating function should not become ill-behaved as the density of sampling points becomes large or small in some regions of the data space
- The algorithm should provide a systematic means for improving the accuracy of its results. It is desirable, but not essential, that the working equations be developed from a variational principle.

- The algorithm should be stable and robust, meaning that it should be able to handle different grid layouts, widely differing function values, etc.
- The algorithm should be broadly applicable to systems of both low and high dimensionality.
- The same algorithm and computer code should be applicable to both interior and edge data points.
- The algorithm should be computationally efficient and the computation time should be reasonable and not grow too rapidly (for example, exponentially) with increasing dimensionality.

The various methods for derivative computation on unstructured grids that will be described later in this chapter are surveyed below.

Finite difference methods. Of course, computing spatial derivatives on structured grids using finite differences is a very common task. However, Perrone et al. [5.5, 5.6] and Liszka and coworkers [5.7–5.10] have described finite difference methods (FDM) for computing partial derivatives on *unstructured grids*. An excellent review has been presented by Orkisz [5.38]. Starting from a Taylor series expansion of the function around a reference grid point, and then using the known function values at the grid points within a surrounding region, a matrix equation can be set up for the partial derivatives at the reference point. (The FDM uses the interpolation equations given in Section 5.2.) The matrix equations are then solved, and all of the partial derivatives are calculated simultaneously. The FDM has the advantage that it can be applied to both an arbitrarily irregular (or regular) grid. This method has been tested on a number of problems, including Poisson's equation and the time-dependent heat equation [5.7]. Because the FDM bears a resemblance to the least squares method, it will not be described separately in the following sections.

Least squares methods. Many of the currently used methods for data fitting on unstructured meshes are based on the least squares algorithm, introduced by Legendre in 1805 (although Gauss reportedly devised the method ten years earlier). In the moving least squares (MLS) variant of this algorithm, a *local* fit is calculated around each of the N data points, so that the algorithm must be called N times. MLS is currently being developed and applied in fluid dynamics and structural mechanics [5.12–5.16]. To enhance the locality of the MLS fit, moving weighted least squares (MWLS) can be used. In this method, local weight functions are included in the original least squares minimization equations. MWLS was introduced by Lancaster and Salkauskas in 1981 [5.2] and reinvented by Nayroles et al. in 1992 from the viewpoint of "diffuse elements" [5.11]. Starting in 1999, MWLS has been used extensively in the quantum trajectory method (see, for example, the studies by Wyatt and Na involving a one-dimensional metastable mode interacting with a multimode harmonic reservoir [5.17]).

As an alternative to MLS and MWLS methods, we also introduce the *dynamic least squares method* (DLS) [5.20]. In most applications of least squares, $g(\vec{r})$ is expanded in a basis set, and the coefficients of each basis function (i.e., the linear parameters to be optimized) are calculated at each time step by solving the corresponding matrix equations. From the DLS viewpoint, the best set of coefficients is found at the minimum of an effective multidimensional potential energy surface, derived from the original minimization equations. To obtain the coefficient values at this minimum, coupled "Newtonian-like" equations of motion are derived for the coefficients, and they are evolved until relaxation occurs into the minimum on the effective potential surface. An informative presentation on the use of this dynamical technique, in the context of least squares fitting, was described by Swanson and Garner [5.19]. In an application of the dynamic least squares method to the propagation of quantum trajectories, Bittner and Wyatt [5.20] studied the scattering of an initial Gaussian wave packet from an Eckart barrier.

Fitting with distributed approximating functionals. Since their introduction in the early 1990s [5.21–5.23], *distributed approximating functionals* (DAFs) have been used extensively for fitting, interpolation, extrapolation, solving differential and partial differential equations in one and two dimensions [5.24], and for the filtering of digital signals. Several types of DAFs have been introduced, including those that use Hermite polynomials, Lagrange interpolation functions, and wavelets [5.25]. The mathematical foundations of DAF theory have been described in two review articles [5.26, 5.27], multidimensional DAFs have been developed for general orthogonal curvilinear coordinates [5.23], and extensions to nonuniform one- and three-dimensional grids have been described [5.22, 5.28]. Like the least squares equations, the DAF working equations can be given a variational derivation [5.27]. In a 2000 study [5.29], DAFs were used to evaluate the spatial derivatives needed to integrate the equations of motion for quantum trajectories undergoing barrier scattering.

Tessellation-fitting method. Another derivative approximation scheme, based on tessellation and function fitting, has been developed by Nerukh and Frederick [5.30]. A *tessellation* is a tiling of a region by *simplex figures* (lines, triangles, and tetrahedra in one, two, and three dimensions, respectively) without gaps or overlaps [5.31]. For example, a *triangulation* of the two-dimensional bounded region occupied by quantum trajectories can be performed by simply connecting adjacent grid points with lines. Following tessellation, an approximate function is obtained around a reference grid point from which the spatial derivatives are then evaluated. This approximate function is calculated locally, using only those elements contained in the corresponding tile (or in an extension, adjacent tiles can be used). This tessellation/fitting method was applied to the scattering of an initial Gaussian wave packet from an Eckart barrier augmented by harmonic potentials in one or two additional degrees of freedom [5.30].

Finite element method. Finite element techniques [5.32] have been used for the evaluation of spatial derivatives in scattering problems in two degrees of freedom [5.33]. This procedure involves mapping a small cluster of distributed grid points from the physical grid to a finite element of simple geometry (such as a square in two dimensions) in computational space. For this finite element, techniques employing low-order basis functions are used to approximate the partial derivatives. In the final step, these derivatives are then mapped from the computational space back into the physical space. This method was used to evolve ensembles of quantum trajectories representing the photodissociation of NO_2 and NOCl [5.33] on electronically excited potential energy surfaces.

This chapter is arranged as follows: In Section 5.2, the MLS method will be described in detail. Later, in Chapter 6, the use of MLS in the quantum trajectory method will be illustrated for one- and two-dimensional problems, and in Chapter 8 applications to multimode systems will be described. Section 5.3 will describe the DLS method, and Section 5.4 describes the use of DAFs for fitting and derivative evaluation. The tesselation/fitting method and the use of finite elements are then described in Sections 5.5 and 5.6, respectively. Lastly, a brief summary of the chapter is presented in Section 5.7.

5.2 Least Squares Fitting Algorithms

A powerful approach that finds frequent application for fitting multidimensional data on unstructured grids is the least squares technique [5.2]. The input data set may represent the entire solution domain or some local region surrounding a reference grid point $\vec{r}_k$. Although the first case leads to the simplest algorithm, the second case is generally preferable, since local solutions are usually much smoother than the global one. For this reason, we will not consider the global least squares procedure. For the local case, the n_p grid points fall within a *stencil* surrounding the reference point.

Around the reference point $\vec{r}_k$, we seek an approximate but accurate *fit* to the local data. From this fit, approximate spatial derivatives will be calculated in this region of the data set. In order to do this, we will assume that the true function $f(\vec{r})$ can be locally expanded in a *basis set* of dimension $(n_b + 1)$. Each individual basis function, denoted by $P_j(\vec{r} - \vec{r}_k)$, where $j = 0, 1, 2, ..., n_b$, depends on the locations of both the observation point $\vec{r}$ and the reference point $\vec{r}_k$. The approximate is then given in the *multiple linear regression model* by the function

$$g(\vec{r}) = \sum_{i=0}^{n_b} a_i P_i(\vec{r} - \vec{r}_k) = P^t(\vec{r})a, \tag{5.1}$$

in which the expansion coefficients $\{a_i\}$ have been arranged into the column vector a (of dimension $n_b + 1$), and the basis functions have been assembled into the vector $P(\vec{r})$. (The coefficients in this expansion also depend on the reference point, but for simplicity, this will not be indicated explicitly.) Because the expansion coefficients are evaluated around each grid point, this method is referred to as the *moving least squares* method. One disadvantage of this method is that a new fit must be calculated around each grid point.

Of course, the choice of the basis is crucial for both representing the data and for speeding the convergence of the expansion. In this book, we will usually focus on *local polynomial basis* sets. When used in any approximate method, this basis has an important property called *polynomial consistency*. This property ensures that a complete polynomial basis of a given order will exactly reproduce the input function if the input function is a polynomial of order less than or equal to the order of the basis set. (The least squares method has a more general consistency feature: *any function* included in the basis set will be *reproduced* by the method.) One commonly used polynomial basis is derived from the Taylor series approximation and has basis functions given by $P_j = (1/j!)\xi^j$, where the displacement from the reference point is denoted by $\xi = x - x_k$. For example, in one dimension a cubic basis set is given by

$$B = \left\{1, \xi, \frac{\xi^2}{2}, \frac{\xi^3}{6}\right\} = \{P_0(\xi), P_1(\xi), P_2(\xi), P_3(\xi)\}. \tag{5.2}$$

In two dimensions, each Taylor basis function is given by $P_{jk} = 1/(j!k!)\xi^j\eta^k$, where $\eta = y - y_k$. As an example, 6-term quadratic basis (a *truncated tensor product space*) would be

$$B = \left\{1, \xi, \eta, \frac{\xi^2}{2}, \frac{\eta^2}{2}, \xi\eta\right\} = \{P_0(\xi,\eta), P_1(\xi,\eta), \ldots, P_5(\xi,\eta)\}. \tag{5.3}$$

In two dimensions, there are 10 functions in the complete cubic basis and 15 in the quartic basis. The cross terms in direct product basis sets, such as the term given by $\xi\eta$, provide for *correlation* among the various degrees of freedom. For the Taylor polynomial basis, *the approximate derivatives evaluated at the reference point are just the expansion coefficients*. For example, in two dimensions, the first derivatives and Laplacian of the approximate are given by

$$\frac{\partial g}{\partial x} = a_1, \quad \frac{\partial g}{\partial y} = a_2, \quad \nabla^2 g = a_3 + a_4. \tag{5.4}$$

Although this polynomial basis is frequently used, it is not without problems. One may think, because of the consistency property, that a high-order polynomial basis should be chosen. However, this is very risky and time-consuming, and we must be reasonably sure that the true function can be represented exactly by some combination of the basis elements which are chosen (which is usually impossible!). If the true function cannot be exactly reproduced, a high-degree polynomial basis will induce oscillations between grid points, thus leading to extremely poor derivative approximations. It is certainly possible to use more general basis sets containing special functions, such as radial basis functions (including Gaussians, multiquadrics, and thin plate splines), rational functions, etc. (see Box 6.2). In addition, polynomial basis sets are sometimes *enriched* with additional functions to give better fits near interesting features that need to be captured (local maxima, minima, and sudden changes associated with shock fronts) [5.1]. Sometimes, search procedures can be implemented to draw in basis functions from a large *dictionary of candidate functions* [5.3]. One strategy for doing this is *forward stepwise selection*: candidate basis functions are added into the summation if they lead to significant improvement in the fit. An opposite strategy, *backward stepwise selection*, deletes candidates from a basis set that starts out "too large". In data mining, an *adaptive parsimonious algorithm* that combines both *forward selection and backward deletion* of candidate functions is MARS (multivariate adaptive regression splines) [5.3].

To derive equations for the expansion coefficients, a *variational criterion* is used. For points within the stencil surrounding point k, the *lack of fit* is measured by the sum of the squares of the errors (the difference between the input data values and the fit values),

$$\Omega_k = \sum_{j=1}^{n_p} \left(f_j - \sum_{i=0}^{n_b} a_i P_i(\vec{r}_j - \vec{r}_k) \right)^2 . \tag{5.5}$$

In regression analysis, this quantity is referred to as the *residual sum of squares* (RSS) [5.3]. In order to make the fit even more local to the reference point, the *weighted errors* can be used,

$$\Omega_k = \sum_{j=1}^{n_p} w_j (f_j - \sum_{i=0}^{n_b} a_i P_i(\vec{r}_j - \vec{r}_k))^2, \tag{5.6}$$

in which the weight associated with each point in the stencil is nonnegative, $w_j \geq 0$. The functional in equation 5.6 can be augmented with *regularization terms*, including those that penalize curvature or roughness in the function (see Chapter 11 in [5.39]).

Usually, the weight functions (also called *window functions* or *kernel functions* [5.3]) are peaked at the reference point and fall off as the distance from the reference point increases. For example, evaluated at point j, a Gaussian weight function would be

$$w_j = w(\vec{r}_j - \vec{r}_k) = e^{-\alpha|\vec{r}_j - \vec{r}_k|^2}, \tag{5.7}$$

in which α is a *bandwidth* or *smoothing parameter*. The region of the stencil where the weight is significant is the *domain of influence* of the reference point $\vec{r}_k$. When various reference points are considered, the patches representing the domains of influence overlap one another, thus providing *grid correlation*. Weight functions other than the Gaussian may also be used, including those with *compact support* (they are *exactly zero* beyond some distance from the reference point). Examples of weight functions with compact support include cubic and quartic splines (see figure 5.2 in [5.36]). Finally, we mention weights that are *almost singular* at each data point, such as the function, $w(\vec{r}) = \exp[-\alpha d^2]/[d^n + \varepsilon]$, where $d = |\vec{r} - \vec{r}_j|$ and ε is a small parameter. This type of weight function is used in the *interpolating moving least squares* method, IMLS, which has been successfully used in fitting potential surfaces to ab initio data [5.35].

The variance defined in equation 5.6 is a *functional of the coefficient vector*, and we could use notation that emphasizes this dependence, such as $\Omega_k\,[a]$. The *variational criterion* is then the following: we seek the "best" vector chosen to minimize the variance as we search the $(n_b + 1)$ -dimensional coefficient space. To find the vector a_{best}, we will first set the partial derivatives of the variance to zero. The derivative of the variance with respect to the coefficient a_p is

$$\frac{\partial \Omega_k}{\partial a_p} = \sum_{j=1}^{n_p} w_j \left(f_j - \sum_{i=0}^{n_b} a_i P_i(\vec{r}_j - \vec{r}_k) \right) \left(-2P_p(\vec{r}_j - \vec{r}_k) \right). \tag{5.8}$$

This expression is then set to zero and rearranged to give the equations ($p = 0, 1, \ldots, n_b$)

$$\sum_{i=0}^{n_b} \left\{ \sum_{j=1}^{n_p} P_p(\vec{r}_j - \vec{r}_k) w_j P_i(\vec{r}_j - \vec{r}_k) \right\} a_i = \sum_{j=1}^{n_p} P_p(\vec{r}_j - \vec{r}_k) w_j f_j. \tag{5.9}$$

We will begin to simplify this system of equations by introducing matrix notation. The P-matrix has rows labeled by grid points and columns labeled by basis functions. For example, the i, j element $P_{i,j} = P_j(\vec{r}_i - \vec{r}_k)$ stores the j-th basis function evaluated at grid point $\vec{r}_i$. This rectangular matrix has n_p rows and $n_b + 1$ columns. In addition, the diagonal $n_p \times n_p$ *weight matrix* has nonzero elements only on the diagonal, $W_{i,j} = w_j \delta_{i,j}$. In terms of these two matrices and the data vector $f(n_p \times 1)$, we obtain the *normal equations* of least squares analysis, given by

$$P^t WPa = P^t Wf. \tag{5.10}$$

The left-hand matrix $P^t WP$ is referred to as the *shape matrix*, or *design matrix*. In this equation, the vector a specifies the function that is being fit in the *spectral representation*, while the vector f specifies the input data in the *grid point representation*.

In equation 5.10, the elements of the left-hand symmetric $(n_b + 1) \times (n_b + 1)$ design matrix are given by

$$\left(P^t WP\right)_{p,q} = \sum_{j=1}^{n_p} P_p(\vec{r}_j - \vec{r}_k) w_j P_q(\vec{r}_j - \vec{r}_k). \tag{5.11}$$

This matrix element is the discrete sum version (rather than integral) of the overlap with respect to the weight function of the two basis functions labeled with p and q. The summation is over the grid points in the stencil. This matrix element is also the discrete version of the *inner product* of the two functions. In addition, the right-hand vector in equation 5.10, having dimensions $(n_b + 1) \times 1$, has the elements

$$\left(P^t Wf\right)_p = \sum_{j=1}^{n_p} P_p(\vec{r}_j - \vec{r}_k) w_j f_j. \tag{5.12}$$

This equation represents the discrete sum form of the overlap integral between basis function p and the *input data vector f*. It is also the *projection* of the input data vector on the p-th basis function.

Formally, the solution vector of equation 5.10 is given by

$$a = \left(P^t WP\right)^{-1} P^t Wf. \tag{5.13}$$

In practice, the matrix $P^t WP$ may become ill-conditioned, and *single value decomposition* (SVD) is needed to solve the normal equations. However, we can continue with the formalism by writing equation 5.13 out in detail for one of the expansion coefficients:

$$a_i = \sum_{j=1}^{n_p} \left\{ \sum_{l=0}^{n_b} \left(P^t WP\right)^{-1}_{i,l} \left(P^t W\right)_{l,j} \right\} f_j. \tag{5.14}$$

When this coefficient is then substituted back into equation 5.1, the *synthesis* of the function can be written as a *discrete transform of the (local) data vector*

$$g(\vec{r}) = \sum_{j=1}^{n_p} S_j(\vec{r} - \vec{r}_k) f_j, \tag{5.15}$$

where the *shape function* (a term borrowed from finite element analysis) is given by

$$S_j(\vec{r} - \vec{r}_k) = \sum_{i,l=0}^{n_b} P_i(\vec{r} - \vec{r}_k) \left(P^t WP\right)^{-1}_{i,l} \left(P^t W\right)_{l,j}. \tag{5.16}$$

Equation 5.15 expresses a mapping from the input data vector on the right side to the (least squares) output function on the left side. The shape functions in this equation are constructed during the analysis, rather than before the computations are initiated. The latter would be case, for example, for finite element computations performed on a fixed grid.

The shape functions have an interesting *normalization property*. If we evaluate the shape function at point j and then sum over the points in the stencil, we obtain

$$\sum_{j=1}^{n_p} S_j(\vec{r}_j - \vec{r}_k) = \sum_{i,l=0}^{n_b} \left(P^t WP\right)^{-1}_{i,l} \left(P^t WP\right)_{l,i}, \tag{5.17}$$

but the right side of this expression is unity. As a result, we obtain

$$\sum_{j=1}^{n_p} S_j(\vec{r}_j - \vec{r}_k) = 1. \tag{5.18}$$

Because of the value expressed on the right side, the left side is said to be a *partition of unity* [5.1].

An alternative method for determination of the coefficient vector, *not the same as least squares*, is to *interpolate the data*. In this case, we force the approximating function on the left side of equation 5.1 to *pass through* n_p of the data points, with the restriction $n_p = n_b + 1$. When this is done, we obtain the $n_b + 1$ equations

$$f_j = \sum_{i=0}^{n_b} a_i P_i(\vec{r}_j - \vec{r}_k) = \sum_{i=0}^{n_b} P_{j,i} a_i, \tag{5.19}$$

or, in matrix notation, $f = Pa$. The coefficient vector is than given by the formal solution to these linear equations is $a = P^{-1} f$. In this case, the square and nonsymmetric matrix P is called the *collocation matrix*. These equations are frequently solved using the SVD algorithm. Interpolation gives a function that, by design, reproduces the data at every point, though the approximate may oscillate in between. Interpolation based on expansion in terms of radial basis functions is described in Box 6.2. Beginning in 1999, Liu and coworkers have developed interpolation methods for problems in structural mechanics and fluid flow (see Chapters 5, 8, and 9 in [5.36]). This method, called PIM (point interpolation method), uses basis sets employing polynomials or radial basis functions. The grids used in these problems are frequently unstructured.

Although the least squares solution does not interpolate the data, it does act as a filter that smoothes the original data, removing high-frequency wiggles that may otherwise contaminate the fitting function. If smoothing is the goal, a low-degree polynomial basis is frequently used, so that the basis functions themselves do not add their own high-frequency wiggles to the function that is being synthesized.

A variation on the least squares theme, referred to as the LRA, *local rational approximation,* has been developed [5.34]. Near the reference point $\vec{r}_k$, the approximate is expressed as a rational function

$$g(\vec{r}) = \frac{a_0 + \sum\limits_{i=1}^{m} a_i P_i(\vec{r} - \vec{r}_k)}{1 - \sum\limits_{i=m+1}^{m+n+1} a_i P_i(\vec{r} - \vec{r}_k)}, \tag{5.20}$$

in which the $m + n + 1$ coefficients $\{a_i\}$ are again determined by the LS procedure. In order to implement this scheme, equation 5.20 is first rearranged into

$$g(\vec{r}) = a_0 + \sum_{i=1}^{m} a_i P_i(\vec{r} - \vec{r}_k) + g(\vec{r}) \sum_{i=m+1}^{m+n+1} a_i P_i(\vec{r} - \vec{r}_k). \tag{5.21}$$

Then, in order to develop a useful scheme, *the input function is substituted on the right side* to give

$$\begin{aligned} g(\vec{r}) &= a_0 + \sum_{i=1}^{m} a_i P_i(\vec{r} - \vec{r}_k) + f(\vec{r}) \sum_{i=m+1}^{m+n+1} a_i P_i(\vec{r} - \vec{r}_k) \\ &= \sum_{i=0}^{m+n+1} a_i Q_i(\vec{r} - \vec{r}_k), \end{aligned} \tag{5.22}$$

in which the new basis set $\{Q_i\}$ incorporates both the original basis functions $\{P_i\}$ and the products of these functions with the input function $\{f \cdot P_i\}$. The coefficients $\{a_i\}$ are then determined using the LS procedure as described earlier in this section. An extension of this procedure leads to a *multilevel least squares algorithm,* with each level based on the use of a rational function [5.34]. This and related LS procedures were used for function fitting and for computation of first and second derivatives for input data given at random test points within three- and six-dimensional hypercubes [5.34]. The procedure is expected to be efficient for high-dimensional problems because only first- and second-order monomials are used within each layer of the hierarchical scheme.

5.3 Dynamic Least Squares

In the least squares method described in the previous section, we were able to approximate spatial derivatives by solving the linear matrix equation given in equation 5.10. To solve the QHEM using this method, we must find a solution to this matrix equation for each of the hydrodynamic fields at each data point at every time step. Though finding coefficient vectors in this manner works in principle, problems are encountered when the P-matrix is ill-conditioned. As mentioned previously, one can use single value decomposition to solve the linear system instead of Gaussian elimination or LU decomposition. Use of the SVD algorithm, however, can be quite time-consuming. In addition to this complication, any matrix-oriented least squares procedure will scale poorly with dimensionality, since the total number of basis functions used in the P-matrix will grow very quickly. To remedy these problems and totally eliminate the need for matrix solutions, we turn to a different version of the least squares procedure, the *dynamic least squares method* (DLS).

To further describe this method, we will consider the *coefficient space* (dimension $n_b + 1$) associated with one of the hydrodynamic fields around one quantum trajectory. We know that the state vector $a(t)$ changes direction and length as this trajectory is guided along by the classical and quantum forces. In the DLS method [5.20, 5.19], we will evolve this state vector in time according to an equation of motion obtained from the variation criteria used in the least squares derivation. In order to derive the appropriate equations of motion, an analogy will be made with the Car–Parrinello (CP) [5.18] technique, which finds frequent applications when molecular dynamics simulations are combined with density functional theory (see Box 5.1).

In order to apply the CP technique, we imagine an imaginary "particle" of mass μ, with the "position vector" $a(t)$ moving under the influence of the effective "potential energy" functional $\Omega[a]$. This functional will be described shortly. For this particle, the Lagrangian can be written as

$$L(\dot{a}, a, t) = \frac{\mu}{2}\sum_{j=0}^{n_b} \dot{a}_j^2 - \Omega[a], \tag{5.23}$$

in which the first term, the kinetic energy, brings in the coefficient "velocities" $\{\dot{a}_j(t)\}$. The action along a trajectory in coefficient space is given by the path integral

$$S\,[a(t)] = \int_0^t L(\dot{a}, a, t)dt. \tag{5.24}$$

The coefficient trajectory for which the first-order variation in the action is zero, i.e., $\delta S = 0$, is a solution of the coupled Newtonian-type equations of motion

$$\mu\ddot{a}_i(t) = -\frac{\partial\Omega[a]}{\partial a_i}, \quad i = 0, 1, \ldots, n_b, \tag{5.25}$$

in which the *force guiding the coefficient-trajectory* is given by the gradient of a potential energy function $\vec{F} = -\vec{\nabla}\Omega$.

The potential energy that governs the dynamics of the coefficient vector should have a minimum at the "best fit" in equation 5.1. This suggests using the variance, which depends on the deviation between the exact and approximate function values at n_p points in the stencil surrounding the fiducial trajectory

$$\Omega[a] = \sum_{j=1}^{n_p} w_j \left[f_j - \sum_{i=0}^{n_b} a_i P_i(x_j - x_k) \right]^2, \tag{5.26}$$

Box 5.1. The Car–Parrinello MD-DFT algorithm

In 1985, Car and Parrinello developed a unified approach to density functional theory (DFT) and classical molecular dynamics (MD) [5.18]. To describe this method, imagine that we are given a set of nuclear positions $\{R_i\}$, some number of external constraints $\{\alpha_i\}$ (such as the volume of the system), and an expression for the DFT ground-state electronic energy $E(\{\psi_i\}, \{R_i\}, \{\alpha_i\})$. Notice that this energy is a functional of the occupied orthonormal orbitals $\{\psi_i\}$. In addition, the electron density is given by $\rho(\vec{r}) = \sum_i |\psi_i(\vec{r})|^2$. Including nuclear motions, we seek to minimize this energy functional with respect to the form of the orbitals, the atomic positions, and possibly the constraint parameters $\{\alpha_i\}$. All of these quantities are time-dependent, and for purposes of developing an optimization scheme, a kinetic energy is defined in the parameter space,

$$K = \sum_i \frac{\mu}{2} \int dv |\dot{\psi}_i|^2 + \sum_i \frac{M_i}{2} \dot{R}_i^2 + \sum_i \frac{\mu_i}{2} \dot{\alpha}_i^2, \tag{1}$$

in which the first term is the electronic contribution, the $\{M_i\}$ are atomic masses, and the parameters μ and μ_i are adjustable "fictitious masses". From this kinetic energy and the electronic energy functional, the Lagrangian in parameter space is given by

$$L = \sum_i \frac{\mu}{2} \int dv |\dot{\psi}_i|^2 + \sum_i \frac{M_i}{2} \dot{R}_i^2 + \sum_i \frac{\mu_i}{2} \dot{\alpha}_i^2 - E(\{\psi_i\}, \{R_i\}, \{\alpha_i\}). \tag{2}$$

This Lagrangian, subject to orthonormality of the orbitals, generates dynamics in parameter space for the orbitals, nuclear positions, and constraint parameters. For example, from the Lagrangian, equations of motion for the nuclear positions are given by $M_i \ddot{R}_i = -\nabla_{R_i} E$, and additional equations of motion are also obtained for $\ddot{\psi}_i$ and $\ddot{\alpha}_i$. The dynamics associated with the latter two quantities are fictitious, and they are introduced only as a tool for performing the optimization of the DFT energy functional. Solving these equations of motion leads to the "equilibrium state", wherein E is a minimum.

in which w_j is the weight attached to the j th point in the stencil. The second summation in this equation is just the approximate value of the function evaluated at the j-th point in the stencil. This effective potential energy is thus the (weighted) sum of the squares of the fitting errors. As time moves along, we want the particle in coefficient space to always seek out the minimum of this time-dependent potential energy. The trajectory that does this satisfies the equations of motion given in equation 5.25. In order to initiate the integration procedure at $t = 0$, values of the expansion coefficients are found by the usual weighted least squares matrix procedure, and the initial velocities $\dot{a}$ are set to zero. Periodically, the coefficients may be "refreshed" by solving the MLS matrix equations, with the intermediate steps handled by the DLS procedure.

The dynamical equations for the coefficients can be solved by the *quenching technique*, similar to the technique used in *simulated annealing* programs for optimization of parameters in complicated nonlinear functions. The procedure is straightforward: solve equations 5.25 using a simple updating scheme in an inner recursion loop. The inner loop is initiated using the coefficient vector from the preceding outer time loop. Then, on some or all of the small inner loop time steps, the velocity vector is scaled, $\dot{a}(t) \rightarrow \lambda \dot{a}(t)$, where the parameter λ controls the quenching rate. With repeated quenches, the kinetic energy gradually vanishes, and the particle settles in toward the minimum of the effective potential. One variation on this scheme is to quench only when the particle is climbing a hill on the effective potential energy surface, i.e., when $d\Omega/dt > 0$.

The dynamic least squares procedure is more efficient than evaluating "fresh" coefficients at each time step. The price to pay is a slight increase in memory requirements and an increase in the complexity of the computer code. However, the procedure does not require inversion of the design matrix at each integration step. In a totally different context, but using procedures similar to those described above, Swanson and Garner gave several examples of dynamic curve fitting [5.19].

Although very few studies have focused on applying this method to the QHEM, it has been used to obtain solutions for the one-dimensional scattering of a wave packet from an Eckart barrier [5.20]. As we shall see, integrating the QHEM for a wave packet scattering from an Eckart barrier is no easy task. In this type of problem, singularities in the equations of motion occur in regions of wave interferences and can cause severe velocity kinks that result in numerical instability. The sources and implications of these singularities will be further discussed in the next chapter. One way of "softening" the effects of a singularity is to use a fitting procedure, such as least squares, to smooth over the kinks. Smoothing also extends the life span of the calculation, and in some cases, it can provide reasonably accurate solutions for long-time propagation. In this study of the Eckart barrier scattering problem, use of the DLS procedure gave a four- to fivefold improvement in CPU time compared to the usual matrix least squares method.

In summary, the DLS method can be a powerful computational approach for fitting discrete data, since it can provide accurate fits while avoiding the direct solution of the MLS matrix equations. To date, we do not believe that this dynamic procedure has been used in other application areas where function fitting is required, though it appears to have considerable utility beyond its use in quantum hydrodynamics.

5.4 Fitting with Distributed Approximating Functionals

In a 2000 study [5.29], distributed approximating functionals (DAFs) were used to evaluate spatial derivatives that are required for integration of the QHEM. To describe this procedure, we will consider the use of DAFs for fitting a function that is known only at discrete points on a uniform grid in one dimension. At reference point x_k, the DAF approximation is obtained from the *discrete convolution* [5.26, 5.27]

$$g(x_k) = \sum_{i=k-m}^{k+m} S(x_i - x_k) f_i, \tag{5.27}$$

in which, for interior points, $2m + 1$ points centered around reference point k are included in the summation. The DAF functions are centered on the grid points, and $S(x_i - x_k)$ denotes the *kernel* evaluated at the i-th grid point.

Although several types of DAFs have been used in previous applications, Hermite DAFs have received the most attention [5.26, 5.27]. These functions have M-degree "shape polynomials" multiplied by a Gaussian envelope that provides for localization around the reference point. The Hermite DAFs are defined by the equation

$$S(\xi) = \frac{\delta_u}{\sqrt{2\pi\sigma^2}} e^{-\xi^2} \sum_{n=0}^{M/2} \frac{(-1)^n}{4^n n!} H_{2n}(\xi), \tag{5.28}$$

in which δ_u is the point spacing on the uniform grid, σ is the width parameter, and $\xi = (x - x_k)/(\sqrt{2}\sigma)$ is the dimensionless displacement. The summation in this equation is restricted to the even Hermite polynomials, while M determines the highest-degree polynomial that is used in the expansion. The two parameters σ and M need not be the same for each point, and they are used to control the degree to which the expansion acts as a smoothing filter. The greater the smoothing, however, the greater the deviation from the function values of the given data set. The DAFs are designed to fit exactly any polynomial of degree M or less. When plotted against the displacement ξ, the Hermite DAF function resembles a "smeared out" approximation to the Dirac δ-function, with a peak at $\xi = 0$ and M real zeros symmetrically distributed on both sides of the central peak. The DAF function approaches the δ-function in either of two limits: $M \to \infty$ at fixed σ or $\sigma \to 0$ at fixed M.

In order to calculate approximate spatial derivatives, the one-dimensional DAF functions in equation 5.27 are directly differentiated, giving

$$g^{(n)}(x_k) = \sum_{i=k-m}^{k+m} S^{(n)}(x_i - x_k) f_i. \tag{5.29}$$

In this equation, the left side denotes the DAF approximation to the n-th derivative evaluated at grid point x_k. On the right side, $S^{(n)}$ is the n-th order differentiating DAF.

In the only study for which the DAF formalism was used in solving the QHEM, the one-dimensional time-dependent scattering of an initial Gaussian wave packet from an Eckart barrier was studied [5.29]. At each time step, the C-amplitude, fluid element positions x, and velocity field v were *mapped from the physical, unstructured grid onto a uniform grid of unit length* using the Jacobian of the transformation (which was also evaluated using the DAF formalism). On this uniform grid, each of these fields was extrapolated onto uniformly spaced grid points in edge regions that extend to the left and right sides of the input data points. In these calculations, 60 fluid elements were used (this is the number of grid points in the central region of the uniform grid), and 15 point extensions were added in each edge region. The DAF parameters were as follows: Gaussian width parameter, $\sigma/\delta_u = 2$ to 3, where δ_u is the grid point spacing on the uniform grid; the highest order of Hermite polynomials was $M = 2$; and the bandwidth for the

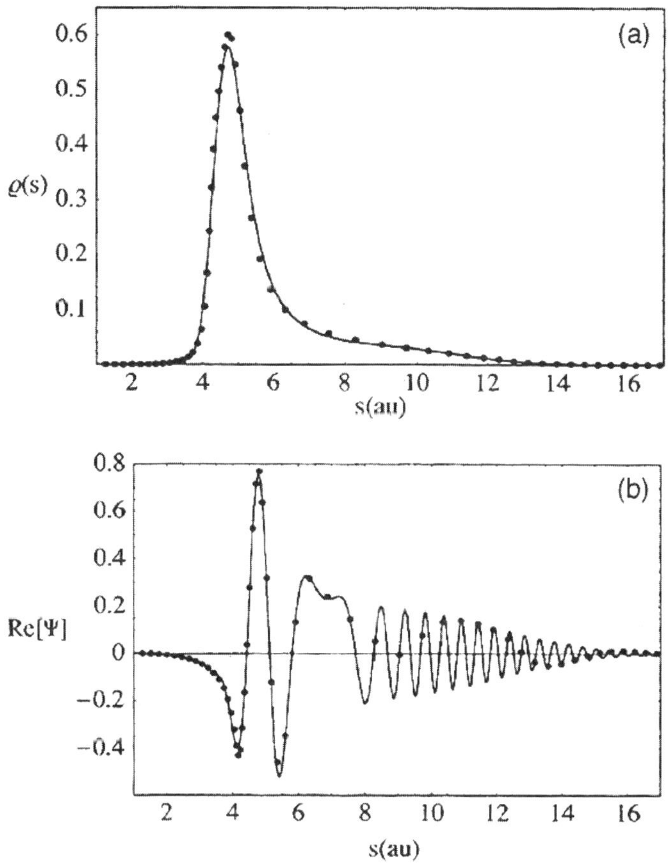

FIGURE 5.1. The probability density (top plot) and real part of the wave function (lower plot) for the scattering of an initial Gaussian wave packet from an Eckart barrier centered at $x_b = 7$ a.u. at time step 460 (44.5 fs) [5.29]. The continuous curves show results from fixed-grid Eulerian calculations, while the dots show results from the QTM using 60 fluid elements.

DAF functions was $m = 12$ to 16. The Eckart barrier, of height 6000 cm^{-1}, was centered at $x_b = 7$ a.u. The initial Gaussian wave packet was centered at $x = 2$ a.u., and the initial translational energy was three fourths of the barrier height.

Figure 5.1 compares the probability density and the real part of the wave function calculated using the QTM/DAF method (dots) with corresponding results obtained by directly solving the TDSE on a large Eulerian grid (continuous curve). The top plot shows the probability density obtained at time step 460 (44.5 fs), while the lower plot shows the real part of the scattering wave function. For the most part, there is excellent agreement between the QTM results and those obtained using the fixed grid. In the lower plot showing the real part of the wave function, there is excellent agreement between the grid results and the QTM results. One remarkable feature is that in the transmitted region given by $x > 7$ a.u., *there is less than one fluid element per de Broglie wavelength*. Even so, the amplitude and phase of the transmitted wave function are accurately reproduced. This is because we are solving the QHEM for smooth hydrodynamic fields rather than for the highly oscillating wave function.

For the QTM/DAF method, after time step 460, errors grow in the reflected region due to amplitude nodes arising from wave interferences. We will soon see that near these nodal regions, adaptive grid techniques of the type described in Chapter 7 are required to circumvent problems associated with grid inflation and compression.

5.5 Derivative Computation via Tessellation and Fitting

Yet another method for approximating the spatial derivatives needed in the QHEM was developed by Nerukh and Frederick in 2000 [5.30]. This novel method is based on function fitting following tessellation of the region occupied by the moving grid points. Because the actual algorithm used by Nerukh and Frederick is sophisticated, we will consider only a simplified two-dimensional example to illustrate this method. To do this, assume that function values are given at the $N = 9$ scattered grid points shown in figure 5.2. As usual, these points will be denoted by $\{x_i, y_i\}$, and the corresponding function values by $\{f_i\}$. We will assume that the partial derivative $\partial f/\partial x$ is required at point P in this figure. In the first stage of this method, a *tessellation* of the bounded region is performed by joining the grid points with lines, as shown in the figure. (There are other tessellations of the same figure that can be used.) Then, using the locations of points 3, 4, and 9, an approximation to the function value can be found at point B along the line that cuts the y axis at the position y_0. Likewise, using points 3, 5, and 6, we can find an approximation at point A along the line at position y_0. Finally, after fitting a polynomial to the

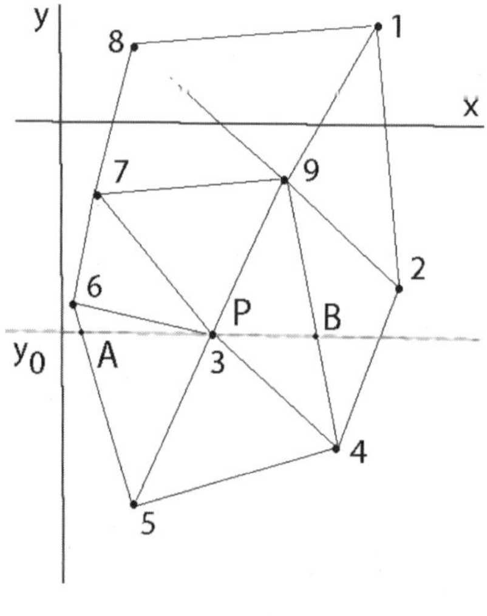

FIGURE 5.2. Triangulation of a planar region containing nine grid points. In this example, the partial derivative $\partial f/\partial x$ is to be calculated at point P.

function values at points A, P, and B, an approximate $\partial f/\partial x$ can be calculated at point P. Along the slice at the position y_0, additional points further from P can be used for more accurate fits using higher-order polynomials, if the information is available. Using this interpolation scheme, partial derivatives can be computed in higher dimensionality. However, the geometric problem of finding the intersection points (A and B in the example) becomes increasingly difficult. In addition, special care is required at points on the edges of the grid, and it is difficult to avoid errors at these points. (An alternative might be to form a multidimensional stencil and do a multivariate polynomial fit, such as described in the least squares section.)

This tessellation/fitting method was applied to the scattering of an initial Gaussian wave packet from an Eckart barrier augmented with uncoupled one- and two-dimensional harmonic potentials [5.30]. For the three-dimensional model, the potential energy was written

$$V(x, y, z) = V_0 \operatorname{sec} h^2 [a (x - x_b)] + b (y - y_b)^2 + c (z - z_b)^2, \tag{5.30}$$

while for the two-degree-of-freedom case, $c = 0$. The parameters used in the three-dimensional calculation were as follows: $V_0 = 3.65$, $a = 1$, $(x_b, y_b, z_b) = (6, 2, 2)$, initial kinetic energy $E_0 = 3.31$, $m = 1$, initial location of the center of the wave packet (2, 2, 2). For the two-dimensional problem, a total of 386 fluid elements of mass $m = 2000$ a.u. were used, and for the three-dimensional problem, a total of 6270 were used. Results for three times for the three-dimensional scattering calculation are shown in figure 5.3. The diameters of the points representing

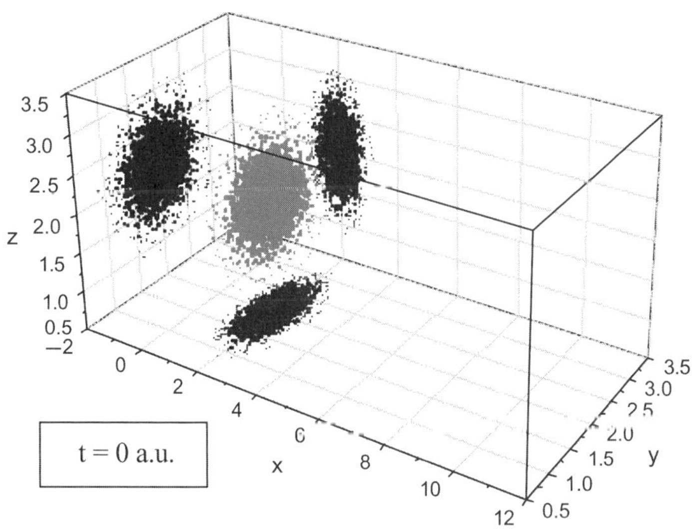

FIGURE 5.3. Propagation of 6270 fluid elements on the three-dimensional Eckart barrier–harmonic oscillator potential surface. Snapshots are shown at three times [5.30]. The diameter of each point is proportional to the density carried by the fluid element. The initial kinetic energy of the wave packet (3.31 a.u.) is slightly lower than the barrier height (3.65 a.u.).

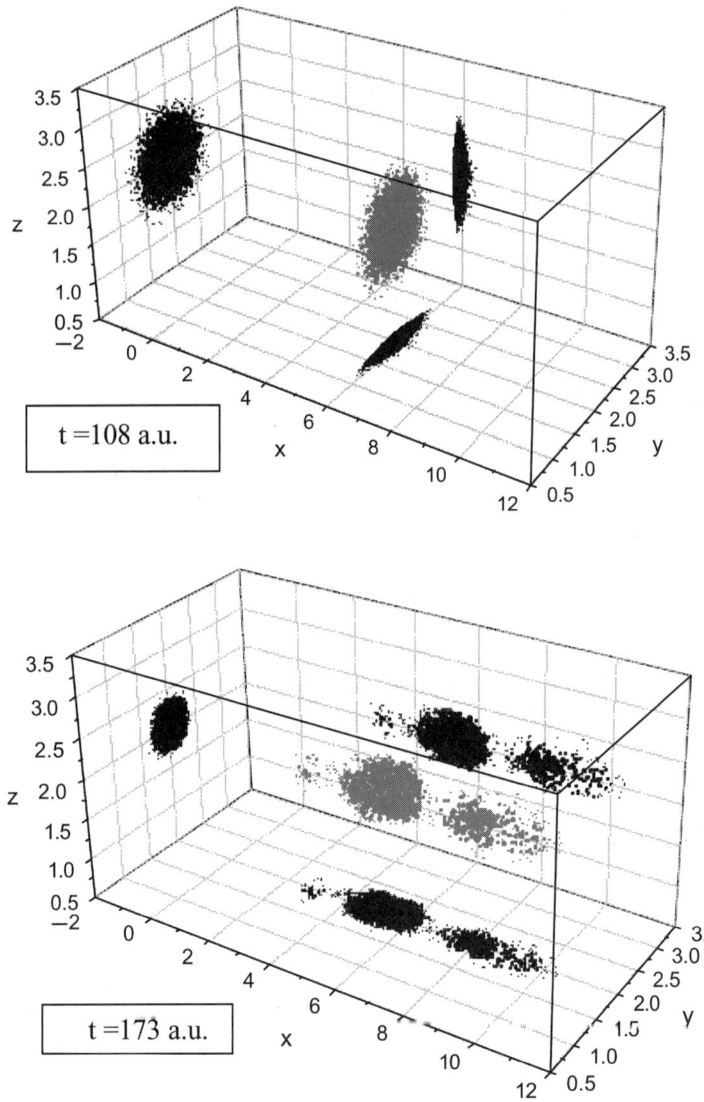

FIGURE 5.3. *Continued*

the fluid elements are proportional to the probability density carried by each of the elements. The distribution starts out spherical and directed toward the right, but as time advances, the distribution flattens just in front of the barrier (centered at the point $x_b = 6$). Soon thereafter, fluid elements pass the barrier and form the transmitted wave packet, shown in part (c). The reflected component to the left of the barrier exhibits interference effects with forward moving components that are still heading for the barrier.

In the two-dimensional version of the same scattering problem, it was found that the transmission probability grew more slowly than "exact" results obtained using an FFT method on a fixed grid. As the wave packet separates into reflected and transmitted components, few fluid elements remain in the barrier region, and numerical errors accumulate due to inaccuracies in the derivative evaluation. However, the tessellation/interpolation method can be generalized to handle unstructured meshes in higher-dimensional problems. In addition, the effectiveness of the algorithm can be improved by adapting algorithms from well-developed areas of computational geometry. Further development and application of this approach would be quite interesting.

5.6 Finite Element Method for Derivative Computation

We will now discuss a method for derivative approximation based on finite elements, the approach used by Sales Mayor et al. to study wave packet evolution in two-dimensional photodissociation problems [5.33]. To describe their method, we will consider a reference point at position (x_k, y_k) and a set of neighboring grid points. The coordinates for these points (including the point k) on the *physical grid* form the set P_k. The technique then involves two main steps, as described below.

(1) *The Mapping Step.* First, the points in the set P_k on the physical grid are mapped to a *computational grid*, where the coordinates of the points (ξ_i, η_i) form the set C_k. This mapping is designed so that the grid points on the computational grid form a standard element with a simple geometry (this type of mapping is described in Chapter 5 of [5.40]). For the element shown in figure 5.4, an unstructured 9-point grid (8 vertices surrounding a central point) in physical space

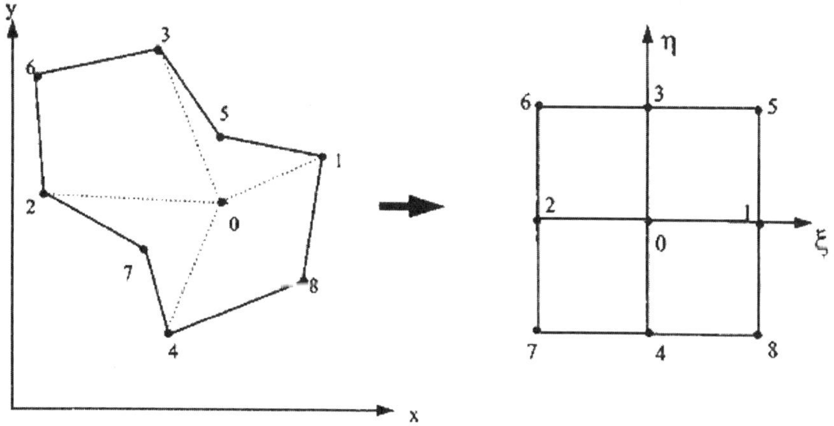

FIGURE 5.4. The mapping of 9 points from the *physical grid* to a square (of edge length $L = 2$) on the *computational grid* [5.33]. The central point, labeled 0, is mapped to the origin of the computational coordinate system (ξ, η).

is mapped to a square computational grid, with the central point (labeled 0 in this figure) located at the origin of the square. The function values and coordinates on the computational grid are denoted by (f_i, ξ_i, η_i). This type of mapping is easily envisioned in two- or three-dimensional spaces, but becomes increasingly difficult in high-dimensional spaces.

(2) *The Interpolation Step.* Next, local interpolation on the finite element in the computational space is performed, and the partial derivatives of the approximate are calculated. For this purpose, it is convenient to introduce the basis set, $B = \{b_1(\xi, \eta), \ldots, b_9(\xi, \eta)\}$, with shape functions having the localization property $b_k(\xi_i, \eta_i) = \delta_{k,i}$. In one dimension, for example, these functions are the Lagrange interpolation polynomials. Expressions for these functions in two- and three-dimensional computational spaces are given in texts such as Zienkiewicz and Taylor, *The Finite Element Method* [5.32].

In the computational space, the interpolant $g(\xi, \eta)$ is then

$$g(\xi, \eta) = \sum_{k=1}^{n} f_k b_k(\xi, \eta), \tag{5.31}$$

in which the expansion coefficients are simply the input function values f_k at the grid points. The partial derivatives at the grid points are then evaluated by differentiating the interpolant,

$$\frac{\partial g}{\partial \xi} = \sum_{k=1}^{n} f_k \frac{\partial b_k}{\partial \xi}. \tag{5.32}$$

If we consider the 9-point example shown in figure 5.4 and the 9 basis functions given by Sales Mayor et al. (see equation A3 in [5.33]), the two approximate partial derivatives at the central point are given by

$$g_\xi = \frac{\partial f}{\partial \xi}|_{(0,0)} = \frac{1}{2}(f_1 - f_2), \quad g_\eta = \frac{\partial f}{\partial \eta}|_{(0,0)} = \frac{1}{2}(f_3 - f_4), \tag{5.33}$$

which are no more than first-order finite difference expressions. In order to transform these partial derivatives from the computational space into the physical space, the 4 derivatives ξ_x, ξ_y, η_x, and η_y, along with the Jacobian J, are required. These can also be obtained by computing derivatives in the computational space.

To cope with edge points, approximate function values were extrapolated at additional grid points outside the grid boundaries. Each original edge point was then transformed into an interior point. Although this technique seemed to work for this problem, extrapolation is always risky, and this method will be difficult to implement in higher dimensionality.

The finite element method was used by Sales Mayor et al. to study the photodissociation of NO_2 and NOCl in two dimensions [5.33]. The initial ground-state vibrational wave function on the ground S_0 electronic potential surface was vertically excited to the S_1 electronic potential surface. For NO_2, this wave packet is created on the inner repulsive wall of the excited surface. On this upper surface, there are shallow wells on both sides of the bisector of the coordinate axes, as noted in the lower panel of figure 5.5. The dynamical part of the problem then

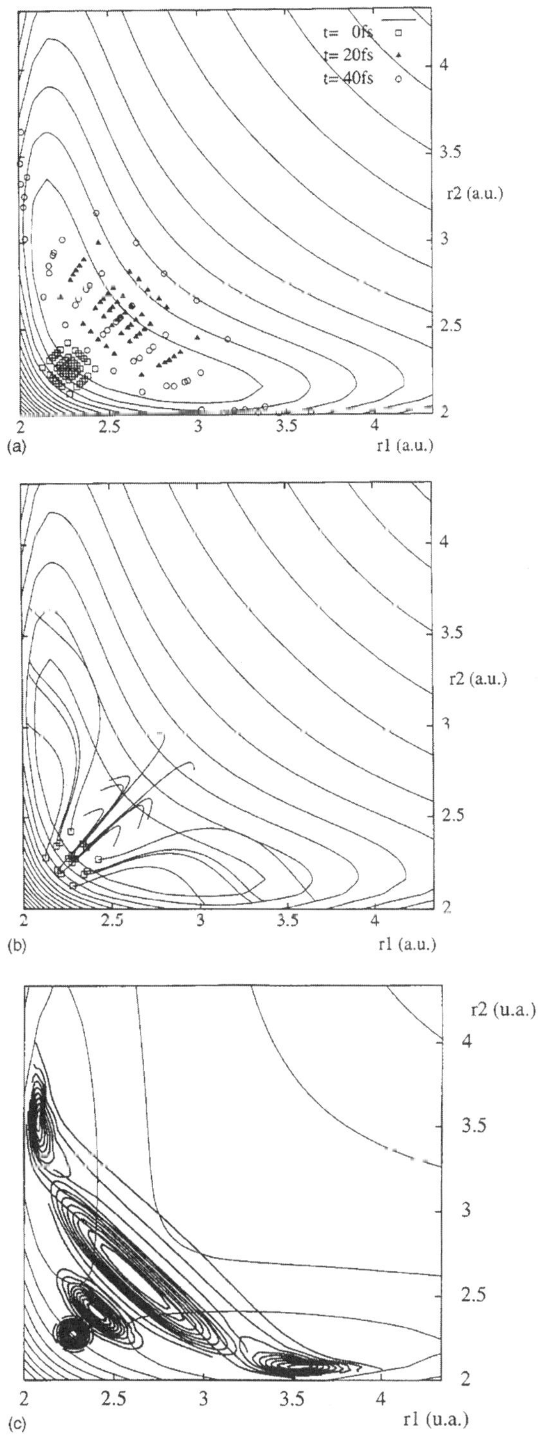

FIGURE 5.5. Quantum trajectory evolution on the NO_2 excited potential energy surface [5.33]. (a) Snapshots at four time steps showing the evolution of 49 fluid elements. (b) Sample trajectories for 16 fluid elements. (c) Equal density contours at three times, $t = 0, 20$, and 40 fs.

involves following the subsequent time evolution of an ensemble of quantum trajectories on the upper electronic potential surface. The initial ensemble spreads as it slides down the inner wall of the surface and then, over the course of time, splits symmetrically into components that evolve into the two NO+O dissociation channels.

In the quantum trajectory studies illustrated in figure 5.5, the NO_2 bending angle was frozen at the equilibrium value on the lower electronic potential surface, $\theta_0 = 133^0$. The two radial distances between each oxygen atom and the nitrogen atom were used as dynamical coordinates. At the initial time, the trajectories were launched from a 19×19 Cartesian grid located around the center of the Gaussian wave packet on the upper potential surface. In the top panel of figure 5.5, the locations of quantum trajectories are displayed at three times, $t = 0, 20$, and 40 fs. In the middle panel, the trajectories followed by 16 of the fluid elements are shown. The lower panel shows contours of the probability density at the same three time steps. The trajectories were propagated to 38 fs, and time correlation functions and dissociation cross sections were computed. Problems that limit long-time propagation arise from the low order of derivative evaluation in the computational space along with the usual difficulties of dealing with edge points.

5.7 Summary

Although each of the algorithms described in this chapter is capable of providing approximate values for partial derivatives on unstructured grids, only the moving least squares (MLS) method has been utilized in quantum trajectory calculations for systems with more than a few degrees of freedom. There is considerable room for further development and application of each of these methods. However, because of its extensive use in multivariate fitting in diverse applications ranging from data mining to fluid dynamics and structural engineering, MLS holds the distinguished position of being the most mature and robust of these algorithms.

Reflecting back to the list of "desirable features" given at the beginning of this chapter, the following comments can be made about MLS:

- The method has a solid foundation based on the minimization of an error functional.
- The method does not assume any particular labeling or arrangement of the input data points.
- Mapping from the physical grid to a computational space with a simpler data layout is not required.
- The same code can be used for interior and edge points, for sparse and dense regions of the data space, and with possible alterations, for both low- and high-dimensional data spaces.
- A high degree of flexibility can be built into both the basis functions and the weight kernel. Individual basis functions can range from simple products of monomials to radial basis functions to rational functions.

- Different basis sets can be used for fitting in various regions of the data space.
- The basis functions can be selected adaptively from a large dictionary, and the stencil size and weight function dilation parameters can also be adaptively determined using cross-correlation techniques.
- The basis set can be expanded in stages, with specific categories of basis functions (such as diagonal or cross terms in polynomial direct product spaces) added within each layer of the hierarchy.

Despite its many charms, the computational requirements of the MLS can be demanding, since it requires refitting to find a new interpolant around each grid point. In fact, in high dimensionality, expansions beyond quadratic forms may not be practical for the matrix-oriented version. Dynamic MLS, however, may reduce the computational time enough to enable the use of larger basis sets for problems in high dimensionality. A challenge for further studies, especially for the propagation of ensembles of quantum trajectories, concerns the use of adaptive algorithms for basis function, kernel, and stencil selection in multivariate applications. In addition, the point interpolation [5.36] and interpolating moving least squares [5.35] methods mentioned in Section 5.2 may also be effective for evaluating the spatial derivatives that are required in the quantum hydrodynamic equations of motion.

References

5.1. T. Belytschko, Y. Krongauz, D. Organ, M. Fleming, P. Krysl, *Meshless methods: An overview and recent developments*, Comp. Methods Appl. Mech. Eng. **139**, 3 (1996).

5.2. P. Lancaster and K. Salkauskas, *Surfaces generated by moving least squares methods*, Math. Comp. **37**, 141 (1981).

5.3. T. Hastie, R. Tibshirani, and J. Friedman, *The Elements of Statistical Learning* (Springer, New York, 2001).

5.4. J.J. Monaghan, *An introduction to SPH* , Comp. Phys. Comm. **48**, 89 (1988).

5.5. N. Perrone and R. Kao, *A general finite difference method for arbitrary meshes*, Computers and Structures **5**, 45 (1975).

5.6. V. Pavlin and N. Perrone, *Finite difference energy techniques for arbitrary meshes applied to linear plate problems*, Int, J. Numer. Meth. in Eng. **14**, 647 (1979).

5.7. T. Liszka and J. Orkisz, *The finite difference method at arbitrary irregular grids and its application in applied mechanics*, Computers and Structures **11**, 83 (1980).

5.8. L. Demkowicz, A. Karafiat, and T. Liszka, *On some convergence results for FDM with irregular mesh*, Comp. Methods in Appl. Mech. and Eng. **42**, 343 (1984).

5.9. T. Liszka, *An interpolation method for an irregular net of nodes*, Int. J. for Numerical Methods in Eng. **20**, 1599 (1984).

5.10. T.J. Liszka, C.A.M. Duarte, and W.W. Tworzydlo, *hp-meshless cloud method*, Computer Methods in Appl. Mech. and Eng. **139**, 263 (1996).

5.11. B. Nayroles, G. Touzot, and P. Villon, *Generalizing the finite element method: diffuse approximation and diffuse elements*, Comput. Mech. **10**, 307 (1992).

5.12. G.A. Dilts, *Moving least squares particle hydrodynamics I. Consistency and stability*, Int. J. Numer. Meth. Eng. **44**, 1115 (1999).

5.13. G.A. Dilts, *Moving least squares particle hydrodynamics II. Conservation and boundaries*, Int. J. Numer. Meth. Eng. **48**, 1503 (2000).
5.14. S.-H. Park and S.-K. Youn, *The least squares meshfree method*, Int. J. Numer. Meth. Eng. **52**, 997 (2001).
5.15. X. Zhang, X.-H. Liu, K.-Z. Song, and M.-W. Lu, *Least squares collocation meshless method*, Int. J. Numer. Meth. Eng. **51**, 1089 (2001).
5.16. X. Zhang, M.-W. Lu, and J.L. Wegner, *A 2-D meshless model for jointed rock structures*, Int. J. Numer. Meth. Eng. **51**, 1089 (2001).
5.17. R.E. Wyatt and K. Na, *Quantum trajectory analysis of subsystem-bath dynamics*, Phys. Rev. E **65**, 016702 (2001).
5.18. R. Car and M. Parrinello, *Unified approach for molecular dynamics and density-functional theory*, Phys. Rev. Lett. **55**, 2471 (1985).
5.19. S. Swanson and J. Garner, *Applications of Newtonian mechanics to curve fitting*, Am. J. Phys. **57**, 698 (1989).
5.20. E.R. Bittner and R.E. Wyatt, *Integrating the quantum Hamilton–Jacobi equations by wavefront expansion and phase space analysis*, J. Chem. Phys. **113**, 8888 (2000).
5.21. D.K. Hoffman, N. Nayar, O.A. Sharafeddin, and D.J. Kouri, *Analytic banded approximation for the discretized free propagator*, J. Phys. Chem. **95**, 8299 (1991).
5.22. D.K. Hoffman M. Arnold, and D.J. Kouri, *Properties of optimal distributed approximating function class propagator for the discretized and continuous wave packet propagations*, J. Phys. Chem. **96**, 6539 (1992).
5.23. D.K. Hoffman and D.J. Kouri, *Distributed approximating function theory for an arbitrary number of particles in a coordinate system-independent formalism*, J. Phys. Chem. **97**, 4984 (1993).
5.24. G.W. Wei, D.S. Zhang, D.J. Kouri, and D.K. Hoffman, *Distributed approximating functional approach to Burgers' equation in one and two space dimensions*, Comp. Phys. Comm. **111**, 93 (1998).
5.25. G.W. Wei, D.J. Kouri, and D.K. Hoffman, *Wavelets and distributed approximating functionals*, Comp. Phys. Comm. **112**, 1 (1998).
5.26. D.K. Hoffman and D.J. Kouri, *Distributed approximating functionals: A new approach to approximating functions and their derivatives*, in *3-rd International Conference on Mathematical and Numerical Aspects of Wave Propagation* (SIAM, Philadelphia, 1995).
5.27. D.K. Hoffman, T.L. Marchioro II, M. Arnold, Y. Huang, W. Zhu, and D.J. Kouri, *Variational derivation and extensions of distributed approximating functionals*, J. Math. Chem. **20**, 117 (1996).
5.28. D.K. Hoffman, A. Fishman, and D.J. Kouri, *Distributed approximating functional approach to fitting multi-dimensional surfaces*, Chem. Phys. Lett. **262**, 393 (1996).
5.29. R.E. Wyatt, D.J. Kouri, and D.K. Hoffman, *Quantum wavepacket dynamics with trajectories: Implementation with distributed approximating functionals*, J. Chem. Phys. **112**, 10730 (2000).
5.30. D. Nerukh and J.H. Frederick, *Mulitdimensional quantum dynamics with trajectories: a novel numerical implementation based upon Bohmian mechanics*, Chem. Phys. Lett. **332**, 145 (2000).
5.31. T.J. Baker, *Delaunay-Voronoi Methods* in J.F. Thompson, B.K. Soni, and N.P. Weatherill, *Handbook of Grid Generation* (CRC Press, New York, 1999).
5.32. O.C. Zienkiewicz and R.L. Taylor, *The Finite Element Method*, Vol. I (Butterworth Heinemann, Boston, 2000).

5.33. F. Sales Mayor, A. Askar, and H. Rabitz, *Quantum fluid dynamics in the Lagrangian representation and applications to photodissociation problems*, J. Chem. Phys. **111**, 2423 (1999).
5.34. X.-G. Hu, T.-K. Ho, and H. Rabitz, *Rational approximation with multidimensional scattered data*, Phys. Rev. E **65**, 035701 (2002).
5.35. G.G. Maisuradze, D.L. Thompson, A.F. Wagner, and M. Minkoff, *Interpolating moving least-squares methods for fitting potential energy surfaces: Detailed analysis of one-dimensional applications*, J. Chem. Phys. **119**, 10002 (2003).
5.36. G.R. Liu, *Mesh Free Methods: Moving Beyond the Finite Element Method* (CRC Press, Boca Raton, 2003).
5.37. G.R. Liu and M.B. Liu, *Smoothed Particle Hydrodynamics* (World Scientific, Singapore, 2003).
5.38. J. Orkisz, *Finite difference methods*, in M. Kleiber (ed.), *Handbook of Computational Solid Mechanics* (Springer, Heidelberg, 1998).
5.39. P. Lancaster and K. Salkauskas, *Curve and Surface Fitting* (Academic Press, New York, 1986).
5.40. E.B. Becker, G.F. Carey, and J.T. Oden, *Finite Elements, An Introduction, Volume I* (Prentice-Hall, Englewood Cliffs, NJ, 1981).

6
Applications of the Quantum Trajectory Method

BY COREY J. TRAHAN

The QTM is applied to four model scattering problems. Plots of quantum trajectories and the quantum potential illustrate wave packet spreading, noncrossing of trajectories, barrier transmission, and instability in nodal regions.

6.1 Introduction

Following its introduction in 1999, the quantum trajectory method (QTM) has been used to obtain numerical solutions to the quantum hydrodynamic equations of motion (QHEM) for a number of one-dimensional and multidimensional problems (see [6.1–6.15] and [6.16, 6.17]). In this chapter, we will examine quantum trajectory solutions for four relatively simple but illustrative examples. In the first example (Section 6.2), the evolution of a free Gaussian wave packet, there are no external forces acting on the trajectories, and the dynamics are governed only by their initial conditions and the evolving quantum force. In the next three examples, we will consider the evolution of a quantum wave packet under the influence of an external potential, which has been calculated beforehand. These examples include wave packet evolution in a two-dimensional anisotropic harmonic oscillator well (Section 6.3), the evolution along a downhill ramp potential curve (Section 6.4), and the one-dimensional scattering from an activation barrier (Section 6.5).

Each of these examples, except for wave packet scattering from an activation barrier, was chosen because the hydrodynamic fields are smooth at all times. We shall see that the QTM is a very reliable computational method when applied to problems whose solutions possess this "smoothness" quality. In the final example, wave packet scattering from an activation barrier, it will be shown that not all problems are so easily solved using the QTM, and in fact, some wave packet scattering problems present severe computational difficulties. We will see that for this problem, instabilities inherent in the QHEM are encountered in the reflected region where nodes occur. It will be these instabilities that motivate us to investigate more sophisticated methods in Chapters 7 and 15.

Before we move to the examples, a few comments must be made concerning details on the numerical methods used in the quantum trajectory algorithm. For the first three examples, the simple first-order Euler time integrator was all that was needed for accurate time stepping. On the other hand, it is well known that in the barrier transmission problem, the QHEM are "stiff", and a fourth-order Runge-Kutta method was used for increased stability. Ideally, an unconditionally stable *implicit time integrator* should be used for robust versions of the QTM, though these methods can be more time consuming than explicit ones. In addition, in evaluating spatial derivatives, the transformation $C(\vec{r}, t) = \ln(R(\vec{r}, t))$ is always made to reduce fitting errors; for more detail, see Box 6.1. For the first two examples, the least squares method (using a local quartic basis set) was used to evaluate the spatial derivatives required in the equations of motion at each time step. For the last two examples, radial basis function (RBF) interpolation was used (see Box 6.2). The actual FORTRAN 90 computer program used for the first three examples is provided in Appendix 2. For the remainder of this chapter, atomic units will be used, unless specified otherwise (see Appendix 1).

Box 6.1. The *C*-space transform

In order to enhance the accuracy in the fitting or interpolation algorithms used to solve the QHEM, we will frequently take advantage of the transformation to the *C*-amplitude $C(\vec{r}, t) = \ln(R(\vec{r}, t))$. It is straightforward to see why this transform is especially useful, since we frequently deal with densities and amplitudes that can be approximated *locally* with Gaussian functions. In *C*-space, this local Gaussian is a quadratic polynomial, which can be accurately fit using the methods presented in the preceding chapter. When used in the QTM, the *C*-space amplitude will be calculated from the *R*-amplitude for the initial wave function and then evolved according to transformed equations of motion involving dC/dt instead of dR/dt. To visualize how this transformation changes the original equations, we can write the original and transformed quantum potentials as

$$Q(R) = -\frac{\hbar^2}{2m}\frac{\vec{\nabla} R}{R}, \qquad Q(C) = -\frac{\hbar^2}{2m}(\nabla^2 C + \vec{\nabla} C \cdot \vec{\nabla} C). \tag{1}$$

Later, when dealing with the derivative propagation method in Chapter 10, we will see that evolving the *C*-amplitude instead of the *R*-amplitude is a necessity.

It should be noted, however, that performing calculations in *C*-space is *not* the solution to all problems connected with derivative evaluation, and there are situations (for example, around wave function nodes) in which use of this transformation can be detrimental. Usually, though, this transform helps to significantly increase the accuracy of the derivative evaluation scheme.

6.2 The Free Wave Packet

The unconstrained free translation of a wave packet is one of the most well known and easily solved problems in quantum dynamics. We start with this example in order to illustrate the effects of the quantum potential on the dynamics of the quantum trajectories and to present methods for calculating and visualizing these quantum trajectories. A similar analysis for the free wave packet was undertaken by Holland in *The Quantum Theory of Motion* [6.18].

Using the C-amplitude, equations of motion for the quantum trajectories are given by

$$\frac{dC_i}{dt} = -\frac{1}{2m}\nabla^2 S_i, \tag{6.1}$$

$$\frac{dS_i}{dt} = \frac{1}{2m}\vec{\nabla}S_i \cdot \vec{\nabla}S_i + \frac{\hbar^2}{2m}(\nabla^2 C_i + \vec{\nabla}C_i \cdot \vec{\nabla}C_i), \tag{6.2}$$

$$\dot{\vec{r}}_i = \vec{v}_i = \frac{1}{m}\vec{\nabla}S_i, \tag{6.3}$$

where i is an index denoting the trajectory at position $\vec{r}_i$. Of course, solutions to these equations are well known and are given analytically for an initial Gaussian in D dimensions [6.18]

$$C(\{x_i\}, t) = -\frac{1}{2}\sum_{i=1}^{D}\left\{\frac{1}{2}\ln\left(\pi\left[\sigma_i^2 + \left(\frac{t}{\sigma_i m_i}\right)^2\right]\right) + \frac{(x_i - x_{0,i} - v_{0,i}t)^2}{\sigma_i^2 + \left(\frac{t}{\sigma_i m_i}\right)^2}\right\}, \tag{6.4}$$

$$S(\{x_i\}, t) = \sum_{i=1}^{D}\left\{p_{0,i}(x_i - x_{0,i}) + \theta_i(t) + \frac{t}{2m_i}\left[\frac{(x_i - x_{0,i} - v_{0,i}t)^2}{\sigma_i^2\left[\sigma_i^2 + \left(\frac{t}{\sigma_i m_i}\right)^2\right]} - p_{0,i}^2\right]\right\}, \tag{6.5}$$

with

$$\theta_i(t) = \frac{1}{2}\tan\left(-\frac{t}{\sigma_i^2 m_i}\right). \tag{6.6}$$

In these equations, m_i, $v_{0,i}$, $p_{0,i}$, $x_{0,i}$, and σ_i respectively are the mass, initial velocity, initial momentum, initial position, and RMS (root mean square) width along coordinate i (atomic units are used in these equations). From this set of equations, it can be seen that the C-amplitude and the phase for the free packet are both quadratic functions in space for all times. Thus, the initial Gaussian distribution remains Gaussian for all times. By taking the gradient of equation 6.5, we find that the free particle velocity field $\vec{v} = (1/m)\vec{\nabla}S(\vec{r}, t)$ is linear in space for all time. These important properties of the free packet enable us to exactly represent the spatial part of C and S by D-dimensional quadratic polynomials. Because of this, quantum trajectories are easily propagated using the QTM.

We will now apply the QTM to an initial one-dimensional initial *stationary* ($v_0 = 0$) Gaussian wave packet. The mass is approximately that of a proton, m = 2000, and the RMS width of the initial Gaussian is $\sigma = 0.18$. Since the free packet solutions are smooth in time, a relatively large time step of 1 a.u. was used for the Euler time integration. A total of 11 trajectories were propagated concurrently.

Figure 6.1 (a) displays the quantum trajectories calculated using the QTM. From this figure, we observe trajectories diverging from the center of the packet and from one another as time moves on, whereas, since $v_0 = 0$ and there are no external forces, classical particles would not move at all. The analytical solution for these quantum trajectories is given by

$$x(t) = v_0 t + x_0 \sqrt{1 + \left(\frac{\hbar t}{2m\sigma^2}\right)^2}. \tag{6.7}$$

The divergent nature of the trajectories is induced by the term in the square root of this equation. We note that this term is purely quantum, since it depends on $\hbar$, and in the classical limit that $\hbar \to 0$, this equation becomes the Newtonian equation for particle translation.

The effect of these divergent trajectories on the density and amplitude fields can be understood if we use the *conservation relation*, which states that along each quantum trajectory, the weight $\rho_i dV_i$ = constant for all time (see equations 4.21–4.23). Since these quantum trajectories are diverging from one another, each of the volume elements $dV_i = dx_i$ is increasing in time. For the conservation property to hold, the densities associated with these volume elements must be decreasing. The result is density spreading as depicted in figure 6.1 (b). This is a well-known attribute of quantum-mechanical wave packets, and if the wave packet remains free, then it will continue to spread until the density is completely delocalized.

To understand the origin of the nonclassical accelerations that lead to wave packet spreading, we write the total force acting on each fluid element as the sum of its classical and quantum components,

$$f_{total} = f_c + f_q = -\frac{\partial}{\partial x}(V + Q) = -\frac{\partial Q}{\partial x}, \tag{6.8}$$

where $f_c = -\partial V/\partial x = 0$ for the free wave packet. The analytical solution for the quantum potential is

$$Q = \frac{\hbar^2}{4m\sigma(t)^2}\left[1 - \frac{(x - v_0 t)^2}{2\sigma(t)^2}\right], \tag{6.9}$$

where the width is given by $\sigma(t) = \sigma_0\sqrt{1 + (\hbar t/2m\sigma_0^2)^2}$. By differentiating this equation, we can obtain the following equation for the quantum force (and total force for this problem):

$$f_q = \frac{\hbar^2}{4m\sigma(t)^4}(x - v_0 t). \tag{6.10}$$

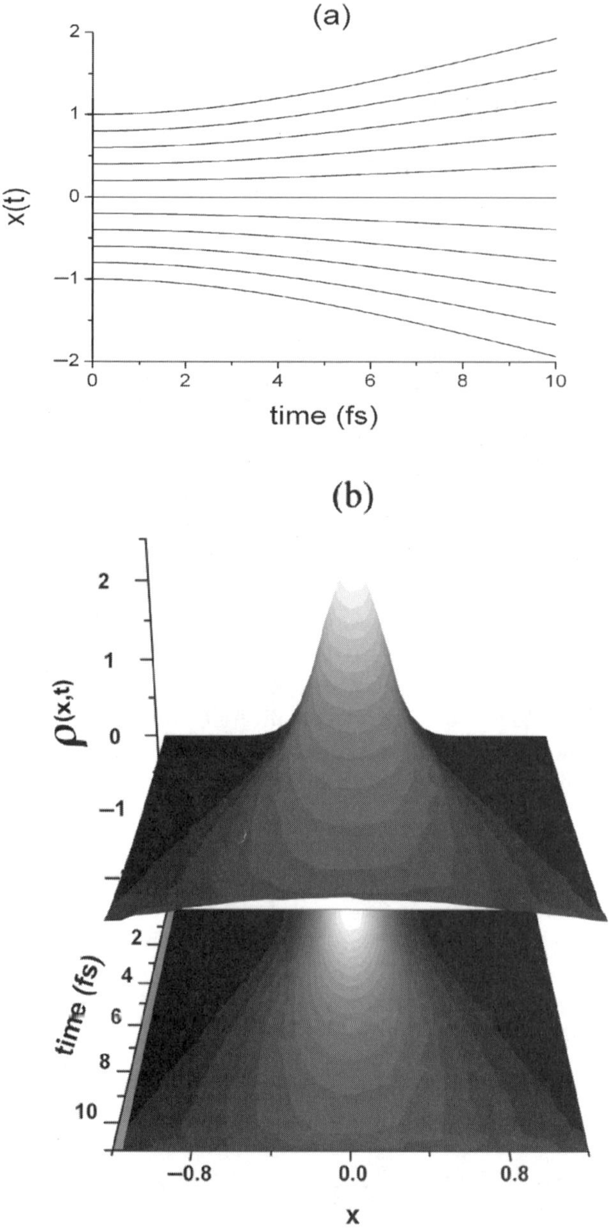

FIGURE 6.1. Time dependence of 11 quantum trajectories (a) and the density (b) for a free one-dimensional Gaussian wave packet (atomic units are used). These solutions were calculated using the QTM.

Plots showing the time dependence of the quantum potential and force are given in figure 6.2. It is the quantum force that spreads the trajectories away from the packet's center. According to this plot and equation 6.10, the quantum force is greater for those fluid elements positioned further away from the wave packet center. At the exact center, the quantum force is zero, and any motion of a fluid element at this location is purely classical. In this problem, all of the fluid elements are initially at rest, and because the quantum force vanishes at the packet's center, the center does not move in time. Another important feature of equation 6.10 is that the magnitude of the quantum force acting on a fluid element at a location $x \neq 0$ decreases as the width of the packet increases. This means that as the wave packet spreads in space, the quantum force decreases in magnitude. This flattening of the quantum force can also be seen in figure 6.2 (b).

It is important to note again that acceleration of the quantum trajectories in this problem is a direct result of the quantum potential. The quantum potential acts as an internal stress that leads to divergence of the particle trajectories and wave packet spreading.

6.3 The Anisotropic Harmonic Oscillator

For the second QTM example, we will propagate an initial Gaussian wave packet on a two-dimensional anisotropic harmonic oscillator potential of the form

$$V(x,y) = k_1x^2 + k_2y^2, \tag{6.11}$$

with force constants $k_1 = 0.009$ and $k_2 = 0.036$. The wave packet will be initially centered at the potential minimum $(0, 0)$ and given initial velocities $v_x = 0.0037$ and $v_y = 0.0048$. The RMS widths for the initial two-dimensional Gaussian are $\sigma_x = \sigma_y = 0.25$, and the mass is $m = 2000$. Once again, analytical solutions to the hydrodynamic equations for this potential have been obtained previously [6.18]. In addition, for the one-dimensional example involving a Gaussian evolving in a harmonic potential, Bowman has given an informative analysis of the dynamics from the viewpoint of Bohmian mechanics [6.33].

In this case, there *is* an external potential present, and the overall force acting on each fluid element is a combination of both the quantum and classical forces, as given in equation 6.8. It is well known that in problems with classical force fields that are linear or constant in space, an initial Gaussian wave packet will remain Gaussian for all times, just as for the free packet. Since this is the case for the anisotropic oscillator, we expect the wave packet center to evolve around the potential well while the amplitude expands or contracts. Consequently, the quantum potential retains an inverted parabolic shape, and the quantum force is a linear function equal to zero at the wave packet center for all times. Since the classical and quantum forces for this problem are both linear, the total force acting on each fluid element is also a linear function in space, just as for the free packet. Once again, the further the fluid elements are from the center of the packet, the

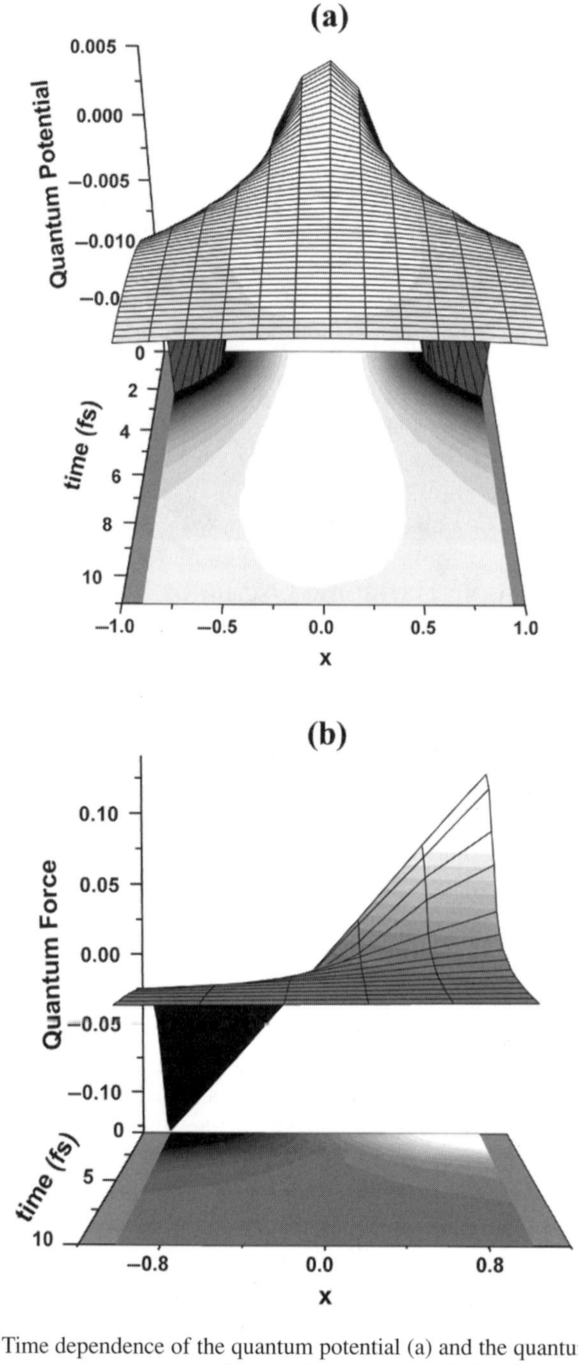

FIGURE 6.2. Time dependence of the quantum potential (a) and the quantum force (b) for a free one-dimensional wave packet. These solutions were calculated using the QTM.

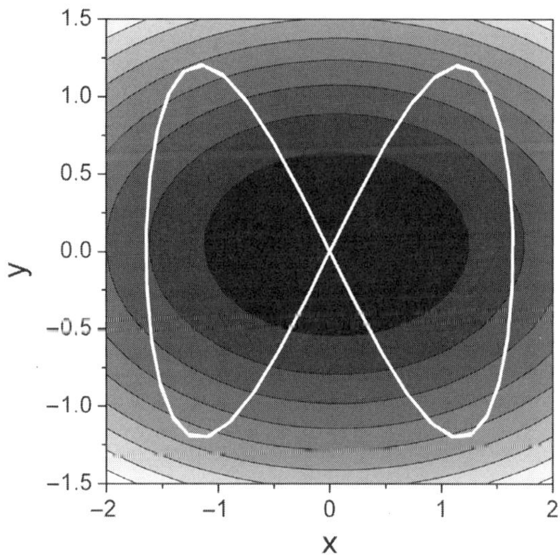

FIGURE 6.3. Time evolution of the quantum trajectory initially located at the center of the Gaussian wave packet. The contour lines represent the anisotropic harmonic oscillator potential. The same trajectory was obtained when the classical equations of motion where integrated for this potential and for the same initial condition.

greater the quantum force acting on the fluid element. The central element will again follow a purely classical trajectory, since the quantum force is zero along this path. The time evolution of the quantum trajectory at the center of the wave packet is illustrated in figure 6.3. Quantum trajectories are plotted in figure 6.4, parts (a) and (b), where two cross-sections are shown to emphasize the anisotropic symmetry of the harmonic well. On comparison, the x-trajectories for the y cross-section have larger amplitudes and a longer period than the y-trajectories for the x cross-section. This was expected since the oscillator potential is "softer" in the x-direction than the y-direction. Once again, if we followed the density along these trajectories, it would reveal an expanding and contracting Gaussian distribution always centered at the central trajectory.

In Section 4.7, we mentioned the noncrossing rule: quantum trajectories are not allowed to cross in space-time. This feature becomes evident when we compare the quantum trajectories shown in figure 6.4 with those obtained classically by integrating Newton's equations of motion (see figure 6.5). (The classical and quantum trajectories were launched from the same initial positions.) All of the classical trajectories cross at two focal points during each oscillatory period. In contrast, the quantum force prevents the quantum trajectories from crossing. In addition to this difference, the maximum and minimum amplitudes of the classical trajectories are significantly smaller than those of the quantum trajectories. This is due to quantum forces pushing the outer fluid elements away from the center of the packet.

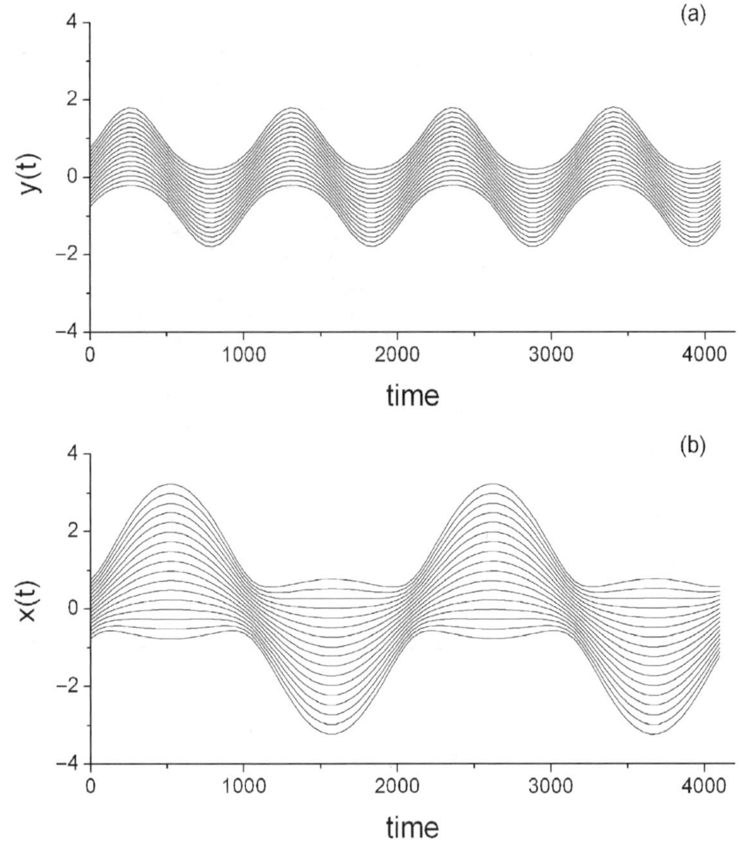

FIGURE 6.4. The time dependence of 17 quantum trajectories calculated using the QTM for the two-dimensional harmonic oscillator. Plot (a) shows the y-dependence for these trajectories in the $x = 0$ cross-section. Plot (b) displays the x-dependence for these trajectories in the $y = 0$ cross-section.

6.4 The Downhill Ramp Potential

We will now use the QTM to follow the evolution of an initial Gaussian wave packet on a one-dimensional downhill ramp potential centered at $x = 1$,

$$V(x) = \frac{V_0}{1 + e^{-2.5(x-1)}}, \tag{6.12}$$

with $V_0 = -0.0068$. The limits on this potential are as follows: as $x \to -\infty$, $V \to 0$; as $x \to \infty$, $V \to -V_0$. This particular model provides insight into the nature of exothermic chemical reactions and some photodissociation processes. For this problem, four wave packets, each initially centered at $x_0 = 0$, were launched with translational energies of 0, 0.0023, 0.0068, and 0.0356 (0, 500, 1500, and 8000 cm^{-1}, respectively). A total of 17 fluid elements were used at each value of the energy. This number is orders of magnitude smaller than the number of fixed

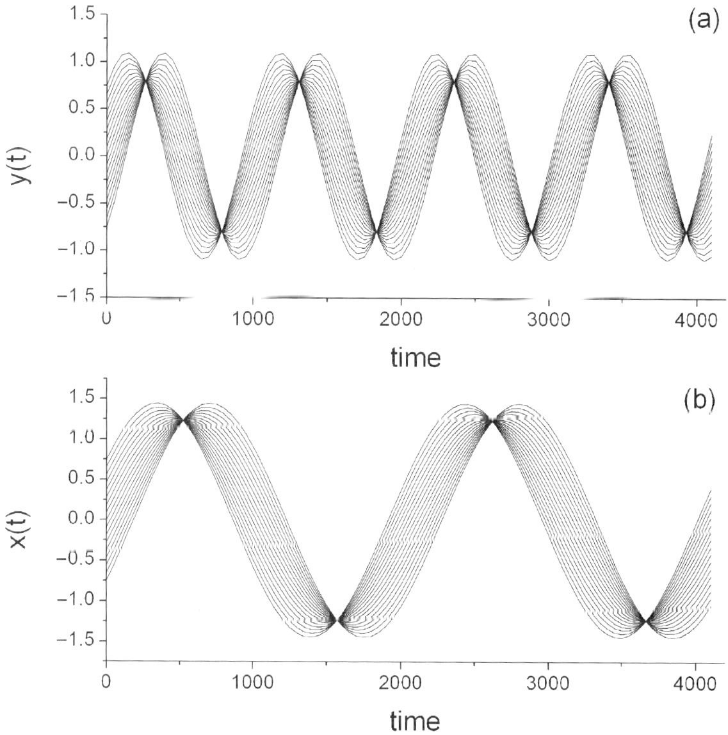

FIGURE 6.5. The time dependence of 17 *classical trajectories* for the two-dimensional harmonic oscillator. Plot (a) shows the y-dependence of trajectories in the $x = 0$ cross-section. Plot (b) displays the x-dependence of trajectories in the $y = 0$ cross-section.

grid points that would be required to solve the time-dependent Schrödinger equation for this problem. The RMS width for each initial one-dimensional Gaussian is $\sigma = 0.25$, and the system mass is m = 2000. For time integration, an Euler integrator with a time step of $dt = 0.1$ was used. In this example, radial basis function interpolation was used to calculate the spatial derivatives needed for the equations of motion. More details on this interpolation procedure are provided in Box 6.2.

One important reason for solving any scattering problem is to obtain time-dependent transmission probabilities. Calculating density transmission for problems such as the downhill ramp may seem a bit odd; however, density reflection can occur whenever there is a change in the potential energy, and this change can be in the vicinity of either a barrier or a drop-off.

The QTM was used to calculate the time-dependent transmission probabilities for the downhill ramp potential, as shown in part (a) of figure 6.6. The solid curves refer to "exact" results obtained by directly solving the time-dependent Schrödinger equation on a space-fixed grid. The QTM and exact results are in excellent agreement. Plot (b) of figure 6.6 shows the time dependence of 17 quantum trajectories for the wave packet having the initial energy $E = 0.0068$ a.u. One important feature shown in this plot is the reflection toward the negative x direction

Box 6.2. Radial basis function (RBF) interpolation

In the last decade or so, radial basis function interpolation has attracted considerable interest due to its ability to interpolate multivariate scattered data relatively accurately [6.19–6.30]. In the typical scenario, a set of n_p discrete function values $\{f(\vec{r}_i), i = 1, 2,.., n_p\}$, is provided at scattered grid point locations. Any interpolation procedure requires that the interpolate exactly reproduce the input function values at the grid points; i.e., $g(\vec{r}_i) = f_i$ for $i = 1, 2,.., n_p$, where $g(\vec{r}_i)$ is called the *interpolate* of the data set $\{f_i\}$. In RBF interpolation, the interpolant has the form

$$g(\vec{r}) = \sum_{j=1}^{n_p} a_j \phi(|\vec{r} - \vec{r}_j|), \tag{1}$$

where $|\cdot\cdot|$ denotes the Euclidean norm, and $\phi(|\vec{r} - \vec{r}_j|)$ are the RBFs. The coefficients $\{a_j\}$ in this equation are found by solving the linear system Φ a = f, where Φ is the $n_p \times n_p$ *collocation matrix* with elements $\phi_{ij} = \phi(|\vec{r} - \vec{r}_j|)$. A few examples of well-known RBFs are

$\phi(r) = (-1)^m(\delta^2 + r^2)^{\beta/2}, (2m - 2 < \beta < 2m),$	Multiquadrics,
$\phi(r) = (\delta^2 + r^2)^{-\beta/2}, (\beta > 0),$	Inverse multiquadrics,
$\phi(r) = (-1)^m(\delta^2 + r^2)^{m-1} \ln(\delta^2 + r^2)^{1/2},$	Shifted thin-plate splines,
$\phi(r) = \exp(-\beta r^2),$	Gaussians.

For the examples in this book, the multiquadric RBF will be used. This particular basis function has three user-supplied parameters, m, β, and δ. For our purposes, we will set the first two parameter values to $m = 1$ and $\beta = 1$.

To date, there are several methods for choosing the δ parameter. Some of these are done "on the fly" while others use static procedures. These methods include Foley and Carlson's scheme for selecting δ by minimizing the average root-mean-square difference between the multiquadric and inverse multiquadric [6.27], Kansa and Carlson's method of selecting local shape parameters (shape parameters that are basis function dependent) [6.26], and lastly, Rippa's method of "*cost*" *minimization* [6.21], which is similar to Goldberg's *method of cross-validation* [6.23]. For the applications described in this chapter [6.13], the shape parameter was chosen by iterating over many values until the *derivative error* in the initial amplitude fit was minimized. This procedure was done once at the initial time, and the selected parameter was then used throughout the time propagation.

The first use of RBF interpolation for propagating quantum trajectories was reported in the study by Hu et al. [6.16]. Using this method, they studied wave packet propagation for the free particle in two and three dimensions, the harmonic oscillator in two dimensions, and for the photodissociation of NOCl on an excited electronic potential energy surface (treated as a problem in two dimensions).

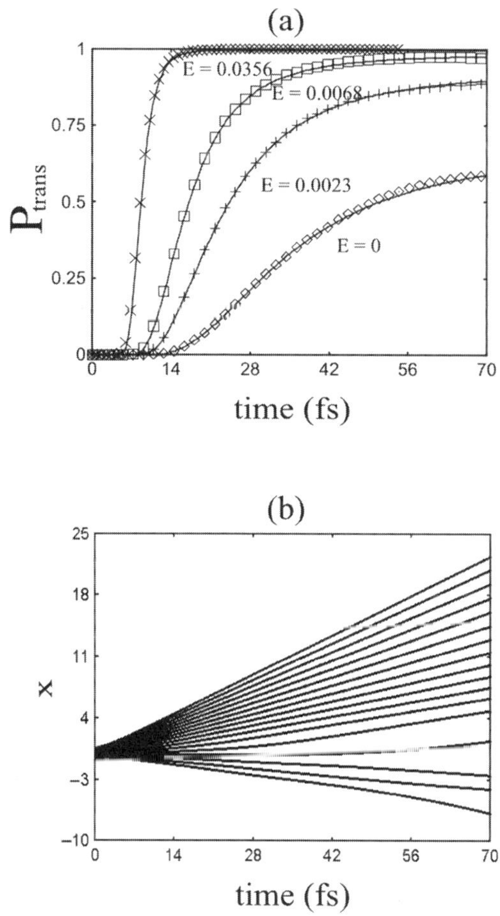

FIGURE 6.6. Part (a) displays the time-dependent transmission probabilities obtained by applying the QTM to the downhill ramp problem. Wave packets with four different energies were propagated. In part (b), 17 quantum trajectories are plotted for the wave packet having an initial energy of 0.0068 a.u. The lowest three curves represent trajectories that move "backwards", away from the downhill ramp.

of the three fluid elements furthest back from the center of the packet. Two asymmetric ensembles are formed during the bifurcation, representing the reflected and transmitted components of the trajectory ensemble. Classically, there is no turning point, and all particles will proceed down the ramp with increasing velocity. However, it is well known that in the quantum case, *above barrier reflection* can occur, resulting in only partial transmission of the amplitude. These unusual accelerations of the reflected fluid elements are due to the quantum potential, which is displayed in figure 6.7. Initially, the quantum potential has the inverted parabolic shape similar to that of a free wave packet, and the quantum force works to spread the packet in space. According to the study by Lopreore and Wyatt [6.6], it is the initial boost in kinetic energy resulting from the quantum force that pushes some fluid elements in the negative x direction and prevents them from transmitting to large values of positive x at later times. Whether the fluid element is transmitted or not depends on its initial location, its initial velocity, and the initial push from the quantum force.

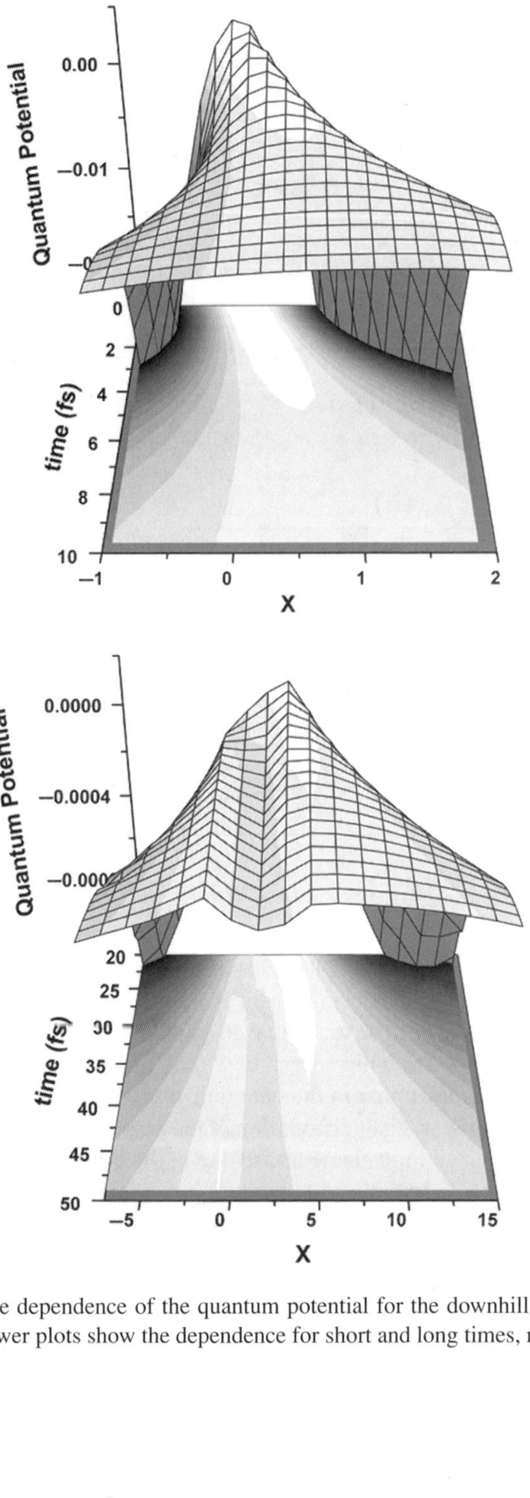

FIGURE 6.7. Time dependence of the quantum potential for the downhill ramp example. The upper and lower plots show the dependence for short and long times, respectively.

6.5 Scattering from the Eckart Barrier

In the examples discussed so far, the QTM is computationally superior to straightforward integration of the time-dependent Schrödinger equation on a fixed grid. This is because very few fluid elements were needed to obtain extremely accurate solutions to the hydrodynamic equations of motion. This is especially so for the downhill ramp potential, since propagating the complex wave function for this problem requires many grid points spread over a large space-fixed lattice. In addition, relatively large time steps can be used to integrate the hydrodynamic equations for these problems. We will now see, however, that the QTM does not always yield such superiority over traditional methods. In fact, for the final application, the simplest version of the QTM fails to provide accurate long-time solutions. (However, *adaptive dynamic grid* techniques discussed in Chapter 7 combined with the *covering function method* introduced in Chapter 15 do provide accurate solutions for this example.)

In this application, we will scatter an initial Gaussian wave packet from the Eckart potential, $V(x) = V_0 \sec h^2 [a(x - x_b)]$. This function is frequently used to model the variation of the potential energy along the reaction path for reactions with an activation barrier. In this example, the barrier height is $V_0 = 0.036$ a.u., the center of the barrier is located at $x_b = 6$, and the width parameter is a $= 0.5$. The initial packet will be launched in the direction of the barrier with the translational energy $E = 0.036$ a.u. The mass and wave packet RMS width are $m = 2000$ and $\sigma = 0.16$, respectively. Just as in the downhill ramp example, radial basis function interpolation will be used to calculate approximate spatial derivatives. For time integration, a fourth-order Runge–Kutta method was used. (For more information on this integration algorithm, see [6.31].)

The quantum trajectory results for this problem are shown in the bottom half of figure 6.8. We note that for the first 45 fs (1850 a.u.), the trajectories are smooth and display expected features. After this time, however, there is trajectory crossing on the left side of the barrier, a feature not allowed for quantum trajectories. The trajectories in the crossing region immediately become unstable, and errors from these regions subsequently propagate throughout the ensemble until the entire calculation is terminated.

In order to analyze the origin of this unstable behavior, the probability density is followed in time up to the point where trajectory crossing occurs. In the top panel of figure 6.8, the density is plotted at the time 47 fs (1920 a.u.). The solid curve in this plot is the density obtained by numerically solving the time-dependent Schrödinger equation. Through examination of this figure, we are able to pinpoint the source of the numerical blow up. Encircled in this figure is a *quasi-node*, a region where the density becomes very small, but not exactly zero. (Nodes and quasi-nodes are described in Section 4.8.) One important attribute of the hydrodynamic equations is that in these nodal regions, the quantum potential and the quantum force have large magnitudes. The effect of the quasi-node on the quantum force is displayed in the bottom panel of figure 6.9. The solid lines in this figure were calculated from the wave function solutions obtained by integrating the Schrödinger equation on a grid.

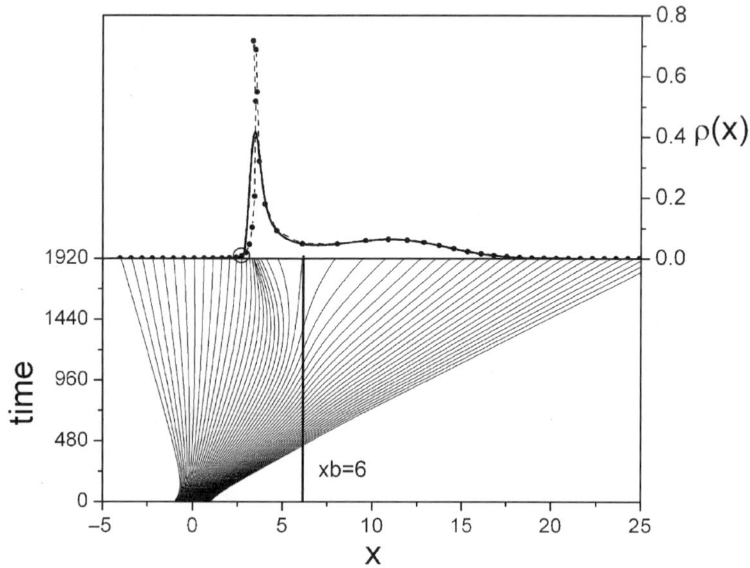

FIGURE 6.8. In the lower panel, QTM trajectories for the Eckart barrier example are plotted against time (vertical axis) until numerical breakdown occurs at 47 fs (1920 a.u.). At the breakdown time, the density is plotted against position (top panel). The solid curve in the top plot is the probability density as calculated from the direct solution to the time- dependent Schrödinger equation, while the dashed curve is the density calculated from the QTM. The solid vertical line in the bottom plot indicates the Eckart barrier maximum. A quasi-node located to the left of the barrier at $x = 2.7$ is encircled.

From the bottom plot we can see the highly localized onset of the singularity in the equations of motion, the source of which is the quantum potential. The unfortunate effects of the developing singularity can be further elucidated by looking closely at the quantum trajectories.

If we follow the quantum trajectories until the breakdown time, we find that for the first 30 fs (1250 a.u.) of the calculation, the fluid elements sweep out very smooth paths. After this time, wave packet bifurcation occurs, and some of the inner trajectories are reflected from the central region of the barrier, represented by the vertical dark solid line in the lower panel of figure 6.8. As times moves on, the quasi-node begins to develop, and fluid elements are forced away from this region as well. After 40 fs (1650 a.u.), some of the trajectories become squeezed between the potential barrier and the quasi-node. Some trajectories become trapped in this region until the quasi-node disappears, after which they are free to move further to the left, away from the barrier. In the trapped region, *compression* occurs as the density of fluid elements increases. However *inflation* occurs near nodes and quasi-nodes, meaning that there is a low density of fluid elements. (These terms were introduced by Wyatt and Bittner in [6.5].) Both of these effects give rise to numerical problems in calculating spatial derivatives using fitting and interpolation algorithms. In addition, because this quasi-node forms rather quickly, the equations

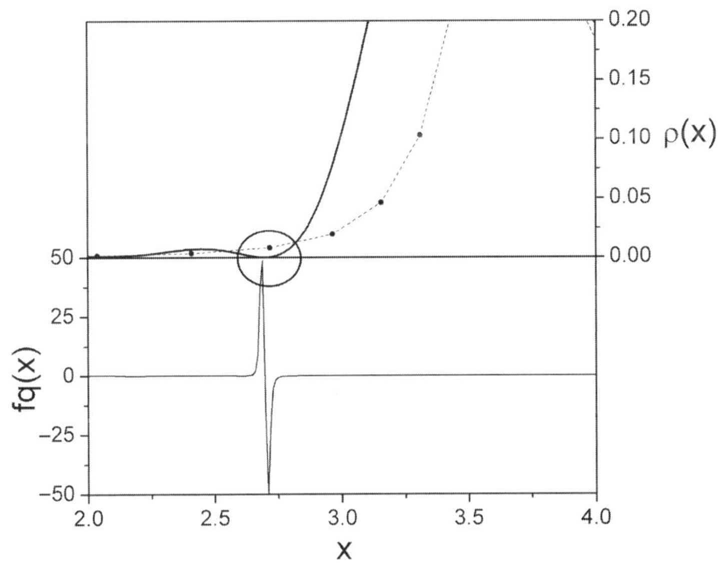

FIGURE 6.9. The region near the quasi-node identified in figure 6.8 is examined further. In the top panel, a close-up on the probability density is given at the breakdown time (1920 a.u.). The dots indicate values obtained from the QTM, and the solid curve displays results obtained by directly solving the TDSE. In the lower panel, the quantum force is plotted at the breakdown time. (Here, the quantum force was calculated from the numerical solution to the Schrödinger equation.) Note that large values of the quantum force are highly localized around the quasi-node.

of motion for C and S are stiff, meaning that only a highly accurate and stable time integrator will capture their dynamics. (Stiff differential equations have solutions involving modes that have vastly different spatial or temporal scales).

In this example, time-stiff trajectories in the compressed region and large-amplitude gradients and curvatures in the region surrounding the quasi-node cause numerical breakdown of the QTM. The Eckart barrier is just one of a number of potential energy surfaces for which problems such as this can arise. Chapters 7 and 15 provide methods to deal with this problem.

6.6 Discussion

In this chapter, the QTM was used to solve the QHEM for four model problems. In the first three of these, excellent results were obtained using fewer grid points and larger time steps than needed for the direct integration of the time-dependent Schrödinger equation on a fixed grid. In these examples, significant wave interference effects were not encountered, and nodes or quasi-nodes did not develop.

In the last example, the QTM was used to study the scattering of a wave packet from an Eckart barrier. This example was included to illustrate the effects of intrinsic singularities in the hydrodynamic equations of motion. These singularities originate in the quantum potential and quantum force near nodal regions. When investigating this problem, we found two ill effects that can lead to numerical breakdown in these regions. One of these was derivative approximation errors resulting from regional compression and inflation, and the second was stiffness in the equations of motion due to velocity kinks. Although the Runge–Kutta method is fourth-order accurate, it is not unconditionally stable, and time integration errors associated with quantum trajectories that make hard turns near the barrier or near nodal regions will quickly lead to numerical blow up. To fully resolve the difficulties near nodal regions associated with inherent instability, derivative approximation, and time integration, the methods introduced in Chapters 7 and 15 must be used.

Finally, we mention that P. Yepes (Department of Physics, Rice University) has developed instructive applets showing Bohmian trajectories for the free Gaussian wave packet, Gaussian wave packet scattering from a square barrier, and the double-slit experiment [6.32]. In addition, David Adcock has produced a colorful animation showing quantum trajectories scattering on a two-dimensional potential surface representing collinear atom-diatomic molecule reactive scattering [6.34].

References

6.1. C. Lopreore and R.E. Wyatt, *Quantum wave packet dynamics with trajectories*, Phys. Rev. Lett. **82**, 5190 (1999).

6.2. R.E. Wyatt, *Quantum wave-packet dynamics with trajectories: wave function synthesis along quantum paths*, Chem. Phys. Lett. **313**, 189 (1999).

6.3. R.E. Wyatt, *Quantum wave packet dynamics with trajectories: Application to reactive scattering*, J. Chem. Phys. **111**, 4406 (1999).

6.4. E.R. Bittner and R.E. Wyatt, *Integrating the quantum Hamilton–Jacobi equations by wave-front expansion and phase space analysis*, J. Chem. Phys. **113**, 8888 (2000).

6.5. R.E. Wyatt and E.R. Bittner, *Quantum wave packet dynamics with trajectories: Implementation with adaptive Lagrangian grids*, J. Chem. Phys. **113**, 8898 (2000).

6.6. C. Lopreore and R.E. Wyatt, *Quantum wave packet dynamics with trajectories: reflections on a downhill ramp potential*, Chem. Phys. Lett. **325**, 73 (2000).

6.7. E.R. Bittner, *Quantum tunneling dynamics using hydrodynamic trajectories*, J. Chem. Phys. **112**, 9703 (2000).

6.8. R.E. Wyatt, D.J. Kouri, and D.K. Hoffman, *Quantum wave packet dynamics with trajectories: Implementation with distributed approximating functionals*, J. Chem. Phys. **112**, 10730 (2000).

6.9. R.E. Wyatt and K. Na, *Quantum trajectory analysis of multimode subsystem–bath dynamics*, Phys. Rev. E. **65**, 016702 (2001).

6.10. C. Lopreore, R.E. Wyatt, and G. Parlant, *Electronic transitions with quantum trajectories*, J. Chem. Phys. **114**, 5113 (2001).

6.11. C. Lopreore and R.E. Wyatt, *Electronic transitions with quantum trajectories. II*, J. Chem. Phys. **116**, 1228 (2001).

6.12. K. Na and R.E. Wyatt, *Quantum hydrodynamic analysis of decoherence: quantum trajectories and stress tensor*, Phys. Lett. A **306**, 97 (2002).

6.13. C. Trahan and R.E. Wyatt, *Radial basis function interpolation in the quantum trajectory method: optimization of the mutiquadric shape parameter*, J. Comp. Phys. **185**, 27 (2003).
6.14. D. Nerukh and J.H. Frederick, *Multidimensional quantum dynamics with trajectories: a novel numerical implementation of Bohmian mechanics*, Chem. Phys. Lett. **332**, 145 (2000).
6.15. R.K. Vadapalli, C.A. Weatherford, I. Banicescu, R.L. Carino, and J. Zhu, *Transient effect of a free particle wave packet in the hydrodynamic formulation of the time-dependent Schrödinger equation*, Int. J. Quantum Chem. **94**, 1 (2003).
6.16. X. Hu, T. Ho, H. Rabitz, and A. Askar, *Solution of the quantum fluid dynamical equations with radial basis function interpolation*, Phys. Rev. E **61**, 5967 (2000).
6.17. F. Sales Mayor, A. Askar, and H.A. Rabitz, *Quantum fluid dynamics in the Lagrangian representation and applications to photodissociation problems*, J. Chem. Phys. **111**, 2423 (1999).
6.18. P. R. Holland, *The Quantum Theory of Motion: An Account of the de Broglie–Bohm Causal Interpretation of Quantum Mechanics* (Cambridge University Press, New York, 1993).
6.19. W. Haussmann, K. Jetter, and M. Reimer (eds.), *Recent progress in multivariate approximation. Proceedings of the 4th International Conference on Multivariate Approximation held at the University of Dortmund* (Birkhäuser Verlag, Basel, 2001).
6.20. T. Hastie, R. Tibshirani, and J. Friedman, *The Elements of Statistical Learning: Data Mining, Inference, and Prediction* (Springer, New York, 2001).
6.21. S. Rippa, *An algorithm for selecting a good value for the parameter c in radial basis function interpolation*, Adv. Comp. Math. **11**, 193 (1999).
6.22. N.J.D. Powell, *A review of methods for multivariable interpolation at scattered data points: The state of the art in numerical analysis* (Oxford University Press, New York, 1997).
6.23. M.A. Goldberg, C.S. Chen, and S.R. Karur, *Improved multiquadric approximation for partial differential equations*, Eng. Anal. with Boundary Elements. **18**, 9 (1996).
6.24. M.A. Goldberg and C.S. Chen, *A bibliography on radial basis function approximation*, Boundary Elem. Commun. **7**, 155 (1996).
6.25. R. Schaback, *Creating Surfaces From Scattered Data Using Radial Basis Functions. Mathematical Methods for Curves and Surfaces* (Vanderbilt Univ. Press, Tennessee, 1995).
6.26. E.J. Kansa and R.E. Carlson, *Improved accuracy of multiquadric interpolation using variable shape parameters*, Comp. Math. Appl. **24**, 99 (1992).
6.27. T.A. Foley and R. Carlson, *The Parameter R2 in Multiquadric Interpolation*, Comp. Math. Appl. **21**, 29 (1991).
6.28. R.L. Hardy, *Theory and applications of the multiquadric-biharmonic method*. Comp. Math. Appl. **19**, 163 (1990).
6.29. E.J. Kansa, *Multiquadrics: A scattered data approximation scheme with applications to computational fluid dynamics. I*, Comp. Math. Applic. **19**, 127 (1990).
6.30. R. Franke, *Scattered data interpolation: A test of some methods*, Math. Comp. **38**, 157 (1982).
6.31. W.H. Press, S.A. Teukolsky, W.T. Vetterling, and B.P. Flannery, *Numerical Recipes in FORTRAN 90*, (Cambridge, University Press, 1996).
6.32. yepes.rice.edu/PhysicsApplets/
6.33. G.E. Bowman, *Bohmian mechanics as a heuristic device: Wave packets in the harmonic oscillator*, Am. J. Phys. **70**, 313 (2002).
6.34. www.cm.utexas.edu/Wyatt/movies/qtm

7
Adaptive Methods for Trajectory Dynamics

BY COREY J. TRAHAN

Three adaptive methods designed to cope with node and derivative evaluation problems are described. These methods include adaptive moving grid design, adaptive smoothing using artificial viscosity, and adaptive hybrid algorithms.

7.1 Introduction

In Chapter 6, we found that as a consequence of inherent singularities in the quantum-hydrodynamic equations of motion (QHEM) that manifest themselves in nodal regions, the equations of motion can become stiff, and spatial gradients or curvatures of the hydrodynamic fields can be very large. These problems are further aggravated when one uses Lagrangian quantum trajectories, since inflation in these nodal regions (the movement of trajectories away from nodes) results in insufficient data for accurate spatial derivative evaluation. Because the singularity dilemma and the associated spatial derivative problem prevent the development of a robust QHEM solver, much effort has been directed toward resolving these numerical issues.

This chapter will focus on three adaptive methods designed to help cope with the singularity and derivative evaluation problems that arise in using the QHEM. The first method deals with repercussions of the singularity problem: inflation and compression. This method is based on use of the *moving-path transform* of the QHEM and will be introduced in Section 7.2. This formulation gives the user *complete control over the paths followed by the grid points*. Because these "non-Bohmian" paths can be chosen arbitrarily "in between" the Lagrangian and Eulerian viewpoints, this method is called the *arbitrary Lagrangian–Eulerian (ALE) method*. In Section 7.3, dynamic grid adaptation and the ALE method will be described and applied to several problems for which the original version of the quantum trajectory method (which used "pure" Lagrangian quantum trajectories) failed. In addition to its use in solving the *moving-path transform* of the QHEM, the ALE method will also be utilized in Section 7.4 to solve the moving-path transform of the TDSE. For this case, the ALE method can be used to decrease the total number of grid points needed for propagating the complex-valued wave function by adaptively guiding grid points to regions where they are most needed according to an *equidistribution principle*. This is in contrast to traditional methods of integrating the TDSE on a fixed lattice, where the same spacing is normally used throughout the entire grid for all times. In Section 7.5, the ALE method will be implemented in conjunction with adaptive smoothing of the force acting on trajectories in nodal regions. In order to do this, an *artificial viscosity* potential is introduced into the quantum Hamilton-Jacobi equation to effectively moderate the large quantum forces experienced by quantum trajectories near nodal regions.

In Section 7.6, we will focus on a different set of adaptive methods that are designed to circumvent problems in nodal regions. These methods combine, in *different spatial regions*, propagation of the TDSE and the QHEM and are therefore called *hybrid methods*. The idea is to solve the QHEM in regions where the hydrodynamic solutions are relatively smooth, while solving the TDSE in nodal regions, where direct integration of the QHEM becomes problematic. A general introduction to these methods will be given, and several successful hybrid applications will be described. In the last section of this chapter, Section 7.7, we will review these adaptive methods and their extensions.

7.2 Hydrodynamic Equations and Adaptive Grids

To alleviate complications arising from inflation and compression, suitable "discipline" must be imposed on the moving grid points. Ideally, we would like to eliminate undersampling in some regions and clustering in other regions while satisfying other goals, such as guiding the grid points to locations that assist with the evaluation of spatial derivatives. It is apparent that this cannot be done using the standard Lagrangian quantum trajectories. It is possible, however, to have *absolute control* over each grid point when using the arbitrary Lagrangian–Eulerian (ALE) method. In this method, *the grid point velocities are not equal to the flow velocity of the probability fluid*, but are user-assigned. These ALE grid velocities can be specified in various ways, including coupling to the velocities of boundary points, adaptive refinement at the end of each time step, or through some combination of these approaches. ALE methods have proven extremely successful in many classical fluid and solid dynamical problems [7.1–7.4].

In order to use the ALE method to solve the QHEM, the *moving-path transforms* of these equations must first be derived. To accomplish this, we begin with the Eulerian versions of the QHEM, given by

$$\frac{\partial S(\vec{r},t)}{\partial t} = -\frac{1}{2}m\vec{v}\cdot\vec{v} - Q(\vec{r},t) - V(\vec{r},t), \tag{7.1}$$

$$\frac{\partial \rho(\vec{r},t)}{\partial t} = -\vec{v}\cdot\vec{\nabla}\rho - \rho\vec{\nabla}\cdot\vec{v}, \tag{7.2}$$

where $Q(r,t)$ is again the quantum potential and $\vec{v} = (1/m)\vec{\nabla}S(\vec{r},t)$ is the flow velocity of the probability fluid. To permit arbitrary grid point motions, a suitable transformation from the Eulerian partial derivative to the total time derivative will be made. By introduction of a *grid velocity* $\dot{\vec{x}}(t)$, which is *not* necessarily equal to the flow velocity of the probability fluid, the relationship between the two time derivatives becomes

$$\frac{d}{dt} = \frac{\partial}{\partial t} + \dot{\vec{x}} \cdot \vec{\nabla}. \tag{7.3}$$

On substitution of this total time derivative, the *moving-path transforms* of the hydrodynamic equations become

$$\frac{dS(\vec{r},t)}{dt} = (\dot{\vec{x}} - \vec{v}) \cdot (m\vec{v}) + \frac{1}{2}m\vec{v} \cdot \vec{v} - Q - V, \tag{7.4}$$

$$\frac{d\rho(\vec{r},t)}{dt} = (\dot{\vec{x}} - \vec{v}) \cdot \vec{\nabla}\rho - \rho\vec{\nabla} \cdot \vec{v}. \tag{7.5}$$

Notice that these equations have been manipulated to isolate the term $\vec{w} = \dot{\vec{x}} - \vec{v}$, called the *slip velocity*. Three cases that depend on the slip velocity can arise:

$$\begin{array}{lll} 1.\ \vec{w} = -\vec{v}, & \dot{\vec{x}} = 0, & \text{Eulerian grid,} \\ 2.\ \vec{w} = 0, & \dot{\vec{x}} = \vec{v}, & \text{Lagrangian grid,} \\ 3.\ \vec{w} \neq -\vec{v} \text{ and } \vec{w} \neq 0, & \dot{\vec{x}} \neq \vec{v} \text{ and } \dot{\vec{x}} \neq 0, & \text{ALE grid.} \end{array} \tag{7.6}$$

For the first case, the grid points are frozen in space, and the original fixed-grid equations of motion for the action and amplitude, equations 7.1 and 7.2, are recovered. Although Eulerian schemes can provide very accurate solutions, they often require a large number of points and are computationally inefficient in regions of low function activity. This problem becomes greatly amplified for unbound problems, high-energy dynamics, and problems with more that a few degrees of freedom.

For the second case, $\vec{w} = 0$ and $\dot{\vec{x}} = \vec{v}$, the grid points (the fluid elements) are locked in concerted motion with the fluid and move along with the fluid's flow velocity. Under this Lagrangian condition, fewer grid points are needed, since the trajectories tend to follow regions of high density and complex dynamics. This approach was used in the QTM. One benefit of using Lagrangian quantum trajectories is that they are governed by a physical law and can be subjected to straightforward physical interpretations. However, as mentioned previously, this method does not always provide a robust algorithm for wave packet dynamics, and the stability and accuracy of the method are almost completely governed by the dynamics of the trajectories.

For the last case, the grid points are not fixed; nor do they follow the fluid's flow. In fact, this condition is the most general of the three and allows for grid velocities to be assigned arbitrarily. There are many ways of determining suitable grid velocities. One frequently used technique is to assign them according to an *equidistribution principle*. Using equidistribution, grid points can dynamically adapt in time according to specified properties of the solution (i.e., the solution's gradient, curvature, etc.). This technique will be described further in Section 7.4. A much simpler alternative, however, is to force the grid points to sweep out a regular grid of uniform spacing. In this approach, the boundary grid points are often assigned Lagrangian velocities so that the uniform grid can translate in space while expanding and contracting according to the Lagrangian boundary grid points. In this case, the grid spacings may change in time, but at any given time, all the grid points are equally spaced. This technique is widely used, since numerical integration and differentiation are much easier and generally more accurate on this type of grid.

In the next three sections, we will discuss applications of the ALE method to both the QHEM and the TDSE. In the first of the two hydrodynamic applications, the ALE method was used to study wave packet evolution on a steep uphill ramp potential. In the second hydrodynamic application, this method was used in conjunction with a method for smoothing over hard density ripples and nodes. In this application, an artificial viscosity term was introduced into the quantum Hamilton–Jacobi equation. This procedure effectively eliminates kinks in the velocity fields around nodes, and a similar procedure has been used in many classical hydrodynamic problems. Using this technique and the ALE method, it is possible to obtain solutions to the QHEM for an Eckart barrier scattering problem where multiple nodes form in the reflected wave packet. Though the artificial viscosity method is an approximation to the exact QHEM, excellent transmission probabilities were obtained, and long-time trajectory propagation was possible.

In addition to solving the QHEM using the ALE method, we will also discuss an application to the TDSE. For this case, the ALE method was used to adapt grid points to the evolving wave function curvature. By doing this, the total number of grid points was reduced relative to Eulerian calculations, while the same solution accuracy was retained.

Throughout this chapter, several one-dimensional examples will be discussed, all of which consider an initial translating Gaussian wave packet of the form,

$$\psi(x, t_0) = \left(\frac{2\beta}{\pi}\right)^{1/4} \exp\left[-\beta(x - x_0)^2 + ik(x - x_0)\right]. \tag{7.7}$$

In this equation, $k = \sqrt{2mE_{trans}}/\hbar$ is the initial translational wave number, and x_0 is the center of the Gaussian. The system mass used for these examples will be $m = 2000$ a.u.

7.3 Grid Adaptation with the ALE Method

In the previous chapter, we found that quantum trajectories can encounter fatal computational problems in nodal regions. Typically, in barrier scattering problems, the wave packet is "squeezed" against the barrier, and thereafter bifurcates into reflected and transmitted components. In many cases, ripples and nodes form in the reflected wave packet, and this is where breakdown occurs. In this section, we describe how the ALE method can be utilized to eliminate trajectory inflation and compression encountered on a "steep" uphill ramp potential [7.5]. This alone does not solve the node problem, but it does give better grid point sampling and more accurate spatial derivatives in these regions.

The uphill ramp potential is given by

$$V(x) = \frac{V_0}{1 + e^{-1.5(x-1)}}, \tag{7.8}$$

where V_0 is the potential maximum at large values of x (values for the barrier height will be mentioned later). In this example, an initial Gaussian wave packet of the form given in equation 7.7 was used with the parameter values (in a.u.) $\beta = 9$, $x_0 = 0$, and $E_{trans} = 0.036$. Numerical solutions to the uphill ramp problem using both the QTM and the ALE method were obtained using 151 grid points. For the QTM, spatial derivatives were calculated using radial basis function interpolation, which was described in Box 6.2 of the previous chapter.

Before propagating the solutions to the hydrodynamic equations, a method for calculating the ALE grid speeds is required. As mentioned previously, there are many ways of doing this. For this problem, the grid speeds were chosen to maintain a grid of uniform spacing. In order for the grid to follow the evolving wave packet density, the grid velocities of the first and last grid points, x_1 and x_N, were assigned to the Lagrangian values, just as in the QTM. The procedure was as follows:

1. First, the boundary grid points were updated using the Lagrangian velocities obtained by the phase gradient, $v = (1/m)\partial_x S(x, t)$, giving $x_1^L(t + \Delta t)$ and $x_N^L(t + \Delta t)$, where the superscript denotes the Lagrangian value for the particle velocity.
2. Next, the updated internal grid points, $x_i(t + \Delta t)$, where $i = 2, 3, \ldots, N - 1$, were obtained by constructing a *uniform grid* between the Lagrangian boundary particles.
3. Lastly, each internal grid velocity was calculated according to the linear difference equation $\dot{x}_i = (x_i(t + \Delta t) - x_i(t)) / \Delta t$. Once again, the grid velocities for the boundary grid points were set to their Lagrangian values, i.e., $\dot{x}_1 = v_1$ and $\dot{x}_N - v_N$.

This procedure can be straightforwardly adapted for use in multidimensional problems, and forcing the grid points to be equally spaced at each time can be numerically beneficial, especially when one usesfinite difference methods to calculate the spatial derivatives.

When the QTM was used to compute transmission probabilities for uphill ramp potentials having the heights $V_0 = 0.027,\ 0.036,\ 0.046,$ and 0.055 a.u., computational breakdown occurred at $t = 76,\ 66,\ 56,$ and 50 fs, respectively. Thus, as the height of the uphill ramp increases, the breakdown time of the QTM decreases. One explanation for this trend is that as the potential is increased, the wave packet is squeezed tighter (smaller width, larger amplitude) as it encounters the potential, and the particle compression is exacerbated. In addition, as the height of the potential is increased, the magnitude and frequencies of the amplitude

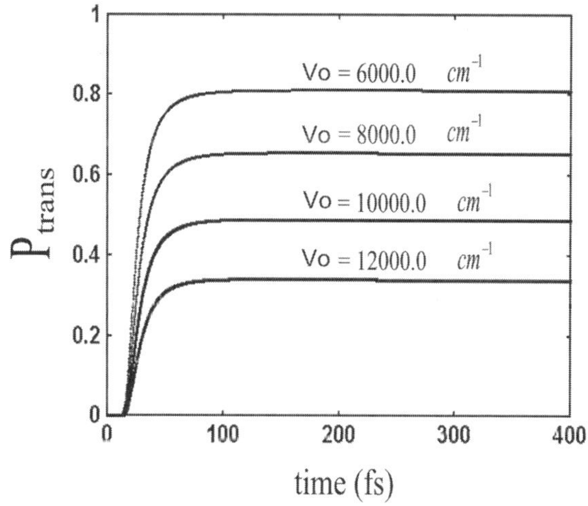

FIGURE 7.1. Transmission probabilities calculated using the ALE method for the uphill ramp with four barrier heights [7.5]. These values were obtained using an initial wave packet translational energy equal to 8000 cm^{-1}.

ripples in the reflected wave packet also increase. The consequences of these amplitude deformations are poor derivative approximation and eventual trajectory crossing.

Problems with computational breakdown *were not encountered* when the ALE method was applied to the *same* uphill ramp models. Figure 7.1 shows transmission probabilities versus time obtained using the ALE method. Although the plot extends for only 400 fs, the transmission probabilities were stable for *over one picosecond*. Such results are a significant improvement over the QTM.

In figure 7.2, the probability density for the uphill ramp potential of height $V_0 = 0.027$ is displayed at $t = 125$ fs. This figure provides one explanation for why the ALE method yields computational results that are both accurate and stable for long times. The most important feature is the location of the grid points. In time, the grid points spread throughout the domain as a uniformly spaced grid, preventing inflation near regions of low density. It is highly probable that most, if not all, of the computational errors in the QTM developed when spatial derivatives were computed in the region of ripple formation in the reflected wave packet $(-16 < x < 0)$. To calculate the spatial derivatives in this critical region, the grid points must be properly positioned to capture the function's local oscillating behavior. Excessive inflation in these regions will result in large-scale errors in the derivatives, probably causing numerical breakdown in the QTM after 76 fs. This is avoided in the ALE method, however, since the grid points are constrained to prevent clustering and inflation.

Lastly, in figure 7.3, it is evident that the solution accuracy increases when the number of grid points in the ALE method is increased from 151 to 251. In this figure, the probability density in the reflected region is plotted at $t = 110$ fs. The accuracy of the action and amplitude is predominantly governed by errors in the interpolation routine, which are greatest in regions of large gradient/curvature. By increasing the number of grid points in these regions, interpolation errors are

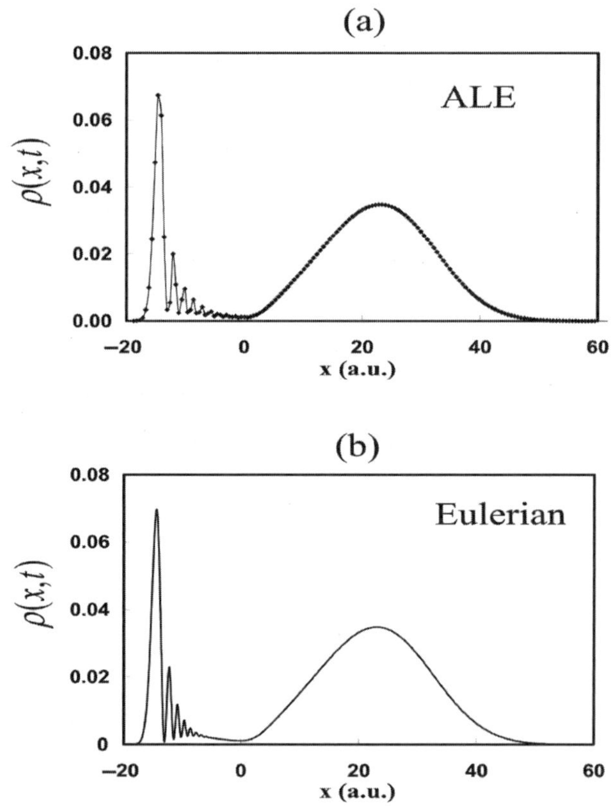

FIGURE 7.2. Density plots at $t = 125$ fs for the uphill ramp potential having the barrier height $V_0 = 6000\ \text{cm}^{-1}$ [7.5]. Part (a) displays the results obtained with the ALE method (connected by linear splines), while part (b) shows results obtained by solving the fixed-grid TDSE.

decreased and more accurate solutions can be obtained. It should be noted that although the number of grid points in the ALE method was as large as 251, this was still only a fraction of the number of grid points (7,150) used to obtain the Eulerian fixed-grid results.

7.4 Grid Adaptation Using the Equidistribution Principle

Although it is very useful to choose the grid velocities so that the points generate a uniform grid at each time, we are not limited to this choice. In this section, we will use a more sophisticated method for choosing the grid point velocities, the *equidistribution principle*. By using this method, we will be able to adapt the grid point paths to the underlying fields *as they evolve in time*. The purpose is not only to guide grid points to regions where there is a significant density, but also to guide

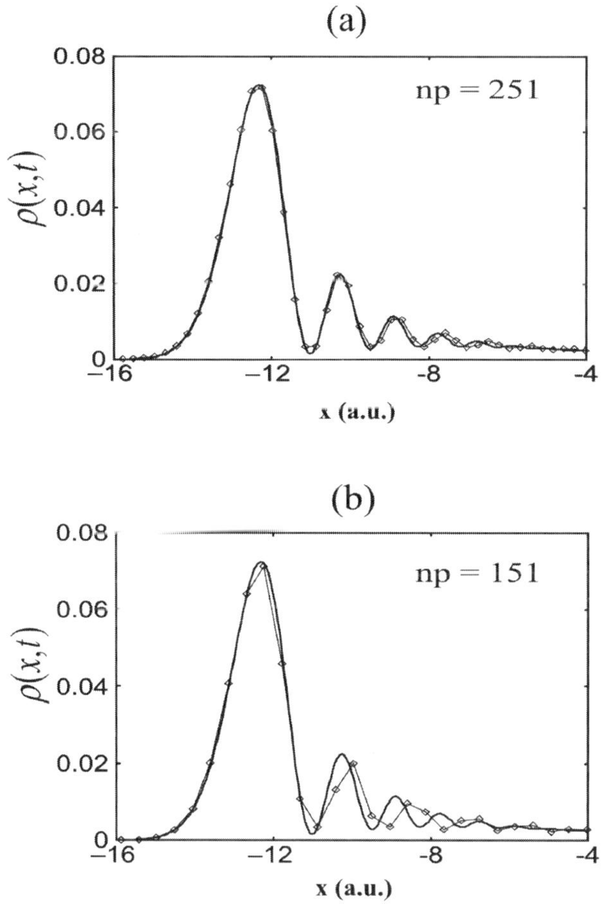

FIGURE 7.3. Density plots at $t = 110$ fs for the uphill ramp potential with a barrier height of 6000 cm^{-1} [7.5]. Parts (a) and (b) show results obtained using the ALE method with 251 and 151 grid points, respectively. The darker solid curve was calculated using fixed-grid integration of the TDSE with 7,150 points.

them to regions where the solution gradients and curvatures are large, so as to reduce fitting errors. We now describe the equidistribution principle.

A number of studies [7.6–7.8] have shown that function approximation errors can be reduced by distributing the grid points so that a positive weight or *monitor function* $M(x)$ is equally distributed over the field,

$$\int_{x_i}^{x_{i+1}} M(x)dx = \text{constant}, \tag{7.9}$$

or, in one-sided discrete form,

$$M_i(x_{i+1} - x_i) = \text{constant}. \tag{7.10}$$

Notice that the above equations are equivalent to the equilibrium conditions for a system of classical springs, where the monitor functions play the role of *spring constants*. It is the monitor function that determines how, when, and to what extent the grid points adapt. Most monitors are designed to detect specific information about the evolving solutions, and subsequently, to use this information to redistribute the grid points. According to equations 7.9 and 7.10, it is the *nonuniformity* in the nearest-neighbor monitor values M_{i-1} and M_{i+1} that cause the grid points to move relative to one another. For example, if the monitor function is gradient- and/or curvature-dependent, then the grid points will be redistributed with a greater density in regions where these quantities are large. If all of the monitor functions have the same value, the grid points will sweep out paths with equal spacing, thus forming an expanding/contracting uniform grid of spacing $h(t)$, as used in the previous sections. Some typical monitor functions are the following: (1) $h^k|u^{k+1}|$; (2) $1 + \beta(\partial u/\partial x)$; (3) $1 + \beta(\partial^2 u/\partial x^2)$; and (4) the truncation error of the solution divided by h (where u^k indicates the k-th x-derivative of the solution, h is the local point spacing, and β is an input parameter).

To obtain the *equilibrium* grid point positions, $\{x_i\}_{i=0}^N$, a homogeneous tridiagonal system of equations of dimension $N \times N$ must be solved. The elements of the spring coefficient matrix will depend on the monitor values at the corresponding grid points. By solving this system of equations, the grid points can be *instantaneously* adapted according to the specific monitor function used. Unfortunately, however, instantaneously adapting grid points does not ensure a smooth dynamic adaptation of the grid, and the grid paths can oscillate and become numerically unstable in time. To eliminate possible erratic behavior, the grid points must be forced to sweep out smooth paths. In addition to this problem, directly solving equation 7.10 can lead to excessive clustering of grid points, since no limit is placed on grid point separation.

One way of overcoming the setbacks encountered with instantaneous dynamic adaptation is to use instead an analogue of the equidistribution principle developed by Dorfi and Drury [7.9], which provides for smooth spatial and temporal adaptation and is given by

$$\frac{1}{M_{j-1/2}}\left(\tilde{n}_{j-1} + \tau\frac{d\tilde{n}_{j-1}}{dt}\right) = \frac{1}{M_{j+1/2}}\left(\tilde{n}_j + \tau\frac{d\tilde{n}_j}{dt}\right), \tag{7.11}$$

with

$$\tilde{n}_j = n_j - \kappa(\kappa+1)(n_{j+1} - 2n_j + n_{j-1}), \tag{7.12}$$

in which $n_j = (\Delta x_j)^{-1}$ is the "grid point concentration", and κ is an input parameter that confines adjacent grid point concentrations to have lower and upper bounds given by

$$\frac{\kappa}{\kappa+1} \leq \frac{\Delta x_j(t)}{\Delta x_{j-1}(t)} \leq \frac{\kappa+1}{\kappa} \text{ for all } j. \tag{7.13}$$

The input parameter τ in equation 7.11 serves as a *temporal smoother* to eliminate grid path oscillations by acting as a delay factor that gives the grid points time to react smoothly to a rapidly changing monitor function. Of course, both smoothing parameters κ and τ are problem-dependent and are determined through trial and error. By noting that

$$\frac{dn_j}{dt} = -n_j^2(\dot{x}_{j+1} - \dot{x}_j), \tag{7.14}$$

We can arrange equation 7.11 into a linear equation of the form $A(x)\dot{x} = B(x)$, where $A(x)$ is an $(N-1) \times (N-1)$ pentadiagonal matrix. In application, this linear system is solved at each time step using the boundary conditions $\dot{x}_0 = \dot{x}_N = 0$, $n_1 = n_0$, and $n_N = n_{N-1}$.

In 2002, Hughes and Wyatt used the equidistribution principle to solve the moving-path transform of the TDSE [7.10], given by

$$i\hbar\frac{d\psi(x,t)}{dt} = i\hbar\left[\frac{\partial}{\partial t} + \dot{x}\frac{\partial}{\partial x}\right]\psi = -\frac{\hbar^2}{2m}\frac{\partial^2\psi}{\partial x^2} + V(x,t)\psi + i\hbar\dot{x}\frac{\partial\psi}{\partial x}. \tag{7.15}$$

In this work, the wave packet dynamics for two problems, the double-well potential and an Eckart barrier, were studied. For adaptation, a monitor function based on the wave function curvature was used, i.e.,

$$M(x) = 1 + \alpha\left|\frac{\partial^2\psi(x,t)}{\partial x^2}\right|, \tag{7.16}$$

where $\alpha \geq 0$ is an adaptivity parameter. Note that if $\alpha = 0$, the grid points will sweep out a grid of uniform spacing. When $\alpha > 0$, however, the grid points are guided into regions of large curvature where the monitor is large. (Remember, a large monitor function means small grid point spacing.) Conversely, grid points are guided away from regions of small curvature, resulting in large particle spacings in these regions.

In the first application, an initial *stationary* wave packet ($k = 0$) centered in the right well (at the position $x_0 = 1.1$) was propagated for the double-well potential, $V(x) = -0.0068x^2 + 0.003x^4$. The system mass and Gaussian width parameter were 2000 a.u. and $\beta = 3$, respectively. The parameters used in equations 7.11, 7.12, and 7.16 are given by $\alpha = 1$, $\kappa = 10$, and $\tau = \Delta t = 0.025$. As the wave packet was propagated, the grid points followed very complicated paths due to interference effects. The adapting grid paths up to $t = 2.5$ ps are shown in figure 7.4, and the wave function amplitude is displayed at eight times in figure 7.5. In this figure, the dashed lines (rarely visible) are results obtained by solving the TDSE on a fixed grid. The agreement between these two methods is excellent.

For the second application, the smoothed version of the equidistribution principle was used to study wave packet scattering from an Eckart barrier given by $V(x) = V_0 \operatorname{sec} h^2[0.4(x-6)]$, where $V_0 = 0.027$ a.u. The initial wave packet parameters were as follows (in a.u.): $m = 2000$, $\beta = 10$, $x_0 = 2$, and $E_{trans} = 0.018$. A total of $N = 249$ grid points were used for this propagation.

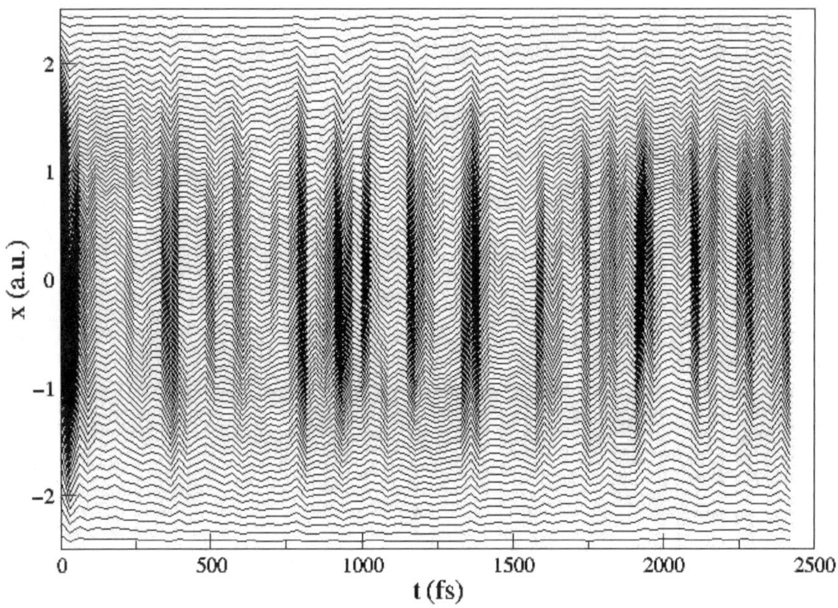

FIGURE 7.4. Grid paths for wave packet propagation in the double-well potential [7.10].

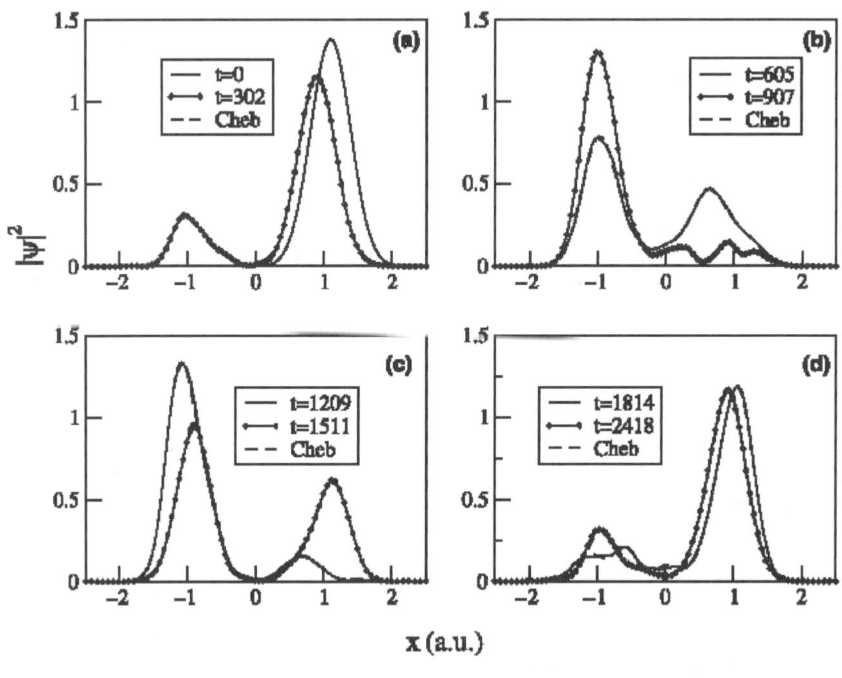

FIGURE 7.5. The solid curves show the time dependence of the probability density for the double-well potential using the grid point equidistribution principle applied to the TDSE. The exact results shown by the dashed curves are barely visible [7.10].

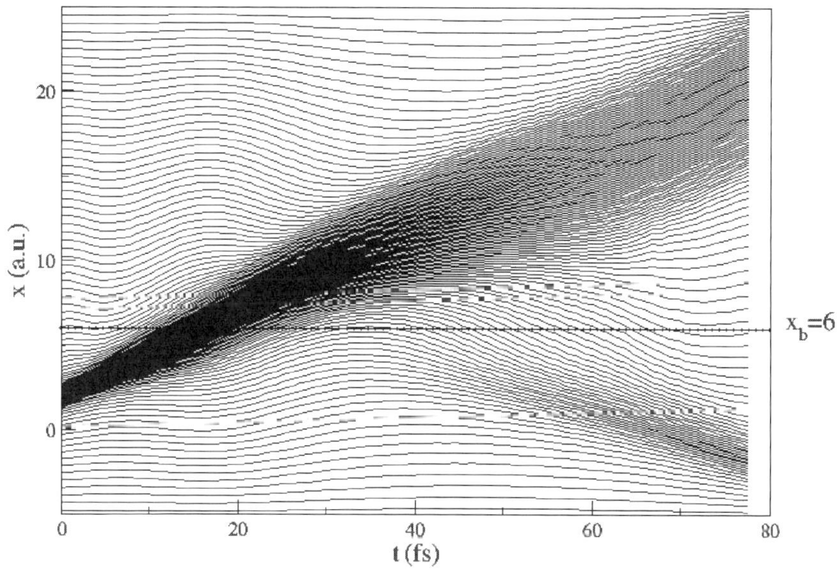

FIGURE 7.6. Adaptive grid paths obtained using the grid point equidistribution principle for the Eckart barrier problem [7.10]. Bifurcation into transmitted and reflected packets occurs in the time range $t = 20$–60 fs.

The adapting grid paths for the Eckart barrier problem are displayed in figure 7.6, which shows that the grid points are initially clustered in the region of significant density (i.e., between $1 \leq x \leq 3$). As time moves on, the wave packet spreads in space and then bifurcates into reflected and transmitted components, as seen by the grid path density. The wave function amplitude at five times is displayed in figure 7.7. In plots (b) and (c), the adaptive results are compared to standard fixed-grid propagation of the TDSE using 599 grid points. Deviation from these results occurs in the regions between the bifurcated packets where the density is small. For regions of significant density, however, there is excellent agreement.

In this section, we have described dynamical grid adaptation based on a smooth version of the equidistribution principle. This method was used to solve the moving-path transform of the TDSE for the double-well and Eckart barrier potentials. It was found that by adapting the grid points to the curvature of the wave function amplitude, the total number of grid points required for solution propagation could be reduced relative to fixed-grid calculations.

7.5 Adaptive Smoothing of the Quantum Force

In this section, we will describe studies that combined the ALE method with the introduction of artificial viscosity in the equations of motion [7.12]. By combining these techniques, it was possible to obtain accurate time-dependent transmission

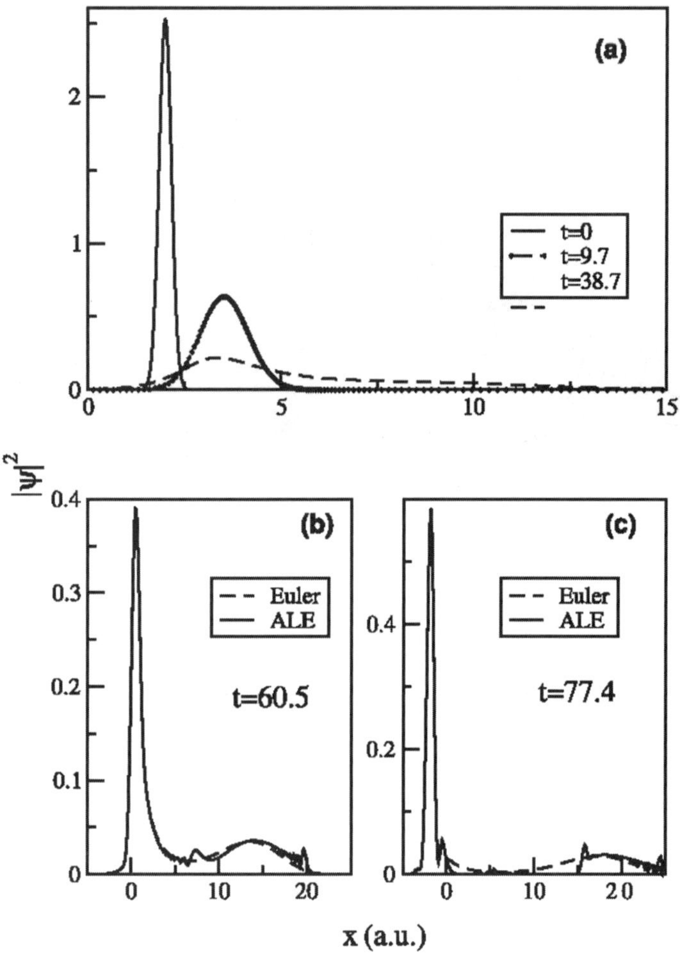

FIGURE 7.7. The density for the scattered wave packet at five different times (in fs) as calculated from the grid point equidistribution principle applied to the moving-path transform of the TDSE [7.10]. In plots (b) and (c), the dashed curves (barely visible) were calculated by propagating the solution to the fixed-grid TDSE.

probabilities for wave packet scattering from a one-dimensional Eckart barrier, thus representing the first time that stable long-time numerical solutions to the QHEM were obtained for this potential. Although the ALE method alone helps to capture the functional form of the solution in nodal regions, solution kinks in these regions may be very large and virtually impossible to capture. In this case, the ALE method alone is not sufficient to handle all problems involved in solving the QHEM.

Coping with the ill effects of singularities and kinks in hydrodynamic fields is not a new topic. Classically, kinks in hydrodynamic solutions can be found when a flowing fluid encounters a barrier or obstacle such that a shock front is created. A standard way to cope with these features is to introduce a new term, the *artificial viscosity,* into the equations of motion. The purpose of this term is to change the effective force the particle feels as it encounters the region where sudden changes occur. When added to the QHEM, artificial viscosity can be used to moderate the strong quantum force in nodal regions by *preventing nodes from fully forming*. This, in turn, provides for much smoother (though approximate) solutions in these regions.

In applications to one-dimensional Eckart barrier scattering, Kendrick introduced "by hand" an artificial viscosity potential V_S given by

$$\begin{aligned} V_S &= c_1 \left|\frac{\partial v}{\partial x}\right| + c_2 \left|\frac{\partial v}{\partial x}\right|^2 \quad \text{for} \quad \frac{\partial v}{\partial x} < 0, \\ V_S &= 0 \quad \text{for} \quad \frac{\partial v}{\partial x} \geq 0, \end{aligned} \tag{7.17}$$

where $v(x)$ is the grid point velocity. The numerical parameters c_1 and c_2 are problem dependent and can be determined by convergence and stability studies. Of course, the larger these parameters become, the greater the deviation between the approximate and exact solutions. Because of this, these parameters should be kept relatively small. In practice, the piecewise cutoff at $\partial v/\partial x = 0$ can lead to numerical problems, so it is best to smooth the viscosity potential using a least squares procedure or by fitting to multiple Gaussians.

When nodes are encountered, a kink occurs in both the quantum force and velocity fields, the latter being an effect of the former. The velocity kink forms due to the quantum force accelerating fluid elements on one side of the node while decelerating those on the opposite side. When artificial viscosity is introduced into the quantum Hamilton–Jacobi equation, an associated viscosity force $f_S = -\partial V_S/\partial x$ acts to squash the kink in the velocity field by counteracting the quantum force in the nodal region. With appropriate viscosity parameters, it is possible to soften the kink enough to permit propagation of quantum trajectories. The results for one crucial time step are displayed in figure 7.8. In plot (a), the amplitude is shown at an early stage during the formation of a node. Although the node is not fully formed, plot (b) displays the velocity kink already associated with it. In plot (c), we see the singularity beginning to develop in the quantum potential. The artificial viscosity potential at this time is also shown in plot (d). By taking the gradient of these two potentials, the quantum force and viscosity forces were obtained, and these are displayed in plots (e) and (f), respectively. From the last two plots, we see that the two forces (quantum and viscous) work to cancel one another in the vicinity of the forming node, resulting in an effective force field that is much smoother than the original quantum force. Because of this, problems associated with node formation in the QHEM are not experienced when this viscosity potential is used.

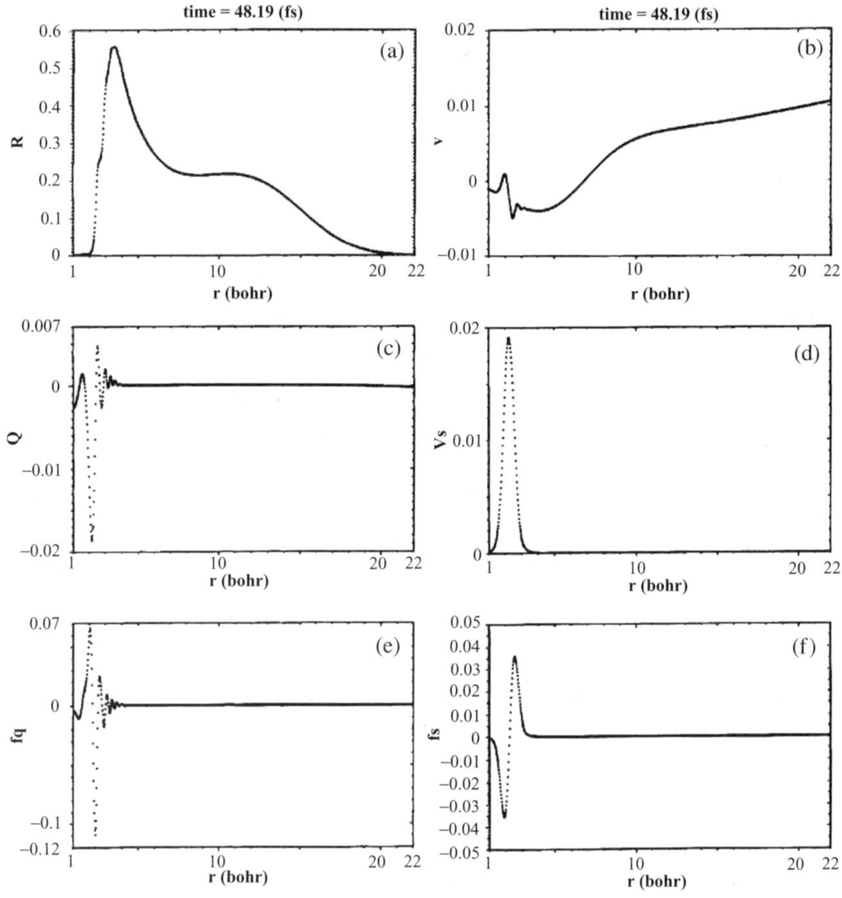

FIGURE 7.8. (a) The R-amplitude [7.12], (b) Lagrangian velocity, (c) quantum potential, (d) viscosity potential, (e) quantum force, and (f) viscosity force at $t = 48.19$ fs. Note the "kink" in the velocity fieldoccurring near $r = 2$ of plot (b).

Further results of the combined use of the ALE method and the artificial viscosity are displayed in figure 7.9, which shows time-dependent transmission probabilities. The solid line was calculated by propagating the wave function on a fixed grid, and the solid dots were obtained using the combined ALE/artificial viscosity method. There is excellent agreement between the two sets of results, though the exact propagation took eight hours of CPU time per energy on a 1.8-GHz Pentium 4 processor, while the QHEM results required only 3 minutes of CPU time per energy [7.12].

In order to investigate how adaptive smoothing affects the amplitude of the wave function in the reflected region, wave packets were scattered from the same Eckart barrier using a range of viscosity parameters. The amplitude at a given time for two sets of parameters and the corresponding exact results are displayed in figure 7.10.

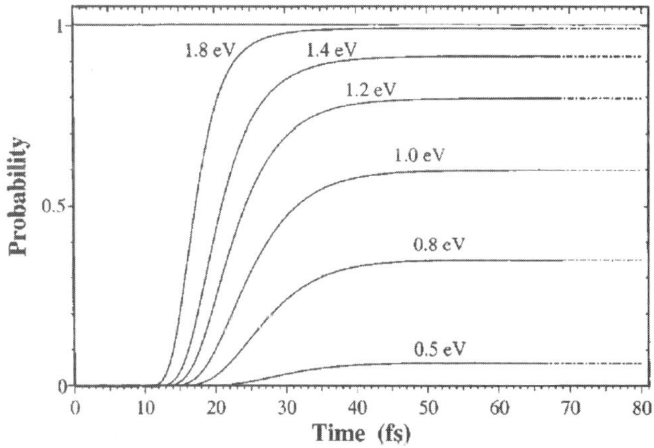

FIGURE 7.9. Time-dependent barrier transmission probabilities for several different initial kinetic energies [7.12]. The solid curves show the exact results, and the long-short, dashed curves were computed using the ALE/artificial viscosity method. The two sets of results are essentially identical.

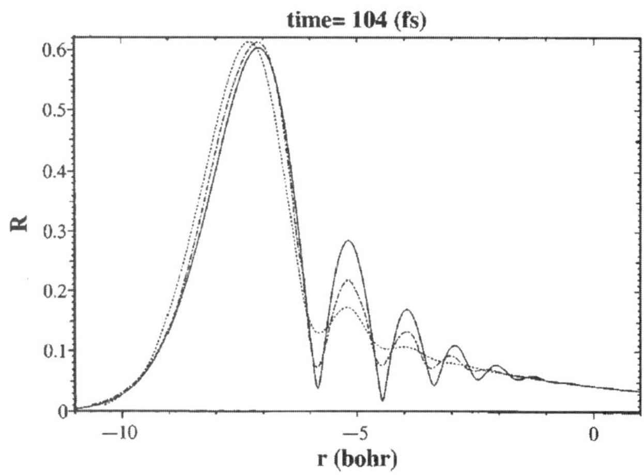

FIGURE 7.10. Amplitude of the wave packet reflected from the Eckart barrier at $t = 104$ fs [7.12]. The solid, dark curve is the exact result. The dashed curve is based on the ALE/artificial viscosity approach using $c_1 = 0.1$, $c_2 = 100$. The long-short dashed curve is also calculated using the ALE/artificial viscosity approach, only this time with the parameter values $c_1 = 0.05$, $c_2 = 5$. Notice that the differences between the QHEM and the exact results decrease as the artificial viscosity parameters decrease.

From this figure, it is clear that as the parameter values increase, the amplitude ripples begin to vanish and the reflected density becomes smooth. An important feature is that even though the hydrodynamic solution differs from the exact one around nodes in the reflected region, the functional form is the same. This is vital, since we do not want the dynamics to completely change with the introduction of V_S. Instead, the viscosity potential eliminates stiffness and kinks that result in numerical instabilities. Another feature not explicitly shown in this plot is that outside of nodal regions (such as in the transmitted region), the hydrodynamic and exact solutions are essentially identical.

In conclusion of this section, an important point should be made regarding the calculation of transmission probabilities. Although the exact solution must be forfeited in adding artificial viscosity to the equations of motion, figure 7.9 confirms that smoothing in nodal regions does not seem to significantly affect the transmission probabilities. Although it is possible that artificial dynamics in the reflected region can affect the dynamics in the transmitted region, the effect does not appear to be significant as long as the magnitude of the viscosity potential is relatively small. Because of this, the combined ALE and artificial viscosity method is very promising, since it allows for the propagation of transmitted trajectories long after nodes form in the true reflected amplitude.

7.6 Adaptive Dynamics with Hybrid Algorithms

Hybrid adaptive methods have been developed to *deal directly* with the intrinsic singularities that arise in nodal regions. In 2000, Wyatt and Bittner [7.13] introduced what would soon become a series of publications based on combined integration of the QHEM and the TDSE. The general idea is to propagate solutions to the hydrodynamic equations in node-free regions while propagating the wave function itself in nodal regions. Although the idea is simple enough, the implementation of hybrid algorithms is not so simple, as we shall soon see. In this section we will discuss the general hybrid algorithm and then investigate successful applications using different types of hybrid algorithms.

An *ideal* hybrid algorithm should do the following: First, the algorithm must be capable of recognizing on the fly when numerical breakdown in the QHEM is about to occur. This can be done, for example, by monitoring for developing nodes. Next, in the vicinity of the forming node, the algorithm must use the hydrodynamic solutions (still accurate, though not for long) to construct the complex-valued wave function. The solution in this region, called a "ψ – patch", should be thereafter *linked to and propagated concurrently* with the remaining hydrodynamic fluid elements outside of the region where nodes develop. Ideally, the algorithm should be able to accommodate the formation of *multiple* nodes, since there is no way of predetermining how many will form in a given problem. Lastly, an ideal hybrid algorithm should have a suitable means for transforming from the wave function in the ψ – patch back into the hydrodynamic representation once the nodes have healed.

The major problem encountered with all hybrid methods concerns *linking the solutions* that are obtained in different spatial regions. In order to propagate solutions in the ψ–patches according to the moving-path transform of the TDSE, boundary conditions are required. In a hybrid method, these conditions are not the usual Dirichlet conditions, wherein $\psi = 0$ at the boundaries. Instead, the wave function at the boundaries must be constructed from the evolving solutions of the QHEM, which are obtained in the region outside of the patch. Because of this, integration of the solution within the ψ –patch is inherently linked to the hydrodynamic solution via its boundaries, which are updated according to the QHEM. Because different algorithms are used to propagate each solution, inconsistencies can develop at the patch boundaries, which can lead to numerical instabilities.

The first example of the use of a hybrid algorithm concerns wave packet dynamics in the double-well potential [7.14]. For this problem, the QTM becomes unstable after the first node starts to form, which occurs shortly after 25 fs. This potential, of the form $V(x) = ax^4 - bx^2$, has two symmetric minima located at $x_{\min} = \pm\sqrt{b/(2a)}$, where the potential at these points is $V_{\min} = -b^2/(4a)$. In this study, the following parameter values were used for the potential (in a.u.): $a = 0.007$, $b = 0.01$. In addition, the initial wave packet had the parameter values $x_0 = -|x_{\min}|$, $\beta = 4.47$, and $E_{trans} = 0.0018$. The QTM was used to propagate the QHEM until nodes began to form, at which time a ψ – patch was created. The moving-path transform of the TDSE, given in equation 7.15, was used in the nodal region. The grid speeds inside the ψ – patch were chosen to sweep out a grid with uniform spacing between the Lagrangian boundary points. By doing this, a finite difference scheme could be used for calculating spatial derivatives within the patch. Each time the Lagrangian fluid elements outside of the patch were evolved one time step, the grid points inside of the patch were evolved anywhere from 20 to 100 time steps in an inner time loop with time step that was a fraction of the hydrodynamic time step. Excellent results were obtained for this problem, as seen from figure 7.11. The solution could be propagated for over 70 fs without difficulty, even though multiple nodes and ripples make their appearance.

In 2003, Hughes and Wyatt developed a different hybrid algorithm [7.15]. Once again, their efforts focused on scattering an initial Gaussian wave packet from an Eckart barrier. The parameters used in the calculation were as follows (in a.u.): $\beta = 10$, $x_0 = 2$, $E_{trans} = 0.018$, $a = 0.4$, $x_b = 6$, and $V_b = 0.027$. In this algorithm, the ALE method with uniform grid spacing was used to propagate the QHEM until a limited number of grid points passed the position of the barrier maximum. When the predetermined "last" allowed hydrodynamic grid point reached the barrier, it was fixed in space, so that two grid domains were created: one on the reactant side of the barrier maximum and the other on the product side. After 60 fs into the propagation, node formation in the reflected packet was detected and a ψ – patch was created, triggered by the quantum potential exceeding a predefined threshold value. Within the patch, the moving-path transform of the TDSE was integrated. This time, instead of assigning the grid speeds to sweep out a patch of uniform spacing, the *equidistribution principle* discussed in Section 7.4 was used to adapt the wave function according to the monitor function

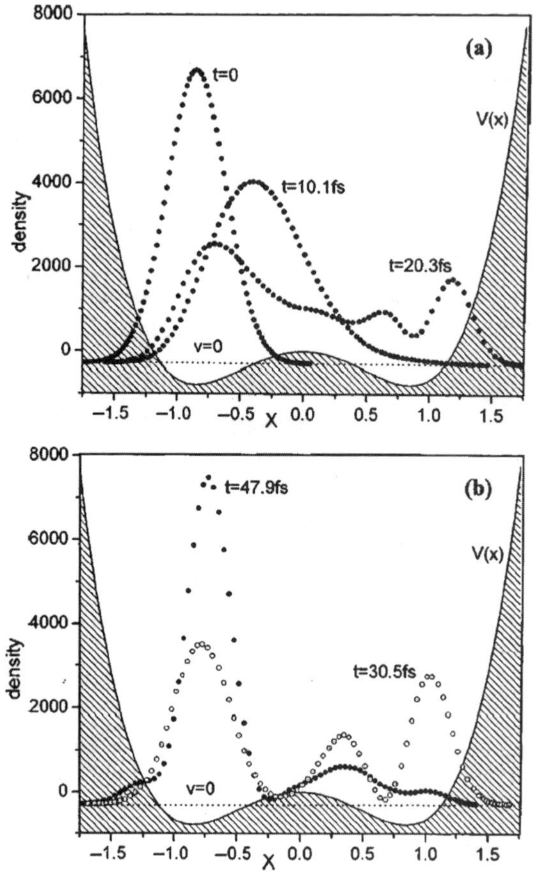

FIGURE 7.11. Wave packet dynamics for the double-well potential [7.14]. The shaded region shows the potential (cm^{-1}). The ground state energy in the harmonic approximation is indicated by the horizontal dotted line. Plots (a) and (b) show the probability densities for five times. For display, these densities have been multiplied by a factor of 4149.

$M(x) = 1 + 0.25\,\left|\partial^2\psi(x,t)/\partial x^2\right|$. Using this method, complete wave packet bifurcation, leading to very accurate transmission probabilities, was obtained. In fact, the transmitted portion of the packet was propagated for over 5 ps! The time dependence of some grid point paths for this problem are plotted in figure 7.12. In addition, by calculating the cross-correlation function (described in Chapter 10), the energy-resolved transmission probability was obtained, and this is shown in figure 7.13. This figure displays the results for calculations using either 20 or 320 moving grid points. The results obtained with 320 grid points are in excellent agreement with the exact fixed-grid results.

Using a simplified algorithm, Babyuk and Wyatt [7.16] investigated the scattering of an initial Gaussian on an Eckart potential. In this study, the parameters for the Gaussian wave packet and potential were given by $\beta = 10$, $x_0 = 2$, $E_{trans} = 0.018$, $a = 0.4$, $x_b = 6$, and $V_b = 0.027$. In this example, the ALE method was used to evolve the hydrodynamic fields on an expanding uniform grid with

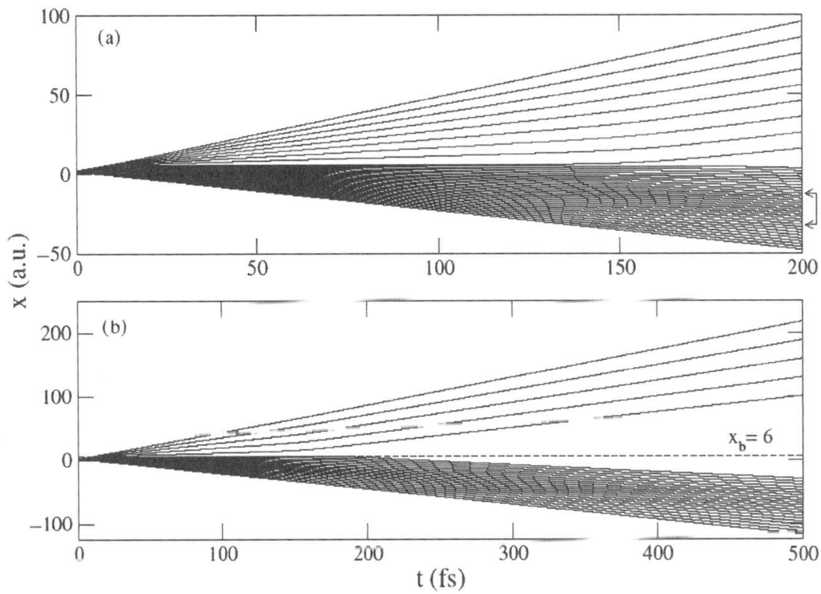

FIGURE 7.12. Grid paths for the Eckart barrier problem [7.15]. Only the paths $j = 0, 8, \ldots, N-8$ are shown in (a) over a 200 fs time scale. Part (b) of this plot illustrates the same propagation over a 500 fs time scale for the paths $j = 0, 16, \ldots, N-16$. In plot (a), grid points have clustered in the reflected region labeled A due to large values of the wave function curvature.

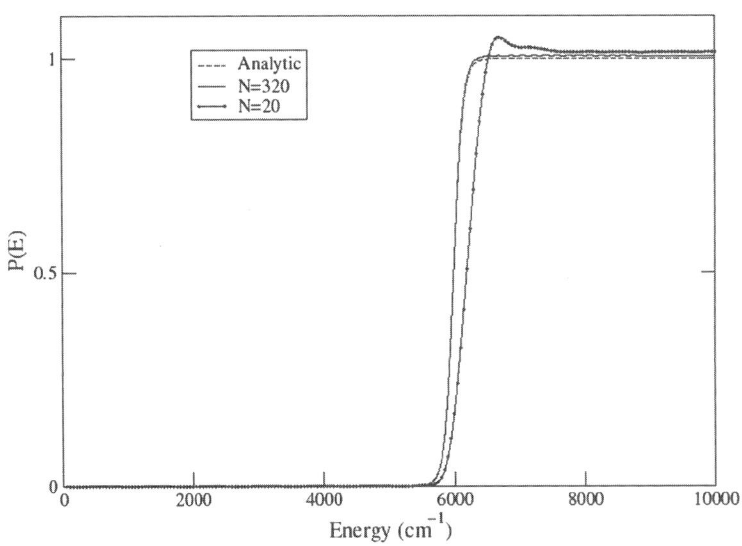

FIGURE 7.13. Energy-resolved transmission probability for the Eckart barrier problem [7.15]. In this plot, N is the total number of grid points.

Lagrangian edge points. When triggered according to the quantum potential, the grid was split into different spatial domains as before. In the reflected region where nodes form, the hydrodynamic solutions were interpolated onto a fine mesh for *fixed-grid integration of the TDSE*, using the fixed-edge particles as boundary points. In the product region to the right side of the barrier, the packet was evolved using the hydrodynamic representation on a uniformly spaced ALE grid. Although the grid points used to propagate the TDSE in the ψ – patch were fixed, the patch was large enough to capture all of the transient nodes that formed during the calculation. One new and important aspect of this algorithm took place after the nodes healed. At this point in time, the reflected packet was converted from the wave function to the hydrodynamic representation. This was done by assembling the piecewise phase into a continuous phase and calculating the amplitude from the expression $R = \sqrt{\psi^*\psi}$. Since fewer grid points were needed in the hydrodynamic picture, fewer than half of the grid points used to propagate the fixed-grid TDSE were retained. This marked the first time that a wave function in a hybrid method was transformed back into the hydrodynamic representation after nodes have healed. The results of the application of this method to the Eckart problem are displayed in figures 7.14 and 7.15. When compared to the solution obtained by a standard fixed-grid propagation of the TDSE, the results are in excellent agreement. Using this method, it was possible to propagate the initial wave packet for over one *nanosecond*!

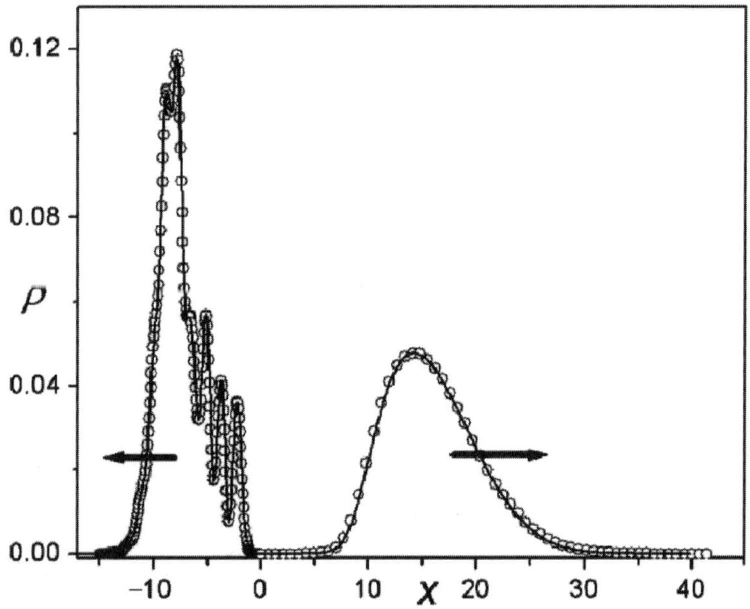

FIGURE 7.14. Probability density calculated with a hybrid method [7.16] at $t = 120.2$ fs. The solid curve was calculated using the fixed-grid TDSE.

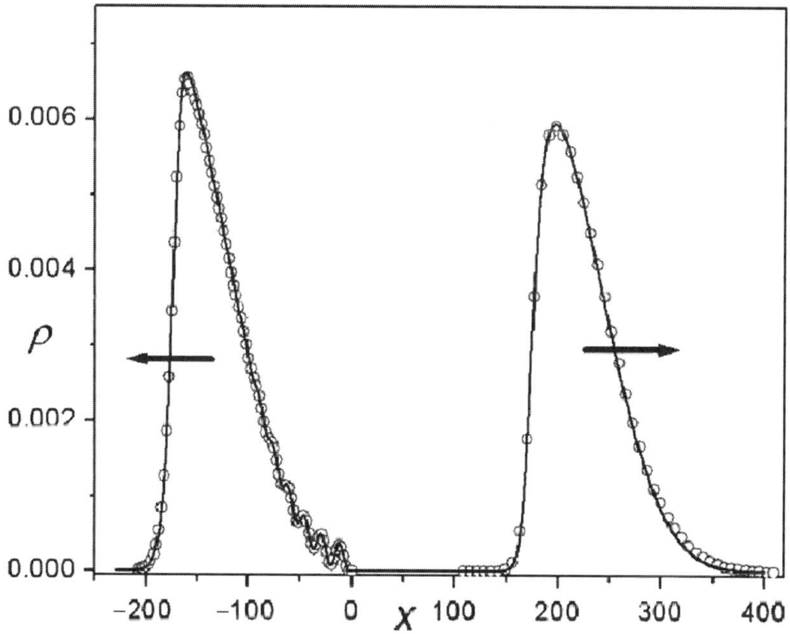

FIGURE 7.15. Probability density at $t = 1.2$ ns calculated by the same hybrid method that was used to calculate the density in figure 7.14 [7.16]. The solid curve was calculated using the fixed-grid TDSE. Note the greatly expanded range of x-values compared with figure 7.14.

7.7 Conclusions

In this chapter, we described three adaptive methods for coping with node problems and derivative evaluation problems that are encountered in solving the QHEM. For wave packet dynamics on simple potential energy surfaces (those without much interference), we saw in Chapter 6 that the original version of the QTM works exceedingly well. However, in regions where nodes and quasi-nodes are encountered, quantum trajectories computed using the simplest version of the QTM become unstable. In the first part of this chapter, we found that by choosing suitable grid point velocities, the ALE method can completely eliminate problems connected with inflation in nodal regions and compression in other regions. One way of choosing these velocities is to force the trajectories to sweep out a grid of uniform spacing. In some cases, such as the uphill ramp potential, this can greatly extend the survival time of the algorithm, even when "hard ripples" form in the reflected wave packet.

We have also seen that the ALE method can be used to adapt grid points to regions where they are most needed. This can help to eliminate extraneous calculations in smooth or insignificant regions where the density is very small. Using a smoothed version of the equidistribution principle, grid points can be guided to regions of large gradients and curvatures, alleviating problems of data inadequacies in derivative approximations. In a multidimensional application of this method [7.17], it was possible to accurately scatter a wave packet from a two-dimensional potential involving an Eckart barrier coupled to a harmonic oscillator potential. Excellent results were obtained, even though grid adaptation was performed only along orthogonal one-dimensional slices of the mesh.

In conjunction with the ALE method, we also discussed adaptive smoothing using artificial viscosity. The introduction of viscous forces can counteract the large quantum forces in nodal regions. Though the solution is no longer exact in these regions, excellent transmission probabilities can still be obtained for barrier scattering problems. To demonstrate the multidimensional capability of this method, Pauler and Kendrick used this method to study wave packet scattering on a two-dimensional coupled Eckart barrier/harmonic oscillator potential [7.18]. Kendrick has generalized this method to reactive scattering on an N-dimensional potential surface [7.19]. A parallelized computer program was used in conjunction with a vibrational decoupling scheme (VDS) to compute reaction probabilities for values of N up to and including 100. In the VDS, the hydrodynamic equations are decoupled into a set of uncoupled one-dimensional problems, and each of these problems is solved separately. In an extension of this method, coupling between these one-dimensional scattering problems can be introduced.

In the final adaptive method discussed in this chapter, several hybrid algorithms that combine coupled integration of the TDSE and the QHEM were discussed. These methods are designed to eliminate problems encountered in the QHEM by switching to the TDSE in nodal regions. In a series of one-dimensional examples, not only was it possible to propagate both equations of motion on an overlapping grid, but it was also possible to revert back to the hydrodynamic representation when nodes healed at later times. Excellent results were obtained for long-time propagation that would have been unreasonable in solving the fixed-grid TDSE.

In Chapter 15, three novel methods for coping with the node problem will be described. Two of these methods involve either expansion of the wave function in node-free counterpropagating components or use of node-free covering functions. The third method uses a modified "quasipolar" form for the wave function.

References

7.1. H. Braess and P. Wriggers, *Arbitrary Lagrangian Eulerian finite element analysis of free surface flow*, Comp. Meth. Appl. Mech. Eng. **190**, 95 (2000).

7.2. T. Belytschko and D. Flanagan, *Finite element methods with user-controlled meshes for fluid–structure interaction*, Comp. Meth Appl. Mech. Eng. **33**, 669 (1982).

7.3. J. Donea, S. Giuliani, and J.P. Halleux, *An arbitrary Lagrangian–Eulerian finite element method for transient dynamic fluid–structure interactions,* Comp. Meth. Appl. Mech. Eng. **33**, 689 (1982).

7.4. T. Hughes and W.K. Liu, *Lagrangian–Eulerian finite element formulation for incompressible viscous flows*, Comp. Meth. Appl. Mech. Eng. **29**, 329 (1981).

7.5. C. Trahan, R.E. Wyatt, *An arbitrary Lagrangian–Eulerian approach to solving the quantum hydrodynamic equations of motion: Equidistribution with "smart" springs*, J. Chem. Phys. **118**, 4784 (2003).
7.6. C. Degand and C. Farhat, *A three-dimensional torsional spring analogy method for unstructured dynamic meshes*, Computers and Structures, **80**, 305 (2002).
7.7. F.J. Blom, *Considerations on the spring analogy*, Int. J. Numer. Meth. Fluids, **32**, 647 (2000).
7.8. B. Palmerio, *A two-dimensional fem adaptive moving-node method for steady Euler flow simulations*, Comp. Meth. Appl. Mech. Eng. **71**, 315 (1988).
7.9. E.A. Dorfi and L.O.C. Drury, *Simple adaptive grids for 1-D initial value problems*, J. Comp. Phys. **69**, 175 (1987).
7.10. K.H. Hughes, R.E. Wyatt, *Wavepacket dynamics on dynamically adapting grids: Application of the equidistribution principle*, Chem. Phys. Lett. **366**, 336 (2002).
7.11. H. Tal-Ezer and R. Kosloff, *An accurate and efficient scheme for propagating the time-dependent Schrödinger equation*, J. Chem. Phys. **81**, 3967 (1984).
7.12. B. Kendrick, *A new method for solving the quantum hydrodynamic equations of motion*, J. Chem. Phys. **119**, 5805 (2003).
7.13. R.E. Wyatt and E.R. Bittner, *Quantum wave packet dynamics with trajectories: Implementation with adaptive Lagrangian grids*, J. Chem. Phys. **113**, 8898 (2000).
7.14. R.E. Wyatt, *Wave packet dynamics on adaptive moving grids*, J. Chem. Phys. **117**, 9569 (2002).
7.15. K.H. Hughes and R.E. Wyatt, *Wavepacket dynamics on arbitrary Lagrangian–Eulerian grids: Application to an Eckart barrier*, Phys. Chem. Chem. Phys. **5**, 3905 (2003).
7.16. D. Babyuk and R.E. Wyatt, *Hybrid adaptive algorithm for wave packet propagation*, Chem. Phys. Lett. **387**, 227 (2004).
7.17. K.H. Hughes, J. Chem. Phys., to be published.
7.18. D.K. Pauler and B.K. Kendrick, *A new method for solving the quantum hydrodynamic equations of motion: Application to two-dimensional reactive scattering*, J. Chem. Phys. **120**, 603 (2003).
7.19. B.K. Kendrick, *Quantum hydrodynamics: Application to N-dimensional reactive scattering*, J. Chem. Phys. **121**, 2471 (2004).

8 Quantum Trajectories for Multidimensional Dynamics

The QTM is applied to three multidimensional problems:
(1) suppression of an interference feature in a 2-mode system;
(2) the decay of a metastable state in an 11-mode system;
(3) electronic energy transfer involving coupled 11-mode potential surfaces.

8.1 Introduction

In Chapter 6, the quantum trajectory method was applied to four one- and two-dimensional model problems. In Chapter 7, dynamic adaptive grid techniques were introduced, and applications to several one-dimensional scattering problems were described. In this chapter, applications of the quantum trajectory method will be made to three wave packet scattering problems that progressively involve increasing complexity in the dynamics and the number of degrees of freedom.

The first example, at the extreme lower end of "multidimensional", concerns wave packet evolution in a *composite system* made up of a subsystem and the environment (the reservoir) to which the system is coupled [8.1]. This composite system is quite simple, but it has just enough complexity to provide a number of insights into the *mechanism for decoherence of a superposition* that is prepared in the subsystem at $t = 0$. We will follow the time-dependent hydrodynamic fields for the composite system for two cases, depending on whether the subsystem–reservoir coupling is "on" or "off". When the coupling is turned off, a prominent interference feature forms in the probability density. However, when the coupling is turned on, growth of this feature is significantly hindered. Hydrodynamic analysis of the suppression of this interference feature will form the focus for this quantum trajectory study.

The second example concerns the time-dependent decay of a metastable state on a potential surface involving a reaction coordinate (the system mode) interacting with a 10-mode harmonic reservoir [8.2]. The total potential energy is the sum of the system anharmonic potential, harmonic oscillator potentials for the M reservoir modes, and bilinear coupling terms linking the system to the reservoir. As in the decoherence example, ensembles of quantum trajectories will be evolved for the composite system. A set of quantum trajectories will be illustrated, and the wave function will be plotted along several of these trajectories.

The third study described in this chapter is an extension of previous quantum trajectory computations on electronic energy transfer [8.3, 8.4]. In these earlier studies, an ensemble of quantum trajectories was launched on the lower of two coupled potential curves. Transfer of amplitude and phase to the upper curve occurred within a localized coupling region. In the present study, the dimensionality of the electronic nonadiabatic problem will be significantly increased [8.5] to allow for the interaction of two 11-dimensional potential surfaces. Each of these surfaces comprises 10 harmonic "bath" modes coupled to a reaction coordinate. Ensembles of quantum trajectories will be launched on each of the two coupled surfaces with the goal of synthesizing the time-dependent wave packets from information carried along by the propagating quantum trajectories.

The outline of this chapter is as follows. The decoherence model is introduced in Section 8.2, and trajectory results are illustrated in Section 8.3. The model used to describe the decay of a metastable state and plots illustrating several quantum trajectories are presented in Section 8.4. The hydrodynamic equations of motion for nonadiabatic dynamics on two coupled potential surfaces are developed in Section 8.5. In Section 8.6, the 11-dimensional potentials and the initial conditions on the quantum trajectories are presented, and the methods used for propagation of the trajectory ensembles are discussed. In Section 8.7, a series of plots is presented to illustrate wave packet propagation on the two surfaces. Finally, a brief summary is presented in Section 8.8.

8.2 Description of the Model for Decoherence

Quantum decoherence plays a central role in a number of interrelated areas [8.6–8.8], including measurement theory [8.6, 8.9], emergent classicality [8.6, 8.10–8.13], quantum computing [8.14], and quantum control theory [8.15]. Several definitions of decoherence have emerged [8.6–8.8], some of which rely on symptoms rather than causes of the effect. For purposes of the model introduced in the following section, *decoherence will be defined operationally* in the following way. First, a *quantum superposition*, a "Schrödinger cat-type state", will be prepared at time $t = 0$ in the specified subsystem. The initial wave function is then something like

$$\psi(x) = a\varphi_1(x) + b\varphi_2(x), \tag{8.1}$$

and the corresponding density matrix is

$$\begin{aligned}\rho(x, x') &= \psi(x)\psi^*(x') = \rho_1(x, x') + \rho_2(x, x') + \rho_{\text{int}}(x, x') \\ &= |a|^2\varphi_1(x)\varphi_1^*(x') + |b|^2\varphi_2(x)\varphi_2^*(x') \\ &\quad + \left(ab^*\varphi_1(x)\varphi_2^*(x') + ba^*\varphi_2(x)\varphi_1^*(x')\right).\end{aligned} \tag{8.2}$$

where ρ_1 and ρ_2 are the *"classical" contributions* and ρ_{int} is the *quantum interference contribution*. If this subsystem is isolated from the environment, then as it evolves, we expect that interference effects will persist. However, if the system is coupled to the environment, then as this state evolves in time, the *suppression or destruction of interference effects is what we mean by decoherence*. After a sufficient *decoherence time*, the system should be well described by a *statistical mixture* (an *incoherent sum*):

$$\rho \xrightarrow{decoherence} \rho_1 + \rho_2 = \rho_{mixture}. \tag{8.3}$$

When this happens, it is said that 'classicality emerges' from the original system.

Several *symptoms of decoherence* include the following:

- approximate *dynamical diagonalization* of the density matrix for the subsystem;
- *destruction of interference ripples* in the Wigner function for the subsystem.

Dynamical diagonalization of the density matrix is most conveniently expressed in terms of the rotated coordinates, $\xi = (x + x')/\sqrt{2}$ and $\eta = (x' - x)/\sqrt{2}$. (These coordinates will be used again in Section 11.10.) The coordinate ξ runs along the diagonal in the (x, x') coordinate space, while η is the "off-diagonal" coordinate. Dynamical diagonalization of the density matrix $\rho(\xi, \eta)$ then means, over the course of time, that this function becomes compressed in the η-direction, approaching a δ-function, but never quite reaching this limit.

In this and the following section, we will study the time-dependent hydrodynamic fields for a *composite system* made up of a subsystem in which the initial superposition is prepared and the environment (the reservoir) to which the system is coupled [8.1]. This superposition is a "2-slit-type state": two well-separated Gaussian wave packets that are given the opportunity to develop interference maxima as they spread into each other. The remainder of the composite system is a single harmonic mode that is bilinearly coupled to the subsystem mode. When the subsystem–reservoir coupling is turned off, a prominent interference peak develops midway between the two initial Gaussians. However, with the coupling turned on, growth of this interference feature is significantly hindered. The hydrodynamic analysis of this retardation will form the focus for this study.

The Hamiltonian for the composite system is decomposed into subsystem, harmonic reservoir, and coupling contributions,

$$H = H_s + H_r + H_c = \frac{1}{2m_0} p_x^2 + \left[\frac{1}{2m} p_y^2 + \frac{1}{2} k y^2\right] + cxy, \tag{8.4}$$

where (x, y) denote the *physical coordinates*. The parameter values are as follows: $\mathrm{m_0 = m = 2000}$ a.u. and $\omega = \sqrt{\mathrm{k/m}} = 0.004556 = 1000\ \mathrm{cm^{-1}}$. In the following section, the dynamics of the coupled system ($c = 0.015$ a.u.) will be compared with those of the uncoupled system ($c = 0$). A contour map of the total potential energy in equation 8.4 is shown in figure 8.1. When the coupling coefficient is positive, the total potential exhibits declining valleys in quadrants two and four of the physical coordinate system. Because of the shape of this potential surface, it

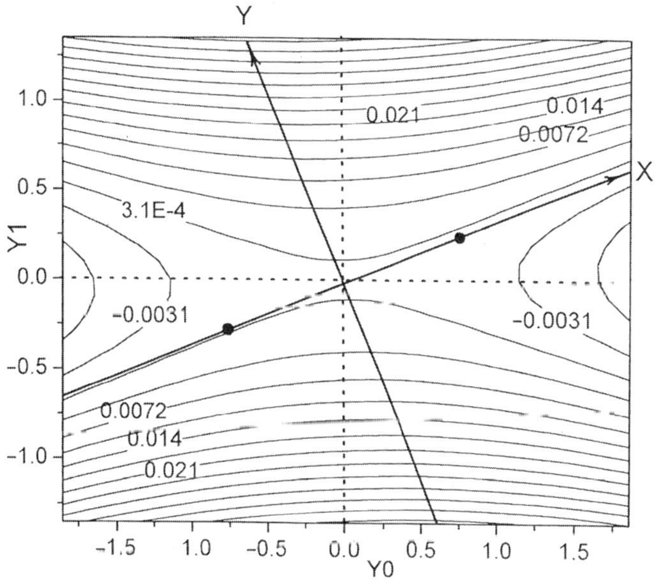

FIGURE 8.1. Contour map of the total potential energy (in a.u.), where (x, y) and (y_0, y_1) are the physical and computational coordinates, respectively [8.1]. The y_0 axis runs along the minimum of the two valleys that decline toward the left and right edges of the figure. The dots on the x axis locate the centers of the two Gaussians that form the initial superposition for the system.

is convenient to introduce two new coordinates: the axis y_0 runs along the valley floor, and y_1 is the perpendicular axis. In these new coordinates, the Hamiltonian is separable, but the initial wave function (specified in the following paragraph) is not. Later, the dynamical equations for the quantum trajectories will be integrated in these coordinates, henceforth referred to as the *computational coordinates*.

At $t = 0$, the wave function is assumed to be the product of a coherent superposition for the subsystem times a ground state harmonic oscillator for the reservoir:

$$\Psi(x, y, t = 0) = \frac{N}{\sqrt{2}}\left[e^{-\beta(x-a)^2} + e^{-\beta(x+a)^2}\right]\varphi_0(y). \tag{8.5}$$

The two Gaussians in the superposition are initially centered at $x = \pm a$, and N is the normalization factor for each Gaussian. The parameters a $= 0.8$ a.u. and $\beta = 4.5$ a.u. ensure that the two displaced Gaussians have a very small initial overlap. In figure 8.1, the dots on the x axis locate the centers of these Gaussians. As time evolves, each Gaussian retains this shape, but the widths along the two principal axes, the twist angle of these axes, and the position and momentum parameters change continuously with time.

In order use the quantum trajectory method to evolve the wave function for the composite system for the coupled case ($c \neq 0$), N fluid elements were launched from a rectangular mesh in the computational (y_0, y_1) coordinate system. For the uncoupled case, the initial particle mesh was set up in the physical (x, y) coordinate system. (The total number of fluid elements was 1215 for the coupled case and 1305 for the uncoupled case.) As time proceeds, the fluid elements no longer form a Cartesian mesh, and the moving weighted least squares method was used to evaluate the spatial derivatives needed for the equations of motion. A ten-term local cubic polynomial basis set was used, with 30 to 35 trajectories in the stencil surrounding the central element.

During the time interval when density is building up in the region between the two Gaussians, some of the fluid elements are forced into close proximity in the interference region. This can cause problems with derivative evaluation on the highly nonuniform mesh. In order to counter this *mesh compression*, the fluid elements were adapted back to a uniform mesh on every time step. All hydrodynamic fields at the next time step were then interpolated onto the new uniform mesh. This procedure stabilizes the calculation and permits integrations to longer times than would otherwise be possible.

8.3 Quantum Trajectory Results for the Decoherence Model

From information carried by the fluid elements, plots may be constructed to illustrate the hydrodynamic fields. The probability density will be illustrated first, followed by flux maps and pictures of the action function. At an early time, $t = 10$ time steps, figure 8.2 shows the probability density and quantum potential for the coupled case. (Each time step corresponds to 2 atomic time units, equal to

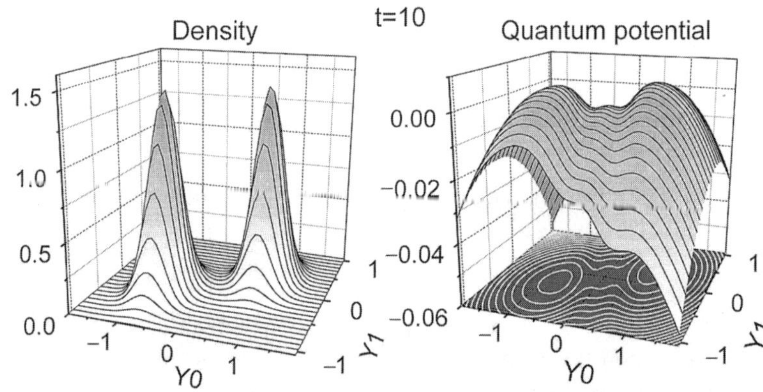

FIGURE 8.2. Probability density (left plot) and quantum potential (right plot) for the coupled case at $t = 10$ time steps [8.1]. These two functions (in a.u.) are plotted in the computational coordinate system. Near the center of each Gaussian, the quantum potential has the shape of an inverted paraboloid, but it "flattens out" in the midregion between the peaks in the density.

0.042 fs.) The twin peaks in the density (left plot) have barely moved from their initial positions. The right peak lies above the y_0 axis while the left peak lies below this axis. An important feature for this case is that the classical force tends to push the center of the left component toward the valley lying along the negative y_0 axis, while at the same time the right component is pushed into the valley that descends along the positive y_0 axis. The classical force thus tends to separate the two components of the initial packet. However, in the central region between the two density peaks, the quantum force acts in the opposite direction. This force pushes fluid elements away from the two density maxima toward the midplane lying between these peaks. Thus in the central region, the classical and quantum forces act in opposite directions, and this has a major influence in preventing density buildup in this region. The situation for the uncoupled case is very different, because the x component of the classical force vanishes. In this case, the quantum force in the central region acts unimpeded in directing fluid elements toward the central interference region near the origin of the coordinate system.

At a later time, $t = 200$ time steps, figure 8.3 shows the density for both the coupled and uncoupled cases. For the uncoupled case (right plot), a large interference peak has formed in the central region between the two initial Gaussians. However, for the coupled case (left plot), the density in the midplane between the two maxima reaches only about $^1\!/_4$ of the value for the uncoupled case. This suppression of the interference density for the coupled case illustrates the phenomenon of "decoherence". At later times, the density in the central region decreases as the two density maxima evolve away from each other, down the two potential valleys shown earlier, in figure 8.1. For the uncoupled case, the density rises slightly higher in the central region and then slowly decays at later times.

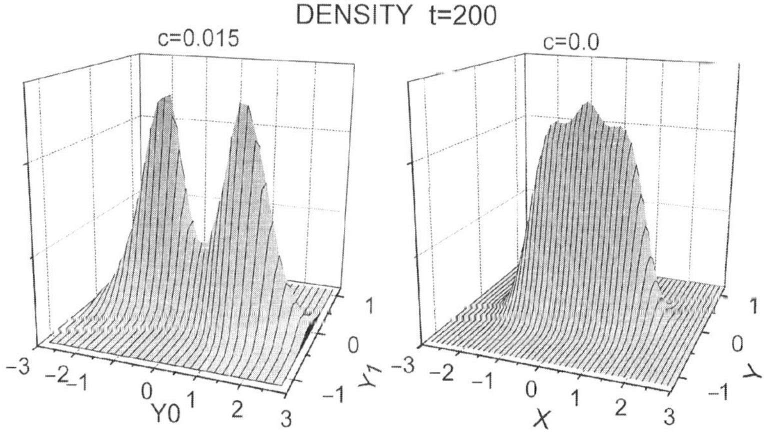

FIGURE 8.3. Probability density for the coupled (left plot) and uncoupled (right plot) cases at $t = 200$ time steps [8.1]. Note that computational coordinates (y_0, y_1) are used for the coupled case, whereas physical coordinates (x, y) are used for the uncoupled case.

The second type of hydrodynamic field that will be illustrated is the probability flux. Figure 8.4 shows flux vector maps and superimposed density contour maps for both the uncoupled and coupled cases. First, for the uncoupled case, the upper two plots show these fields for two time steps. When $t = 10$ time steps, the flux map shows a remarkable feature: all vectors are directed away from two vertical axes at $x = \pm 0.8$. When $|x| < 0.8$, these vectors are directed toward the $x = 0$ midplane, while for $|x| > 0.8$, all of these vectors are directed away from the

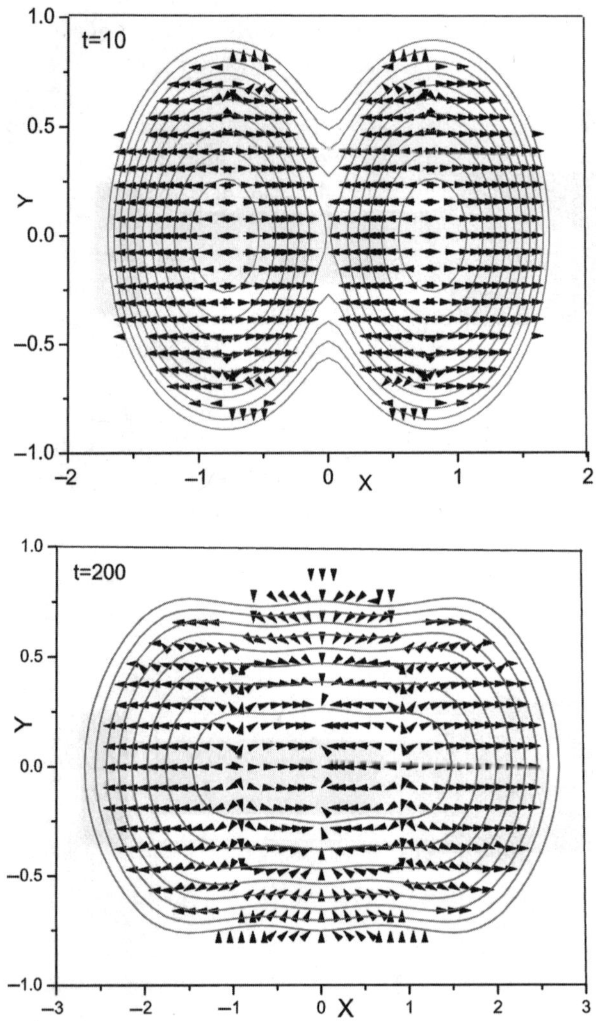

FIGURE 8.4. Flux maps and superimposed density contour maps for the *uncoupled case* (top two plots) and the *coupled case* (bottom two plots) at two time steps, $t = 10$ and $t = 200$ [8.1]. The dotted lines in the lower two plots for the coupled case are attractors for flux vectors.

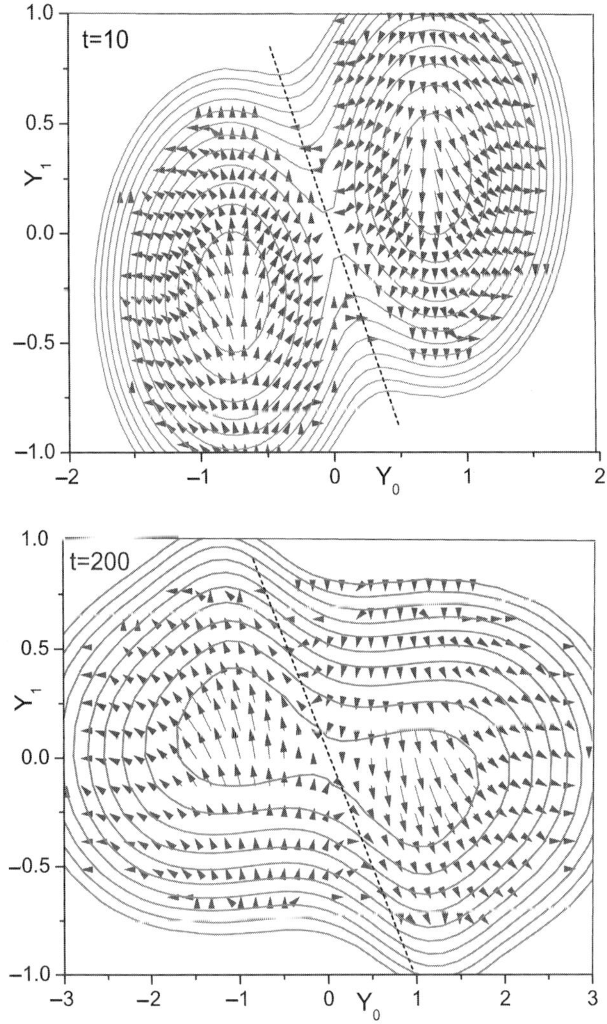

FIGURE 8.4. *Continued*

central region. The "collision" of flux vectors arriving from opposite directions at the $x = 0$ midplane leads to the buildup of density in this region at later times. At $t = 200$ time steps, the midplane at $x = 0$ continues to act as an attractor for flux vectors, while the two vertical axes near $x = \pm 0.8$ continue to act as repellers. The flux maps for the coupled case (lower two plots in figure 8.4) show significant differences from those for the uncoupled case. Turning first to the map for the early time, $t = 10$ time steps, almost all of the flux vectors point nearly parallel or antiparallel to the y_1 axis. Not surprisingly, the longest vectors are found near the two maxima in the density. At later time, $t = 200$ time steps, the flux is directed away from the origin, which is where the large interference feature arises for the uncoupled case.

The action function $S(x, y, t)$, which determines the phase of the wave function for the composite system, also determines the direction and magnitude for the flow velocity of the probability fluid. The action function at one time, $t = 200$, is shown in figure 8.5 for both the coupled and uncoupled cases. For the uncoupled case (top plot), the action function is symmetric under reflection in both the x and y axes. The action is largest near the left and right edges of the figure and displays a ridge along the $x = 0$ midplane. The lower plot in this figure shows the action

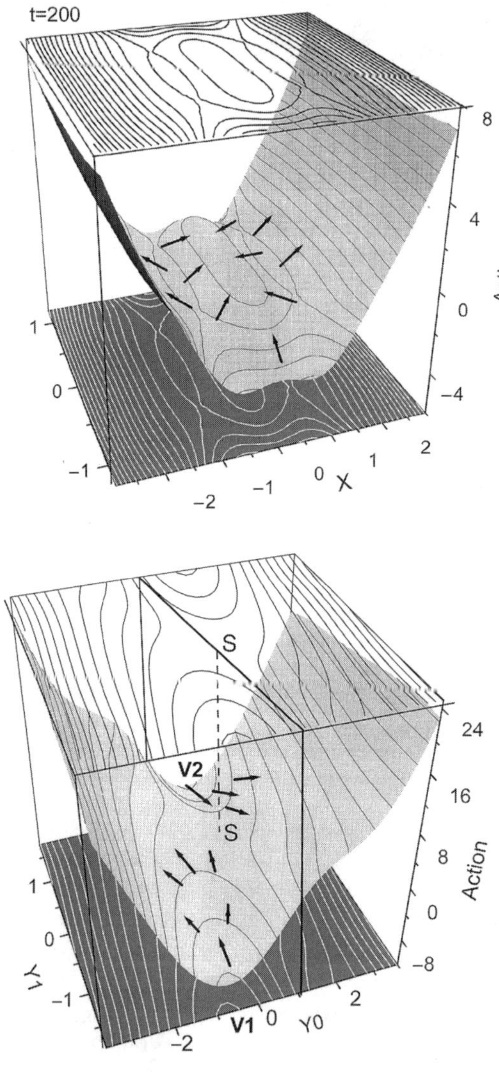

FIGURE 8.5. Wave function phase ($S/\hbar$ in radians) at $t = 200$ for the uncoupled (top plot) and coupled (bottom plot) cases [8.1]. Several gradient vectors are shown on each plot. In the lower plot, the saddle point (S) and valleys (V_1 and V_2) are indicated.

function for the coupled case. The reflection symmetry with respect to both axes has been broken, but there is inversion symmetry in the (y_0, y_1) coordinate system. In addition, the saddle point at the origin (marked S) separates the two valleys V_1 and V_2. In each of these plots, the local velocity vectors are parallel to the gradient vectors, some of which are shown.

We mentioned in the preceding section that commonly used indicators for decoherence include the dynamical diagonalization of the reduced density matrix and the modulation and annihilation of ripples in the reduced Wigner function for the subsystem. The reduced density matrix for the subsystem is found by tracing (averaging) over the reservoir coordinate, and the reduced Wigner function is found by Fourier transforming this function. The relevant transformations are given by

$$\rho_s(x, x', t) = \int \rho(x, y, x', y, t)dy,$$

$$W_s(x, p_x, t) = \frac{1}{2\pi\hbar}\int_{-\infty}^{\infty} \rho_s(x + \zeta/2, x - \zeta/2, t)e^{ip_x\zeta/\hbar}. \tag{8.6}$$

Both of these functions were computed from information carried by the trajectories, and plots are presented in [8.1]. When $t = 0$, the reduced density matrix displays large off-diagonal interference peaks located symmetrically on either side of the diagonal axis. For the uncoupled case, as time advances, the off-diagonal peaks persist, although they become significantly broadened. However, for the coupled case, the off-diagonal peaks are gradually damped, so that what survives lies close to the diagonal. The subsystem Wigner function at the initial time has large peaks centered at $x = \pm a$ and a series of interference ripples lying between these peaks and along the p_x axis. As time advances, the hills broaden and the negative valleys fill in, thus leading to a Wigner function that is positive and smooth everywhere. These results thus conform to the canonical measures for decoherence. One additional feature is that at later times, the reduced density matrix for the subsystem becomes very similar to that for a mixture of independently evolving components.

In Section 13.4, we will return to the same decoherence model and describe the classical and quantum components of the stress tensor, which enter the quantum Navier–Stokes equation describing the rate of change in the momentum density (probability density times momentum).

8.4 Quantum Trajectory Results for the Decay of a Metastable State

The quantum trajectory method has been applied [8.2] to wave packet dynamics in a composite system involving a one-dimensional reaction coordinate (the system mode) interacting with a multimode harmonic reservoir. Computational results were obtained for $M = 1$, 10, and 15 bath modes. As in the decoherence model, ensembles of quantum trajectories were evolved for the composite system, and the wave function was synthesized along individual quantum trajectories. The system and bath coordinates are denoted by y_0 (the *reaction coordinate*) and $\{y_1, y_2, \ldots, y_M\}$, respectively. The total potential energy is the sum of the system anharmonic potential, harmonic oscillator potentials for the M reservoir modes, and bilinear coupling terms linking the system to the reservoir,

$$V(y_0, \{y_i\}) = V_0(y_0) + \frac{1}{2}\sum_{i=1}^{M} f_i y_i^2 + y_0 \sum_{i=1}^{M} c_i y_i. \tag{8.7}$$

The reservoir modes are not coupled among themselves, but each is coupled to the system mode. For the 10-bath-mode-model, the system potential is an exponential repulsive term augmented by a Gaussian barrier centered on the reaction coordinate at the position $y_0 = y_b$,

$$V_0(y_0) = Ce^{-\delta y_0} + De^{-\eta(y_0 - y_b)^2}. \tag{8.8}$$

(The parameters for this potential are listed in Table II in [8.2].) The 10 reservoir mode frequencies varied from 1700 to 2400 cm^{-1}. For this study, an ensemble containing a very small number of fluid elements, $N = 110$, was propagated in the 11-dimensional coordinate space.

At the initial time, the total wave function was assumed to be the product of M ground state harmonic oscillator functions for the reservoir modes times a displaced harmonic oscillator function for the system mode. In addition, the translational wave packet for the system mode was given an initial thrust in the $+y_0$ direction. If the number of reservoir modes is small, $M \leq 3$, then at $t = 0$ it is convenient to start out the coordinates for each fluid element on a Cartesian grid. However, for a larger number of bath modes, it is not necessary to do this. For the studies reported in this section, an N-point uniform grid was first set up along the y_0 axis. For each of these points on the reaction coordinate, a random value was selected for each of the M reservoir coordinates. However, each reservoir coordinate was restricted to the interval $-y_i^{\max} \leq y_i \leq y_i^{\max}$, where the edge values were chosen so that the wave function for each mode was larger than a predetermined minimum value. By doing this, each fluid element was assigned $M + 1$ initial coordinates, forming an *unstructured grid* in the $(M + 1)$-dimensional coordinate space.

The least squares fitting algorithm was used to determine the spatial derivatives of the hydrodynamic fields required for time propagation. Within the stencil surrounding each fluid element, a local quadratic basis set was used for each of these fields. This basis included constant, linear, and quadratic terms in the displacement coordinates for all $M + 1$ degrees of freedom:

$$B = \left\{1, \xi_0, \xi_1, \ldots, \xi_0^2, \xi_1^2, \ldots, \xi_0\xi_1, \ldots\right\}. \tag{8.9}$$

The dimension of this basis set is given by

$$n_b = 1 + 2(M + 1) + M(M + 1)/2. \tag{8.10}$$

For example, when $M = 10$, there are 78 terms in the basis, and when $M = 15$, there are 153 terms.

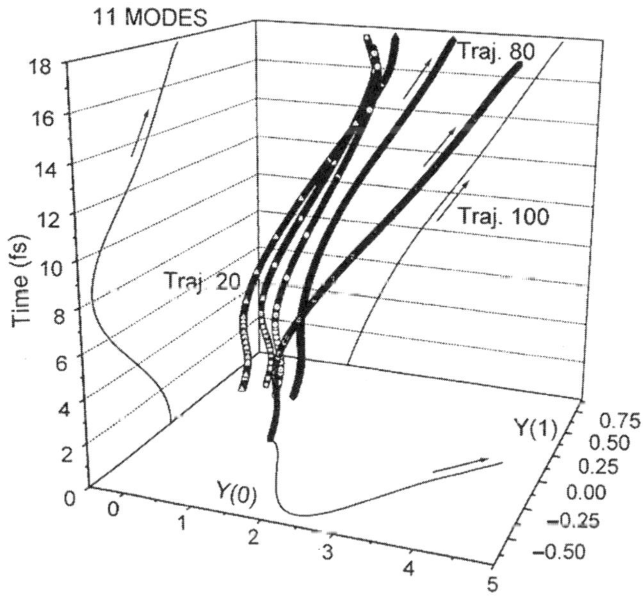

FIGURE 8.6. Time dependence of trajectories 20, 40, 60, 80, and 100 for the 11-mode system [8.2]. Trajectories 20 and 100 are near the trailing and leading edges of the trajectory ensemble, respectively. The projections of trajectory 100 on the three coordinate planes are also shown. The barrier maximum is located at $y_b = 2.0$ a.u. along the system coordinate y_0.

For the 11-mode system, the time evolution of five quantum trajectories is shown in figure 8.6. The time dependence for trajectories 20, 40, 60, 80, and 100 is shown in the (y_0, y_1) plane in the 11-dimensional coordinate space. In addition, the projection of trajectory 100 on the three coordinate planes is also shown. Trajectories 20 and 100 are launched from near the back and front ends of the ensemble, respectively. It is evident that trajectories 20, 40, and 60 decelerate as they approach the barrier region (this occurs at about $t = 15$ fs) and that trajectories 80 and 100 accelerate quickly to larger values of the reaction coordinate (y_0) after crossing the barrier.

Since the time-dependent C-amplitude and the action function S are computed along each trajectory, the complex-valued wave function can be easily synthesized. The plots displayed in figure 8.7 show the time dependence of the wave function for two quantum trajectories selected from the ensemble. The vertical axis denotes the propagation time, and the horizontal axes show the real and imaginary parts of the wave function. In each horizontal time slice, the radial distance from the vertical axis is the R-amplitude, and the twist angle around this axis is determined by the phase of the wave function, $S/\hbar$. In part (a), for trajectory number 72, there is gradually increasing amplitude as the trajectory moves toward the barrier maximum, which is reached for times near the top of the figure. In part (b), for trajectory number 45, there is an increase in density near 15 fs, but the density

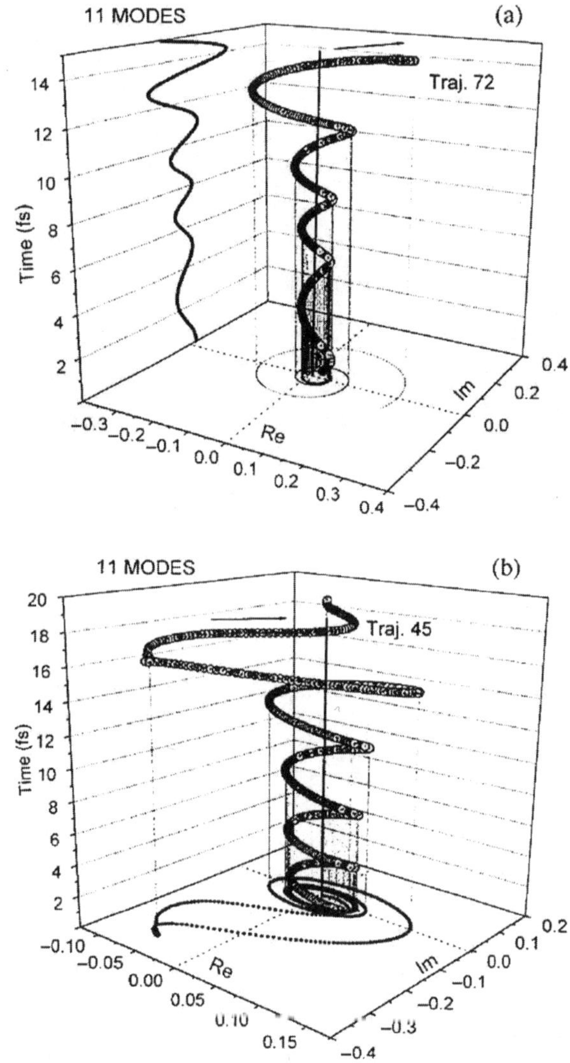

FIGURE 8.7. Time dependence of the wave function for trajectories 72 (top plot) and 45 (bottom plot) [8.2]. The real and imaginary parts of the wave function are plotted versus time. At each time, the distance from the vertical time axis is $\rho^{1/2}$, and the phase angle around this axis is $S/\hbar$. The barrier maximum is located at $y_b = 2.0$ a.u. along the reaction coordinate.

starts to decrease as the trajectory accelerates on the downhill side of the barrier (near $t = 20$ fs).

A number of additional results were obtained [8.2], including (1) modification of the bilinear system–reservoir coupling terms in equation 8.7 to include higher-order terms; (2) a comparison of effects on the trajectory dynamics of systematic versus random system–reservoir coupling terms; (3) energy partitioning among the system and reservoir modes along individual quantum trajectories for a model involving 15 reservoir modes.

8.5 Quantum Trajectory equations for Electronic Nonadiabatic Dynamics

In this and the following two sections, the dynamics of swarms of quantum trajectories moving on two coupled 11-dimensional potential surfaces will be investigated. Each of the potential surfaces is a downhill ramp along the reaction coordinate, and the coupling matrix element is a Gaussian centered just before the downhill region of each potential surface. On each of the potential surfaces, hydrodynamic equations of motion are integrated to find the probability density, acceleration, and action function along each of the trajectories. In order to improve the numerical stability of the integration scheme, periodic remeshing is used. The number of trajectories evolving on each surface is constant, although both amplitude and phase are transferred smoothly from one surface to the other during the evolution. In Section 8.7, wave packet evolution at several time steps will be displayed in the form of surface and gray-scale density plots.

We have seen in Chapters 2 and 4 that there are three hydrodynamic equations of motion for quantum trajectories evolving on a single electronic potential surface. These equations allow us to update the probability density $\rho(x,t)$, trajectory location $x(t)$, flow momentum $p(x,t) = \partial S(x,t)/\partial x$, and action function $S(x,t)$ along each trajectory. These equations were summarized in Box 2.5.

When we develop the equations of motion for electronic nonadiabatic problems, there are several starting points that depend on the representation used for the electronic potential surfaces and associated wave functions. These two starting points, the *adiabatic and diabatic representations*, are described in Box 8.1. For two coupled electronic potential energy surfaces, the total number of quantum trajectory equations of motion doubles to six. These equations were first described in the study by Wyatt, Lopreore, and Parlant [8.3], and an outline of the derivation is presented in Box 8.2. These six equations, expressed in the Lagrangian frame, will be summarized below.

In each of the six equations of motion given below, the last term on the right side is the contribution to the rate due to intersurface coupling. The first two equations of the set are continuity equations modified for gain or loss of probability due to coupling to the other potential surface:

$$\begin{aligned} d\rho_1/dt &= -\rho_1 \nabla \cdot v_1 - \lambda_{12}, \\ d\rho_2/dt &= -\rho_2 \nabla \cdot v_2 - \lambda_{21}. \end{aligned} \tag{8.11}$$

The second pair of equations are of Newtonian form, in which the right side gives the total force acting on the fluid element:

$$\begin{aligned} dp_1/dt &= -\nabla (V_{11} + Q_1 + Q_{12}) = F_1, \\ dp_2/dt &= -\nabla (V_{22} + Q_2 + Q_{21}) = F_2. \end{aligned} \tag{8.12}$$

Box 8.1. Diabatic and adiabatic electronic representations

The diabatic electronic representation and its connection to the familiar adiabatic representation were described by Felix Smith in 1969 [8.16]. A thorough presentation of these relationships is presented in the text by Tannor [8.17]. The starting point is the familiar electronic Schrödinger equation in the Born–Oppenheimer approximation (BOA), where at each fixed value for the nuclear coordinate (a single nuclear coordinate will be assumed here), the following eigenproblem is solved:

$$H_e(e;x)\chi_j(e;x) = E_j(x)\chi_j(e;x), \tag{1}$$

where H_e is the x-dependent electronic Hamiltonian, $\chi_j(e;x)$ is the electronic wave function, and $E_j(x)$ is the electronic energy. The electronic Hamiltonian includes the electronic kinetic energy, electron–electron repulsion, electron–nuclear attraction, and nuclear–nuclear repulsion terms,

$$H_e(e;x) = T_e(e) + V_{ee}(e) + V_{ne}(e,x) + V_{nn}(x). \tag{2}$$

The eigenfunctions in equation 1, the *adiabatic states*, diagonalize the electronic Hamiltonian. Within the BOA, the *total wave function* is the product of nuclear motion and electronic functions,

$$\Psi_j(x,e) = \xi_j^0(x)\chi_j(e;x). \tag{3}$$

In this approximation, the electronic eigenvalue (the *potential curve*) enters the nuclear motion Schrödinger equation as a potential energy

$$\left[-\frac{\hbar^2}{2m}\frac{\partial^2}{\partial x^2} + E_j(x)\right]\xi_j^0(x) = E_j(x)\xi_j^0(x). \tag{4}$$

In order to go beyond the adiabatic approximation, we will expand the total wave function in the complete set of electronic basis functions (at each value for x)

$$\Psi(x,e) = \sum_{j=1}^{N} \xi_j(x)\chi_j(e;x). \tag{5}$$

Equations for the nuclear motion functions can be derived by substituting this expansion into the TDSE. If the first and second derivatives of the electronic functions with respect to x are neglected, then equation 4 is recovered for each approximate nuclear motion wave function. However, if these derivatives are retained, we obtain the following system of coupled differential equations for the nuclear motion functions:

$$\left[-\frac{\hbar^2}{2m}\frac{\partial^2}{\partial x^2} + E_j(x) - E\right]\xi_j(x) = \frac{\hbar^2}{2m}\sum_{k=1}^{N}\left[2T_{j,k}^{(1)}(x)\frac{\partial}{\partial x} + T_{j,k}^{(2)}(x)\right]\xi_k(x), \tag{6}$$

in which the derivative matrix elements are

$$T_{j,k}^{(1)}(x) = \langle\chi_j|\partial/\partial x|\chi_k\rangle, \quad T_{j,k}^{(2)}(x) = \langle\chi_j|\partial^2/\partial x^2|\chi_k\rangle. \tag{7}$$

These matrix elements are frequently small except near avoided crossings between pairs of adiabatic potential curves. In these regions, another electronic basis set is frequently used, these being the diabatic functions.

In the *diabatic representation*, the electronic matrix elements in equation 7 are zero, or at least they are very small. There are an infinite number of routes to the diabatic representation. The rather trivial way is to choose the diabatic functions to be equal to the adiabatic ones *at a fixed value of* x, denoted by x_0, which could be in the asymptotic region. In this basis, the diabatic functions are denoted by $\varphi_j(e) = \chi_j(e; x_0)$, and the coupling matrix elements linking the electronic states are given by

$$V_{i,j}(x) = \langle \varphi_i | V(e, x) - V(e, x_0) | \varphi_j \rangle, \tag{8}$$

in which integration (over electronic coordinates only) is over the change in total potential energy relative to the reference geometry.

Comment about notation:

	Adiabatic representation	Diabatic representation
Electronic wave function	χ	φ
Nuclear motion wave function	ξ	ψ

A more general way to relate the adiabatic basis to the diabatic one is to consider the unitary transformation (written here for the transformation from two diabatic functions to two adiabatic ones)

$$\begin{bmatrix} \chi_1 \\ \chi_2 \end{bmatrix} = \begin{bmatrix} A_{1,1} & A_{1,2} \\ A_{2,1} & A_{2,2} \end{bmatrix} \begin{bmatrix} \varphi_1 \\ \varphi_2 \end{bmatrix}, \tag{9}$$

or, written as a transformation of column vectors, $\chi = A\varphi$. In the diabatic basis, the new nuclear motion functions are represented by the column vector ψ. These functions satisfy the matrix equation

$$A\frac{\partial^2 \psi}{\partial x^2} + 2\left(\frac{\partial A}{\partial x} + T^{(1)}A\right)\frac{\partial \psi}{\partial x} + \left(\left(\frac{\partial^2}{\partial x^2} + 2T^{(1)}\frac{\partial}{\partial x} + T^{(2)}\right)A \right.$$
$$\left. - \frac{2m}{\hbar^2}\left(U(x) - E\right)A\right)\psi(x) = 0, \tag{10}$$

in which $U(x)$ is the *diagonal* adiabatic electronic energy matrix $U_{i,i}(x) - E_i(x)$.

One way to choose the transformation matrix $A(x)$ is to set the term multiplying $\partial\psi/\partial x$ to zero,

$$\frac{\partial A}{\partial x} + T^{(1)}A = 0, \tag{11}$$

and this condition leads to

$$\frac{\partial^2 A}{\partial x^2} + 2T^{(1)}\frac{\partial A}{\partial x} + T^{(2)}A = 0. \tag{12}$$

The solution to equation 11 for the transformation matrix $A(x)$ was discussed by Baer in 1975 [8.18] and by Mead and Truhlar in 1982 [8.19].

Equations 11 and 12 then lead to a much simplified matrix equation for the nuclear motion functions $\psi(x)$:

$$-\frac{\hbar^2}{2m}\frac{\partial^2\psi(x)}{\partial x^2} + A^{\dagger}UA \cdot \psi(x) = E\psi(x). \tag{13}$$

We will now define the coupling matrix in the diabiatic representation, $V = A^{\dagger}UA$, but it is possibly more informative to rewrite this equation as $U(x) = A(x)V(x)A(x)^{\dagger}$. This version shows that diagonalization of the electronic coupling matrix in the diabatic representation gives the adiabatic electronic potential energy curves and that the eigenvector matrix is just the transformation matrix $A(x)$.

Box 8.2. Quantum trajectory equations for electronic nonadiabatic processes

We will consider two diabatic electronic basis functions (real-valued and orthonormal at each value of x, the reaction coordinate), $\varphi_1(e;x)$ and $\varphi_2(e;x)$, where e denotes the set of electronic coordinates. The origin for x is within the coupling region, with positive x directed toward the product channel. The Hamiltonian can be expressed as the sum of the kinetic energy operator for motion along the reaction coordinate plus the electronic Hamiltonian

$$H(x,e) = T(x) + H_e(x,e) = -\hbar^2/(2m)\partial^2/\partial x^2 + H_e(x,e). \tag{1}$$

In this representation, the electronic coupling matrix elements are $V_{i,j}(x) = \langle\varphi_i|H_e|\varphi_j\rangle$. The two diagonal elements of the matrix $V(x)$ are the (diabatic) potential energy curves, $V_{1,1}(x)$ and $V_{2,2}(x)$, while the off-diagonal elements $V_{1,2}(x) = V_{2,1}(x)$ couple the two potential curves. It will be assumed that the matrix elements of the nuclear momentum and kinetic energy are zero in this representation; for example, $\langle\varphi_i|\partial/\partial x|\varphi_j\rangle = 0$.

At each time, the total (nuclear motion plus electronic) wave function can be expanded in the diabatic electronic basis set,

$$\Psi(x,e,t) = \psi_1(x,t)\varphi_1(e;x) + \psi_2(x,t)\varphi_2(e;x). \tag{2}$$

The first goal is to obtain equations of motion for the nuclear motion components of the total wave function. These are readily derived by substituting this total wave function in equation 2 into the TDSE, followed by projection on each of the diabatic electronic functions. The two resulting coupled equations for the nuclear motion functions may be expressed in matrix form (see equation 13 in Box 8.1)

$$-\frac{\hbar^2}{2m}\frac{\partial^2}{\partial x^2}\begin{bmatrix}\psi_1(x,t)\\ \psi_2(x.t)\end{bmatrix} + \begin{bmatrix}V_{1,1}(x) & V_{1,2}(x)\\ V_{2,1}(x) & V_{2,2}(x)\end{bmatrix}\begin{bmatrix}\psi_1(x,t)\\ \psi_2(x,t)\end{bmatrix} = i\hbar\frac{\partial}{\partial t}\begin{bmatrix}\psi_1(x,t)\\ \psi_2(x,t)\end{bmatrix}. \tag{3}$$

The electronic coupling matrix $V(x)$ serves as input, and this is assumed to be known as a function of x. The initial conditions on these functions are that we start with a wave packet in one of the channels in the reactant region (large negative values for x), say channel 1, and direct this packet in toward the coupling region. The goal of the calculation is typically to evaluate the probability for occupying channel 2 in the product region.

The next step is to substitute the polar forms for each nuclear motion function, given by $\psi_j(x,t) = R_j(x,t)\exp\left[iS_j(x,t)/\hbar\right]$, into equation 3. After some algebra, first carried out by Gerard Parlant, we obtain the six coupled hydrodynamic equations given by equations 8.11–8.14. In these equations, the density on each electronic surface is $\rho_j(x,t) = R_j(x,t)^2$, and the flow velocity for the probability fluid is $v_j(x,t) = (1/m)\partial S_j(x,t)/\partial x$.

The total force acting on a fluid element is made up of three components: the classical force, the quantum force arising from other trajectories evolving on the same potential surface, and a coupling force arising from intersurface linkage.

The final pair of equations gives the rate of change in the action functions. The right side of each equation brings in the quantum Lagrangian along with the rate of change in the action due to coupling with the other surface

$$\begin{aligned} dS_1/dt &= (1/2)mv_1^2 - (V_{11} + Q_1 + Q_{12}) = L_1 - Q_{12}, \\ dS_2/dt &= (1/2)mv_2^2 - (V_{22} + Q_2 + Q_{21}) = L_2 - Q_{21}. \end{aligned} \tag{8.13}$$

The two coupling terms are (λ_{12} has units of a rate and Q_{12} is a potential energy):

$$\lambda_{12} = (2V_{12}/\hbar)\cdot(\rho_1\rho_2)^{1/2}\sin(\Delta), \quad Q_{12} = V_{12}(\rho_2/\rho_1)^{1/2}\cos(\Delta), \tag{8.14}$$

in which the phase shift (or *coherence function*) is defined by $\Delta = (S_1 - S_2)/\hbar$ and where $(\rho_1\rho_2)^{1/2}$ is the *transition density*. The coupling terms in equation 8.14 obey the symmetry relations $\lambda_{12} = \lambda_{21}$ and $\rho_1 Q_{12} = \rho_2 Q_{21}$. In the above equations, the quantum potential depends on the curvature of the amplitude on each surface $Q_j = (-\hbar^2/(2mR_j))\nabla^2 R_j$, where as usual, $R_j = \sqrt{\rho_j}$.

The quantum trajectory method for electronic nonadiabatic collisions is embodied in equations 8.11–8.14. No approximations have been made in deriving these equations from the TDSE. Additional details, including calculations for two coupled potential curves, are provided in previous studies [8.3, 8.4].

Starting with a wave packet launched on the lower potential surface, the goal is to compute the resulting transmission and reflection of amplitude on surface 1, as well as the transmission and reflection of the wave packet on surface 2, after it is created in the coupling region. However, since the coupled trajectory equations of motion contain terms (λ_{ij} and Q_{ij}) involving division by the densities, we must always have nonnegligible density *on both surfaces* at the positions of the fluid elements. For this reason, two separate wave packet propagation sequences are run, and the time-evolved wave functions on the two surfaces will then be combined to produce the results that would be obtained when a wave packet is launched on *only one* of the two surfaces [8.3, 8.4].

At the initial time, Gaussian wave packets are launched on each of the potential surfaces (these are denoted by G_1 and G_2), and two separate wave packet propagations are then executed, the first starting with the following two wave packets:

$$\begin{aligned} &\text{Surface 1:} \quad G_1 e^{ik(x-x_0)}, \\ &\text{Surface 2:} \quad G_2 e^{ik(x-x_0)}. \end{aligned} \tag{8.15}$$

These two functions later evolve into the wave packets $\Psi_1^{(1)}$ and $\Psi_2^{(1)}$ on the two surfaces (the superscripts refer to "calculation number 1").

The second calculation starts with the same two spatial wave packets, except that the phase of the upper packet is shifted by π. This has the effect of multiplying G_2 by (-1). As a result, the two wave packets launched in the second calculation are given by

$$\begin{aligned} &\text{Surface 1:} \quad G_1 e^{ik(x-x_0)}, \\ &\text{Surface 2:} \quad G_2 e^{ik(x-x_0)} e^{i\pi}. \end{aligned} \tag{8.16}$$

and these functions later evolve into the two wave packets $\Psi_1^{(2)}$ and $\Psi_2^{(2)}$. At each time, the wave functions on the two surfaces that would have evolved from a *single* wave packet launched on surface 1 are given by the superpositions

$$\begin{aligned} &\text{Surface 1:} \quad (1/2)\left[\Psi_1^{(1)} + \Psi_1^{(2)}\right], \\ &\text{Surface 2:} \quad (1/2)\left[\Psi_2^{(1)} + \Psi_2^{(2)}\right]. \end{aligned} \tag{8.17}$$

and these combinations lead to $G_1 e^{ik(x-x_0)}$ as the wave packet launched at $t = 0$ on surface 1, along with a "zero-density" or "virtual" wave packet starting on surface 2. In this way, we are able to circumvent problems in the hydrodynamic equations of motion associated with division by "zero-density".

The quantum trajectory method for electronic nonadiabatic processes conserves the number of quantum trajectories evolving on each potential surface, although the amplitude and phase of the wave function on each surface changes smoothly and continuously. This is very different from the approximate trajectory method introduced by Tully and Preston in 1971 [8.20]. In the *trajectory surface hopping* (TSH) algorithm (see Box 8.3), *classical trajectories* evolve on each surface. An approximate probability function tells the trajectory whether or not to hop to another potential surface at a particular position in the coupling region.

Rather than developing the quantum trajectory equations of motion as described in Box 8.2, an alternative is to expand the total wave function in adiabatic electronic functions and then use a "pseudopolar" form for the nuclear motion functions. In this approach, pursued by Burant and Tully, the final working equations were *not* solved using quantum trajectories [8.22]. This approach is explored further in Box 8.4.

Box 8.3. Trajectory surface hopping

Trajectory surface hopping (TSH), introduced by Tully and Preston in 1971, is a phenomenological extension of classical dynamics to nonadiabatic processes [8.20]. In the simplest version, we begin with two adiabatic potential curves $E_1(x)$ and $E_2(x)$ with an avoided crossing centered at the position denoted by x_0. Starting with an ensemble of classical trajectories on curve 1, the goal is to calculate the probability of ending up on curve 2. The simplest version of the TSH algorithm is based on the following sequence of events:

- a classical trajectory is followed on curve 1 up to the avoided crossing;
- a transition probability is assigned for making the hop from curve 1 to curve 2;
- if the particle hops, the velocity is adjusted so that the total energy is conserved;
- the classical trajectory is then followed on curve 2.

Thus, except at the crossing point, the particle moves on *either* curve 1 or curve 2.

There are several semiclassical methods for estimating the hopping probability $P(x, v)$, which depends on the velocity (v) of the particle. For example, the Landau–Zener (LZ) approximation for the transition probability depends on the energy gap between the adiabatic curves at the crossing point, $A = E_2(x_0) - E_1(x_0)$, and the difference in the slopes of these curves, $B = F_2 - F_1$ (we will assume that $F_2 > F_1$). The LZ transition probability is then given by

$$P(x_0, v) = \exp\left[-2\pi A^2/(\hbar B v)\right]. \tag{1}$$

A second method for estimating this probability is to integrate coupled equations of motion for the amplitudes to occupy the adiabatic states. The matrix elements of the interaction potential, which depend on the nuclear coordinate x, can be evaluated using the classical path approximation for $x(t)$.

The next step is to allow for *branching* of an ensemble of independent classical trajectories heading for the crossing region. The "ants procedure" is one way to handle the branching [8.20]. A column of N ants, the classical trajectories, march in along curve 1 toward the crossing region. The column splits into two groups at the crossing point, with $N_2 = P(x_0, v)N$ (converted to an integer) following along curve 2 and the remainder continuing along curve 1. The ants that hop to curve 2 adjust their velocities to conserve the total energy (electronic plus nuclear motion). The trajectories moving on either potential curve might encounter additional crossing points, but the same procedure is applied at each point. This procedure is effective when the number of crossings is relatively small. Other procedures may be used when there are multiple crossings.

Some features of the TSH method are as follows: the method is based on probabilities, not amplitudes; there is a lack of continuity (the trajectory changes electronic state in one time step); localized avoided crossings are assumed.

Because of the first item in this list, quantum effects connected with the nuclear motion (interferences, zero-point energies, tunneling, phase coherences) are neglected.

In an extension of the basic TSH method, an algorithm has been developed to allow for extended regions of nonadiabatic coupling [8.21]. Individual trajectories may hop many times between the coupled potential surfaces, but the number of switches is minimized subject to approximately maintaining the correct distribution of final states.

Box 8.4. Alternative decomposition of the wave function

An alternative way to decompose the total wave function, using two electronic states as an example, is to begin with the expansion in adiabatic electronic states

$$\Psi(x, e, t) = \sum_j \xi_j(x, t)\chi_j(e; x). \tag{1}$$

Each of the two nuclear motion functions is then represented in pseudopolar form

$$\xi_j(x, t) = A_j(x, t)e^{iS_j(x,t)/\hbar}, \tag{2}$$

where, *in contrast to the polar decomposition in equation 2.2, it is not assumed that the amplitudes A_j are real-valued.* When equations 1 and 2 are substituted into the coupled Schrödinger equations for the nuclear motion wave functions, Burant and Tully derived the following equations of motion [8.22]:

$$-\frac{\partial S_j}{\partial t} = \frac{1}{2m}\left(\frac{\partial S_j}{\partial x}\right)^2 + E_j - \frac{\hbar^2}{2m}\frac{1}{A_j}\frac{\partial^2 A_j}{\partial x^2}, \tag{3}$$

$$\begin{aligned}\frac{\partial A_j}{\partial t} &= -\left(v_j\frac{\partial}{\partial x} + \frac{1}{2}\frac{\partial v_j}{\partial x}\right)A_j \\ &\quad - \sum_i \left(T_{ji}^{(1)}A_i v_i - \frac{i\hbar}{m}T_{ji}^{(1)}\frac{\partial A_i}{\partial x} + \frac{i\hbar}{2m}T_{ji}^{(2)}A_i\right)e^{i\Delta_{ij}/\hbar}.\end{aligned} \tag{4}$$

in which $mv_i = \partial S_i/\partial x$, $\Delta_{ij} = S_i - S_j$, and the other notation is in accord with that in Box 8.2. The last term in the Hamilton–Jacobi equation, equation 3, is analogous to the quantum potential for the wave packet on the j-th potential surface. It is important to note that all coupling between the adiabatic surfaces has been incorporated in the equation of motion for A_j. As a result, the Hamilton–Jacobi equation, equation 3, is *formally* identical to that derived earlier in Chapter 2 for motion on a single potential surface.

By dropping certain terms, Burant and Tully proposed several approximations in equations 3 and 4 [8.22]. First, the final term in equation 3 was dropped, thus leaving the *classical* Hamilton–Jacobi equation

$$-\frac{\partial S_j}{\partial t} = \frac{1}{2m}\left(\frac{\partial S_j}{\partial x}\right)^2 + E_j. \tag{5}$$

The trajectory solutions to this equation are "pure" classical trajectories with momenta given by $p_j(x) = \partial S_j/\partial x = (2m[E - E_j(x)])^{1/2}$. As a second approximation, the $\hbar$-dependent electronic coupling terms within the second set of parentheses in equation 4 were also dropped, thus leaving the much simpler equation

$$\frac{\partial A_j}{\partial t} = -\left(v_j \frac{\partial}{\partial x} + \frac{1}{2}\frac{\partial v_j}{\partial x}\right) A_j - \sum_i T_{ji}^{(1)} A_i v_i e^{i\Delta_{ij}/\hbar}. \qquad (6)$$

These two equations form the basis of the *velocity coupling approximation* (VCA) to equations 3 and 4, the latter being termed the *classical limit Schrödinger equation* (CLSE).

The interpretation of the CLSE equations is straightforward. Equation 5 provides for purely classical propagation on each potential surface, while equation 6 yields continuously and smoothly changing complex-valued amplitudes on each surface. Thus while there is wave packet-like motion on each surface, the actual dynamics are classical.

The VCA and a related approximation were applied to several avoided crossing models for which there are no classical turning points or tunneling corrections. The classical momentum given previously was fed into equation 6, which was then solved on a space fixed grid. Transition probabilities computed using these models agreed well with exact results and in some cases were considerably better than those obtained using the trajectory surface hopping model.

8.6 Description of the Model for Electronic Nonadiabatic Dynamics

Each of the two 11-dimensional potential surfaces used in this model is a downhill ramp along the reaction coordinate (denoted by x) with harmonic oscillator potentials for the remaining 10 degrees of freedom (the displacement coordinates are denoted by y_j) [8.5]. The lower potential surface involves a *steep downhill ramp* centered near x_r, which is given by

$$V_{11}(x, \{y_j\}) = V_{rc}(x) + V_h(\{y_j\}) - \frac{c_{11}}{2}[\tanh(x_r - x) - 1] + \frac{1}{2}\sum_{j=1}^{10} k_j y_j^2, \qquad (8.18)$$

in which the first term is the potential along the reaction coordinate. The reaction coordinate potential has the following limits: as $x \to \infty$, $V_{rc} \to -c_{11}$; as $x \to -\infty$, $V_{rc} \to 0$; and $V_{rc}(x = x_r) = -c_{11}/2$. The force constants may be specified in terms of the effective mass and the frequency for each mode, $k_j = m_j\omega_j^2$. The upper potential surface V_{22} is a *shallow downhill ramp* with the same functional form as V_{11}, except that c_{22} is used in place of c_{11} ($c_{22} < c_{11}$). The vibrational components of these two surfaces are identical. The *coupling matrix element* that links these two surfaces is an 11-dimensional Gaussian

$$V_{12}(r) = c_{12} \cdot \exp\left\{-\beta\left[(x - x_c)^2 + \sum_{j=1}^{10} y_j^2\right]\right\}. \tag{8.19}$$

The coupling potential is centered at the position x_c along the reaction coordinate, and this point lies just before the position x_r where the reaction coordinate potential reaches the halfway point on its descent into the product valley. Figure 8.8 shows plots of the three potentials V_{11}, V_{22}, and V_{12}, versus x and y_1, where the latter coordinate is the displacement for vibrational mode number 1.

As mentioned in the previous section, at the initial time, Gaussian wave packets were launched on each of the two potential surfaces. The wave packet on the lower surface is given by the product of 11 one-dimensional Gaussians:

$$G_1(x, \{y_j\}) = (2\alpha_0/\pi)^{1/4}\, e^{-\alpha_0(x-x_0)^2} \prod_{j=1}^{10} \left(2\alpha_j/\pi\right)^{1/4} e^{-\alpha_j y_j^2}. \tag{8.20}$$

The first term on the right is the translational wave packet centered at position x_0 in the reactant valley of the potential surface. The initial wave packet on the upper surface has the same form, with the same parameter values. In order to implement the QTM for this problem, each initial wave packet is discretized in terms of N fluid elements, where for these calculations, $N = 110$. In order to locate each fluid element, 11 coordinates are required, and these are denoted by $\{x, y_1, y_2, \ldots, y_{10}\}$. The x-coordinates for the fluid elements are uniformly spaced about x_0 as the center, and the nearest-neighbor separations are $\Delta x = 0.04$ a.u. For each of these fluid elements, the vibrational coordinates are chosen at random from within the range $[-y_0, y_0]$, with y_0 chosen to be large enough so that some of the trajectories lie in the nonclassical part of the vibrational potential. The initial translational energy along the reaction coordinate is 6000 cm^{-1}, and other wave packet parameters were listed in ref. [8.5].

The quantum trajectories were propagated for 1400 time steps with the first-order Euler scheme, using a step size of $\Delta t = 2.0$ a.u. (0.048 fs). Stability of the propagation scheme was enhanced by periodically remeshing the Lagrangian grid points. At some propagation time t, the grid points along the reaction coordinate fall between the limits x_{min} and x_{max}. The new x-coordinates for the fluid elements were then equally distributed between these limits, and the y-coordinates were reset to their original values. When viewed just after each of these remeshing steps, the swarm of mesh points expands along x but is rigidly constrained along each of the y-directions. The remeshing procedure was performed every 20 time steps. The hydrodynamic fields were then least-squares interpolated onto the new mesh after the coordinates were adapted.

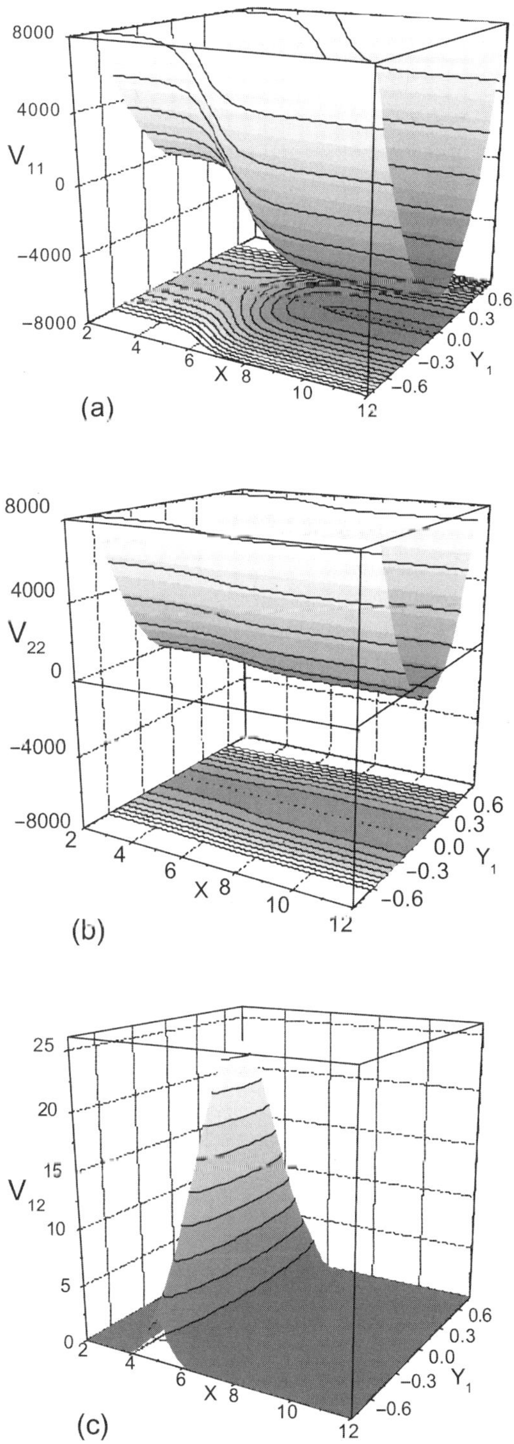

FIGURE 8.8. Electronic matrix elements plotted in the (x, y_1) subspace of the 11-dimensional coordinate space [8.5]. In (a) and (b), slices through the V_{11} and V_{22} potential surfaces are shown, and in (c) the coupling potential V_{12} is plotted. In each part, the vertical (energy) axis is in cm^{-1}.

8.7 Nonadiabatic Dynamics From Quantum Trajectory Propagation

In this section, wave packet dynamics on the two coupled potential surfaces will be illustrated. At several time steps, on each potential surface, the real part of the wave function was synthesized on the irregular grid defined by the positions of the quantum trajectories. Then, as a prelude to plotting, least squares interpolation was used to construct this function on a regular Cartesian grid in the (x, y_1) subspace. A grey-scale surface map of the wave packet on surface 1 at t = 200 a.u. (4.8 fs) is shown in figure 8.9. Initially, this wave packet was centered at $x = 2$ along the reaction coordinate, and at the current time step, the leading edge of the packet is approaching the coupling region, which is centered at $x = 5$. The wave packet on surface 2 will be created as the leading edge of this wave packet enters the coupling region.

The wave packets on the two potential surfaces at t = 600 a.u. (14.5 fs) are shown in figure 8.10. The wave packet on surface 1 shown in part (a) is almost centered in the coupling region, and the leading edge has moved over the steep downhill portion of the potential. Part (b) shows the newly created wave packet on surface 2. Part of this packet eventually reflects back toward smaller values of x, but most of it will progress toward the product direction at larger values of x. At this time step, the amplitude of the wave function on surface 2 is about 1/200 of the maximum on the lower surface.

A grey-scale map of the real part of the wave function on surface 1 at t = 1000 a.u. (24 fs) is shown in figure 8.11. The decreasing de Broglie wavelength associated with the steep downhill ramp is evident between $x = 5$ and $x = 8$. The large white dots locate 63 of the quantum trajectories at this time step; others lie outside of the plotting region. What is remarkable is that the complex oscillatory structure

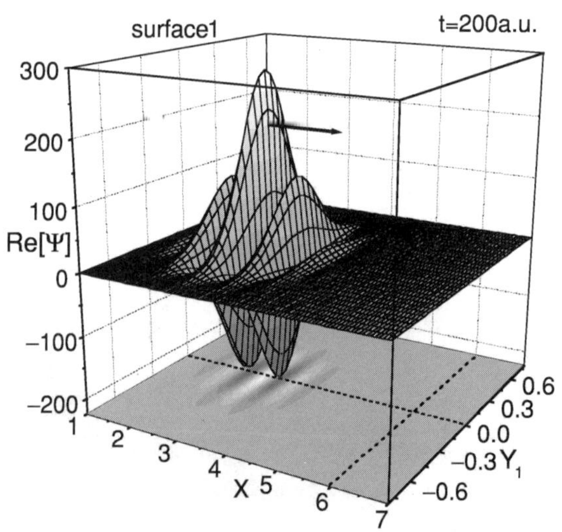

FIGURE 8.9. Wave packet on surface 1 at $t = 200$ a.u. (4 fs). The real part of the wave function is plotted in the (x, y_1) subspace [8.5]. The coupling region is centered at $x = 5$ along the reaction coordinate, and the ramp is halfway downhill when $x = 6$. The information needed to make this plot was interpolated onto a uniform mesh from the trajectory locations.

FIGURE 8.10. Wave packets on surfaces 1 (part a) and 2 (part b) at t = 600 a.u. (12 fs) [8.5]. The real part of the wave function is plotted in the (x, y_1) subspace. Note the different scales on the two vertical axes.

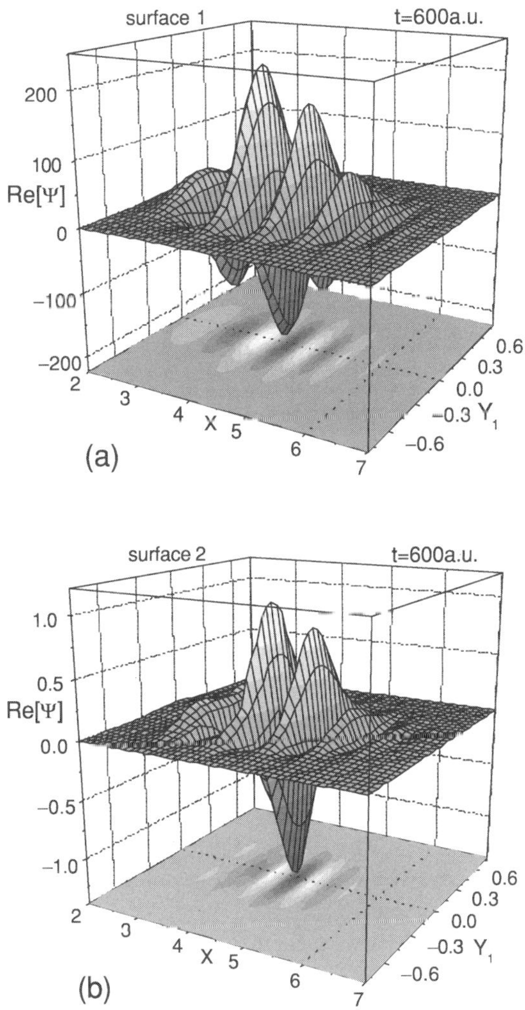

displayed in this figure was generated from information carried at a small number of moving "information centers".

8.8 Conclusions

Applications of the quantum trajectory method were made to three wave packet scattering problems. The first example involved wave packet evolution in a composite system involving a subsystem coupled to a single-mode environment [8.1]. When the coupling was turned off, a prominent interference feature formed in the probability density. However, when the coupling was turned on, growth of this feature was significantly hindered. The environment thus acted as a "decohering

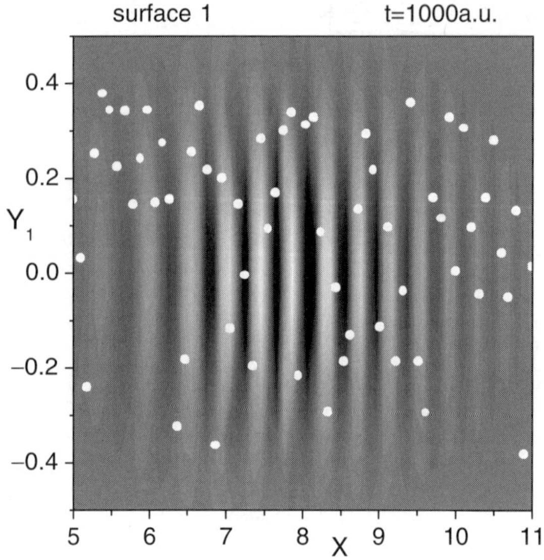

FIGURE 8.11. Grey-scale map in the (x, y_1) subspace showing the real part of the wave function on surface 1 at $t = 1000$ a.u. (24 fs) [8.5]. The large white dots locate some of the quantum trajectories at this time step (additional trajectories are located outside of the plotting region).

agent". Hydrodynamic analysis of the suppression of this interference feature formed the focus for this quantum trajectory study. The second example involved the time-dependent decay of a metastable state on a potential surface involving a reaction coordinate coupled to a 10-mode harmonic reservoir [8.2]. Quantum trajectories were illustrated and the wave function was plotted along several of the trajectories. Using the analytical method for hydrodynamic computations (in which the precomputed wave function is used), Nagomi et al. have calculated Bohmian trajectories for the decay of an initial state confined near the origin by a δ-function potential [8.23].

In the third example, electronic nonadiabatic dynamics were studied by propagating ensembles of quantum trajectories on two coupled 11-dimensional potential energy surfaces [8.5]. Along each quantum trajectory, the scattering wave function was synthesized, and for plotting purposes, the wave function was interpolated onto a regular grid. Plots showed the evolution of the wave packet on the lower surface toward the coupling region followed at later times by the gradual growth of the wave packet on the upper potential surface. These results demonstrate that for some multidimensional systems, quantum trajectories can provide useful dynamical information, even though a small number of trajectories were propagated.

References

8.1. K. Na and R.E. Wyatt, *Quantum hydrodynamic analysis of decoherence*, Physica Scripta, **XX**, 1 (2003).

8.2. R.E. Wyatt and K. Na, *Quantum trajectory analysis of multimode subsystem-bath dynamics*, Phys. Rev. E **65**, 016702 (2002).

8.3. R.E. Wyatt, C.L. Lopreore, and G. Parlant, *Electronic transitions with quantum trajectories*, J. Chem. Phys. **114**, 5113 (2001).
8.4. C.L. Lopreore and R.E. Wyatt, *Electronic transitions with quantum trajectories II*, J. Chem. Phys. **116**, 1228 (2002).
8.5. R.E. Wyatt, *Electronic nonadiabatic dynamics with quantum trajectories: Wavepackets on two coupled 11-dimensional potential surfaces*, presented at ITAMP (Institute for Theoretical Atomic, Molecular and Optical Physics) Workshop, Harvard University, May 9–11, 2002.
8.6. W.H. Zurek, *Decoherence and the transition from quantum to classical*, Physics Today, p. 36 (1991).
8.7. E. Joos, *Decoherence through interaction with the environment*, in D. Giulini, E. Joos, C. Kiefer, J. Kupsch, I.-O. Stamatescu, and H.D. Zeh (eds.), *Decoherence and the Appearance of a Classical World in Quantum Theory* (Springer, New York, 1998).
8.8. H.D. Zeh, in P. Blanchard, D. Giulini, E. Joos, C. Kiefer, and I.O. Stanatescu (eds.), *Decoherence: Theoretical, Experimental, and Conceptual Problems* (Springer, New York, 2000).
8.9. E. Joos and H.D. Zeh, *The emergence of classical properties through interaction with the environment*, Z. Phys. B **59**, 223 (1985).
8.10. J.P. Paz and W.H. Zurek, *Enviornment-induced decoherence, classicality, and consistency of quantum histories*, Phys. Rev. D **48**, 2728 (1993).
8.11. W.H. Zurek, *Preferred states, predictability, classicality and the environment-induced decoherence*, Prog. Theor. Phys. **89**, 281 (1993).
8.12. W.H. Zurek, *Decoherence, chaos, quantum-classical correspondence, and the algorithmic arrow of time*, Physica Scripta T **76**, 186 (1998).
8.13. S. Habib, K. Shizume, and W.H. Zurek, *Decoherence, chaos, and the correspondence principle*, Phys. Rev. Lett. **80**, 4361 (1998).
8.14. C.P. Williams and S.H. Clearwater, *Explorations in Quantum Computing* (Springer-Verlag, New York, 1998).
8.15. C. Brif, H. Rabitz, S. Wallentowitz, and I.A. Walmsley, *Decoherence of molecular vibrational packets: Observable manifestations and control criteria*, Phys. Rev. A **63**, 063404 (2001).
8.16. F.T. Smith, *Diabatic and adabiatic representations for atom–molecule collisions*, Phys. Rev. **179**, 111 (1969).
8.17. D.J. Tannor, *Introduction to Quantum Mechanics: A Time Dependent Perspective* (University Science Books, New York, 2004).
8.18. M. Baer, *Adiabatic and diabatic representations for atom–molecule collisions: Treatment of the collinear arrangement*, Chem. Phys. Lett. **35**, 112 (1975).
8.19. C.A. Mead and D.G. Truhlar, *Condition for the definition of a strictly diabatic electronic basis for molecular systems*, J. Chem. Phys. **77**, 6090 (1982).
8.20. J.C. Tully and R.K. Preston, *Trajectory surface hopping approach to nonadiabatic molecular collisions: The reaction of H^+ with D_2*, J. Chem. Phys. **55**, 562 (1971).
8.21. J.C. Tully, *Molecular dynamics with electronic transitions*, J. Chem. Phys. **93**, 1061 (1990).
8.22. J.C. Burant and J.C. Tully, *Nonadiabatic dynamics via the classical limit Schrödinger equation*, J. Chem. Phys. **112**, 6097 (2000).
8.23. Y. Nogami, F.M. Toyama, and W. van Dijk, *Bohmian description of a decaying quantum system*, Phys. Lett. A **270**, 279 (2000).

9
Approximations to the Quantum Force

Methods are described for computing approximations to the quantum force. Either the density or the log-derivative of the density is fit to a parameterized function.

9.1 Introduction

In the previous chapters, we first derived equations of motion for quantum trajectories and then presented a number of applications involving wave packet scattering. We also outlined a methodology for computing these trajectories by using least squares fitting to estimate the quantum potential and quantum force required to propagate the trajectories. In this chapter, several methods for estimating the quantum potential and the quantum force will be described. These methods fit a parameterized function to data carried by the trajectories. From the fit that is obtained, approximations to the quantum potential and the quantum force may be computed.

In the work of Maddox and Bittner [9.1], *statistical analysis* is used to obtain a smooth probability function that is *most likely* to represent the data carried by the trajectories. (The text by Gershenfeld [9.7] provides excellent background information on statistical fitting techniques.) This scheme is built on a *parameterized Gaussian model* for the probability density. An iterative procedure, *expectation-maximization* [9.7] is used to find a set of Gaussian parameters that best approximates the true density function. This fitted density is then used to compute an approximate quantum force that drives the ensemble of trajectories. As a nontrivial example, it will be demonstrated that this approach can be used to determine the *ground state density* and energy for a two-degree-of-freedom anharmonic potential representing the stretch-bend states of methyl iodide.

Garashchuk and Rassalov [9.2, 9.3] used a different global method, least squares fitting to a sum of Gaussians, to develop approximations to the probability density and the quantum potential. This method has been applied to wave packet scattering from an Eckart barrier, where different numbers of Gaussians were used to compute the energy-resolved transmission probability. In addition, it was found that the transmission probability obtained using a single Gaussian is essentially equivalent to the result obtained using a well-known semiclassical approximation. In these global Gaussian fitting methods, the total energy is not strictly conserved. This defect provided at least part of the motivation to develop a different fitting method that can be used to compute an approximation to the quantum potential.

Rather than fitting the density directly, approximations can be developed by fitting the *log derivative of the density* [9.4–9.6]. Both global and local fitting procedures have been implemented, of which the latter are based on the use of overlapping *domain functions*. These procedures are more sensitive to local features of the density, including ripples and nodes, and they have the important advantage of *conserving the total energy* in a closed system, independent of the quality of the fit. For an assumed linear fit to the log derivative, the parameters are obtained in terms of first and second moments of the density. Applications have been made to the collinear $H+H_2$ exchange reaction and to the photodissociation of ICN.

Section 9.2 introduces the statistical approach for fitting the density to a sum of Gaussians. In Section 9.3, the expectation-maximization algorithm for determining the Gaussian parameters is described. This statistical fitting procedure is then applied in Section 9.4 to the determination of the vibration-bend ground state of the methyl iodide molecule. The use of an alternative algorithm for fitting the density to a sum of Gaussians, the least squares procedure, is presented along with applications in Section 9.5. Rather than fitting the density directly, an alternative technique involving global or local fits to the log derivative of the density is described in Sections 9.6 and 9.7. Conclusions are presented in Section 9.8.

9.2 Statistical Approach for Fitting the Density to Gaussians

In this section, we will assume that the density can be approximated by summing contributions from M Gaussian components. In the work of Maddox and Bittner [9.1], the decomposition of the density is expressed as a sum of *joint probabilities,*

$$\rho(r) = \sum_{m=1}^{M} p(r, c_m), \tag{9.1}$$

where $p(r, c_m)$ is the probability that a randomly chosen fluid element is located at position r and "belongs to" the m-th Gaussian, labeled c_m. Each of these Gaussians is parameterized by a *weight* $p(c_m)$, a *mean position vector* μ_m, and a vector of *variances* σ_m. We can also replace the variance vector with a full *covariance matrix* C_m if necessary. The expansion weights $p(c_m)$ are non-negative and sum to unity.

By definition [9.7], each joint probability in equation 9.1 is related to a pair of *conditional probabilities* according to the relation

$$p(r, c_m) = p(r|c_m)p(c_m) = p(c_m|r)\rho(r), \tag{9.2}$$

where the following two conditional probabilities have been introduced. (The notation is as follows: $p(x|y)$ is the probability for event x given a specific value for event y.)

- $p(r|c_m)$ is the *forward probability* that a randomly chosen Gaussian c_m "has" the configuration point r. (Two examples of this probability will be given shortly.)
- $p(c_m|r)$ is the *posterior probability* that the configuration point r is a "member" of the Gaussian function labeled c_m.

Substituting the first equality of equation 9.2 into equation 9.1 we have

$$\rho(r) = \sum_{m=1}^{M} p(r|\, c_m)p(c_m). \tag{9.3}$$

The forward probability $p(r|c_m)$ can be specified with either of two different Gaussian models. The *first model* assumes that each multidimensional Gaussian is *completely separable* and takes the form of a product over the D-dimensional configuration space:

$$p(r|c_m) = \Pi_{d=1}^{D} \left[1/(2\pi\sigma_{m,d}^2)\right]^{1/2} \exp\left[-(r - \mu_{m,d})^2/(2\sigma_{m,d}^2)\right]. \tag{9.4}$$

The *second model* explicitly takes into account *nonseparable correlations* and incorporates the full *covariance matrix:*

$$p(r|c_m) = \left[||C_m^{-1}||/(2\pi)^D\right]^{1/2} \exp\left[-(r - \mu_m)^t C_m^{-1}(r - \mu_m)/2\right]. \tag{9.5}$$

In comparison with the separable case, the fully covariant model can represent more complicated density features with fewer functions; however, this is at greater computational expense. For low-dimensional systems, it is advantageous to use the fully covariant model, but in high dimensionality, it is much more efficient to use a larger number of separable Gaussians.

9.3 Determination of Parameters: Expectation-Maximization

Now that the Gaussian fitting model has been established, the goal is to determine the fitting parameters $p(c_m)$, μ_m, and σ_m (or C_m). The mean position vector and covariance matrix of the Gaussian components are defined in terms of the first and second moments of the forward probabilities:

$$\mu_m = \int r\; p(r|c_m)dr, \tag{9.6}$$

$$C_m = \int (r - \mu_m)^t\, (r - \mu_m)\; p(r|c_m)\, dr. \tag{9.7}$$

For the separable case, the variances are given by the diagonal elements $\sigma_{m,i}^2 = (C_m)_{ii}$. After substituting $p(r|c_m)$ from equation 9.2 into equations 9.6 and 9.7, we can express the integrands in terms of the density:

$$\mu_m = \int r\, p(c_m|r)\rho(r)/p(c_m) \cdot dr, \tag{9.8}$$

$$C_m = \int (r - \mu_m)^t\, (r - \mu_m)\; p(c_m|r)\rho(r)/p(c_m) \cdot dr. \tag{9.9}$$

These expressions can be approximated by *sums over the N fluid elements*

$$\mu_m = \frac{1}{Np(c_m)} \sum_{n=1}^{N} r_n p(c_m|r_n), \tag{9.10}$$

$$C_m = \frac{1}{Np(c_m)} \sum_{n=1}^{N} (r_n - \mu_m)^t (r_n - \mu_m) p(c_m|r_n). \tag{9.11}$$

The expansion weights can also be written as a sum over fluid elements:

$$p(c_m) = \frac{1}{N} \sum_{n=1}^{N} p(c_m|r_n). \tag{9.12}$$

We still need a method to determine the *posterior probabilities* $p(c_m|r_n)$ in equations 9.10–9.12. These quantities can be evaluated directly from the *forward probabilities* according to one of the principal equations of *Bayesian statistics* (this expression is described in Section 10.2 in [9.7]):

$$p(c_m|r_n) = \frac{p(r_n|c_m)p(c_m)}{\sum_{m=1}^{M} p(r_n|c_m)p(c_m)}. \tag{9.13}$$

The numerator, denominator, and ratio in this expression have the following interpretations:

- *Numerator:* A measure of how well *the Gaussian* c_m describes the fluid element with the configuration r_n.
- *Denominator:* A measure of how well the fluid element at r_n is described by *all of the Gaussians*.
- *The ratio:* The chance that the fluid element at r_n is described by the Gaussian with the label c_m. Hence, the Gaussian that best describes r_n will have the largest posterior probability for that point

The circular structure given by equations 9.1–9.13 provides the framework for an iterative procedure, the *expectation-maximization* (EM) algorithm [9.7], that seeks a set of Gaussian parameters yielding the best estimate for the density. The parameters are evaluated by *maximizing the log-likelihood* (L) of the Gaussian fitting function, when it is evaluated at the set of input data points (this algorithm is described in Section 14.3 of [9.7])

$$L = \log \Pi_{n=1}^{N} \rho(r_n). \tag{9.14}$$

The log-likelihood measures how well the density model describes the fit at the ensemble of data points. The EM algorithm works very much like a variation principle, in that there is a likelihood equation defined over parameter space, $\nabla L = 0$, such that L *is a maximum for parameter sets that are effective in describing the ensemble's distribution.* Furthermore, it can be shown that the update rules in equations 9.10–9.12 move the Gaussians through parameter space in the direction that improves the density estimate. The cycle of estimating the expected distribution function and maximizing the log-likelihood is repeated iteratively until a satisfactory estimate of the density is achieved.

One problem that needs to be addressed concerns the number of Gaussians (M) used in the density estimate. For a Gaussian wave packet evolving in a parabolic potential the answer is known, but in general, we will never really know how many Gaussians to use. When a wave packet bifurcates at a potential barrier, it will often develop complicated oscillations and nodal structures that are impossible to capture with a small number of Gaussians. Though there are statistical methods for "guessing" the number of components in a data set, these will not be used for the application described in the next section. Instead, we simply try to use the minimum number of Gaussians that give reasonable results.

The overall Gaussian fitting approximation and EM algorithm is as follows: First we generate the ensemble of fluid elements, usually a Gaussian density, via some appropriate sampling technique. The EM algorithm is initialized by choosing a set of parameters for a preset number of Gaussians. Typically, the initial Gaussians are given a uniform weight, $p(c_m) = 1/M$. The mean position vectors are randomly selected from the domain of the ensemble. The initial variances are chosen to be large enough to encompass the entire ensemble, and the cross terms are set to zero. We then cycle through the EM routine until the parameters converge to an acceptable estimate for the density. Convergence can be evaluated in a number of ways, such as by monitoring the Gaussian parameters, the conditional probabilities, or the log-likelihood.

9.4 Computational Results: Ground Vibrational State of Methyl Iodide

We now turn to a problem with physical merits. With an estimate for the density given by equations 9.4 or 9.5, it is straightforward to compute an approximate quantum force in terms of the Gaussian parameters. The quantum and classical forces are then used to drive the ensemble of trajectories over a small time step. For nontrivial problems, the quantum density will generally exhibit a very complicated structure. Clearly, the Gaussian approximation will not be able to capture the intricacies of a realistic quantum density, especially if there are multiple ripples and nodes present. Consequently, it is not feasible, using the present formulation of this methodology, to obtain numerically accurate quantum densities for complex nonstationary systems.

Ground state quantum densities, on the other hand, are characteristically much simpler than their excited state and nonstationary counterparts. As a result, this fitting approach is most useful for determining the ground state properties of high dimensional systems. For stationary systems, the quantum force exactly counterbalances the classical force, and the ensemble of quantum trajectories does not evolve in time. The ground state can then be realized from an initial nonstationary state by adding a small *damping term* to the Newtonian-type equation of motion,

$$m\dot{v} = -\nabla(V + Q) - \gamma m v. \tag{9.15}$$

This fictitious friction term, $-\gamma m v$, causes the fluid elements to lose a small amount of kinetic energy at each time step. For a classical ensemble, the distribution would collapse to a delta function(s) centered at the minimum energy point(s) of the potential surface. For the quantum trajectory ensemble, however, as the distribution becomes increasingly narrow, the quantum force becomes very strong and requires the ensemble to maintain a minimum finite width. When equilibrium is reached at longer times, the resulting distribution is representative of the ground state quantum density. The corresponding ground state energy can be resolved to within the statistical error of a Monte Carlo integration over the ensemble elements.

In order to illustrate these features, we will follow the evolution of an initial Gaussian ensemble toward the ground state for the stretch-bend modes of the lowest electronic state of methyl iodide. The vibrational system is treated as a reduced mass (m = 20,000 a.u.) evolving on a two-dimensional anharmonic potential energy surface [9.8]. Contour lines for this potential surface are depicted (gray lines) in the four parts of figure 9.1. In part (c) of this figure, we illustrate the numerically exact ground state density obtained by diagonalizing the stretch-bend Hamiltonian using the discrete variable representation (DVR). The grid points displayed in this figure provide the minimum number that are needed to converge the lowest-energy eigenvalue. Clearly, a much larger grid would be necessary to perform a wave packet calculation on this potential surface.

Parts (a) and (b) of figure 9.1 illustrate the density evolution for both the separable and fully covariant Gaussian fitting models, respectively. The black ovals represent the half-width contours of the Gaussians used to fit the density. There are four Gaussians in the separable case and two for the fully covariant model. The various contour plots labeled (1), (2), and (3) correspond to snapshots of the density at different times during the evolution of the trajectory ensemble. For both models, the initial density at step (1) is Gaussian, and all but one of the Gaussians is redundant. However, as the ensemble is propagated, the individual Gaussians behave differently from one another. The contours at step (2) show the quantum density at an intermediate time after roughly 10,000 time steps ($\Delta t = 1$ atomic time unit). At longer times, equilibrium is achieved, and the contours at step (3) are representative of the quantum ground state.

Part (d) in figure 9.1 shows the energy of the system relative to the minimum of the potential well as a function of the number of time steps. The exact ground state energy (591 cm^{-1}) serves as a benchmark and is indicated by the dashed horizontal line. The dotted and solid energy curves are for the separable and nonseparable

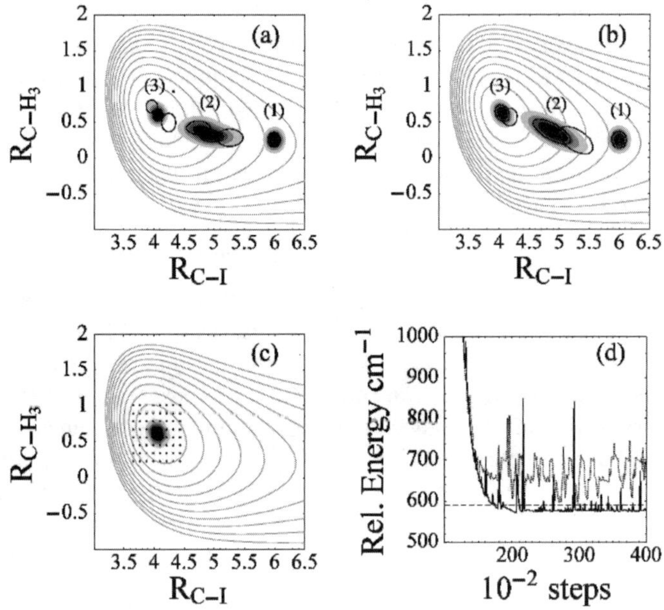

FIGURE 9.1. Plots (a) and (b) show the relaxation of an initial Gaussian wave packet in an anharmonic potential well for both the separable and fully covariant models, respectively [9.1]. The gray contours reflect the potential energy curves for the stretch-bend dynamics of CH_3I. The shaded contours indicate the shape of the approximated density after (1) 0, (2) 10,000, and (3) 40,000 time steps, respectively. The solid curves represent the half-width contours of the Gaussians. Plot (c) shows the numerically accurate quantum-mechanical ground state and the associated grid of quadrature points. Plot (d) shows the energy of the estimated density as a function of the number of time steps. The dotted and solid curves correspond to the separable and nonseparable models, respectively, while the dashed horizontal line represents the numerically exact ground state energy.

Gaussian models, respectively. Discarding the first 20,000 time steps, the average energy for the separable case is 679 $\pm$ 28 cm^{-1}, which is well above the exact energy. The average energy for the nonseparable model falls within reach of the exact energy at 601 $\pm$ 24 cm^{-1}. The sharp energy spikes for the nonseparable calculation are due to anomalous changes in the Gaussian parameters, such as a sudden jump in μ_{m} or rotation of C_m. These effects do not pose a significant problem, since the Gaussian parameters respond within a few time steps to correct the abnormalities. Filtering out these sporadic deviations improves the accuracy of the ground state energy estimate and significantly reduces the statistical variation, thus yielding 593 $\pm$ 5 cm^{-1}. For comparison, the same calculation has been performed using four fully covariant Gaussians. The average energy for the equilibrated system improves slightly to 593 $\pm$ 3 cm^{-1}. However, this is at the cost of greater computational effort and slower convergence.

9.5 Fitting the Density Using Least Squares

A different computational procedure was followed by Garashchuk and Rassalov [9.2, 9.3] to fit a sum of Gaussians to the values of the density at the trajectory locations. The global approximation for the density was written in terms of a sum over M Gaussians,

$$\rho(x) \approx f(x) = \sum_{j=1}^{M} c_j^2 e^{-a_n^2(x-X_n)^2}. \tag{9.16}$$

This approximation is evaluated once at each time step and can be made arbitrarily accurate by increasing the number of terms in the summation. For each Gaussian, there are three fitting parameters to be determined, $\{c_j, a_j, X_j\}$. Consequently, the total number of parameters is $3M$. These parameters are determined by minimization of the least squares error functional

$$J(s) = \int (\rho(x) - f(x;s))^2\, dx, \tag{9.17}$$

where s denotes the set of parameters. This functional is optimized by using the gradient conditions $G_j = \partial J(s)/\partial s_j = 0$ and these equations are then solved using an iterative quadratic method that makes use of the matrix of second derivatives. The procedure starts with a single Gaussian, and then M can be gradually increased up to a maximum allowed value. In principle, in the limit of large M, accurate estimates of the quantum potential and force can be obtained from the fitted density.

As an example of this global fitting procedure, we return once again to wave packet scattering from an Eckart barrier [9.2, 9.3]. The initial Gaussian wave packet is centered at $x_0 = -3$, with the initial momentum $p_0 = 6.0$ (parameter values are given in atomic units, with the unit of time given by 918 a.u.), and the barrier is centered at the origin. (Other parameter values are listed in the cited references.) In these calculations, an ensemble of 199 quantum trajectories was evolved, but additional trajectories were added in the barrier region when the separation between adjacent trajectories exceeded a threshold value. Figure 9.2 shows the fit to the density at $t = 0.9$ for calculations using either 2 or 16 Gaussians. The solid curve shows the accurate quantum-mechanical density at this time step. The fit with two Gaussians reproduces the bifurcation of the wave packet, but does not describe the quasi-node that develops near $x = -1.6$ a.u. However, the calculation with 16 Gaussians is able to approximately capture this feature. In general, when a small number of Gaussians are used in the fitting procedure, ripples in the density cannot be accurately reproduced. Adding more Gaussians improves the fit to the density, but adds to the computational expense. In addition, long-time stability becomes a major issue because errors in the quantum potential and quantum force accumulate over time, and this may eventually destroy the calculation.

From the time dependence of the trajectory ensemble, we can readily obtain the time-dependent transmission probability $P(t)$. A related quantity of prime importance in reactive scattering is the energy-resolved transmission probability $P(E)$,

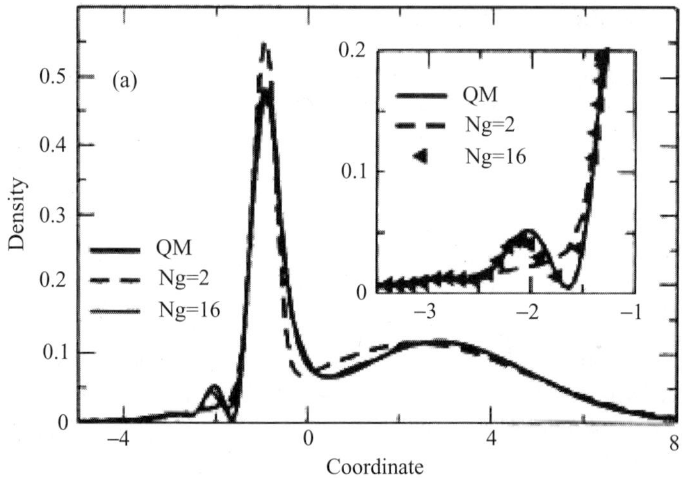

FIGURE 9.2. Wave packet scattering from an Eckart barrier centered at the origin [9.3]. Approximate quantum potentials were computed by fitting the density with either 2 or 16 Gaussians. The insert shows an enlargement of the region to the left of the barrier where interference effects create ripples in the density.

that can be obtained by Fourier transforming the time-dependent cross-correlation function $C(t)$. In Chapter 10, the equations relating $C(t)$ to $P(E)$ will be presented in more detail, but for now, we will describe what is involved in calculating the cross-correlation function. This quantity is the overlap between two wave functions: the time-evolving wave packet $\psi(x,t)$, which was launched from the reactant (left) side of the barrier, and a stationary monitor function $\phi(x)$ located on the product (right) side of the barrier. (For the calculations reported later, $\phi(x)$ is centered at $x_0 = 3.0$, and the initial Gaussian wave packet is centered at $-x_0$.) The correlation function, defined by the matrix element $C(t) = <\phi|\psi(t)>$, is then transformed to give the energy-resolved transmission probability

$$P(E) = \eta(E)|\int_0^\infty C(t)e^{iEt/\hbar}dt|^2, \tag{9.18}$$

where $\eta(E)$ is an energy dependent factor (see equation 10.15).

Figure 9.3 shows $P(E)$ for calculations where the density was fit using 1, 2, or 4 Gaussians. Starting at the top, the three panels in this figure show results for broad, moderate (similar to the $H+H_2$ reaction), and narrow barriers, all of which have the same barrier height ($V_0 = 16$ in reduced units). In addition, the analytical $P(E)$ curve is shown with filled circles and results obtained using a semiclassical method are shown with triangles (this method uses the Herman–Kluk propagator with the semiclassical initial value representation). For all three barrier widths, there is remarkable agreement between the semiclassical results and those obtained when *only one Gaussian* was used to fit the density. The agreement between this pair of results and the analytic curve improves as the barrier becomes

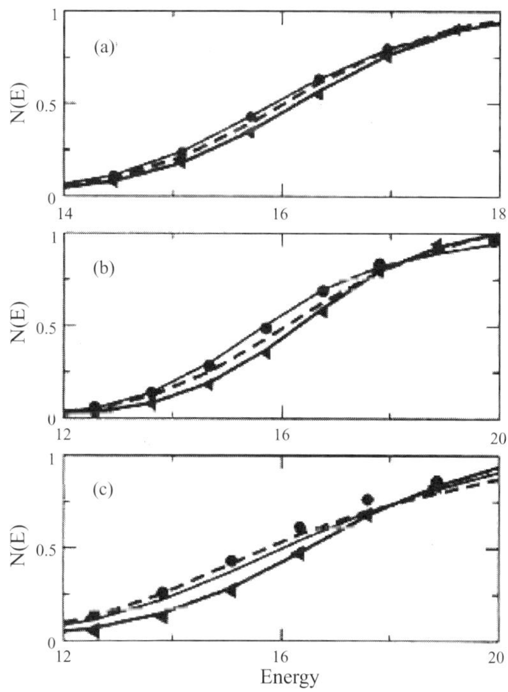

FIGURE 9.3. Energy-resolved reaction probability obtained by fitting the density with 1 (thick solid curve), 2 (dashed curve), or 4 (thin solid curve) Gaussians [9.3]. The analytical and semiclassical results are shown with filled circles and triangles, respectively. The top, middle, and bottom panels show results for wide, moderate, and thin barriers. The barrier height is $V_0 = 16$ in these reduced-energy units. Note the change in energy range between the top panel and the two lower panels.

thicker (more "classical"). In addition, the results obtained with 2 or 4 Gaussians are closer to the analytic result than for the case in which a single Gaussian was used. These calculations demonstrate, for all barrier thicknesses, that very good energy-resolved transmission probabilities can be obtained when just a few Gaussians are used in the fitting procedure. Additional comparisons in systems of higher dimensionality would be of interest.

9.6 Global Fit to the Log Derivative of the Density

In the previous section, *global fitting of the density* in terms of a superposition of Gaussians was the first step toward computation of an approximate quantum force for use in the QHEM. However, in regions of low density, such as around wave function nodes, this method may provide poor estimates for the quantum force. In addition, these global fitting methods do not strictly conserve the total energy. An alternative way of approximating the quantum force has been developed by Garashchuk and Rassolov (GR) [9.4, 9.5]. This method may be more sensitive to local features such as to ripples and nodes. In the *linear quantum force* (LQF) method, no predetermined form is imposed on the density itself. Rather, the procedure centers on obtaining the fit to a function of the density, the *nonclassical component of the momentum operator* (see Box 13.2).

When the momentum operator acts on the polar form of the wave function, we obtain

$$\hat{p}\psi = \left[\frac{\partial S}{\partial x} - \frac{i\hbar}{2}\frac{1}{\rho}\frac{\partial \rho}{\partial x}\right]\psi = m\left[\frac{1}{m}\frac{\partial S}{\partial x} - i\frac{\hbar}{2m}\frac{1}{\rho}\frac{\partial \rho}{\partial x}\right]\psi = m(v + iu)\psi, \tag{9.19}$$

where v is the flow velocity and $u = -D\partial \ln(\rho)/\partial x$ is the Einstein *osmotic velocity* (the quantum diffusion coefficient is $D = \hbar/(2m)$). (The osmotic velocity and the imaginary component of the complex-valued velocity are also mentioned in Boxes 2.6 and 13.2, and near equations 13.7 and 13.11–13.12.) In the work of GR, the *log derivative of the density* $\mu(x) = \partial \ln(\rho)/\partial x$ is fit with a function $g(x; s)$, where $s = \{s_1, s_2, \ldots, s_m\}$ is a set of time-dependent parameters. Linear fits have the advantage that they are quickly determined and can be readily extended to high dimensionality. However, in regions where there are ripples in the density and around wave function nodes, linear fits cannot be accurate. In this case, higher-order fits using extended basis sets might be used, albeit at higher computational expense. In terms of the exact log derivative of the density, the quantum potential is given by

$$Q(x) = -\frac{\hbar^2}{2m}\frac{R(x)''}{R(x)} = -\frac{\hbar^2}{8m}\left[\mu(x)^2 + 2\mu(x)'\right]. \tag{9.20}$$

An *approximate quantum potential* (AQP) is generated when an approximation is substituted for $\mu(x)$ on the right side of this expression.

The fitting procedure for $\mu(x)$ can be based on either *global* or *local* approximations. The *global approach* will be considered first, and then the local approximation will be described in the following section. In the global approach, a least squares error functional is defined by

$$J(s) = \int (\mu(x) - g(x; s))^2 \rho(x) dx. \tag{9.21}$$

Minimization of this functional then leads to the system of equations $\partial J/\partial s_i = 0$, $i = 1, 2, \ldots, m$. In the simplest case, the two-parameter *linear fit* is given by $g(x; s_1, s_2) = s_1 + s_2 x$, and the variational criterion leads to the pair of equations

$$s_1 + M_1 s_2 = \int \mu(x)\rho(x) dx,$$

$$M_1 s_1 + M_2 s_2 = \int x\mu(x)\rho(x) dx. \tag{9.22}$$

In these equations, the *first and second moments* of the density are given by

$$M_1 = \int x\rho(x) dx, \quad M_2 = \int x^2 \rho(x) dx. \tag{9.23}$$

Each integral in equations 9.22 and 9.23 can be evaluated using trajectory locations and weights $w_i = \rho(x_i) dx_i$. For example, the second moment is evaluated from the following trajectory discretization:

$$M_2 = \sum_{i=1}^{N} w_i x_i^2. \tag{9.24}$$

Once the parameters have been determined, the AQP is given by (compare with equation 9.20)

$$Q(x) = -\frac{\hbar^2}{8m}\left[g(x;s)^2 + 2g(x;s)'\right]. \tag{9.25}$$

Using the linear fit for g mentioned above, the quantum force evaluated from this potential is proportional to g itself:

$$f_q(x) = \frac{\hbar^2}{4m} s_2(s_1 + x s_2). \tag{9.26}$$

It is important to remark that linear fitting gives the exact time evolution for a Gaussian wave packet in a quadratic potential. In addition, in some cases, the linear quantum force may provide the dominant quantum correction even when the dynamics are not strictly Gaussian.

To summarize some features of the global fitting procedure for the log derivative of the density, the optimization problem has a simple analytic solution requiring only first and second moments of the trajectory distribution. In addition, in multidimensional problems, parameter optimization can be performed separately for each degree of freedom. The method conserves the total energy, and the average value over the trajectory ensemble of the quantum force is zero.

The global fitting method described above has been applied to several scattering problems, including the following: transmission through the Eckart barrier in one dimension, the photodissociation cross section of ICN in two dimensions, and the collinear $H+H_2$ reaction (two degrees of freedom). The latter example will be described in more detail in the remainder of this section [9.5]. Jacobi coordinates for reactants R (reactant atom to the center of mass of the reactant molecule) and r (vibrational coordinate in the reactant molecule) were used to describe the dynamics. These trajectory results were compared with results from quantum-mechanical grid calculations and with purely classical calculations for which the quantum potential was set to zero.

Figure 9.4 shows some details for a single initial relative translational energy. Part (a) compares exact and linear quantum force calculations for the time-dependent reaction probability, and part (b) shows the time dependence of the average position of the wave packet. The average positions agree well until the wave packet turns the corner on the potential surface (at $t = 0.35$ time units) and bifurcates into transmitted and reflected components. Energy-resolved reaction probabilities (not shown here) were also compared, and the overall shapes of the curves were similar. When compared with the linear force approximate results, the maximum value of the exact reaction probability curve was about 10% smaller and was shifted to lower energy by about 0.08 eV.

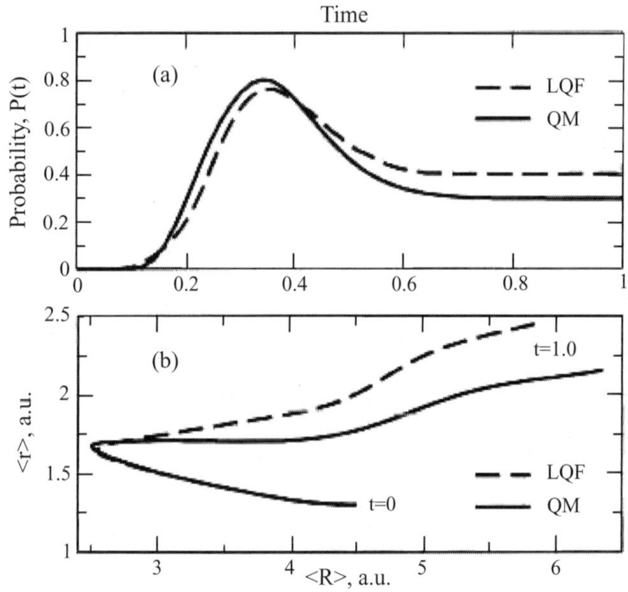

FIGURE 9.4. Quantum trajectory results for the collinear H+H_2 reaction [9.5]: (a) time-dependent reaction probability; (b) time dependence of the average position parameters. In each plot, the exact quantum-mechanical results (QM) are compared with results from the linear force approximation (LQF). The initial relative momentum is 9.5 a.u. and one unit of time is 918 a.u.

9.7 Local Fit to the Log Derivative of the Density

In the *local fitting approach* to the determination of an approximate quantum force, the first step is to divide the coordinate space into a series of *domains* (labeled $l = 1, 2, \ldots, L$) [9.6]. The log derivative of the density is then fit separately in each domain. Associated with each domain is a space-fixed "*domain function*" $\Omega_l(x)$, that may be chosen to reach a relatively large value in the specified domain, but which may extend into other domains. As an example, Gaussian functions can be used: $\Omega_l(x) = \exp(-\beta(x - q_l)^2)$. The parameters in the function may be based on the shape of the density or on physically defined regions of a potential surface (such as reactants, products, and transition state). Similar to the shape functions used in both least squares and finite element analysis, the domain functions have the *partition of unity property*

$$\sum_{l=1}^{L} \Omega_l(x) = 1. \tag{9.27}$$

A simple way to ensure that this property is satisfied is to choose one of the functions, the *complementary function* in domain L, to have the property

$$\Omega_L(x) = 1 - \sum_{l=1}^{L-1} \Omega_l(x). \tag{9.28}$$

Even when the domain functions in equation 9.27 are Gaussians, the complementary function $\Omega_L(x)$ will *not* have the Gaussian shape.

The proper use of domain functions in evaluating the quantum potential is suggested by examining matrix elements of the kinetic energy operator. For functions vanishing at $x = \pm\infty$, integration by parts can be used to convert the usual "second derivative" form of the matrix element into another form where the integrand is a "double first derivative"

$$\frac{\hbar^2}{2m}\int \phi_i^*(x)\phi_j(x)''dx = \frac{\hbar^2}{2m}\int \phi_i'(x)^*\phi_j'(x)dx. \tag{9.29}$$

There is a modification of this relationship (again obtained by integration by parts) that comes into play when a domain function is incorporated:

$$\int \phi_i'(x)^*\phi_j'(x)\Omega_l(x)dx = -\int \Omega_l(x)\phi_i^*(x)\left[\frac{\partial^2}{\partial x^2} + \frac{\Omega_l'(x)}{\Omega_l(x)}\frac{\partial}{\partial x}\right]\phi_j(x)dx. \tag{9.30}$$

Using this result as a guide, the form of the quantum potential for one domain is given by

$$Q_l(x,t) = \frac{\hbar^2}{2m}\frac{1}{R(x,t)}\left[\frac{\partial^2}{\partial x^2} + \frac{\Omega_l(x)'}{\Omega_l(x)}\frac{\partial}{\partial x}\right]R(x,t). \tag{9.31}$$

The parameters in the fitting function $g_l(x;s)$ are then evaluated *separately* for each domain. These parameters are determined by minimizing the functional

$$J_l(s) = \int (\mu(x) - g_l(x;s))^2\, \Omega_l(x)\rho(x)dx, \tag{9.32}$$

in which $\mu(x)$ is again the log derivative of the density. The approximate quantum potential on this domain is then given by

$$Q(x) = -\frac{\hbar^2}{8m}\left[\left(2g(x;s)' + g(x;s)^2\right)\Omega_l(x) + 2g(x;s)\Omega_l'(x)\right], \tag{9.33}$$

and the total quantum potential is the sum over contributions from the overlapping domains

$$Q(x,t) = \sum_{l=1}^{L} Q_l(x,t). \tag{9.34}$$

In practice, linear fitting functions with different parameters can be used in each of a small number of domains, but this will not provide high accuracy near nodes.

The domain method for computation of the quantum force will be applied to a familiar example: the scattering of a Gaussian wave packet from an Eckart barrier.

In this study, viewed as a simplified model for the H_3 system, scaled atomic units are used in which the reduced mass of the hydrogen molecule is set to unity, but distances are measured in atomic units. In these units, the parameter values $V_0 = 16$ and $\alpha = 1.36$ are used in the Eckart potential $V(x) = V_0 \cosh(\alpha x)^{-2}$. The wave packet was launched from the initial position $x_0 = -2$ a.u., and the barrier was centered at the origin. The initial translational energy of the wave packet was about one-half of the barrier height.

In one representative calculation, five equally spaced Gaussian domain functions were centered at the positions $q_1 = -2.0, \quad q_2 = -1.5, \ldots q_5 = 0.0$. (Adjacent domain functions have an overlap of about 0.55.) In figure 9.5 (a), the sum of

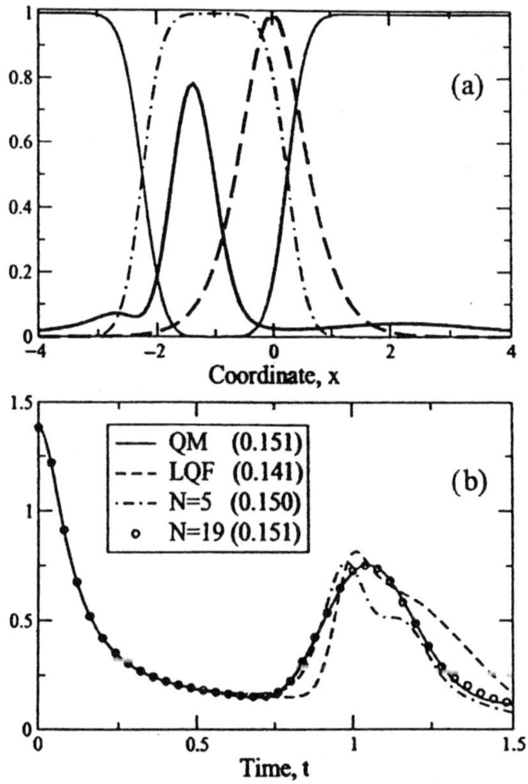

FIGURE 9.5. Results for scattering of an initial Gaussian wave packet from an Eckart barrier [9.6]. (a) The barrier, centered at the origin, is shown by the dashed curve. The sum of five domain functions is shown by the dash–dot curve, and the complementary domain function (approaching unity at large displacements from the barrier) is shown by the thin solid curve. The density associated with the scattering wave function (centered near the point $x = -1.5$) at $t = 0.8$ is shown by the thick solid curve. (b) The time correlation function (density overlap) is shown for the exact quantum calculation, for calculations employing either 6 or 20 domain functions, and for the global linear quantum force (LQF) calculation. The numbers in parentheses are the reaction probabilities for these four calculations.

these five domain functions is shown along with the complementary function $\Omega_6(x)$. The scattering probability density at one time step is also shown. The first five domain functions are concentrated on the left side of the barrier, where ripples form in the density at later times due to reflection from the barrier. The complementary function, clearly not having a Gaussian shape, approaches unity in the limits $x \to \pm\infty$.

Two quantities were calculated from the density carried by the quantum trajectories. The first of these, the *reaction probability* (the total probability on the product side of the barrier), was computed from the equation

$$P = \lim_{t\to\infty} \sum_{i=1}^{N} w_i H(x_i(t)), \tag{9.35}$$

in which w_i is the trajectory weight (see Section 4.5), and $H(x)$ is the Heaviside step function ($H(x) = 1$ if $x \geq 1$ and $H(x) = 0$ otherwise). For calculations using the six domain functions described above, the reaction probability was 0.150, in good agreement with the exact quantum-mechanical value of 0.151. When 20 Gaussian domain functions were used, the reaction probability agreed with the exact result, $P = 0.151$. These results are better than those obtained using a global fit to the linear quantum force, where the reaction probability was 0.141.

The second quantity computed from the trajectory information was the *density correlation function*, defined by $C(t) =< \rho(0)|\rho(t) >$. This function measures the overlap of the time-evolving density with the initial density on the left side of the barrier. The time dependence of this function is shown in part (b) of figure 9.5. This correlation function drops from an initial value of unity and then experiences a recurrence near $t = 1$ as the reflected wave packet passes over the region occupied by the initial packet. The result obtained with 20 domain functions is in excellent agreement with the exact quantum result. The correlation function obtained using the global fit to the linear quantum force also shows a recurrence peak, although the result is not as accurate as those obtained using multiple domain functions.

9.8 Conclusions

In this chapter, several new strategies were explored for approximately determining the quantum potential and quantum force associated with an ensemble of quantum trajectories. In order to model the quantum force, these methods fit a parameterized function to either the density or the log derivative of the density. Moreover, because these fitting procedures are formulated in terms of simple sums over information carried by the quantum trajectories, they can be extended to higher dimensionality and implemented on parallel computers. It would be of interest to explore the use of higher order-fitting models that can possibly account for more complicated features in the density, such as ripples and nodal structures.

A local fitting method based on the use of a set of domain functions was also described in this chapter. This method has been applied to the collinear hydrogen exchange reaction and to several one-dimensional problems involving anharmonic potentials. An important feature is that the fitting can be performed independently for each domain and for each degree of freedom. Extensions to wave packet scattering in higher dimensionality would be of interest.

As demonstrated for methyl iodide, adding a small viscous drag to the equations of motion for the quantum trajectories slowly removes excess kinetic energy from the system. After a sufficient equilibration time, the ensemble settles into the ground state density and can be used to compute properties such as the zero-point energy and other expectation values. This approach may be useful for computing the ground state vibrational density in high-dimensional systems, such as weakly bound van der Waals clusters and Bose–Einstein condensates.

It would also be worthwhile to investigate the use of Gaussian fitting methods in connection with "smoothed" phase space distributions, such as the Husimi distribution (which is described later, in Section 11.3). Since Husimi functions are generally less structured than the corresponding density matrices and Wigner functions, the Gaussian density approximation may provide an accurate representation for the phase space dynamics. Such a scheme could be used for examining mixed states and quantum dissipation in phase space.

Before leaving the main topic of this chapter, we mention that LaGattuta has used what he refers to as a "drastic approximation" to the quantum potential for Coulombic systems to study the dynamics of systems of atoms, ions, and electrons, all interacting with one another [9.9].

References

9.1. J.B. Maddox and E.R. Bittner, *Estimating Bohm's quantum force using Bayesian statistics*, J. Chem. Phys. **119**, 6465 (2003).

9.2. S. Garashchuk and V.A. Rassolov, *Semiclassical dynamics based on quantum trajectories*, Chem. Phys. Lett. **364**, 562 (2002).

9.3. S. Garashchuk and V.A. Rassolov, *Semiclassical dynamics with quantum trajectories: Formulation and comparison with the semiclassical initial value representation propagator*, J. Chem. Phys. **118**, 2482 (2003).

9.4. S. Garashchuk and V.A. Rassolov, *Quantum dynamics with Bohmian trajectories: Energy conserving approximation to the quantum potential*, Chem. Phys. Lett. **376**, 358 (2003).

9.5. S. Garashchuk and V.A. Rassolov, *Energy conserving approximations to the quantum potential: Dynamics with linearized quantum force*, J. Chem. Phys. **120**, 1181 (2004).

9.6. V.A. Rassolov and S. Garashchuk, *Bohmian dynamics on subspaces using linearized quantum force*, J. Chem. Phys. **120**, 6815 (2004).

9.7. N. Gershenfeld, *The Nature of Mathematical Modeling* (Cambridge University Press, Cambridge, 1999).

9.8. M. Shapiro and R. Bersohn, *Vibrational energy distribution of the CH_3 radical photodissociated from CH_3I*, J. Chem. Phys. **73**, 3810 (1980).

9.9. K. LaGattuta, *Introduction to the quantum trajectory method and to Fermi molecular dynamics*, J. Phys. A **36**, 6013 (2003).

10
Derivative Propagation Along Quantum Trajectories

Two methods are introduced for propagating spatial derivatives along quantum trajectories. Truncation of the infinite hierarchy of equations of motion leads to approximate quantum trajectories that can be run one at a time.

10.1 Introduction

The applications of the hydrodynamic formulation described in the preceding chapters used various fitting techniques (whose accuracy may be difficult to assess and control) to evaluate spatial derivatives of the fields around each fluid element in the evolving ensemble. In this chapter, two related but distinct methods [10.1, 10.12] are described that *circumvent the use of fitting techniques to evaluate spatial derivatives*. The first of these methods, reported by Trahan, Hughes, and Wyatt in 2003 [10.1], develops a set of exact analytic equations of motion for the spatial derivatives of C and S that appear in the Madelung–Bohm polar form for the wave function $\psi(r, t) = \exp(C(r, t) + iS(r, t)/\hbar)$. The various orders of derivatives are coupled together in an infinite hierarchy, but low-order truncations of this set may yield useful and accurate approximations. This method is referred to as the DPM, which stands for *derivative propagation method*. The second method, developed by Liu and Makri in 2004 [10.12], has a different starting point. Spatial derivatives of the density are derived from the conservation relation for the weight along the quantum trajectory (the weight, defined at the end of Section 4.5, is the product of the probability density and the volume element). These derivatives are evaluated in terms of density derivatives at the starting point for the trajectory, x_0, and the *coordinate sensitivities* $\partial^m x(t)/\partial x_0{}^m$, which in turn are evaluated from the *trajectory stability matrix* and its spatial derivatives. As in the DPM, the various orders of spatial derivatives are coupled together in an infinite hierarchy, and low-order truncations lead to approximate quantum trajectories. This method is referred to as the TSM, which stands for *trajectory stability method*.

Because function fitting is eliminated, use of the DPM or the TSM leads to orders of magnitude reduction in the propagation time compared with that for an ensemble of linked trajectories. Significantly, *quantum effects can be included at various orders of approximation, there are no basis set expansions, and single quantum trajectories (rather than correlated ensembles) may be propagated, one at a time*. For some problems, including barrier transmission, excellent computational results are obtained. However, both methods run into problems with interference effects, such as nodal regions on the "back side" of barrier transmission problems.

In the computational fluid mechanics literature, there is one method that bears a (small) overlap with these two derivative propagation methods. Aoki has developed a scheme, called IDO (interpolated differential operator) [10.2, 10.3], in which spatial derivatives are evaluated by applying the operator $\partial^n/\partial x^n$ to a Hermite interpolation function. Rather than introducing Lagrangian trajectories, the equations of motion are usually solved in the Eulerian frame. However, in one application, the simulation of an ensemble of falling leaves (no joke, this is a difficult problem!), the Poisson equation for the pressure was solved using the IDO scheme in a moving frame.

In Section 10.2, the one-dimensional version of the hydrodynamic equations of motion will be reviewed. In Section 10.3, the hierarchy of coupled differential equations for the spatial derivatives of C and S will be developed. In Section 10.4, details concerning computational implementation of the DPM will be described. Two examples involving wave packet scattering will be presented in Section 10.5, and comments concerning multidimensional extensions are in Section 10.6. In Section 10.7, the TSM is introduced, and this is followed by an example in Section 10.8. Comparisons between these methods and additional remarks are presented in Section 10.9.

10.2 Review of the Hydrodynamic Equations

Before proceeding to the DPM, we will first review some features of the hydrodynamic equations. We have already seen that nonlinear coupled partial differential equations for the amplitude C and the action function S are given by

$$\frac{\partial C}{\partial t} = -\frac{1}{2m}[\nabla^2 S + 2\nabla C \cdot \nabla S],$$
$$\frac{\partial S}{\partial t} = -\frac{1}{2m}\nabla S \cdot \nabla S + \frac{\hbar^2}{2m}[\nabla^2 C + \nabla C \cdot \nabla C] - V(r,t). \qquad (10.1)$$

The first equation is a version of the continuity equation, and the second is the quantum Hamilton–Jacobi equation (QHJ). In the QHJ equation, the term involving $\hbar^2$ is the negative of the Bohm quantum potential, which introduces all quantum effects into the dynamics. In addition, the flow momentum associated with the evolving density is given by the gradient $p(r,t) = \nabla S(r,t)$, and the probability density is $\rho(r,t) = \exp(2C(r,t))$. A final point is that these two equations are expressed in the Eulerian picture: an observer monitors the flow from a fixed vantage point. The view as seen by a moving observer will be described later.

For later use in Section 10.3, the equations of motion in one dimension are given by

$$\begin{aligned}\frac{\partial C}{\partial t} &= -\frac{1}{2m}\left[S_2 + 2C_1 S_1\right],\\ \frac{\partial S}{\partial t} &= -\frac{1}{2m}(S_1)^2 + \frac{\hbar^2}{2m}\left[C_2 + (C_1)^2\right] - V,\end{aligned} \tag{10.2}$$

where the order of the spatial derivative is denoted by the subscript, $S_n = \partial^n S/\partial x^n$, etc.

Trajectory representations, including the QTM, are obtained by discretization of the initial wave packet $\Psi(r, 0)$ into N fluid elements followed by evolution of the ensemble of coupled trajectories. The trajectories are developed through integration of the equation of motion $dr/dt = (1/m)\nabla S(r, t)$. We have already seen that the spatial derivatives involved in the gradients and curvatures in equation 10.1 are frequently computed through least squares polynomial fitting of the fields C and S on the unstructured mesh defined by the locations of the trajectories. In all previous studies, an ensemble of correlated trajectories was integrated in a lockstep fashion; correlation between the fluid elements was induced by the quantum potential, which explicitly incorporates nonlocality into the dynamics. We are now ready to introduce an alternative method for evaluating the spatial derivatives required for the equations of motion.

10.3 The DPM Derivative Hierarchy

An alternative to propagation of an ensemble of correlated fluid elements is obtained by propagation of the required spatial derivatives along individual trajectories. This derivative propagation method was described by Trahan, Hughes, and Wyatt in 2003 [10.1]. In the DPM, the equations of motion for the spatial derivatives are obtained by *direct differentiation* of equations 10.1, with repeated application of the chain rule for differentiating the product of two functions. In order to provide a simple illustration, we will directly differentiate the hydrodynamic equations of motion in one dimension. By differentiating equations 10.2 two times with respect to x, we obtain

$$\begin{aligned}\frac{\partial C_1}{\partial t} &= -\frac{1}{2m}\left[S_3 + 2C_1 S_2 + 2C_2 S_1\right],\\ \frac{\partial C_2}{\partial t} &= -\frac{1}{2m}\left[S_4 + 2C_1 S_3 + 4C_2 S_2 + 2C_3 S_1\right],\\ \frac{\partial S_1}{\partial t} &= -\frac{1}{2m}\left[2S_1 S_2\right] + \frac{\hbar^2}{2m}\left[C_3 + 2C_1 C_2\right] - V_1,\\ \frac{\partial S_2}{\partial t} &= -\frac{1}{2m}\left[2S_1 S_3 + 2(S_2)^2\right] + \frac{\hbar^2}{2m}\left[C_4 + 2C_1 C_3 + 2(C_2)^2\right] - V_2.\end{aligned} \tag{10.3}$$

In these equations, not only is there back-coupling from higher derivatives to lower derivatives, but there is additional up-coupling; for example, $\partial C_2/\partial t$ is coupled to the higher derivatives S_4, S_3, and C_3. As a result, unless the higher derivatives vanish for both C and S (as would be the case for a Gaussian wave packet launched on a harmonic potential, for example), the four equations given above form the beginning of an infinite hierarchy of coupled equations for the spatial derivatives. That is the bad news. The good news is that these equations are easy to derive, just by applying $\partial/\partial x$ over and over, and they are exact. However, there remains the issue of how to deal with an infinite number of coupled equations.

Before describing how to deal with this hierarchy of equations, we will develop general equations for spatial derivatives of any order. In order to differentiate terms involving the product of two functions in equations 10.2, we employ the Leibniz theorem. This theorem gives the n-th derivative of the product of two functions, $H(x) = F(x)G(x)$:

$$H_n = \sum_{j=0}^{n} b(n, j) F_{(n-j)} G_j, \tag{10.4}$$

where the binomial coefficient is given by $b(n, j) = n!/\left[j!(n-j)!\right]$. As an example, the fourth derivative of this product function is given by the sum of six terms:

$$H_4 = F_0 G_4 + 4F_1 G_3 + 6F_2 G_2 + 4F_3 G_1 + F_4 G_0. \tag{10.5}$$

Using the Leibniz relation, we obtain from equations 10.2 the following equations of motion for the spatial derivatives:

$$\begin{aligned}
\frac{\partial C_n}{\partial t} &= -\frac{1}{2m}\left[S_{(n+2)} + 2\sum_{j=0}^{n} b(n, j) S_{(1+j)} C_{(1+n-j)}\right], \\
\frac{\partial S_n}{\partial t} &= -\frac{1}{2m}\left[\sum_{j=0}^{n} b(n, j) S_{(1+j)} S_{(1+n-j)}\right] \\
&\quad + \frac{\hbar^2}{2m}\left[C_{(n+2)} + \sum_{j=0}^{n} b(n, j) C_{(1+j)} C_{(1+n-j)}\right] - V_n.
\end{aligned} \tag{10.6}$$

Again, we note the presence of both up-coupling and down-coupling in both equations.

In order to make progress in dealing with the hierarchy, we will now assume that the hydrodynamic fields are smooth enough to be well approximated by low-order polynomial expansions in the displacements around a trajectory. If we let ξ denote the displacement from a trajectory at time t, then we are assuming that C and S can be approximated by

$$C(\xi, t) = \sum_{k=0}^{K} \frac{1}{k!} c_k(t) \xi^k, \qquad S(\xi, t) = \sum_{k=0}^{L} \frac{1}{k!} s_k(t) \xi^k. \tag{10.7}$$

Substitution of these equations into equations 10.3 followed by taking the limit $\xi \to 0$ then gives a finite number of equations of motion, $(K + 1)$ or $(L + 1)$, for the expansion coefficients. For example, quadratic expansions ($K = L = 2$) lead to a system of six coupled equations. It is not necessary that $K = L$; for example, zero order in C and first order in S will give a time-dependent WKB-type approximation in which the fluid element follows a classical trajectory along which both C and S are independent of $\hbar$.

In order to be more specific, assume quadratic expansions for both C and S,

$$\begin{aligned} C(\xi, t) &= c_0(t) + c_1(t)\xi + \frac{1}{2}c_2(t)\xi^2, \\ S(\xi, t) &= s_0(t) + s_1(t)\xi + \frac{1}{2}s_2(t)\xi^2. \end{aligned} \tag{10.8}$$

substitute these expressions into equations 10.2 and 10.3 and then let $\xi \to 0$. The six coupled equations for the expansion coefficients are

$$\begin{aligned} \frac{\partial c_0}{\partial t} &= -\frac{1}{2m}\left[s_2 + 2c_1 s_1\right], \\ \frac{\partial s_0}{\partial t} &= \frac{1}{2m}(s_1)^2 + \frac{\hbar^2}{2m}\left[c_2 + (c_1)^2\right] - V, \\ \frac{\partial c_1}{\partial t} &= -\frac{1}{2m}\left[2c_1 s_2 + 2c_2 s_1\right], \\ \frac{\partial s_1}{\partial t} &= -\frac{1}{2m}\left[2s_1 s_2\right] + \frac{\hbar^2}{2m}\left[2c_1 c_2\right] - V_1, \\ \frac{\partial c_2}{\partial t} &= -\frac{1}{2m}\left[4c_2 s_2\right], \\ \frac{\partial s_2}{\partial t} &= -\frac{1}{2m}\left[2(s_2)^2\right] + \frac{\hbar^2}{2m}\left[2(c_2)^2\right] - V_2. \end{aligned} \tag{10.9}$$

These equations have exactly the same form as the starting equations, equations 10.2 and 10.3, after the higher derivatives C_n and S_n (for $n \geq 3$) are set to zero. *Truncation of the derivative hierarchy is thus equivalent to assuming polynomial smoothness at some level.*

A few additional comments can be made about equations 10.2 and 10.9. These are coupled, nonlinear differential equations expressed in the Eulerian frame. Conversion of the time derivatives to those appropriate for motion along an arbitrary path $r(t)$ is made through the relation $(d/dt) = (\partial/\partial t) + \dot{r}(t) \cdot \nabla$, where $\dot{r}$ in the last (convective) term is the *path velocity*. This equation has been used in some of the previous chapters to convert derivatives to the moving frame. For example, from equation 10.2 for the one-dimensional case, the n-th partial derivatives of C and S along the path $x(t)$ change at the rates

$$\begin{aligned} \frac{dC_n}{dt} &= -\frac{1}{2m}\left[S_{2+n} + 2(C_1 S_1)_n\right] + \dot{x}(t)C_{1+n}, \\ \frac{dS_n}{dt} &= -\frac{1}{2m}((S_1)^2)_n + \frac{\hbar^2}{2m}\left[C_{2+n} + ((C_1)^2)_n\right] - V_n + \dot{x}(t)S_{1+n}. \end{aligned} \tag{10.10}$$

For the special but important case of a Lagrangian path, the velocity is gradient driven and matches that of the fluid, $\dot{r}(t) = (1/m)\nabla S$.

With quadratic expansions for $C(x, t)$ and $S(x, t)$, the wave function synthesized around each fluid element is a local Gaussian. However, this does not imply that the global wave function is of Gaussian form. Beginning with the studies of Heller that began over 25 years ago [10.4], frozen or thawed (fixed or variable width, respectively) Gaussian wave packets have been used in many semiclassical studies of time-dependent processes. A significant difference between the latter studies and the DPM is that we use quadratic expansions (or higher, if necessary) for the amplitude and phase of the wave function *propagated along each trajectory*, rather than for the global wave function. In this sense, the DPM is an extension of Heller's earlier studies.

Within the hydrodynamic formulation, a different infinite hierarchy of equations has recently been described [10.5, 10.6]. In the studies of Burghardt and Cederbaum, position space equations of motion were derived for momentum moments of the Wigner function. This approach to the hydrodynamic equations was described earlier, in Chapter 3. For pure states, the hierarchy terminates at the second moment, but the formalism is also applicable to mixed states, where (in general) all moments are coupled. The hierarchy described in these studies is different from that developed in the DPM in that we propagate coupled equations for spatial derivatives rather than momentum moments of a phase space distribution function.

10.4 Implementation of the DPM

In order to build a computer program to run the DPM, the following steps can be followed. To be specific, assume that we are operating at the quadratic level, so that the six functions and derivatives, denoted by $\Phi = \{C, C_1, C_2, S, S_1, S_2\}$, are computed along each trajectory at each time step. We will specify the initial conditions at the end of this section. For now, assume that one trajectory has been followed for n-time steps, where the time step is Δt. At the current time step, the position and momentum for the trajectory are x and $p = \nabla S = S_1$, respectively. In order to advance one time step, the following procedure can be followed:

Step 1. Using information provided by the set Φ, compute the set of Eulerian time derivatives $\partial\Phi/\partial t$. Equations 10.2 and 10.3 are used for this purpose.

Step 2. Convert the time derivatives to the moving frame using equations 10.10.

Step 3. Using the current functions Φ and the derivatives in the moving frame, update the set Φ. In the simplest integration scheme, first-order Euler, for example, $C_n(t + \Delta t) = C_n(t) + (dC_n/dt)\Delta t$. A higher-order integrator should be used for generating the new set Φ from the values at the beginning of the time step.

Step 4. Update the trajectory. Again, for simplicity, using the first-order Euler integrator, the new values are given by

$$x_{new} = x_{old} + (p/m) \cdot \Delta t \quad \text{and} \quad p_{new} = p_{old} + F \cdot \Delta t, \tag{10.11}$$

where the force is $F = -(\partial V/\partial x)_{xnew}$. Again, a higher-order integrator should be used for this update.

Step 5. Both the trajectory and the set of functions and derivatives Φ have been updated; now return to *Step 1* to advance one more time step.

The conditions on the functions and derivatives at the start of the trajectory are determined from the initial wave function. Assuming a normalized Gaussian wave packet given by

$$\psi(x, t = 0) = (2\beta/\pi)^{1/4}\, e^{-\beta(x-x_0)^2} e^{ik_0 x}, \tag{10.12}$$

the C and S amplitudes are given by

$$\begin{aligned} C(x, t = 0) &= \ln(2\beta/\pi)^{1/4} - \beta(x - x_0)^2, \\ S(x, t = 0) &= \hbar k_0 x. \end{aligned} \tag{10.13}$$

As a result, the only nonzero x-derivatives are

$$\begin{aligned} C_1 &= -2\beta(x - x_0), \\ C_2 &= -2\beta, \\ S_1 &= \hbar k_0. \end{aligned} \tag{10.14}$$

Given the values in equations 10.13 and 10.14, the trajectory may be launched from position x at $t = 0$.

10.5 Two DPM Examples

In order to demonstrate some features of the DPM, computational results will be presented for two model problems. The first of these concerns the decay of a wave packet launched from the quasi-bound region of the potential, $V(x) = (1/2)kx^2 - \gamma kx^3$. (This model was solved recently using an ensemble of coupled trajectories in phase space [10.7]). The force constant $k = 4.05 \cdot 10^{-3}$ a.u. is chosen so that the harmonic term reaches 2000 cm^{-1} when $x = 1.5$. In addition, the value $\gamma = 0.14$ was used (unless specified otherwise, atomic units are used). The resulting potential, plotted in figure 10.1, displays a near-harmonic bowl around $x = 0$ and reaches a local maximum near $x = 2.4$, where the potential is 1680 cm^{-1}. The potential drops to $V = 0$ at $x_0 = 3.57$ and becomes increasingly negative as x increases. The center of the initial Gaussian is $x_0 = -0.5$, the width parameter multiplying $(x - x_0)^2$ in the exponent is $\beta = 6.0$, the initial translational energy is set to zero, and the mass is $m = 2000$.

In this example, the aim is to compute the time correlation function $C(t) = < \psi(t)|\phi >$, where the test function $\phi(x)$ is chosen as a delta function at x_0. In effect, we are computing the time dependence of the wave function at a single fixed point. For this purpose, we will integrate the equations of motion for the derivatives in the Eulerian, fixed in space, representation. Using the third-order DPM, eight coupled equations of motion are integrated for C and S and their first three spatial derivatives. The resulting correlation function shown in figure 10.2 is

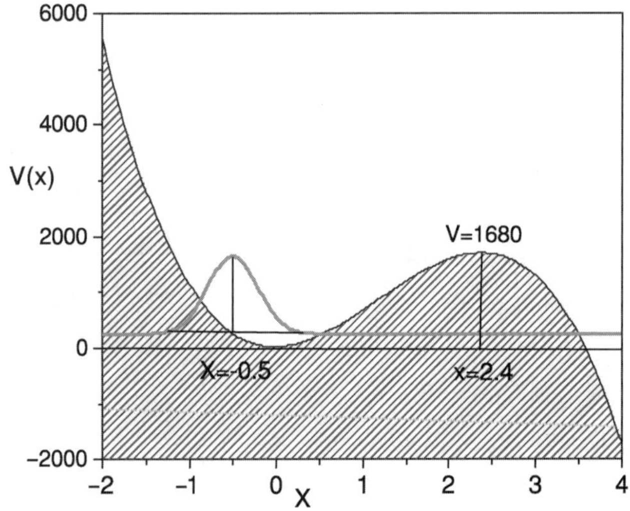

FIGURE 10.1. The quadratic plus cubic potential used to study the decay of a metastable state (the potential is in cm^{-1}). The initial Gaussian wave function (multiplied by 10^3 and then shifted up by 253 cm^{-1})is also shown.

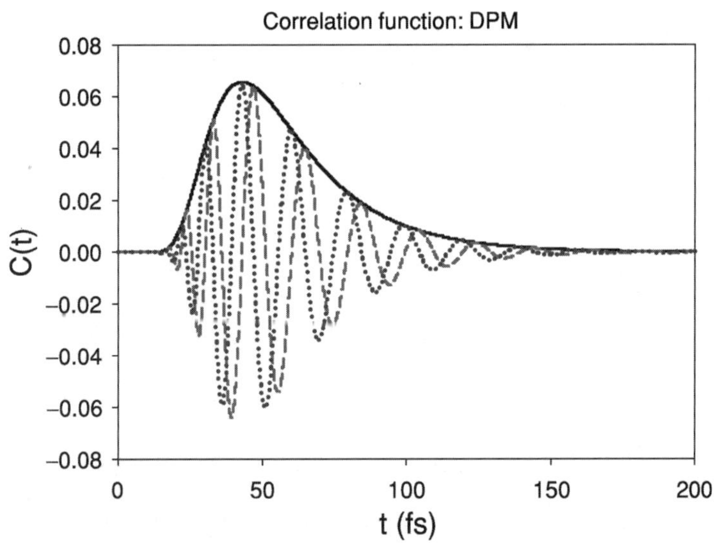

FIGURE 10.2. Correlation function calculated at one point by the Eulerian version of the DPM [10.1]. The initial Gaussian wave packet evolves on the potential shown in figure 10.1. The real and imaginary parts of $C(t)$ are shown by dashed and dotted lines, respectively.

in excellent agreement with results obtained using traditional global wave packet propagation on a fixed grid of 3000 points. Use of the DPM thus permitted the correlation function to be obtained through the ultimate compactification of the Eulerian grid to a single point (at the position of the test function)! However, this type of calculation does not always work at one test point.

In the second application of the DPM, the energy-resolved transmission probability $P(E)$ will be computed from the time-dependent scattering of a wave packet from a repulsive barrier. Equations relating time-dependent scattering to energy-resolved quantities were developed and applied by Tannor and Weeks [10.8]. This topic is also described in the books by Tannor [10.9] and by Zhang [10.10]. Tannor and Weeks showed that $P(E)$ may be computed from the Fourier transform of the cross-correlation function

$$P(E) = \left(\frac{k}{m}\right)^2 \sqrt{\frac{\beta}{2\pi}} e^{(k-k_0)^2/(2\beta)} \left| \int_0^\infty C_{\alpha\beta}(t) e^{iEt} dt \right|^2 . \tag{10.15}$$

The wave number corresponding to the translational energy E is given by $k = \sqrt{2mE/\hbar^2}$, k_0 is the wave number corresponding to the initial wave packet translational energy E_0, and the width parameter for the initial Gaussian is β. The *cross-correlation function* $C_{\alpha\beta}(t)$ between the wave packet $\Psi_\alpha^{(+)}(x,t)$ launched from the reactant region ($x \to -\infty$) and a stationary target function $\phi_\beta^{(-)}(x) = \delta(x - x_0)\exp(-ikx)$ located at position x_0 on the product side of the barrier (which is centered at $x = 0$) is given by $< \Psi_\alpha^{(+)} | \phi_\beta^{(-)} >$. In this example, an Eckart potential is used, $V(x) = V_0/\cosh^2(\kappa x)$, with the barrier height $V_0 = 3000$ cm^{-1} and the width parameter $\kappa = 1.2$. The initial Gaussian wave packet, with translational energy $E_0 = 3000$ cm^{-1}, was centered at $x_c = -6.0$, and the test function used to monitor the wave function in the product region was located at $x_0 = 5.0$.

Quantum trajectories were fired, one at a time, toward the barrier from initial values of x in the interval $x^* \le x \le x_{front}$, where x_{front} is a point on the leading edge of the wave packet where the density is very low. The starting point for each successive trajectory was then moved back, away from the barrier, until the bifurcation point x^* was found. For all starting positions $x > x^*$, the trajectories make it over the barrier. However, for starting positions $x < x^*$, the trajectories evolve to form the nonreactive portion of the wave packet. For these calculations, $x_{font} = -3.5$, and it was found that $x^* \approx -6.0001$. The trajectories launched from within this reactive zone yielded the cross correlation function over the time interval 14.0 fs to 147.0 fs.

Using second-order DPM trajectories, the cross-correlation function for this model was computed and this is shown in figure 10.3. The peak of the transmitted wave packet passes the monitor point in the product channel about 70 fs after the packet was launched from the reactant side of the barrier. Then, from the Fourier transform of $C_{\alpha\beta}(t)$, $P(E)$ was calculated using equation 10.15. The analytic transmission probability for the Eckart barrier (derived in the text by

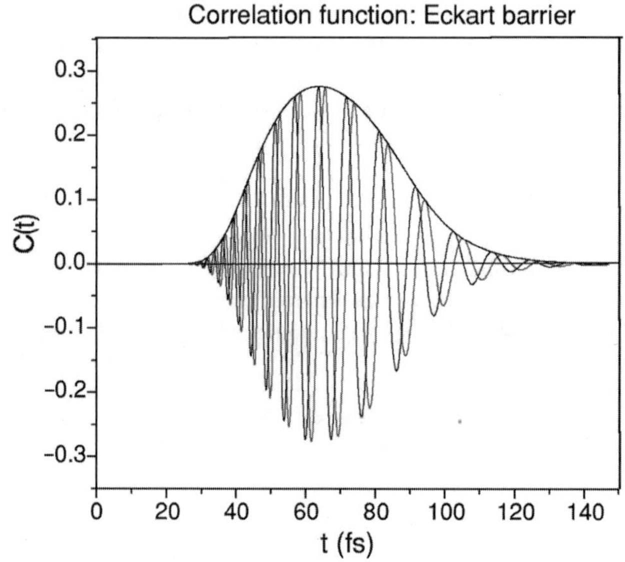

FIGURE 10.3. Time-dependent cross-correlation function for the Eckart barrier calculated using the second-order DPM [10.1]. The trajectories were monitored at $x_0 = +5.0$ after being launched from the reactive zone, $-6.0001 \leq x \leq -3.5$, on the reactant side of the barrier. The two oscillating curves show the real and imaginary parts, and the envelope shows the magnitude of the correlation function.

Landau and Lifshitz [10.11]) is shown in figure 10.4. This figure also shows the DPM results for both second- and third-order expansions. The DPM curves capture the energy dependence of the transmission probability, including the low-energy tunneling region, but the second-order DPM results slightly overestimate $P(E)$ in the energy range 3200–4500 cm^{-1}. However, the curve obtained using third order DPM is in good quantitative agreement with the analytic result.

10.6 Multidimensional Extension of the DPM

The DPM may be readily extended to evolution in higher dimensionality, although the resulting equations are necessarily more complicated than those presented earlier. For example, in two degrees of freedom, the partial derivative of the dot product of two gradients, $\nabla S \cdot \nabla C = S_{(1,0)}C_{(1,0)} + S_{(0,1)}C_{(0,1)}$, is given by

$$(\nabla S \cdot \nabla C)_{(n,m)} = \sum_{j=0}^{n}\sum_{k=0}^{m} b(n,j)b(m,k)[S_{(1+j,k)}C_{(1+n-j,m-k)} + S_{(j,1+k)}C_{(n-j,1+m,k)}], \tag{10.16}$$

where the following notation is used for the partial derivatives: $S_{(n,m)} = \partial^n \partial^m S(x_1, x_2, t)/\partial x_1^n \partial x_2^m$. In terms of the derivative in equation 10.16, equations

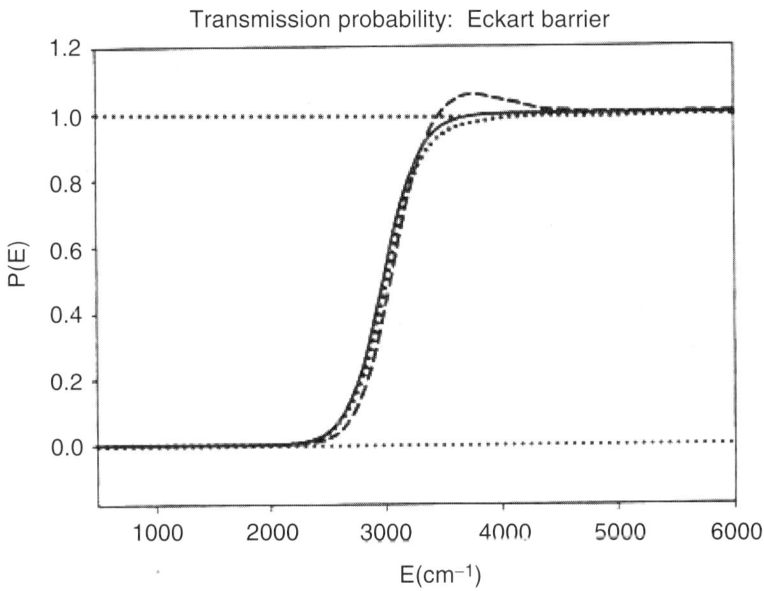

FIGURE 10.4. Energy-resolved transmission probabilities for the Eckart barrier (barrier height $V_0 = 3000\,\text{cm}^{-1}$) [10.1]. Second-order (dashed curve) and third-order (dotted curve) DPM results are compared with the analytic result (solid curve).

of motion for the derivatives of C and S are

$$\begin{aligned}
\partial_t C_{(n,m)} &= -\frac{1}{2m}\left[S_{(2+n,m)} + S_{(n,2+m)} + 2(\nabla S\cdot\nabla C)_{(n,m)}\right],\\
\partial_t S_{(n,m)} &= -\frac{1}{2m}(\nabla S\cdot\nabla S)_{(n,m)} + \frac{\hbar^2}{2m}\left[C_{(2+n,m)} + C_{(n,2+m)} + (\nabla C\cdot\nabla C)_{(n,m)}\right]\\
&\quad -V_{(n,m)}.
\end{aligned} \tag{10.17}$$

These equations, and their multidimensional extensions, are readily programmed.

Of course, because of the increasing number of derivatives, the computational cost for each trajectory increases with the dimensionality. For example, using quadratic expansions for C and S in D dimensions, there are $(D+1)(D+2)/2$ equations of motion for each function and its derivatives. As an example, we will list the 14 derivatives for C in $D = 4$ dimensions:

$$\begin{aligned}
&C_{(1,0,0,0)},\ C_{(0,1,0,0)},\ C_{(0,0,1,0)},\ C_{(0,0,0,1)},\\
&C_{(2,0,0,0)},\ C_{(0,2,0,0)},\ C_{(0,0,2,0)},\ C_{(0,0,0,2)},\\
&C_{(1,1,0,0)},\ C_{(1,0,1,0)},\ C_{(1,0,0,1)},\\
&C_{(0,1,1,0)},\ C_{(0,1,0,1)},\\
&C_{(0,0,1,1)}.
\end{aligned}$$

At the second-order level, there will thus be 15 equations of motion for C and its 14 derivatives.

The use of DPM expansions beyond the quadratic, or possibly the cubic, level is probably not practical in high dimensionality because of the large number of derivatives that need to be propagated. However, implementation at the "cheap" quadratic level dresses what would otherwise be a bare classical trajectory with an approximate quantum potential and its derivatives. Propagation at the quadratic level is feasible, and as an example, this approach has been implemented in a scattering program that handles 10 degrees of freedom.

10.7 Propagation of the Trajectory Stability Matrix

In the DPM, we have seen that the wave function amplitude and phase along with their low order spatial derivatives are propagated along individual quantum trajectories. From these quantities, the complex-valued wave function may be synthesized. A distinct, though related, derivative propagation scheme has been implemented by Liu and Makri [10.12]. The probability density is propagated along quantum trajectories, but the action function is not explicitly considered and the wave function is not synthesized. In order to propagate the density and its spatial derivatives, the Jacobian and its derivatives are evaluated along individual trajectories. The latter quantities are evaluated through integration of the *trajectory stability matrix* and its derivatives. In order to propagate the quantum trajectory, the quantum force is required, and this in turn is also evaluated in terms of elements of the stability matrix. In common with the DPM, this scheme leads to an infinite hierarchy of coupled equations that are truncated at low order. The low order scheme approximately captures quantum effects and allows for one-at-a-time propagation of quantum trajectories. Overall, this method and the DPM have in common both positive and negative features; these aspects will be listed later, in Sec. 10.9.

The starting point for development of what we will term the *trajectory stability method* (TSM) is the *conservation relation for the weight evaluated along the trajectory*

$$\rho(x,t)dx(t) = \rho(x_0,0)dx(0) = \rho_0(x_0)dx(0). \tag{10.18}$$

Along the trajectory $x(t)$, launched from the starting point x_0, the increment $dx(t)$ may increase or decrease depending on whether neighboring trajectories are diverging or converging. From this equation, we obtain the evolved density in terms of the initial density and the inverse of the Jacobian evaluated along the trajectory

$$\rho(x,t) = \rho(x_0,0)\frac{\partial x(0)}{\partial x(t)} = \rho(x_0,0)J(x(t),x(0))^{-1}. \tag{10.19}$$

Since the quantum force may be expressed in terms of spatial derivatives of the density, our first task in to evaluate the derivatives of the density. In order to do this, we repeatedly apply $\partial/\partial x$ to both sides of equation 10.19, thus leading to the first few spatial derivatives

$$\rho(x,t)' = \rho_0(x_0)' \left(\frac{\partial x(0)}{\partial x(t)}\right)^2 + \rho_0(x_0)\frac{\partial^2 x(0)}{\partial x(t)^2}, \tag{10.20}$$

$$\rho(x,t)'' = \rho_0(x_0)'' \left(\frac{\partial x(0)}{\partial x(t)}\right)^3 + 3\rho_0(x_0)'\frac{\partial x(0)}{\partial x(t)}\frac{\partial^2 x(0)}{\partial x(t)^2} + \rho_0(x_0)\frac{\partial^3 x(0)}{\partial x(t)^3}. \tag{10.21}$$

In these equations, spatial derivatives of the density at time t are evaluated in terms of derivatives of the density *at the starting point,* along with derivatives of the Jacobian. The derivatives of the initial density are known from the specified form of the initial wave packet.

The derivatives of the initial position with respect to the final position are eval uated in terms of the inverse of the Jacobian and its derivatives. The first few of these derivatives are listed below:

$$\begin{aligned} \frac{\partial x(0)}{\partial x(t)} &= \left(\frac{\partial x(t)}{\partial x(0)}\right)^{-1} = J^{-1}, \\ \frac{\partial^2 x(0)}{\partial x(t)^2} &= \left(\frac{\partial x(t)}{\partial x(0)}\right)^{-3} \frac{\partial^2 x(t)}{\partial x(0)^2} = -\frac{1}{J^3}\frac{\partial J}{\partial x(0)}. \end{aligned} \tag{10.22}$$

Evaluation of these quantities thus involves calculation of derivatives of the final position reached by the trajectory with respect to the initial position. This brings us to consideration of the *trajectory stability matrix* (TSM).

The TSM is well known in classical dynamics, where it is used to evaluate *sensitivities* of the final position and momentum with respect to variations of the initial position and momentum. The TSM is frequently used to evaluate trajectory stability around fixed points, to see whether they diverge, converge, or oscillate about this point. There are four sensitivity coefficients, and these are arranged to form the 2×2 *stability (or monodromy) matrix*

$$M(t) = \begin{bmatrix} \frac{\partial x(t)}{\partial x(0)} & \frac{\partial x(t)}{\partial p(0)} \\ \frac{\partial p(t)}{\partial x(0)} & \frac{\partial p(t)}{\partial p(0)} \end{bmatrix}, \tag{10.23}$$

in which each matrix element involves the rate of change of the "new" coordinate with respect to the initial coordinate. Especially important is that the (1,1) element of this matrix is the Jacobian that we are seeking.

The stability matrix evolves in time according to the first-order matrix differential equation

$$\frac{dM(t)}{dt} = T(t)M(t). \tag{10.24}$$

The matrix T is given by

$$T(t) = \begin{bmatrix} 0 & \partial^2 H/\partial p^2 \\ -\partial^2 H/\partial x^2 & 0 \end{bmatrix} = \begin{bmatrix} 0 & 1/m \\ d(x,t) & 0 \end{bmatrix}, \tag{10.25}$$

in which $H = p^2/(2m) + V + Q$. The initial condition on the stability matrix is $M(0) = 1$. The *force derivative* in equation 10.25 is given by

$$d(x,t) = f(x,t)' = -\partial^2(V+Q)/\partial x^2. \tag{10.26}$$

The next step is to differentiate equation 10.24 repeatedly with respect to x_0. The first few derivatives of this equation are as follows (where $x_0 = x(0)$):

$$\frac{d}{dt}\frac{\partial M(t)}{\partial x_0} = T(t)\frac{\partial M(t)}{\partial x_0} + \frac{\partial T(t)}{\partial x_0}M(t), \tag{10.27}$$

$$\frac{d}{dt}\frac{\partial^2 M(t)}{\partial {x_0}^2} = T(t)\frac{\partial^2 M(t)}{\partial {x_0}^2} + 2\frac{\partial M(t)}{\partial x_0}\frac{\partial T(t)}{\partial x_0} + \frac{\partial^2 T(t)}{\partial {x_0}^2}M(t). \tag{10.28}$$

The (1,1) elements of the two matrices $\partial M/\partial x_0$ and $\partial^2 M/\partial {x_0}^2$ provide the two derivatives $\partial^2 x(t)/\partial {x_0}^2$ and $\partial^3 x(t)/\partial {x_0}^3$ that we are seeking. The only interesting terms in the derivatives of the matrix $T(t)$ occur in the (2,1) position. The first few derivatives of the force term with respect to x_0 are given by

$$\frac{\partial d(x,t)}{\partial x_0} = \frac{\partial d(x,t)}{\partial x(t)}\frac{\partial x(t)}{\partial x_0}, \tag{10.29}$$

$$\frac{\partial^2 d(x,t)}{\partial {x_0}^2} = \frac{\partial^2 d(x,t)}{\partial x(t)^2}\left(\frac{\partial x(t)}{\partial x_0}\right)^2 + \frac{\partial d(x,t)}{\partial x(t)}\frac{\partial^2 x(t)}{\partial {x_0}^2}. \tag{10.30}$$

With the last set of derivatives, the system is closed; all needed derivatives are provided by this infinite hierarchy.

The preceding equations for evaluating spatial derivatives are summarized as follows:

- Use equations 10.19–10.21 to update $\rho(x,t)$ and its derivatives.
- Use equation 10.22 to find the derivatives of $\partial x(0)/\partial x(t)$.
- Use equations 10.24, 10.27, and 10.28 to evaluate $M(t)$ and its derivatives.
- Use equations 10.29 and 10.30 to update $d(x,t)$ and its derivatives.

Although equations have been provided for only the lowest few derivatives, corresponding equations for higher derivatives may be derived. All of the preceding equations are integrated along the quantum trajectory, which satisfies the now familiar equations

$$\frac{dx(t)}{dt} = v(t), \tag{10.31}$$

$$m\frac{dv(t)}{dt} = -\frac{\partial}{\partial x(t)}\left[V(x) + Q(x,t)\right]. \tag{10.32}$$

We note again that the density is evolved along the trajectory, but (unlike the DPM) the action function and its derivatives (except for $v(t)$) *do not* make an appearance.

In the simplest version of this scheme, the derivative $\partial x(t)/\partial x_0$ is evaluated along the trajectory and all higher derivatives are, in effect, set to zero. As a result, only six first-order differential equations are integrated. These are:

- the two trajectory equations, equations 10.31 and 10.32;
- the four equations embedded in equation 10.24.

This system is referred to by Liu and Makri as the *second-order scheme*, an example of which will be presented in the next section.

Several comments are in order regarding the TSM:

- The two key equations are (1) the conservation relation for the weight along the trajectory, equation 10.18; (2) The matrix trajectory stability equation, equation 10.24.
- All spatial derivatives are evaluated with respect to the starting point x_0.

10.8 Application of the Trajectory Stability Method

Using second-order TSM trajectories, energy-resolved transmission probabilities have been computed for the Eckart barrier scattering problem. Regarding the Eckart potential, the barrier height is $V_0 = 0.016$ a.u., the width parameter is $a = 1.3624$ a.u., and the mass used in the calculations is $m = 1016$ a.u. A Gaussian initial wave packet was used with width parameter $\beta = 8$ a.u. In figure 10.5, the

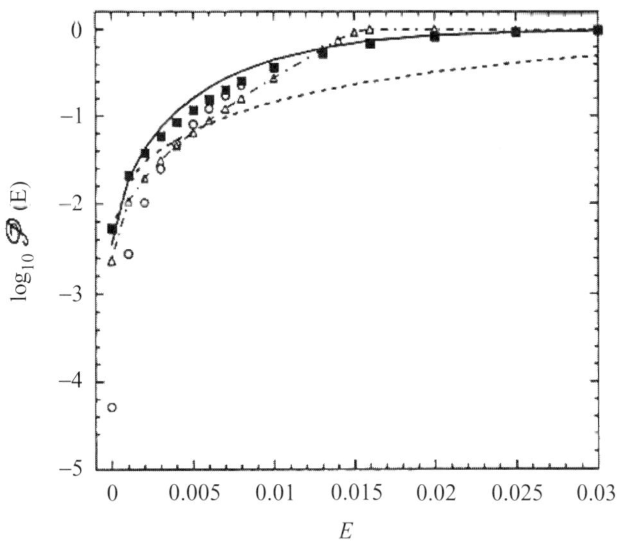

FIGURE 10.5. Energy resolved transmission probabilities [10.12] as a function of energy (in a.u.): second-order TSM (solid squares); accurate quantum-mechanical result (continuous curve); classical model obtained by integration over initial momentum distribution (hollow circles); alternative classical model as described in this reference (triangles and dash–dot curve). The transmission probability for a parabolic barrier with the same curvature at the barrier maximum as the Eckart barrier is also shown (dashed curve).

TSM energy-resolved transmission probability (solid squares) is compared with the exact quantum-mechanical result (continuous curve). The TSM results, which vary by a factor of over 300 over the energy range shown in this figure, are in reasonable agreement with the exact results. Also shown in this figure is the transmission probability curve for a parabolic barrier that has the same curvature at the barrier maximum as the Eckart barrier. It is clear that anharmonicities present in the Eckart potential lead to significant deviations from the transmission probabilities obtained for the parabolic barrier. This figure also shows the transmission curve for a classical model for barrier transmission that involves integration over the momentum distribution in the initial wave packet (hollow circles). At low energies, these results are several orders of magnitude below the TSM results.

10.9 Comments and Comparisons

The two methods introduced in this chapter both lead to infinite systems of coupled equations for the spatial derivatives of hydrodynamic fields along individual quantum trajectories. For the DPM, spatial derivatives of the wave amplitude and phase, the functions C and S, are propagated. In the TSM, spatial derivatives of the density $\rho(x, t)$ and the Jacobian $\partial x(t)/\partial x(0)$ are propagated. The use of low-order truncated sets of equations allows the trajectory to monitor and adapt to the surrounding fields. These fields are not brought into all orders of the derivatives; hence these methods dress what would otherwise be a bare classical trajectory with approximations to the quantum potential and its derivatives. This "tube" built around each trajectory brings in what we will term *regional nonlocality*; the trajectory may not be aware of the hydrodynamic fields and their changes beyond a limited horizon. In general, only very high order versions would allow for more distant features (such as ripples or nodes) to influence the approximate quantum trajectory. However, high-order calculations are not practical because of trouble with round-off errors and other instabilities. An example that brings out one way in which these approximate quantum trajectories can fail is the familiar two-slit diffraction experiment. DPM or TSM trajectories launched at the same time from different slits cannot be expected to adjust their motions so as to land on the detection screen in the higher-intensity bands. For this reason, the propagation of ensembles of correlated trajectories, as described in the earlier chapters, still plays a significant role.

One additional point is worth elaborating. When the moving least squares method that was described in Chapter 5 is used for fitting the hydrodynamic fields within a *local stencil*, only low-order derivatives are evaluated from the fit. Yet, this method brings in *global nonlocality* because the local stencils formed around the various fluid elements overlap one another. By the time that we advance to the next time step, each fluid element is "aware" of the others, even the ones far away. In contrast, although local derivatives are also evaluated around each trajectory in the DPM or the TSM, there is no communication through "overlapping stencils" of the type that occurs in evolving an entire ensemble using local least squares fitting. Thus, the DPM and TSM are not able to build in long-range correlation that would, for example, allow a trajectory to respond to inflation associated with a node forming some distance away.

After combining the DPM and the initial value representation (described earlier, in Section 4.6) to calculate survival probabilities for wave packets, Bittner made several interesting observations about DPM quantum trajectories [10.13]. Given the initial condition specified at $t = 0$, each approximate DPM trajectory $x_k(t)$ is guided by an underlying approximate wave function within which are embedded truncated C-amplitude and action functions,

$$\psi(x,t) = \exp\left[\sum_{j=0}^{n} \left(C_j^n + iS_j^n/\hbar\right)(x - x_k(t))^j \,/j!\right]. \tag{10.33}$$

This wave function may be accurate only within a local patch surrounding the trajectory. It would be exact, or at least very accurate, if we had the exact expansion coefficients to substitute into the exponent, but of course, we lack this information. Because the coefficients in the exponent of this wave function are approximate, neighboring trajectories lose their correlation as the coefficients in the exponent wander off into different regions of parameter space, as if guided by different initial wave functions. As a result, truncation of the derivative hierarchy leads to dephasing and decoherence in the dynamics. From this viewpoint, DPM-type dynamics will be accurate only when the *"truncation-induced" decoherence* time is relatively long compared with the observation time for interesting observables. In spite of this, derivative propagation schemes do work very well on some problems, including barrier transmission. Clearly, there are opportunities for improving our understanding of when these schemes work well and when they do not.

Several additional comments will be made about the DPM:

(1) In practice, some of the DPM trajectories may propagate for a while, then "blow up" over the next few time steps. However, a trajectory launched from a neighboring position may propagate to long times without difficulties. There are several reasons why some trajectories may fail. When a trajectory gets close to a node or quasi-node, the C-amplitude becomes very negative, and low-order DPM becomes inaccurate. In addition, the truncated system of DPM differential equations becomes stiff, with different scales and rates of change for the functions and their various derivatives. If this is the case, then special numerical integration algorithms for stiff systems (i.e., implicit integrators) are required.

(2) Even when the derivatives of the potential truncate, such as for a potential of polynomial form, the derivatives of C and S do not necessarily go to zero after a certain cutoff value. We do not have a way to predict what order of DPM should be used to generate results of a specified accuracy. Experience so far indicates that it is better to use low orders, usually meaning less than fifth order. When higher orders are brought in, the results may get progressively worse. These problems may arise in part because of the way that the DPM equations were truncated; rather than simply setting the higher-order spatial derivatives to zero, a "soft" truncation involving approximations for these derivatives would be preferable.

However, a soft truncation scheme will not solve the failure of DPM trajectories to "know about" events occurring some distance from the trajectory. These DPM trajectories still have a limited event horizon.

In summary, we will list some common *positive features* shared by the DPM and the TSM:

- Both methods lead to an infinite hierarchy of coupled equations, which, for a practical scheme, must be truncated at a low derivative order.
- All spatial derivatives are evaluated through analytic equations.
- In effect, approximations to the quantum potential (or quantum force) are integrated along the trajectory.
- Truncation at the second order leads to trajectory equations that are exact for quadratic potentials.
- Quantum effects are approximately included, so that barrier penetration can be modeled (in some cases very accurately).
- Single trajectories can be run one at a time.
- The methods can be readily extended to high dimensionality.
- The computer codes can be easily parallelized.

In addition, low-order versions of the DPM and the TSM share common *defects*, including the following:

- Propagation near nodes in the wave function is inaccurate (in fact, the trajectories may "blow up").
- Since only low-order spatial derivatives are propagated, these methods bring regional nonlocality ("local nonlocality"!) into the dynamics.
- Interference effects leading to ripples and nodes are not accurately incorporated (these effects may be "smoothed over").

The DPM will reappear in Chapter 11 (Sections 11.6–11.8), when trajectory approaches to the quantum dynamics of open systems (governed by Caldeira–Leggett-type equations) are described. For some phase space problems, such as the decay of a metastable state, DPM trajectories lead to very good predictions of the decay probabilities [10.14].

References

10.1. C.J. Trahan, K. Hughes, and R.E. Wyatt, *A new method for wave packet propagation: Derivative propagation along quantum trajectories*, J. Chem. Phys. **118**, 9911 (2003).
10.2. T. Aoki, *Interpolated differential operator (IDO) scheme for solving partial differential equations*, Comp. Phys. Comm. **102**, 132 (1997).
10.3. T. Aoki, *3D simulation of falling leaves*, Comp. Phys. Comm. **142**, 326 (2001).
10.4. E.J. Heller, *Time-dependent approach to semiclassical dynamics*, J. Chem. Phys. **62**, 1554 (1975).
10.5. I. Burghardt and L.S. Cederbaum, *Hydrodynamic equations for mixed quantum states. I. General formulation*, J. Chem. Phys. **115**, 10303 (2001).

10.6. I. Burghardt and L.S. Cederbaum, *Hydrodynamic equations for mixed quantum states. I. Coupled electronic states*, J. Chem. Phys. **115**, 10312 (2001).
10.7. A. Donoso and C.C. Martens, *Quantum tunneling using entangled classical trajectories*, Phys. Rev. Lett. **87**, 223202 (2001).
10.8. D.J. Tannor and D.E. Weeks, *Wave packet correlation function formulation of scattering theory: The quantum analog of classical S-matrix theory*, J. Chem. Phys. **98**, 3884 (1993).
10.9. D.J. Tannor, *Introduction to Quantum Mechanics: A Time Dependent Perspective* (University Science Books, New York, 2004).
10.10. J.Z.H. Zhang, *Theory and Application of Quantum Molecular Dynamics* (World Scientific, Singapore, 1999).
10.11. L.D. Landau and E.M. Lifshitz, *Quantum Mechanics: Non-relativistic Theory* (Pergamon Press, London, 1958).
10.12. J. Liu and N. Makri, *Monte Carlo Bohmian dynamics from trajectory stability*, J. Phys. Chem. A **108**, 5408 (2004).
10.13. E.R. Bittner, *Quantum initial value representations using approximate Bohmian trajectories*, J. Chem. Phys. **119**, 1358 (2003).
10.14. C.J. Trahan and R.E. Wyatt, *Classical and quantum phase space evolution: fixed-lattice and trajectory solutions*, Chem. Phys. Lett. **385**, 280 (2004).

11
Quantum Trajectories in Phase Space

Evolution of the quantum phase space distribution function and the density matrix for open systems (a system coupled to a heat bath) is described in terms of trajectories influenced by both local and nonlocal forces.

11.1 Introduction

The quantum trajectories developed in the preceding chapters were computed and analyzed in position space. For *open quantum systems*, those coupled to an environment, such as a thermal reservoir, many of the formulations and applications are developed in phase space. For a one-degree-of-freedom system, for example, the goal is to describe the evolution of the distribution function for the quantum system $W(x, p, t)$ in the two-dimensional phase space characterized by the pair of coordinates $\{x, p\}$. In addition to providing informative pictures of the dynamics, various average values can be calculated directly in terms of this distribution function.

For an *isolated classical system*, the Liouville equation that was introduced in Chapter 3 describes the evolution of the phase space distribution function. This equation, and the characteristics of phase space flow that are associated with it, are described in Section 11.2. When the classical system is allowed to interact through friction terms with a thermal bath, the distribution function evolves according to the Kramers equation, also described in the same section. We saw in Chapter 3 that for an *isolated quantum system*, the Wigner–Moyal equation governs the time evolution of the Wigner phase space distribution function. This function is used in many areas, including the analysis of systems with underlying classically chaotic dynamics and in quantum optics. Because this function may develop negative basins, smoothed positive semidefinite distribution functions are also frequently used. The Husimi function, for example, is obtained from the Wigner function by smoothing with a Gaussian kernel. The Wigner and Husimi equations are described in Section 11.3. When the quantum subsystem is coupled to an environment, the equations of motion become more complicated. For *open quantum systems*, the Caldeira–Leggett equation is a widely studied evolutionary equation for the quantum distribution function. This equation and a variant that has two additional smoothing terms, the Diosi equation, are described in Section 11.4. The derivation and study of quantum equations of motion for open systems continues to be an active area of research.

It has only been within the past few years that quantum trajectories have been used to evolve phase space distribution functions. Trahan and Wyatt used a version of the DPM, which was introduced in Chapter 10, to propagate single trajectories that evolve under the influence of both local and nonlocal (density-dependent) terms in the equations of motion [11.1]. This work was stimulated by earlier research in which Donoso and Martens [11.2–11.4] developed a method for propagating ensembles of "entangled trajectories". The density-dependent derivative coupling terms in the equations of motion were evaluated by fitting a Gaussian function to the density surrounding each trajectory. The propagation method used by Donoso and Martens is described in Section 11.5. Use of the DPM for phase space dynamics is described in Section 11.6. Conversion of the equations of motion derived originally in the Eulerian frame into the moving with the fluid (Lagrangian) frame is described in Section 11.7. In Section 11.8, plots are shown to illustrate the evolution of quantum trajectories for the decay of a metastable state and for scattering from an Eckart barrier. In Section 11.9, we return to the momentum moments of the Wigner function, a topic that was introduced earlier, in Chapter 3. In this section, these moments and coupled rate equations for their time evolution will be considered for mixed state systems undergoing dissipative dynamics. The phase space distribution function is connected through a Fourier transform to another important function, the density matrix $\rho(x, x', t)$. Novel trajectory methods for evolving the density matrix are described in Section 11.10, and two applications are presented in Section 11.11. Finally, concluding remarks are in Section 11.12.

11.2 The Liouville, Langevin, and Kramers Equations

In classical dynamics, we saw in Chapter 3 that the evolution of an ensemble of N trajectories can be described in several ways. For example, we can begin by integrating the classical equations of motion to find the coordinate and momentum $\{x(t), p(t)\}$ for each of the trajectories. Each trajectory can then be plotted in the two-dimensional phase space having the coordinates $\{x, p\}$. As each trajectory is integrated, the orbit can be plotted in this space, so that at time t we have N trajectories, each leading from an initial point $\{x_0, p_0\}$ at $t = 0$ to the current value at time t. In a small box of area $\Delta x\ \Delta p$ around the point $\{x, p\}$, the number of trajectories is $\Delta N(x, p, t)$, and the fraction of the total number of trajectories located in this box is $\Delta N(x, p, t)/N$. The probability density is phase space is then defined by the limit $\Delta x \to 0,\ \Delta p \to 0$:

$$W(x, p, t) = \lim \left[\frac{\Delta N(x, p, t)}{N \cdot \Delta x \Delta p} \right]. \tag{11.1}$$

It follows directly from this relation that the density is normalized

$$\int_{-\infty}^{\infty}\int_{-\infty}^{\infty} W(x,p,t)\,dx\,dp = 1. \tag{11.2}$$

The flow of this classical probability fluid in phase space was addressed by Liouville. He derived an equation, now called the *Liouville equation* [11.5], for the rate of change in the density at a fixed point in phase space:

$$\frac{\partial W}{\partial t} = -\frac{p}{m}\frac{\partial W}{\partial x} + \frac{\partial V}{\partial x}\frac{\partial W}{\partial p}. \tag{11.3}$$

With assistance from the classical equations of motion, $\dot{x} = p/m$ and $\dot{p} = -\partial V/\partial x$, we will define two components of the (time-dependent) velocity vector $\vec{V}$ in the two-dimensional phase space: $v_x = \dot{x}$ and $v_p = \dot{p}$. In addition, the gradient operator $\vec{\nabla}$ in this space has components given by $\{\partial/\partial x, \partial/\partial p\}$. Using these relations, the Liouville equation may be written in the compact form

$$\frac{\partial W}{\partial t} = -\vec{V}\cdot\vec{\nabla}W. \tag{11.4}$$

Equation 11.4 expresses the Liouville equation in the Eulerian frame. Another version is obtained by converting this equation into the Lagrangian frame. This transformation begins with

$$\frac{dW}{dt} = \frac{\partial W}{\partial t} + v_x\frac{\partial W}{\partial x} + v_p\frac{\partial W}{\partial p} = \frac{\partial W}{\partial t} + \vec{V}\cdot\vec{\nabla}W. \tag{11.5}$$

If we now substitute the Eulerian time derivative from equation 11.4, we obtain the result that *the density does not change along the flow,*

$$dW/dt = 0, \tag{11.6}$$

or stated another way, the density at time t along the flow has the same value that was imposed by the initial conditions

$$W(t) = W(t_0). \tag{11.7}$$

The results expressed in the two preceding equations are known as *Liouville's theorem*. Some consequences of Liouville's theorem include the following interrelated statements: (1) the Jacobian evaluated along the flow is invariant, $dJ/dt = 1$; (2) the flow is volume preserving (isovolumetric); (3) the flow is incompressible, $\vec{\nabla}\cdot\vec{v} = 0$. Before concluding this section, we will point out that Liouville's theorem, along with the three auxiliary statements, is *not valid* for quantum-mechanical phase space flow.

Liouville's equation governs the evolution of the classical phase space density in an *isolated system*. We now turn to *open systems*, those that can exchange energy with the surroundings. One example is a system in contact with a *heat bath* that is maintained at the temperature T. A classical trajectory evolving in such a system is subject to *frictional effects*, the damping of the velocity due to coupling between the system and the bath degrees of freedom. The friction force is given by $F_{friction} = -\gamma m v$, in which γ (units of 1/time) is the *phenomenological friction coefficient*. This term acting alone would cause the particle to settle to zero velocity in a minimum on the potential surface.

Historical comment. Joseph Liouville (1809–1882) was for a number of years professor of analysis and mechanics at the Ecole Polytechnique, in Paris. He is especially remembered for his contributions in partial differential equations, electrodynamics, and heat transfer. His work was extremely wide-ranging, from physics to astronomy to pure mathematics. He was also involved in politics, having been elected to the Assembly in 1848, just before he began lecturing at the College of France in 1851.

In addition, the system evolves under the influence of two additional contributions to the total force: as usual, there is the classical force arising from the potential $V(q)$, plus an additional *random force* due to interaction with the bath. The latter is also called the *stochastic force* $F(t)$. For identically prepared ensembles that then evolve in time, the ensemble average of the stochastic force is assumed to be zero, $\langle F(t)\rangle = 0$. However, the *two-time correlation function* (*force autocorrelation function*) is not necessarily zero:

$$\langle F(t)F(t')\rangle = C(|t-t'|). \tag{11.8}$$

This expression relates the force at time t' to the force at some other time t. An important case is that in which the correlation function is nonzero only when $t' = t$, so that the right side is proportional to a δ-function, $\langle F(t)F(t')\rangle \approx \delta(t-t')$. For a system coupled to a thermal bath, it can be shown that the friction coefficient and the correlation function are related through the *fluctuation-dissipation theorem:*

$$\langle F(t)F(t')\rangle = 2m\gamma k_B T\delta(t-t'), \tag{11.9}$$

where k_B is Boltzmann's constant. This theorem is discussed in textbooks on statistical mechanics [11.5] and will not be derived here. When the stochastic force is δ-correlated, it is also referred to as *white noise*, because the Fourier transform of equation 11.9 is flat in frequency space. *Colored noise* refers to the more general case in which the correlation function is not δ-spiked, so that the Fourier transform acquires some frequency dependence.

The *stochastic process* for a particle influenced by the three force terms mentioned previously is governed by the *Langevin equation*, which is used to describe *classical Brownian motion in phase space*,

$$m\frac{dv}{dt} = -\frac{\partial V}{\partial x} - \gamma m v + F(t), \tag{11.10}$$

where $\dot{x} = v$. Rather than viewing γ and $F(t)$ in phenomenological terms, the friction coefficient and the random force may be derived from Hamiltonian models involving a subsystem interacting through bilinear coupling terms with a harmonic bath [11.6].

Historical comment. In addition to his work on diffusive transport, French physicist Paul Langevin (1872–1946) developed an equation relating paramagnetism and absolute temperature. He also studied the properties of ionized gases and Brownian motion in gases. During World War I, he worked on submarine detection using ultrasonic vibrations, a precursor to sonar detection that was used during World War II.

One approach to modeling system-bath dynamics is to treat the fluctuating force as additive Gaussian noise and then solve equations 11.10 numerically for each member in an ensemble of N trajectories. Another approach to the stochastic dynamics of an ensemble is to *propagate the density* $W(x, p, t)$ in phase space using an appropriate equation of motion. By doing this, the stochastic term in the Langevin equation, $F(t)$, is replaced by a *deterministic* term proportional to the second-order derivative of $W(x, p, t)$ with respect to the momentum.

In 1940, Kramers derived an important equation [11.7] that governs the evolution of the density for a subsystem in contact with a heat bath at the equilibrium temperature T. Kramer's model for a condensed phase chemical reaction again involves a probability distribution moving in a one-dimensional potential with the remaining degrees of freedom for both the reacting and solvent molecules constituting the heat bath. The Kramers equation is given by

$$\frac{\partial W}{\partial t} = -\frac{p}{m}\frac{\partial W}{\partial x} + \frac{\partial V}{\partial x}\frac{\partial W}{\partial p} + \gamma\frac{\partial}{\partial p}\left[p + mk_B T\frac{\partial}{\partial p}\right]W, \tag{11.11}$$

where the first of the two terms involving γ is the *dissipative term*, and the second one leads to *momentum diffusion*. The first two terms on the right side are recognized as the convective terms from the Liouville equation. A straightforward and informative derivation of this equation is given in the chemical kinetics text by Billing and Mikkelsen [11.8]. Kramers obtained steady-state solutions of this equation for two limiting cases, weak and strong friction. In each case, he solved an approximate diffusion equation, in configuration space for strong friction and in energy space for weak friction. In more recent times, the Kramers rate equation has been subjected to rederivation from Hamiltonian models [11.9–11.10]. A number of analyses have focused on determination of the escape rate of a particle initially trapped in a metastable well as a function of both temperature and the friction coefficient. These and other results are described in the comprehensive review article by Hanggi et al. [11.11].

Historical comment. Dutch physicist Hendrick A. Kramers, 1894–1952, served as an assistant to Bohr in Copenhagen during the period 1916–1926. Bohr was known by some as the "Pope" while Kramers was "His Eminence". Bohr almost never calculated anything, while Kramers did most of the calculations during this period. In addition to the "Kramers problem" described above, he was one of the three developers of the WKB semiclassical approximation, he introduced the idea of taking the "thermodynamic limit", and he developed several dispersion relations that bear his name. There are two biographies [11.12, 11.13], one appropriately titled, *H.A. Kramers, Between Tradition and Revolution*. In addition, an article in *Physics Today* summarized Kramers's contributions to statistical mechanics [11.14].

11.3 The Wigner and Husimi Equations

In the quantum mechanics of *closed systems*, one of several equations that governs the evolution of a phase space distribution function is the Wigner–Moyal (WM) equation [11.15, 11.16, 11.31], which was introduced in Section 3.3:

$$\frac{\partial W}{\partial t} = -\frac{p}{m}\frac{\partial W}{\partial x} + \frac{\partial V}{\partial x}\frac{\partial W}{\partial p} - \frac{\hbar^2}{24}\frac{\partial^3 V}{\partial x^3}\frac{\partial^3 W}{\partial p^3} + \frac{\hbar^4}{1920}\frac{\partial^5 V}{\partial x^5}\frac{\partial^5 W}{\partial p^5} + O(\hbar^6). \tag{11.12}$$

Additional quantum corrections on the right side bring in the higher-order odd momentum derivatives of the density multiplied by the odd coordinate derivative of the potential. For a potential expressible as a polynomial, the quantum corrections terminate with the highest nonzero derivative of the potential. The Wigner function itself may be generated through a Fourier transform of the density matrix $\rho(x, x', t)$ for either a pure or a mixed state [11.16]. We note that the first two terms on the right side have the form of the classical Liouville equation.

Unlike the nonnegative classical distribution functions that solve the Liouville and Kramers equations, the Wigner function may become negative in some regions of phase space, even when $W(x, p)$ is initially nonnegative. These *negative basins* arise due to interference effects. It has been proven that a necessary and sufficient condition for the Wigner function to be nonnegative is that the corresponding wave function must be the exponential of a quadratic form, $\psi(x) = \exp[-(ax^2 + bx + c)]$ [11.42, 11.43]. Needless to say, most Wigner functions will develop one or more negative basins during the course of their time evolution. An example of negative basins in a Wigner function is described in Box 11.1.

The use of trajectories to evolve phase space distributions, especially for open quantum systems, forms the subject of the remainder of this chapter. Before about 2000, trajectories were used on occasion in semiclassical studies of phase space dynamics. One of these approaches, the Lee–Scully *Wigner trajectory* method, is treated in more detail in Box 11.2. The trajectory approaches developed by Donoso and Martens [11.2–11.4] in 2001–2002 and by Trahan and Wyatt [11.1] in 2003 are described in the following sections. There are challenging problems that must be solved before robust computational methods can be developed for evolving quantum phase space distributions; one such problem encountered with trajectory methods is considered in Box 11.3.

Box 11.1. Example of a Wigner function with negative basins

The Wigner function for a superposition of two Gaussian functions has negative interference ripples in the region between the two Gaussian peaks. The Wigner function for this example is obtained from the (unnormalized) superposition

$$\psi(x) = \sqrt{2\beta/\pi}\,e^{-\beta(x-x_0)^2} + \sqrt{2\beta/\pi}\,e^{-\beta(x+x_0)^2}, \tag{1}$$

in which the two Gaussians are separated by the distance $\Delta x = 2x_0$. The Wigner function for this state is given by

$$\begin{aligned} W(x,p) &= (1/\pi\hbar)e^{-[2\beta(x-x_0)^2+p^2/(2\beta\hbar^2)]} + (1/\pi\hbar)e^{-[2\beta(x+x_0)^2+p^2/(2\beta\hbar^2)]} \\ &\quad +2\cos(\Delta x\cdot p/\hbar)\cdot e^{-[2\beta x^2+p^2/(2\beta\hbar^2)]}. \end{aligned} \tag{2}$$

The first two terms in this expression are the Gaussian densities for each of the two components of the wave function. The third term is the interference term, which oscillates along the p-axis with wavelength $\lambda_p = 2\pi\hbar/\Delta x$. For the parameter values $x_0 = 1$ and $\beta = 1$, the Wigner function is plotted below (atomic units are used here). Two relatively deep negative basins are located along the p-axis, between the Wigner densities for the two "pure" Gaussians (which are centered at $x_0 = \pm 1$).

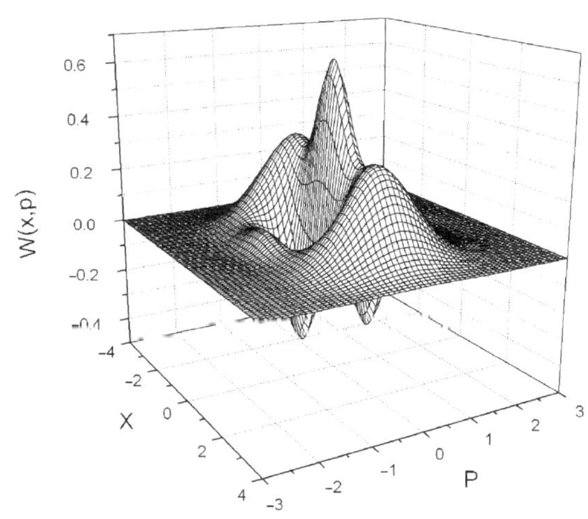

Box 11.2. The Lee–Scully Wigner trajectory method

A semiclassical trajectory approach has sometimes been used to estimate state-to-state transition probabilities in molecular collisions. In the version developed by Lee and Scully [11.17, 11.18], the following four steps were proposed. First, given the wave function for the initial state, the Wigner transform is computed. Second, initial conditions for trajectories are sampled from this Wigner distribution. Third, *classical trajectories* are propagated, given the space of initial conditions obtained in the preceding step. Finally, the trajectory ensemble is analyzed using the Wigner function for the final collision state. For example, if the Wigner function for the final state is denoted by $W^{(f)}(x, p)$ and the distribution function that evolved from the initial state is $W^{(i)}(x, p, t)$, the $i \rightarrow f$ transition probability is given by the overlap integral

$$P_{i\rightarrow f} = Tr\left(W^{(f)}W^{(i)}(t)\right), \tag{1}$$

in which $Tr(..)$, the *trace operation*, is an integral over phase space. Thus, quantum mechanics enters only through the Wigner functions for separated reactants and products. The dynamics linking the initial and final states are purely classical.

Heller has given an analysis of errors that can arise when classical trajectories are used to propagate the quantum-mechanical Wigner function [11.19, 11.20]. In addition, McLafferty has discussed trajectory approximations to the "Wigner path integral" [11.39].

Box 11.3. Phase space trajectories and negative regions of *W*(*x*, *p*, *t*)

In an insightful study, Daligault [11.34] made a number of comments, and expressed some concerns, about the use of Lagrangian trajectories to propagate quantum phase space distributions. One issue that he discussed, on which we will elaborate slightly, is the following. How can methods be developed to allow for the creation or annihilation of trajectories in regions where the distribution function becomes negative? Since the exact Lagrangian trajectories cannot cross nodal surfaces defined by $W(x, p, t) = 0$, how can the dynamics be properly modeled when only positive-weight trajectories are followed at earlier times? The entangled trajectory method used by Donoso and Martens [11.2, 11.4] and the DPM used by Trahan and Wyatt [11.1] do not permit negative regions of $W(x, p, t)$ to form. Stochastic hopping models, somewhat like trajectory surface hopping methods, might possibly be devised for this purpose. Takabayasi proposed a stochastic model in his 1954 study [3.1]. In addition, Maddox and Bittner [11.32, 11.33] have developed trajectory methods for propagation of the complex-valued density matrix $\rho(x, x', t)$, which in principle can be transformed into phase space. Computationally, this may be difficult, because information is available at a limited number of trajectory locations.

(Section 11.10 will be concerned with trajectory methods for evolving the density matrix.) The issue of trajectory propagation and negative basins in quantum phase space needs further exploration.

We have emphasized that *Lagrangian trajectories* cannot evolve across nodal surfaces in phase space. What is the situation for *non-Lagrangian trajectories*? These trajectories *can* penetrate into negative basins. The propagation of these ALE trajectories is explored further in Box 11.4.

In addition to these methods, Shifren and Ferry have developed a quantum Monte Carlo method that employs the time-evolving Wigner function [11.44]. An additional quantity, the *affinity* A_i, is assigned to each trajectory. This quantity, which can take on negative values, has a magnitude bounded by unity, $|A_i| \leq 1$. The Wigner distribution at each phase space point is then obtained through the trajectory sum

$$W(x, p) = \sum \delta(x - x_i)\delta(p - p_i)A_i. \tag{1}$$

Use of the affinity permits the Wigner distribution to become negative. Sample calculations were performed for an initial Gaussian wave packet scattering from an Eckart barrier. On the "back side" of the barrier, negative basins formed in the Wigner function. A different method, developed by Wong [11.45], is also capable of generating negative basins.

Given an ensemble of evolving Bohmian trajectories, can the quantum phase space distribution be generated from information carried by these trajectories? Surprisingly, the answer is yes, at least formally. Dias and Prata (see Section 4 in [11.35]) have derived an expression for the quantum phase space distribution in terms of the (complicated) transform of the hydrodynamic phase space distribution $W_{hydro}(x, p, t) = R(x, t)^2\delta(p - \partial S/\partial x)$. Computational investigations of this connection would be very informative.

Box 11.4. Non-Lagrangian phase space trajectories

The Eulerian version of the equation of motion for the phase space distribution function can be written

$$\frac{\partial W}{\partial t} = \hat{L}W, \tag{1}$$

in which $\hat{L}$ is a linear operator. This operator, for example, could be the Wigner or Caldeira–Leggett operator. This equation of motion will now be transformed to the ALE moving frame. (The ALE method was introduced in Section 7.2) The ALE grid point velocity vector is given by $U = (\dot{x}, \dot{p})$, where the two velocity components are arbitrary. In terms of the phase space gradient operator $\nabla = (\partial/\partial x, \partial/\partial p)$, the rate of change in the distribution function is given by

$$\frac{dW}{dt} = \frac{\partial W}{\partial t} + U \cdot \nabla W = \hat{L}W + U \cdot \nabla W, \tag{2}$$

The operator $\hat{L}$ will now be split into the Liouville ("classical-type") term, and a second component which may include quantum and dissipative terms

$$\hat{L} = \hat{L}_0 + \hat{L}_1 = -V \cdot \nabla + \hat{L}_1. \tag{3}$$

In this equation, the Lagrangian velocity vector is given by

$$V = \left(\frac{p}{m}, -\frac{\partial V}{\partial x} \right). \tag{4}$$

Substitution of equation 3 into equation 2 then gives the equation of motion for W in the ALE frame,

$$\frac{dW}{dt} = (U - V) \cdot \nabla W + \hat{L}_1 W, \tag{5}$$

in which $(U - V)$ is the slip velocity. In the special case that $U = V$, the equation of motion is expressed in the Lagrangian frame. In addition, if $\hat{L}_1 = 0$, we then recover Liouville's theorem.

The important point about equation 5 is that the non-Lagrangian term $(U - V) \cdot \nabla W$ permits the density to become negative along the ALE phase space path given by $(x(t), p(t))$. ALE dynamical frames have apparently not been applied to trajectory propagation in phase space.

Historical comment. Hungarian physicist Eugene Paul Wigner (1902–1995) immigrated to the United States in 1930. His main contribution was in applying group theory to quantum mechanics. In addition, his work on the strong nuclear force led to his sharing the Nobel Prize in physics in 1963 with Maria Goeppert Mayer and Hans Jensen.

Positive definite coarse-grained (smoothed) transforms of the Wigner distribution, such as the *Husimi function* (convolution of the Wigner function with Gaussian functions), have been developed to average over the negative valleys in the underlying Wigner function [11.16]. The Husimi function $H(x, p, t)$ is defined through an integral transform of the Wigner function,

$$H(x, p, t) = \int_{-\infty}^{\infty} \int_{-\infty}^{\infty} K(x, p; x', p') W(x', p', t) dx' dp', \tag{11.13}$$

where the *smoothing kernel* is a normalized two-dimensional Gaussian function

$$K(x, p, x', p') = (\pi\hbar)^{-1} \exp\left[-(x - x')^2/\alpha - \alpha(p - p')^2/\hbar^2\right], \tag{11.14}$$

in which α sets the scale for smoothing. In equation 11.13, the smoothing kernel draws in values of the Wigner function from within an elliptic region surrounding the "observation point" $\{x, p\}$.

The Husimi distribution obeys the complicated equation of motion

$$\frac{\partial H}{\partial t} = -\frac{p}{m}\frac{\partial H}{\partial x} - \frac{\hbar^2}{2m\alpha}\frac{\partial^2 H}{\partial x \partial p} + \sum_{\lambda=1,3,..}^{(odd)} \sum_{\mu=0,1,..}^{\infty} \sum_{\kappa=0,1,..}^{(\mu-2\kappa)\geq 0} C_{\lambda\mu\kappa} \frac{\partial^{\lambda+\mu} V}{\partial x^{\lambda+\mu}} \frac{\partial^{\lambda}\partial^{\mu-2\kappa} H}{\partial p^{\lambda} \partial x^{\mu-2\kappa}}, \tag{11.15}$$

where $C_{\lambda\mu\kappa} = (i\hbar)^{\lambda-1}\alpha^{\mu-\kappa}/\left[2^{\lambda+\mu-1}\lambda!\kappa!(\mu-2\kappa)!\right]$. The triple summation in equation 11.15 is tedious to evaluate, but for potentials expressed as polynomials, termination occurs after a finite number of terms. For example, for a potential containing no higher than cubic terms, equation 11.15 becomes

$$\frac{\partial H}{\partial t} = -\frac{p}{m}\frac{\partial H}{\partial x} + \left[\frac{\partial V}{\partial x} + \frac{\alpha}{4}\frac{\partial^3 V}{\partial x^3}\right]\frac{\partial H}{\partial p} - \left[\frac{\hbar^2}{2m\alpha} - \frac{\alpha}{2}\frac{\partial^2 V}{\partial x^2}\right]\frac{\partial^2 H}{\partial x \partial p} + \frac{\alpha^2}{8}\frac{\partial^3 V}{\partial x^3}\frac{\partial^3 H}{\partial x^2 \partial p} - \frac{\hbar^2}{24}\frac{\partial^3 V}{\partial x^3}\frac{\partial^3 H}{\partial p^3}. \tag{11.16}$$

Four of the seven terms on the right side depend on the smoothing parameter, while the remaining three are identical to those appearing in equation 11.12 for the Wigner function. One aspect of the Husimi distribution that blurs its interpretation is that unlike the Wigner distribution, integration over x does not yield the momentum space probability density, and integration over p does not yield the position space probability density (see equations 3.6 and 3.7).

A comprehensive review of quantum phase space distribution functions may be consulted for more details [11.16]. Plots of Wigner and Huisimi functions for stationary states of the quartic oscillator and the double-well potential were presented by Novaes [11.41].

11.4 The Caldeira–Leggett Equation

The inclusion of dissipative effects into the dynamics of an open quantum system is a topic with a rich history [11.21]. The well-known *Caldeira–Leggett* (CL) equation [11.30] applies in the small-friction-coefficient, high-temperature limit ($k_B T \gg E_0$, where E_0 is the zero-point energy of the uncoupled system). The Hamiltonian used in the derivation of this equation is described in Box 11.5, and Box 11.6 gives additional details about the derivation of the CL equation. The phase space version of the CL equation has the same form as the Wigner equation with the addition of the same two friction terms that appear in the Kramers equation. Including only the lowest-order quantum term, this equation has the form

$$\frac{\partial W}{\partial t} = -\frac{p}{m}\frac{\partial W}{\partial x} + \frac{\partial V}{\partial x}\frac{\partial W}{\partial p} - \frac{\hbar^2}{24}\frac{\partial^3 V}{\partial x^3}\frac{\partial^3 W}{\partial p^3} + \gamma\frac{\partial (pW)}{\partial p} + \gamma m k_B T\frac{\partial^2 W}{\partial p^2}. \tag{11.17}$$

Equation 11.17, regarded as the *semiclassical version of the Kramers equation*, has been used to estimate quantum effects on thermally activated barrier crossings [11.22].

Historical comment. Anthony Leggett, born in 1938 in Great Britain, started his college career as a philosophy major before converting to physics. He made the correct choice. He has done pioneering work in condensed matter theory, especially regarding normal and superconducting helium liquids. In 2003, he shared the Nobel Prize in physics with Alexei Abrikosov and Vitaly Ginsburg for work on superfluidity. Leggett was associated with the University of Sussex when that work was done, but since 1983 he has been a professor at the University of Illinois at Urbana-Champaign.

Box 11.5. The system-bath Hamiltonian

This discussion of the system-bath Hamiltonian is based on the presentation appearing in the book by Weiss [11.21]. The Hamiltonian for the composite system is frequently decomposed into a Hamiltonian for the subsystem, the bath Hamiltonian, plus the interaction term

$$H = H_s + H_b + H_i, \tag{1}$$

where for a single-mode system and an N-mode harmonic bath

$$H_s = \frac{p^2}{2m} + V(x), \tag{2}$$

$$H_b = \sum_{k=1}^{N} \left[\frac{p_k^2}{2m_k} + \frac{1}{2} m_k \omega_k^2 q_k^2 \right]. \tag{3}$$

Assuming bilinear coupling between each mode of the bath and the system, the interaction term is

$$H_i = -\sum_{k=1}^{N} c_k q_k x + \Delta V(x), \tag{4}$$

where the *counterterm* $\Delta V(x)$ is introduced to adjust the position of the minimum of the total potential at each value of x.

The *total potential* is

$$V_t = V(x) + \frac{1}{2} \sum_{k=1}^{N} m_k \omega_k^2 q_k^2 - \sum_{k=1}^{N} c_k x q_k + \Delta V(x). \tag{5}$$

For a fixed value of x, the minimum in V_t occurs when the bath coordinate has the value

$$q_k = \frac{c_k x}{m_k \omega_k^2}, \tag{6}$$

and the total potential evaluated at the minimum is given by

$$V_t = V(x) - \frac{1}{2} \sum_{k=1}^{N} \frac{c_k^2 x^2}{m_k \omega_k^2} + \Delta V(x). \tag{7}$$

The usual choice for ΔV is to "counter" the second term in the above expression, thus restoring the total potential (evaluated at this minimum) back to $V(x)$

$$\Delta V(x) = \frac{1}{2} \sum_{k=1}^{N} \frac{c_k^2 x^2}{m_k \omega_k^2}. \tag{8}$$

Using this result, the total potential may be written in terms of the adjusted bath coordinates,

$$V_t = V(x) + \frac{1}{2}\sum_{k=1}^{N} m_k\omega_k^2\left[q_k - \frac{c_k}{m_k\omega_k^2}x\right]^2, \tag{9}$$

so that the Hamiltonian then becomes

$$H = \frac{p^2}{2m} + V(x) + \frac{1}{2}\sum_{k=1}^{N}\frac{p_k^2}{2m_k} + \frac{1}{2}\sum_{k=1}^{N} m_k\omega_k^2\left[q_k - \frac{c_k}{m_k\omega_k^2}x\right]^2. \tag{10}$$

This is the form of the Hamiltonian that was used by Caldeira and Leggett and by many others who have studied system-bath dynamics.

The density operator $\hat{\rho}(t)$ is required to be *positive semidefinite* (also referred to as "simple positivity"). This means that when the density operator is represented in a Hilbert space basis set, and brought into diagonal form, there are no negative eigenvalues. A stronger condition than this is *complete positivity* (CP), a mathematical property that is defined by Lindblad [11.24] and by Pechukas [11.23]. The CL equation does not satisfy the conditions for CP. However, quantum master equations for $\partial\hat{\rho}/\partial t$ that can be expressed in *Lindblad form* [11.25] have time-dependent solutions that are *guaranteed* to remain completely positive. In a sense, the CL equation for the density operator is missing some terms that are needed to guarantee that the solution satisfies the CP condition. Possibly it is worth emphasizing that the Lindblad form specifies a *mathematical structure* for quantum master equations that guarantees solutions that are CP. However, merely satisfying this condition does not imply that the physics is realistic. Box 11.7 provides additional information about dissipation operators and the *Lindblad form for the quantum master equation*.

Box 11.6. On the derivation of the Caldeira–Leggett equation

Starting from the system-bath model Hamiltonian described in Box 11.3, Caldeira and Leggett used a real-time path integral approach to derive an equation of motion for the reduced density matrix for the system. They first set up a path integral expression for the total density matrix for the composite system and then averaged (traced) over the bath modes. At the initial time, it was assumed that the total density operator could be written as a separable product, the density operator for the system times the canonical density operator (at temperature T) for the bath. During the derivation, they took the continuum limit for the bath, and assumed a specific form for the product of the density of bath modes and the square of the coupling coefficient, given by $\rho(\omega)c(\omega)^2$. During the derivation, the weak-coupling, high-temperature limit was assumed.

Box 11.7. Density operator evolution with dissipation

The Hermitian total *density operator* for an isolated composite system evolves according to the *Liouville–von Neumann* (LvN) equation

$$\partial\hat{\rho}(t)/\partial t = -i/\hbar\left[\hat{H}, \hat{\rho}(t)\right] = \hat{L}\hat{\rho}(t), \tag{1}$$

where $[\hat{A}, \hat{B}]$ denotes the commutator of the two operators and where $\hat{L}$ defines the *quantum Liouville operator*. This equation of motion is straightforward to derive. For simplicity, if we assume a pure state, $\hat{\rho}_t = |\psi(t)\rangle\langle\psi(t)|$, the time derivative is

$$\partial\hat{\rho}_t/\partial t = |\psi\rangle\langle\dot{\psi}| + |\dot{\psi}\rangle\langle\psi|. \tag{2}$$

From the Schrödinger equation we obtain $|\dot{\psi}\rangle = (1/(i\hbar))\hat{H}|\psi\rangle$ and $\langle\dot{\psi}| = (-1/(i\hbar))\langle\psi|\hat{H}$. Using these equations to evaluate $\partial\hat{\rho}_t/\partial t$, we immediately obtain the LvN equation.

For a *mixed state*, we would start with the expansion in pure states

$$\hat{\rho}_t = \sum_j \omega_j|\psi_j\rangle\langle\psi_j| \tag{3}$$

(with weights w_j having the properties $\omega_j \geq 0$ and $\sum_j \omega_j = 1$) and then proceed as before. The total density operator must be Hermitian and positive semidefinite (i.e., no negative eigenvalues).

The evolution of the *reduced density operator* for the subsystem can be written

$$\partial\hat{\rho}_s/\partial t = \hat{L}_s\hat{\rho}_s + \hat{L}_D\hat{\rho}_s, \tag{4}$$

where $\hat{L}_s = -i/\hbar[H_s, \ldots]$ is the Liouville operator for the isolated subsystem and $\hat{L}_D$ is the *dissipation operator*, which provides linkage between the subsystem and the bath.

The Lindblad form for the dissipation operator is

$$\hat{L}_D\rho_s = \frac{1}{2}\sum_j\left\{\left[\hat{L}_j\hat{\rho}_s, \hat{L}_j^\dagger\right] + \left[\hat{L}_j, \hat{\rho}_s\hat{L}_j^\dagger\right]\right\}, \tag{5}$$

in which $\hat{L}_j$ is one of a set of *Lindblad operators*, or *Lindblads*. For dissipation operators of this form, the trace of the density operator is conserved, and positivity of the density operator is guaranteed. Hermitian Lindblads are used to represent the effect of measurements, and non-Hermitian variants produce dissipative interactions. Examples of Lindblads are given in the book by Percival [11.25].

A more complete discussion of positivity of the density operator and the Lindblad form for the dissipation operator is provided in the study by Kohen, Marsten, and Tannor [11.26].

In a careful rederivation of the CL equation, Diosi [11.27, 11.28] found two additional smoothing terms that bring the modified equation into the Lindblad form. As a consequence, the solutions satisfy the CP condition. The phase space version of the *modified CL equation* has the following form:

$$\frac{\partial W}{\partial t} = -\frac{p}{m}\frac{\partial W}{\partial x} + \frac{\partial V}{\partial x}\frac{\partial W}{\partial p} - \frac{\hbar^2}{24}\frac{\partial^3 V}{\partial x^3}\frac{\partial^3 W}{\partial p^3} + \gamma\frac{\partial (pW)}{\partial p} + \gamma m k_B T\frac{\partial^2 W}{\partial p^2}$$
$$+ D_{xx}\frac{\partial^2 W}{\partial x^2} + D_{px}\frac{\partial^2 W}{\partial p \partial x}, \tag{11.18}$$

where $D_{xx} = 1/(2m) \cdot [(\gamma\hbar^2)/(6k_B T)]$ and $D_{px} = (\Omega/\pi) \cdot [(\gamma\hbar^2)/(6k_B T)]$, in which Ω is the cut-off frequency for the harmonic bath. In the derivation of this equation, three conditions were placed on the parameters: (1) the bath response time $1/\Omega$ is short compared to characteristic system times; (2) weak coupling to the reservoir $\gamma << \Omega$; (3) moderate to high temperature, $k_B T \geq \hbar\Omega$. Note that the two Diosi smoothing terms become very small at high temperature, which suggests that under certain conditions (at high temperature), the CL equation has protection against loss of positivity.

Warning: It should *not be assumed* that a *positive density operator* necessarily leads to phase space distributions $W(x, p, t)$ that lack negative basins. For example, for a free particle, a superposition of two Gaussians in position space leads to a distribution that has negative interference ripples in some regions of phase space.

11.5 Phase Space Evolution with Entangled Trajectories

A novel method for evolving both classical and quantum phase space distributions, $W(x, p, t)$, through use of *ensembles of linked trajectories* has been developed by Donoso and Martens [11.2–11.4]. The trajectories are *entangled* through coupling terms in the equations of motion that depend on x and p derivatives of the local density around each trajectory. The coupling terms, in both classical and quantum mechanics, introduce *nonlocality and contextuality* (dependence on the initial state) into the equations of motion. The probability fluid evolves through phase space as a unified whole with each part both influencing, and dependent on, the dynamics of every other part. As stated by Donoso and Martens [11.3], “quantum effects arise as a breakdown of the statistical independence of the trajectories”.

In the computational algorithm used by Donoso and Martens, the nonlocal density-dependent force acting on a specific trajectory was determined by fitting one Gaussian to the known density values at a set of nearby trajectories. A moment method was employed to evaluate the parameters in the Gaussian exponent. From this fit, derivatives of the density were evaluated, and these were then used to calculate the nonlocal terms in the equations of motion for each of the entangled trajectories. The local nature of the fit allows multiple extrema in the density to be captured. Over the course of time, errors develop in the trajectories due to unavoidable inaccuracies arising from the fitting procedure. In spite of this, good results were obtained for evolution of trajectory ensembles for several model problems. These include the solution of the Kramers equation for the double-well potential [11.4] and evolution of the Wigner density for the decay of a metastable state [11.2, 11.3].

11.6 Phase Space Evolution Using the Derivative Propagation Method

In this section, a different method for propagation for classical or quantum entangled phase space trajectories will be described. Rather than step-by-step propagation of an ensemble of linked trajectories, *analytic equations of motion for the partial derivatives of the distribution function with respect to x and p will be derived, and these quantities will be propagated along the trajectories concurrently with the density itself.* This method is an application of the DPM, the derivative propagation method, which was described in Chapter 10. The DPM has been used to evolve trajectories for both classical and quantum open systems [11.1].

An enormous benefit of the DPM is that *single trajectories may be propagated, one at a time*, and fitting is no longer required to compute the spatial derivatives that are required in the equations of motion. The derivatives themselves introduce *regional nonlocality* into the trajectory dynamics, thus bypassing the step-by-step propagation of an ensemble comprising a large number of entangled trajectories. In effect, each trajectory acquires a "width" (due to the derivative terms) that permits it to monitor the surrounding hydrodynamic fields. The various orders of partial derivatives are coupled together in an infinite hierarchy, but low-order truncations of this system yield useful approximations. A significant benefit is that because function fitting is eliminated, the computation time is reduced by orders of magnitude. However, in spite of these benefits, DPM trajectories are approximate in the sense that *they do not build in long-range correlations* that might be important in determining the local dynamics.

In order to pave the way for use of the DPM to solve the phase space transport equations, we will recast these equations in terms of the *C-density*, defined by $W = \exp(C)$. The reason for introducing this exponential transform is that the C-density can frequently be represented by low-order polynomial expansions around a given trajectory, and this may not be true for W itself. For example, a Gaussian form for W is associated with a quadratic polynomial in (x, p) for the C-density. Of course, this transformation tacitly assumes that W is nonnegative. For solutions of the Kramers and Husimi equations, this is not an issue, but it is for solutions of the Wigner, Caldeira–Leggett, and modified Caldeira–Leggett equations. Following the spirit of the work by Donoso and Martens for the Wigner function, we might identify the computed $W(x, p, t)$ as a nonnegative transform of the underlying rippled (sometimes with negative basins) quantum distribution function.

Before presenting equations of motion for the C-density, we will introduce the following notation for the various partial derivatives with respect to x and p:

$$C_{(m,n)} = \frac{\partial^m \partial^n C(x, p, t)}{\partial^m x \partial^n p}. \tag{11.19}$$

The partial derivatives of W are readily expressed in terms of these partial derivatives; for example,

$$\frac{\partial W}{\partial x} = C_{(1,0)}W, \quad \frac{\partial^2 W}{\partial p^2} = (C^2_{(0,1)} + C_{(0,2)})W. \tag{11.20}$$

Using these relations, the phase space equations of motion for the density can be transformed into equations of motion for the C-density. For example, from equation 11.11 we obtain the transformed Kramers equation

$$\partial_t C_{(0,0)} = -\frac{p}{m}C_{(1,0)} + V_{(1,0)}C_{(0,1)} + \gamma\left[1 + pC_{(0,1)} + mk_BT\left(C_{(0,2)} + C^2_{(0,1)}\right)\right]. \tag{11.21}$$

The transformed version of the modified CL equation is obtained from equation 11.18:

$$\begin{aligned}\partial_t C_{(0,0)} = &-\frac{p}{m}C_{(1,0)} + V_{(1,0)}C_{(0,1)} - \frac{\hbar^2}{24}V_{(3,0)}\left[C_{(0,3)} + 3C_{(0,1)}C_{(0,2)} + C^3_{(0,1)}\right] \\ &+\gamma\left[1 + pC_{(0,1)} + mk_BT\left(C_{(0,2)} + C^2_{(0,1)}\right)\right] + D_{xx}\left(C^2_{(1,0)} + C_{(2,0)}\right) \\ &+D_{px}\left(C_{(1,0)}C_{(0,1)} + C_{(1,1)}\right).\end{aligned} \tag{11.22}$$

The derivative propagation method that we will use to evolve the x and p derivatives of C along individual trajectories will now be described.

In the DPM, the spatial derivatives are propagated in time along trajectories according to *exact* equations of motion. As shown previously, in Chapter 10, the equations of motion for these derivatives are straightforward to derive. As an example, in order to formulate the DPM for a generic evolutionary partial differential equation given by

$$\frac{\partial W(x,t)}{\partial t} = F\left(x, t; W, \frac{\partial W}{\partial x}, \frac{\partial^2 W}{\partial x^2}, \ldots\right), \tag{11.23}$$

we simply need to differentiate both sides with respect to x and then switch the order of the x and t partial derivatives. By doing this, the *analytic equation of motion* for an arbitrary n-th order spatial derivative becomes

$$\frac{\partial W_{(n)}}{\partial t} = \frac{\partial}{\partial t}\left(\frac{\partial^n W(x,t)}{\partial x^n}\right) = \frac{\partial^n F(x,t)}{\partial x^n}. \tag{11.24}$$

In order to demonstrate how this procedure leads to an infinite hierarchy of coupled equations, consider the one-dimensional diffusion equation

$$\frac{\partial W(x,t)}{\partial t} = -\frac{\partial}{\partial x}\left\{D(x)\frac{\partial W}{\partial x}\right\} = -\frac{\partial D}{\partial x}\frac{\partial W}{\partial x} - D\frac{\partial^2 W}{\partial x^2}, \tag{11.25}$$

In order to differentiate the right side with respect to x, we will again use the Leibniz identity for the n-th derivative of a product function $H(x) = F(x)G(x)$:

$$H_{(n)} = \sum_{j=0}^{n} b(n,j)F_{(n-j)}G_{(j)}, \tag{11.26}$$

where $b(n, j)$ is the binomial coefficient. Using this relation, from equation 11.25

we obtain the equation of motion for the n-th spatial derivative:

$$\partial_t W_{(n)} = -\sum_{j=0}^{n} b(n, j) D_{(n-j+1)} W_{(j+1)} - D W_{(n+2)}. \tag{11.27}$$

It can be seen from this equation that not only is there *up-coupling* to higher-order derivatives, but there is also *down-coupling* in the summation. It is the up-coupling, however, that leads to the infinite hierarchy. For arbitrary evolutionary equations, this chain of equations is impossible to solve exactly. However, in favorable situations a suitable truncation can be imposed.

Returning to the phase space equations of motion, the DPM equations are obtained by application of the operator $\partial^{(n+m)}/\partial x^n \partial p^m$ (n and m each taking on values 0, 1, 2, ...). The equation for the (n, m) partial derivative has the general form

$$\frac{\partial C_{(n,m)}}{\partial t} = G(x, p; C, C_{(1,0)}, C_{(0,1)}, \dots), \tag{11.28}$$

where the right side brings in coupling to both lower and higher derivatives. These algebraic equations are usually not written down explicitly; rather, they are generated directly within recursion loops in the computer program that solves these equations. Most results so far have been obtained with DPM orders $n + m$ of 2, 3, and 4, corresponding to assumed quadratic, cubic, and quartic local polynomial expansions of the C-density around each trajectory. Keeping the order as low as possible is vital, since in multidimensional problems one must propagate not only the "pure" spatial derivatives in each coordinate, but the various cross terms as well.

Use of the DPM equations brings up the issue of *algorithmic nonlocality*: in order to approximate the spatial derivatives at specific trajectory locations, information riding along nearby trajectories in a surrounding stencil is needed. It appears that all grid points must be propagated simultaneously as a correlated ensemble, with the result that if one trajectory goes bad, this can terminate propagation of the entire ensemble. The quantum trajectory method described in Chapter 4 is based on this form of lockstep propagation of an ensemble of trajectories, as is the method of Donoso and Martens that was described in Section 11.5. The DPM follows a different strategy: the hydrodynamic fields and their derivatives are propagated along individual trajectories, and these may be run one at a time.

11.7 Equations of Motion for Lagrangian Trajectories

The equations of motion described previously were all expressed in the Eulerian frame, i.e., from the viewpoint of an observer fixed at the phase space point (x, p). We will now transform these equations into the Lagrangian frame, wherein an observer follows along a trajectory with the local fluid velocity. In order to do this, we will follow Donoso and Martens [11.2–11.4] and note that each phase space equation of motion can be written in the form of a *continuity equation* involving the divergence of the flux vector $\vec{J}$:

$$\frac{\partial W}{\partial t} = -\vec{\nabla} \cdot \vec{J}, \tag{11.29}$$

where the phase space gradient operator is again $\vec{\nabla} = (\partial/\partial x, \partial/\partial p)$. If we require that the flux be the product of the density and the velocity, $\vec{J} = W\vec{V}$, where $\vec{V} = (V_x, V_p)$ is the phase space velocity vector, then the velocity components are defined when the right side of each equation of motion is written as a divergence. Except for the modified CL equation, the x-velocity is $V_x = p/m$. The velocity component in the p-direction can be decomposed into a *classical (local) component* along with a *nonlocal density-dependent term*,

$$V_p = dp/dt = F_{local} + F_{nonlocal}(W), \tag{11.30}$$

in which F_{local} includes both the classical force $-\partial V/\partial x$ and the dissipative term $-\gamma p$, if it is required from the original equation of motion. As examples of p-velocities, this quantity for the Kramers and modified CL equations, expressed in terms of the C-density, is given by the following expressions:

$$\begin{aligned} Kramers: \quad & V_p = -V_{(1,0)} - \gamma p - \gamma m k_B T C_{(0,1)}, \\ ModCL: \quad & V_p = -V_{(1,0)} - \gamma p + \frac{\hbar^2}{24} V_{(3,0)} \left(C_{(0,2)} + C^2_{(0,1)}\right) - \gamma m k_B T C_{(0,1)} \\ & \quad -D_{pq} C_{(1,0)}. \end{aligned} \tag{11.31}$$

In these equations, the first term is the classical force, the second term $(-\gamma p)$ leads to momentum relaxation, and the remaining nonlocal terms bring in x and p derivatives of the density.

These velocity components also find use in the transformation from the Eulerian to the Lagrangian frame. The rate of change in the function $F(x, p, t)$, as seen by an observer moving at velocity $\vec{V}$, is

$$\frac{dF}{dt} = \frac{\partial F}{\partial t} + \vec{V} \cdot \vec{\nabla} F, \tag{11.32}$$

where the last term is the convective contribution. When the function F refers to the density, this equation becomes the Lagrangian form of the continuity equation

$$\frac{dW}{dt} = -\vec{\nabla} \cdot (\vec{V} W) + \vec{V} \cdot \vec{\nabla} W = -W\vec{\nabla} \cdot V. \tag{11.33}$$

This equation can be integrated along the trajectory to give the new density in terms of the value at the initial time,

$$W(q(t), p(t), t) = \exp\left(-\int_0^t \vec{\nabla} \cdot \vec{V}(\tau) d\tau\right) W(q(0), p(0), 0). \tag{11.34}$$

Except for the classical Liouville equation, the other evolutionary equations in Sections 11.2–11.4 lead to *non-Hamiltonian dynamics* in which the flow is generally compressible (expansive or contractive), with a nonzero value for the divergence, $\vec{\nabla} \cdot \vec{V} \neq 0$. As a result, elementary volume elements may contract or expand along the flow, and the flow itself is state-dependent. Daligault [11.34] has made interesting comments about these features of phase space flow; see Box 11.3.

For the derivatives of the C-density, the Lagrangian derivative, equation 11.32, becomes

$$\frac{dC_{(n,m)}}{dt} = \frac{\partial C_{(n,m)}}{\partial t} + V_x C_{(n+1,m)} + V_p C_{(n,m+1)}, \tag{11.35}$$

and this brings in the next higher order of derivative with respect to both x and p. The $\vec{V} \cdot \vec{\nabla} F$ terms always cancel some of the terms in the Eulerian derivative, so that the Lagrangian derivative is actually simpler that the one calculated by a fixed observer. For example, for the Kramers equation, the Lagrangian time derivative of the C-density is given by

$$\frac{dC_{(0,0)}}{dt} = \gamma + \gamma m k_B T C_{(0,2)}, \tag{11.36}$$

and this is much simpler than the Eulerian derivative given earlier in equation 11.21.

11.8 Examples of Quantum Phase Space Evolution

In this section, two computational examples will be described to illustrate the evolution of phase space quantum trajectories for open quantum systems. In the first example, we will consider the decay of a metastable state using a potential of the same functional form that was shown earlier, in figure 10.1. The metastable well potential $V(x) = (1/2)m\omega^2 x^2 - (1/3)bx^3$ was used with the parameter values (unless specified otherwise, all quantities are in atomic units) $m = 2000$, $\omega = 0.01$, and $b = 0.2981$ (see also the results by Donoso and Martens in [11.2]). The harmonic period for the potential well centered at $x = 0$ is $\tau_{vib} = 2\pi/\omega = 628$ a.u. $= 15$ fs, the barrier maximum occurs at $x^* = 0.67$, the barrier height is $V^* = 0.015$ (3292 cm^{-1}), and the barrier momentum is $p^* = \sqrt{2mV^*} = 7.745$. In addition, the following parameter values were used in the equations of motion: $\gamma = 5x10^{-4}$, $k_B T = 0.025$, and $\Omega = 0.05$ (bath cutoff frequency).

Using the third-order DPM, trajectories were evolved for the modified CL equation. These trajectories were launched one at a time from initial phase space points chosen as follows. The center of the initial minimum uncertainty distribution given earlier, in equation 11.14, was located at $x_0 = -0.3$, $p_0 = 0.0$ (these quantities are denoted by (x', p') in equation 11.14), and the width was set by the parameter $\alpha = \hbar/(m\omega) = 0.05$. When plotted in phase space, the elliptic contour having the value 10% of the maximum value ($W_{\max} = 1/\pi = 0.318$) of this initial density stretches about the center to ± 6.7 along the p axis and ± 0.3 along the x axis. Around this central point, trajectories were selected from points within the elliptic boundary on which the density is $W = 10^{-4}$. From within this outer contour, trajectories were launched from 9281 initial positions located on a rectangular mesh. Figure 11.1 shows the locations of the initial trajectories (black dots) superimposed on a grey-scale map of the density.

Starting from the initial phase space points illustrated in figure 11.1, the time evolution of the ensemble was generated using the DPM to propagate single trajectories. Figure 11.2 shows the evolving probability distribution at five times: 4, 8, 12, 16, and 20 fs. The trajectories launched from the initial distribution rotate in a

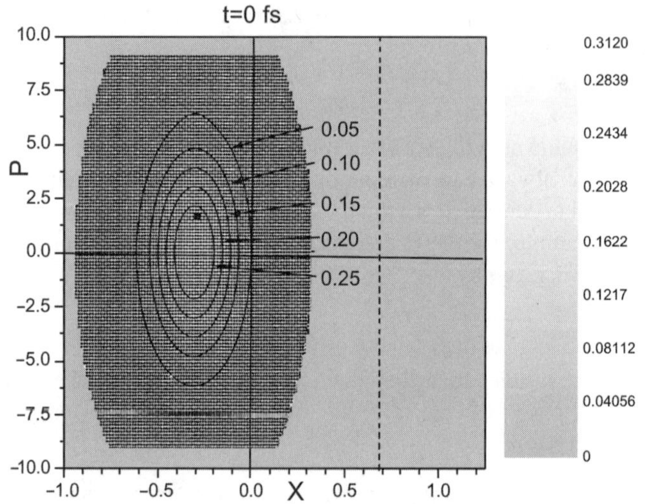

FIGURE 11.1. Trajectory locations (black dots) at $t = 0$ superimposed on a map of the density. The elliptic contour lines cut through the density at the values indicated. The vertical dashed line at $x^* = 0.67$ locates the barrier maximum.

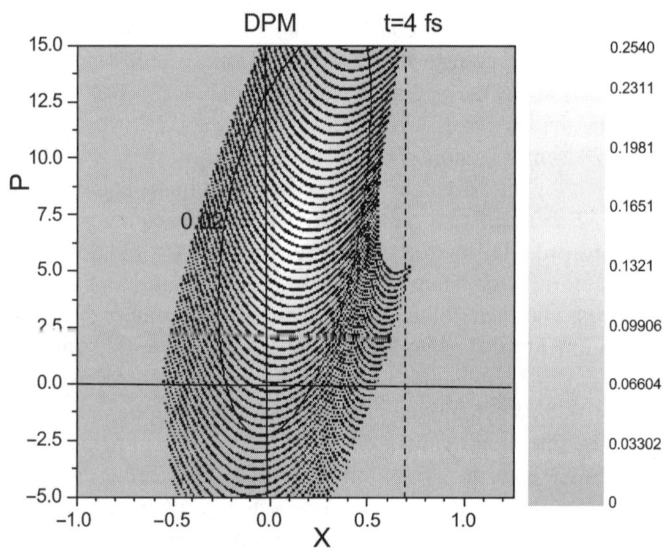

FIGURE 11.2. Trajectory locations (black dots) at five times superimposed on grey-scale maps of the density. The contour lines cut through the density at the value that is shown in each figure. The vertical dashed line at $x^* = 0.67$ locates the barrier maximum.

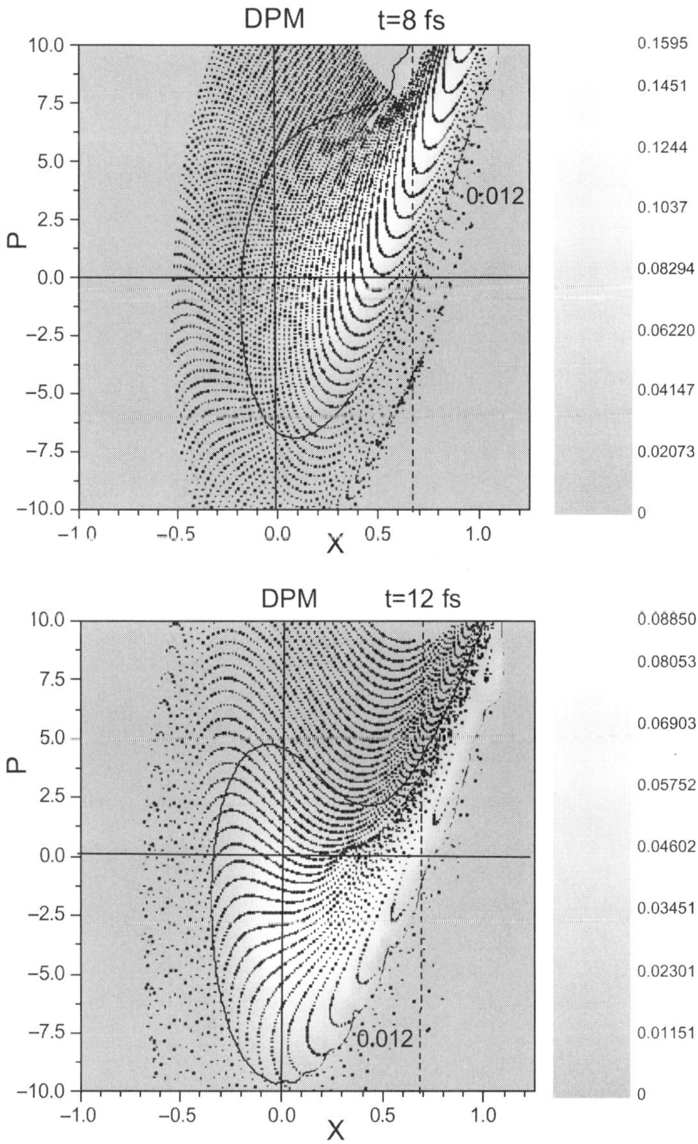

FIGURE 11.2. *Continued*

clockwise sense about the origin, and during each cycle some of these trajectories surge over the barrier to form the diffuse tube that extends toward the upper right of each figure (except for the early time plot at 4 fs). As time proceeds, the central part of the distribution gradually relaxes toward the origin.

By 4 fs, the center of the distribution has moved to about $x = 0.20$, $p = 7.5$ and shows rapid spreading in the momentum direction. By 8 fs, near the half-period of the motion around the bottom of the well, a tube of trajectories is evolving over the

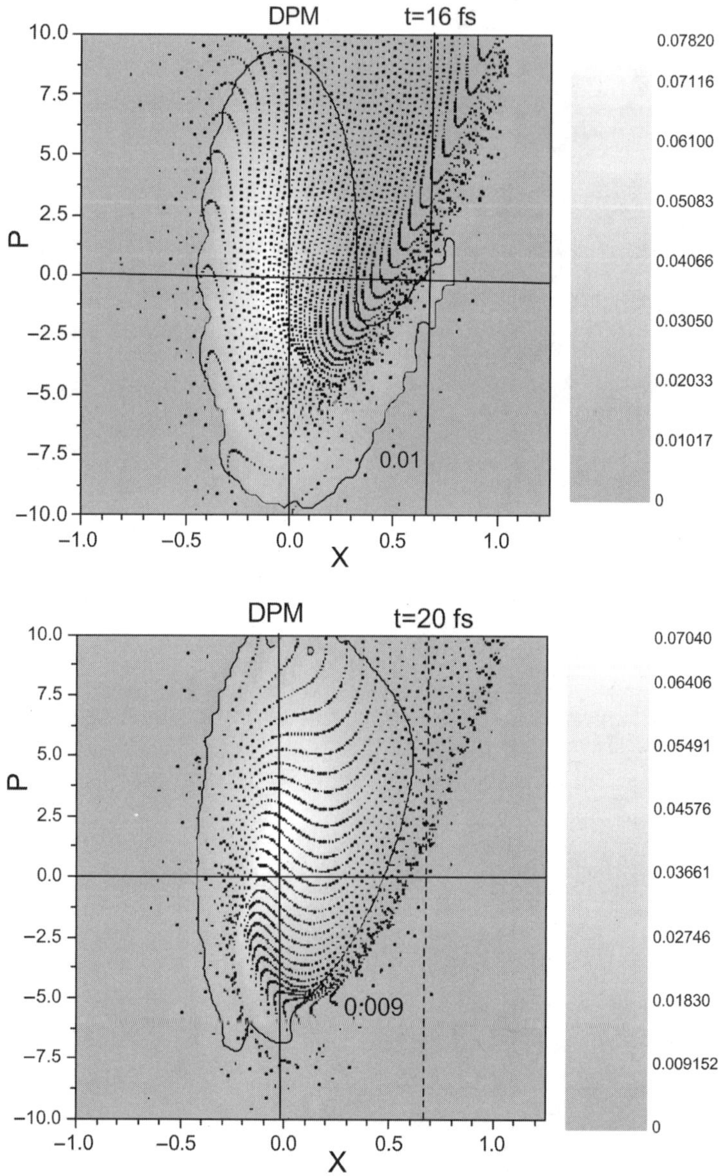

FIGURE 11.2. *Continued*

barrier toward the upper right side of the figure. The trajectories evolve over the barrier through what Skodje et al. described as a "flux-gate" [11.36]. At 12 fs, the center has moved to about $x = 0.25$, $p = -3.5$, and a few trajectories continue to escape over the barrier. At 16 fs, approximately one period of motion, the center is near $x = -0.15$, $p = 0$. At 20 fs, the center has twisted further in the clockwise sense to $x = 0.1$, $p = 3$, and the second surge of trajectories is starting to cross the barrier. At later times, the central part of the density continues to spiral in toward the origin, and on each revolution a smaller number of trajectories make it over the barrier.

In order to assess the accuracy of the approximate DPM solutions that were shown in figure 11.2, we will compare these results with accurate solutions obtained on a lattice of space-fixed points. After setting up the lattice in phase space and discretizing the differential operators that appear in the mod CL equation of motion, a system of ordinary differential equations may be integrated to find the time evolution of the solution at the lattice points. This procedure is described in more detail in Box 11.8. The grid solution at one time, $t = 16$ fs, is displayed in figure 11.3, and this should be compared with the DPM solution at the same time in figure 11.2. First, we note that the two figures are qualitatively similar: the maxima in the density are at almost the same locations, and in both figures there is a narrow tube of density extending toward the upper right. The maximum density for the grid solution is 0.0798, and this is within 2% of the DPM value. However, there is one important qualitative difference: the solution obtained using the fixed-grid becomes slightly negative (minimum value -0.008) in the banana shaped region just above the tube of density that extends over the barrier toward the upper right side of the figure. The trajectory ensemble has no way of generating negative basins in the density. However, for the parameter values (friction coefficient and temperature) used to generate these figures, the negative basins are not very significant. At other times, there may be several shallow negative basins, always located on the back side of the density tube (i.e., toward higher values of the momentum) that extends over the barrier.

The final result that will be presented for the modified Caldeira–Leggett equation concerns the effect of increasing the friction parameter. Figure 11.4 shows the density obtained by integrating this equation using a value for the friction coefficient that is twice as large as the value that was used previously (the new value is $\gamma = 0.0010$), but the temperature was not changed. Comparing the density in figure 11.4 with the density at $t = 16$ fs in figure 11.2, we note several differences. First, the maximum density, 0.086, is higher than the maximum value for the previous density (0.078). In addition, the integrated density in the internal region (to the left of the barrier maximum), 0.54, is 15% higher than for the previous case. Thus, at this temperature, increasing γ leads to slightly enhanced localization in the internal region.

The last computational result concerns a comparison between the distribution function obtained by integrating the Kramers equation (no quantum effects) with the result obtained by integrating the modified Caldeira–Leggett equation. Figure 11.5 shows the density obtained by integrating the Kramers equation (using the same values for the friction coefficient and the temperature that were used to obtain the quantum results). Comparing the density in figure 11.5 with the density at $t = 16$ fs in figure 11.2, we note several differences. First, the maximum value

Box 11.8. Phase space dynamics on a lattice

For problems with just a few degrees of freedom, it is possible to solve the partial differential equation for the phase space distribution function on a space-fixed lattice [10.14]. The evolutionary equation will be denoted by $\partial W/\partial t = \hat{L}W$, where $\hat{L}$ denotes the operator containing the classical force and the various differential operators. As an example, we will assume a single degree of freedom, so that the phase space is characterized by coordinates (x, p). First, the accessible region between $x_{\min} = x_0$ and $x_{\max} = x_m$ is subdivided into m cells, each of length $\Delta x = (x_m - x_0)/m$. Likewise, the momentum axis between $p_{\min} = p_0$ and $p_{\max} = p_n$ is subdivided into n cells of length $\Delta p = (p_n - p_0)/n$. Rather than finding the continuous function $W(x, p, t)$, we will settle for obtaining approximate values at the lattice points. The solution at the lattice point (x_j, p_k) is denoted by $W_{j,k}$. Then, around this point, the differential operators appearing in the equation of motion are approximated by finite difference expressions. For example, the second-order derivatives can be represented by the usual expressions (which have fourth-order error terms)

$$\frac{\partial^2 W_{j,k}}{\partial x^2} = \frac{1}{\Delta x^2}\left[W_{j+1,k} - 2W_{j,k} + W_{j-1,k}\right] + O(\Delta x^4), \tag{1}$$

$$\frac{\partial^2 W_{j,k}}{\partial p^2} = \frac{1}{\Delta p^2}\left[W_{j,k+1} - 2W_{j,k} + W_{j,k-1}\right] + O(\Delta p^4), \tag{2}$$

where the values of the distribution function in the stencil surrounding the point of interest enter the expressions for the derivatives. These expressions are valid for interior lattice points; similar equations are used to evaluate derivatives on the boundary. On the boundary, the solution is forced to vanish, $W_{j,k} = 0$. In order to prevent back reflections from the boundary from spoiling the solution in the internal region, at each time step the solution near the boundary is usually damped with a Gaussian function.

The evolutionary equation at each grid point may now be expressed in the form $dW_{j,k}/dt = F(W_{j,k}, W_{j+1,k}, \ldots)$, where the right side depends on a few of the discretized solution values. Given the initial values of the solution at $t = 0$, this system of ordinary differential equations may then be integrated using a standard algorithm, such as fourth-order Runge–Kutta.

for the Kramers density, 0.140, is almost a factor of two higher than the maximum value for the quantum density, and the location of the maximum for the Kramers case occurs for a lower (more negative) value of the momentum. In addition, the integrated density in the internal region (to the left of the barrier maximum), 0.66, is 42% higher than for the quantum density. Thus, turning on the quantum terms in the equation of motion leads to a significant decrease in the density trapped in the internal region to the left of the barrier. We may be tempted to attribute this to tunneling in the quantum case, but we should remember that there may

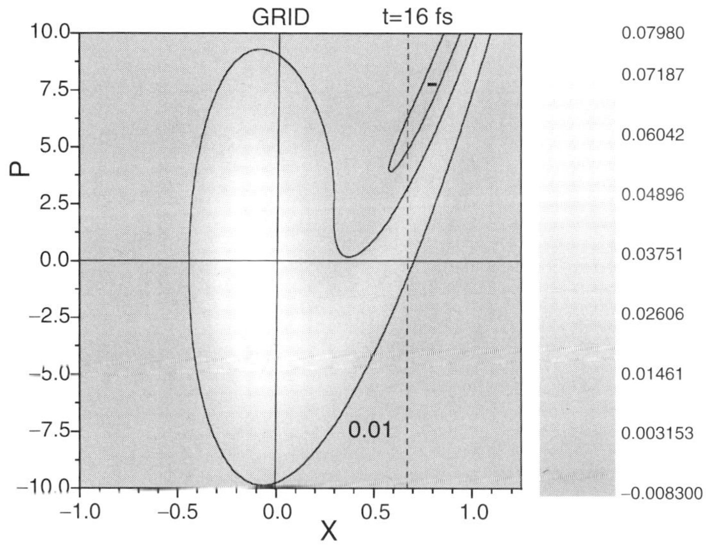

FIGURE 11.3. Solution of the modified Caldeira–Legget equation obtained on a space-fixed lattice in the two-dimensional phase space. The solution becomes slightly negative in the banana-shaped region near the upper segment on the vertical dashed line, which marks the location of the barrier maximum. This density should be compared with the DPM solution at the same time that was shown in figure 11.2.

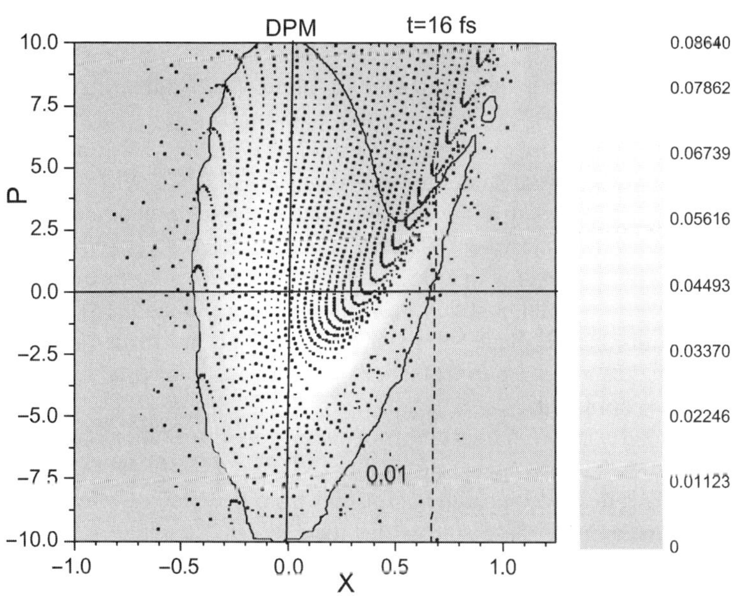

FIGURE 11.4. Trajectory locations (black dots) superimposed on a grey-scale map of the density. The contour line cuts through the density at the value that is shown in the figure. The vertical dashed line at $x^* = 0.67$ locates the barrier maximum. Compared with the parameters used to generate the results shown in figure 11.2, the value of the friction coefficient has been increased by a factor of two (to the value $\gamma = 0.001$), but $k_B T$ retains the same value.

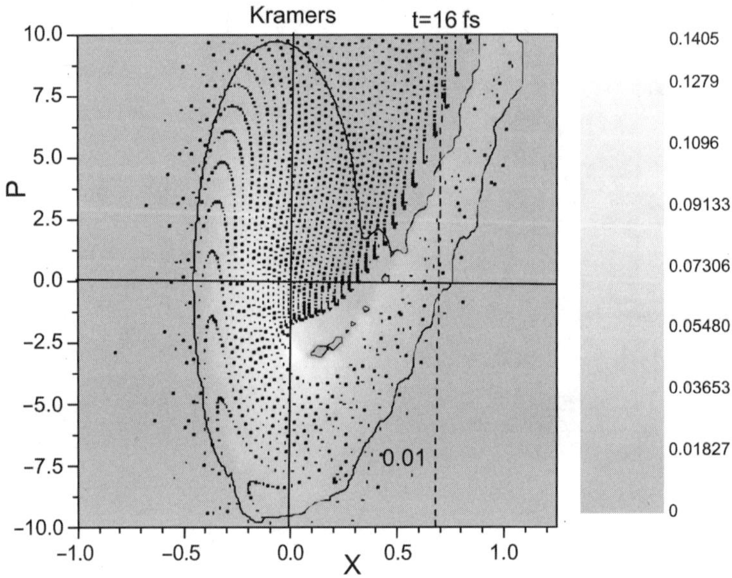

FIGURE 11.5. Trajectory locations (black dots) superimposed on a grey-scale map of the probability density. Rather than the modified Caldeira–Leggett equation that was used to generate the results in figures 11.2–11.4, the Kramers equation was integrated using the DPM.

be additional quantum effects (interference effects) that influence the quantum evolution.

In the earlier study by Trahan and Wyatt [11.1], the DPM was applied to the Kramers, Wigner, Husimi, Caldeira–Leggett, and modified Caldeira–Leggett equations of motion. The decay of a metastable state in the same anharmonic potential was studied, for different values of the friction coefficient. In 1989, Skodje et al. [11.36] presented a detailed hydrodynamic analysis of phase space evolution for the double-well potential, and this was followed by a similar analysis in 1991 of the periodically kicked plane rotor (the quantum standard map) [11.37]. In the latter two studies, the wave function was precomputed before the hydrodynamic analysis was performed.

In the second example, the scattering of phase space distributions from an Eckart barrier were studied. In the work by Hughes and Wyatt [11.40], DPM transmission probabilities were also compared with results from fixed-lattice calculations. The Eckart barrier used in these studies has the following parameter values: $V_0 = 3000\ \text{cm}^{-1}$, $q_b = 3$ a.u. (barrier location), and $\alpha = 2.5$ a.u. (width parameter). The initial distribution was centered at $x = 0$ and $p = 7.39$ ($E_{trans} = 3000\ \text{cm}^{-1}$) and the mass is $m = 2000$ (atomic units are used, unless specified otherwise).

For the DPM calculations, 18,113 trajectories were launched one at a time from initial locations on a uniform grid. These trajectories at $t = 0$ are shown by the swarm of dots in figures 11.6 (a) and 11.7 (a). The subsequent time evolution for

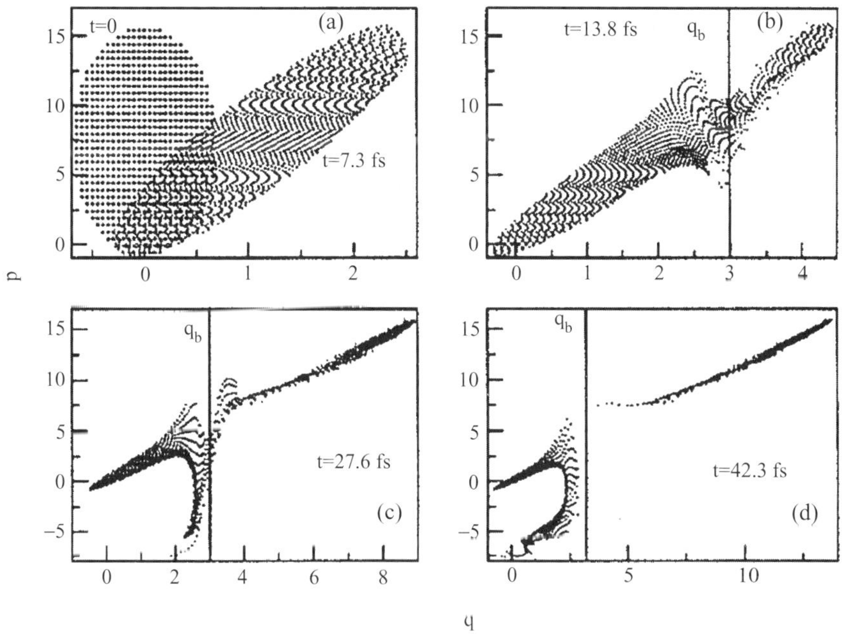

FIGURE 11.6. Evolution of third-order DPM trajectories as they encounter an Eckart barrier centered at $q_b = 3$ a.u. [11.40]. The trajectory equations of motion were derived from the Wigner–Moyal equation. The dots show the trajectory locations at five times. Initially, there were 18,113 trajectories in the ensemble.

these trajectories is shown in figure 11.6 for the Wigner–Moyal equation and in figure 11.7 for the modified CL equation. At early times, shown in part (a) of each figure, the distribution rotates clockwise and stretches because the higher-momentum components move faster than the lower-momentum components. Closer to the potential barrier, shown in part (b) of each figure, the higher-momentum components are slowed down before crossing the barrier maximum at $q = 3$. The trajectory ensemble then bifurcates in the barrier region as the higher momentum components cross to form the transmitted distribution. For the dissipative system shown in figure 11.7, there is a sinking of the distribution into negative regions of p space and, although only slightly apparent, there is a slowing of the whole distribution as it moves along the q direction. These observations are especially apparent in figures 11.7 (c) and (d). In addition, compared with the nondissipative case, both the transmitted and reflected distributions are blurred into rather fuzzy distributions. Finally, in figure 11.8, the time-dependent transmission probability obtained for the Wigner function is compared with exact results from finite difference fixed-grid calculations. From this figure, it is seen that there is excellent agreement between the two sets of results. Other results (density matrix elements, average energies, and transmission probabilities as a function of friction coefficient and temperature) are described in [11.40].

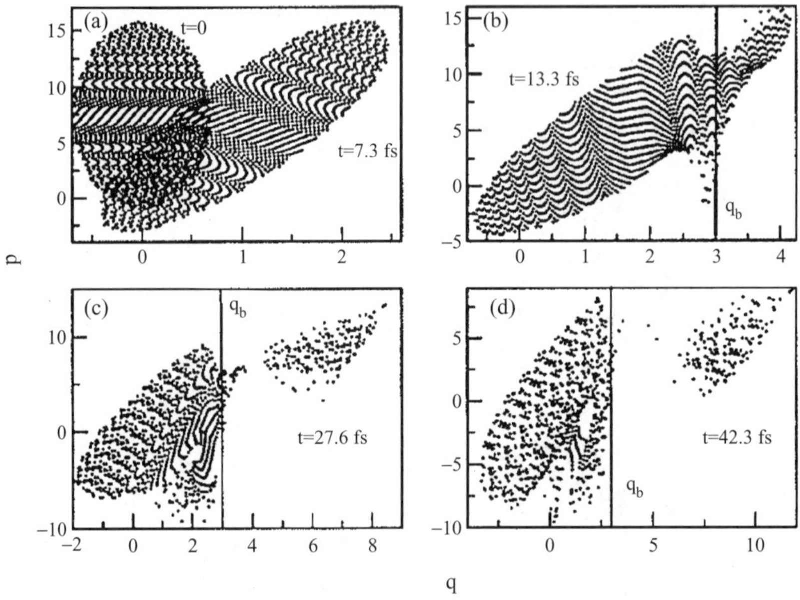

FIGURE 11.7. Evolution of third-order DPM trajectories as they encounter an Eckart barrier centered at $q_b = 3$ a.u. [11.40]. The trajectory equations of motion were derived from the modified CL equation. Trajectory locations are shown at five times. Initially, there were 18,113 trajectories in the ensemble. The parameter values used in the equations of motion were as follows: $\gamma = 5 \times 10^{-4}$ and $k_B T = 1390$ cm^{-1}.

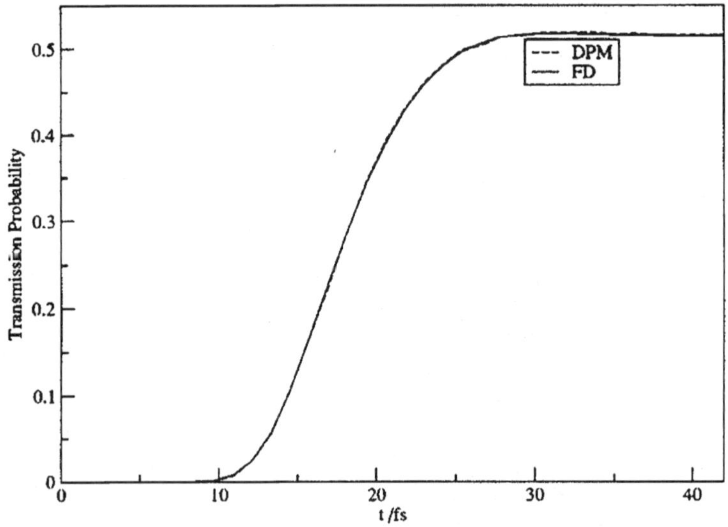

FIGURE 11.8. Time-dependent transmission probabilities for Eckart barrier scattering [11.40]. For the DPM results (dashed curve), the trajectory equations of motion were derived from the Wigner–Moyal equation. The fixed grid results (continuous curve) are almost indistinguishable from the DPM results.

11.9 Momentum Moments for Dissipative Dynamics

In the phase space route to the hydrodynamic equations described earlier, in Chapter 3, we began by defining the *momentum moments* $\bar{p}_n(x, t)$ of the phase space distribution function $W(x, p, t)$. Then, from the equation of motion for the distribution function (the Wigner–Moyal equation was used), coupled equations of motion for these moments were derived. In general, this system of equations forms an infinite hierarchy, with the rate of change in the n-moment coupled to the next higher moment and to a sequence of lower moments. For a system described by a wave function, a *pure state*, the 2-moment and all higher moments can be expressed in terms of the 0-moment (the position space probability density) and the 1-moment (which is related to the flux). For these states, the rate equation hierarchy can thus be replaced with a closed set of rate equations involving just the lowest two moments. For *mixed states*, the rate equation hierarchy in general does not terminate exactly. However, for specific assumed forms for the distribution function, the hierarchy can be made to terminate.

A relatively simple but informative example was described by Burghardt and Moller [11.29]. Using the Caldeira–Leggett (CL) equation of motion for the phase space distribution function, they computed momentum moments for a time-dependent Gaussian distribution evolving in a harmonic oscillator potential. In spite of the dissipative effects, an initial Gaussian retains this functional form at all later times, but the average position, momentum, and width parameters are time-dependent. Because this potential is quadratic, the $\hbar$-dependent terms in the CL equation (which bring in third and higher derivatives of the potential) all vanish. However, the moments themselves bring in "quantum corrections" from the initial distribution function.

From the CL equation, Burghardt and Moller [11.29] derived the following equations of motion for the first three moments:

$$\frac{\partial \rho(x,t)}{\partial t} = -\frac{1}{m}\frac{\partial \bar{p}_1(x,t)}{\partial x}, \tag{11.37}$$

$$\frac{\partial \bar{p}_1(x,t)}{\partial t} = -\frac{1}{m}\frac{\partial \bar{p}_2(x,t)}{\partial x} - \frac{\partial V}{\partial x}\rho(x,t) - \gamma \bar{p}_1(x,t), \tag{11.38}$$

$$\frac{\partial \bar{p}_2(x,t)}{\partial t} = -\frac{1}{m}\frac{\partial \bar{p}_3(x,t)}{\partial x} - 2\frac{\partial V}{\partial x}\bar{p}_1(x,t) - 2\gamma \bar{p}_2(x,t) + 2\gamma m < E(T) > \rho(x,t). \tag{11.39}$$

In the second and third of these equations, γ is again the friction coefficient, which links the open quantum subsystem to the thermal bath. In the continuity equation, equation 11.37, thermal or friction effects do not appear explicitly. However, the rate equations for the 1-moment and the 2-moment do involve frictional relaxation terms. In equation 11.39, the *average thermal energy* for the quantum harmonic oscillator is given by

$$\langle E(T)\rangle = \frac{\hbar\omega}{2}\coth\left[\frac{\hbar\omega}{2k_B T}\right]. \tag{11.40}$$

At low temperature, $k_B T \ll \hbar\omega$, the average energy approaches the zero-point value, $E_0 = \hbar\omega/2$, while at high temperature, $k_B T \gg \hbar\omega$, the thermal energy approaches the classical limit, $k_B T$. For a *Gaussian density*, the 3-moment in equation 11.39 can be expressed in terms of the three lower moments,

$$\bar{p}_3(x,t) = -2p(x,t)^3\rho(x,t) + 3p(x,t)\bar{p}_2(x,t), \tag{11.41}$$

where $p(x,t)$ is the average momentum. Because of this, the coupled equations for the first three moments form a closed set. For other systems where the density could assume a non-Gaussian shape, truncation at this order would serve as an approximation.

Equation 11.38 can be recast into the form of equations 3.30 and 3.31, which involve the hydrodynamic force

$$\frac{dp(x,t)}{dt} = -\frac{\partial V(x)}{\partial x} - \gamma p + F_{hydro}(x,t) = -m\omega^2 x - \gamma p + F_{hydro}(x,t), \tag{11.42}$$

in which

$$F_{hydro} = -\frac{1}{m\rho(x,t)}\frac{\partial\sigma(x,t)}{\partial t} = \frac{D(t)}{m\sigma_{xx}^2(t)}(x - \langle x(t)\rangle). \tag{11.43}$$

The *momentum variance* of the phase space distribution function (see equation 3.19) can also be expressed, $\sigma(x,t) = \bar{p}_2(x,t) - p(x,t)^2\rho(x,t)$. For the Gaussian density evolving in the harmonic potential, the hydrodynamic force becomes $F_{hydro}(x,t) = \kappa(t)\,[x - \langle x(t)\rangle]$, where $\kappa(t)$, see equation 11.43, has been expressed in terms of the Gaussian parameters that appear in equation 11.44 (next paragraph). This force vanishes at the time-dependent average position and is linear in the displacement from the average position.

For this example, another approach can be followed in order to verify the preceding equations. Because the phase space density is always Gaussian, it can be written in the form

$$W(x,p,t) = \frac{1}{2\pi\sqrt{D}}\exp\left\{-\frac{1}{2D}\left[\sigma_{pp}\delta_x^2 + \sigma_{xx}\delta_p^2 - 2\sigma_{xp}\delta_x\delta_p\right]\right\}, \tag{11.44}$$

where the x and p deviations from the mean values are $\delta_x = x - \langle x\rangle$ and $\delta_p = p - \langle p\rangle$, respectively. Correlation between the x and p motions is brought in by the term involving σ_{xp}. When this correlation parameter is nonzero, the cigar-shaped density has its principal axes at an angle with respect to the (x,p) coordinates. The two mean values $\langle x\rangle$ and $\langle p\rangle$ along with the three width parameters $\{\sigma_{xx}, \sigma_{pp}, \sigma_{xp}\}$ are all time-dependent quantities. Also, the quantity D in this equation and in equation 11.43 is the time-dependent determinant of the variance matrix, $D = \sigma_{xx}\sigma_{pp} - \sigma_{xp}^2$. (For a pure state, this determinant becomes independent of time, $D = (\hbar/2)^2$). A system of five coupled differential equations can be set up and integrated to provide values for these parameters. It is possible to express the momentum moments of this phase space density in terms of the same set of Gaussian parameters. As a consistency check, the moments calculated this way can then be compared with those obtained by directly integrating the coupled rate equations for the momentum moments.

For times longer than the relaxation time, the distribution settles into a thermal equilibrium state characterized by time-independent values for the five parameters in the Gaussian function. In this case, the average position and momentum are both zero, $\langle x \rangle = \langle p \rangle = 0$, and the three width parameters are $\sigma_{xx} = \langle E(T) \rangle / (m\omega^2)$, $\sigma_{pp} = m\langle E(T) \rangle$, and $\sigma_{xp} = 0$. Using these parameter values, the phase space density for the *thermal equilibrium state* becomes

$$W_{thermal}(x, p) = \frac{\omega}{2\pi \langle E(T) \rangle} \exp\left[-H(q, p)/\langle E(T) \rangle\right], \tag{11.45}$$

where $H(q, p)$ is the Hamiltonian for the classical harmonic. The position space density evaluated from this equation becomes

$$\rho(x)_{thermal} = \left(\frac{m\omega^2}{2\pi \langle E(T) \rangle} \right)^{1/2} \exp\left[-\frac{(m\omega^2 x^2/2)}{\langle E(T) \rangle} \right]. \tag{11.46}$$

In addition, the hydrodynamic force becomes, $F_{hydro} = m\omega^2 x$, and this exactly balances the classical force $F_{cl} = -m\omega^2 x$, so that the total force acting on each quantum trajectory is zero in the equilibrium state.

In figures 11.9 and 11.10, quantum trajectory results are shown for an initial Gaussian distribution relaxing toward an equilibrium distribution. Initially, the average position and momentum are $\langle x \rangle = -0.75$ a.u. and $\langle p \rangle = 0$, and the other parameter values are given in the caption to figure 11.9. Figure 11.9 shows the average values for the coordinate and momentum as the distribution relaxes toward the thermal equilibrium values, $\langle x \rangle = \langle p \rangle = 0$. Figure 11.10 shows the time dependence of five quantum trajectories as the distribution gradually

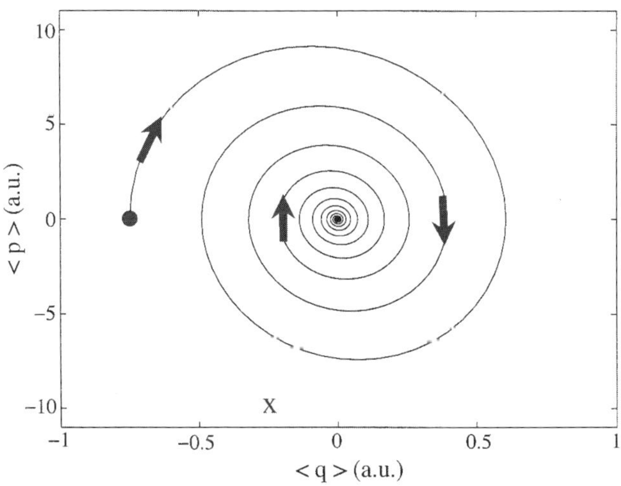

FIGURE 11.9. Mean values for x and p for a time evolving Gaussian distribution governed by Caldeira–Leggett dissipative dynamics [11.29]. The parameter values are: $m = 10$ a.u., $\omega = 0.000741$ a.u. (which gives the period, 205 fs), $\gamma = 0.0001$ a.u., and $k_B T = 0.0016$ a.u.

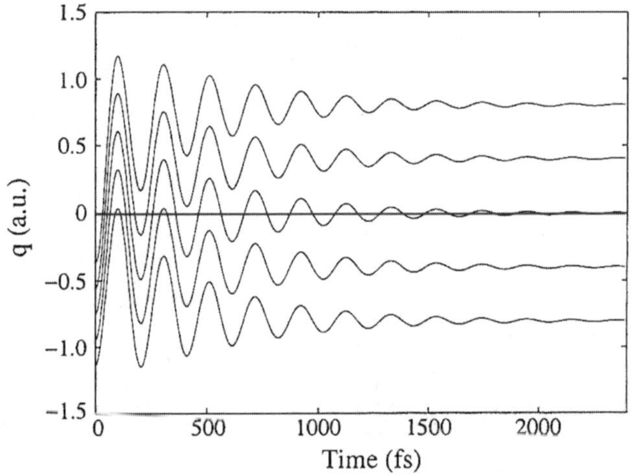

FIGURE 11.10. A set of five hydrodynamic trajectories associated with a time-evolving Gaussian distribution that is governed by Caldeira–Leggett dissipative dynamics [11.29]. The parameter values are given in the caption to figure 11.9.

approaches (over a time interval of about 2000 fs) the thermal equilibrium state. During the first 100 fs, the distribution spreads, and then the trajectories undergo damped oscillations (but they remain in phase with one another) about the equilibrium values. The trajectories never cross paths, but move "parallel" to one another. For the central trajectory of this cluster, the hydrodynamic force vanishes at all times, and this trajectory (equivalent to a classical trajectory) gradually settles in on the average values given by $\langle x \rangle = \langle p \rangle = 0$.

11.10 Hydrodynamic Equations for Density Matrix Evolution

In the hydrodynamic approach to density matrix evolution, equations of motion are integrated to yield the time evolution of quantum trajectories [11.32, 11.33]. From amplitude and phase information carried along these trajectories, diagonal and off-diagonal elements of the density matrix can be synthesized. An advantage of this approach is that the trajectory locations define a moving grid that adapts to the flow. This moving grid eliminates the need to have grid points in regions where there is little activity, but if the density needs to build up in a formerly vacant region, the grid faithfully follows the flow. In the first part of this section, the equations of motion for the hydrodynamic fields are derived for a *pure state* not coupled to an environment. Then, in the remainder of this section, density matrix evolution is considered for an *open quantum system* in contact with a thermal reservoir. In the latter case, there are many possible starting points, but in this section the equations of motion will be derived from the Caldeira–Leggett equation. Two applications to modeling open dissipative quantum systems will then be discussed in Section 11.11.

Matrix elements of the density operator in the position representation, denoted by $\rho(x, y, t) = \langle x|\hat{\rho}(t)|y\rangle$, define the density matrix. When the values for x and y are equal, the diagonal elements measure *populations*, $\rho(x, t) = \langle x|\hat{\rho}(t)|x\rangle$, and when $x \neq y$, the density matrix element refers to *coherence* between the two position eigenstates, $|x\rangle$ and $|y\rangle$. For an isolated quantum system in a pure state, the density matrix at all times can be factored, $\rho(x, y, t) = \langle x|\psi(t)\rangle\langle\psi(t)|y\rangle$, and the density matrix evolves according to the *Liouville–von Neumann equation of motion*

$$i\hbar\frac{\partial\rho(x, y, t)}{\partial t} = [H(x) - H(y)]\,\rho(x, y, t). \tag{11.47}$$

Written in more detail, this equation of motion becomes

$$\begin{aligned} i\hbar\frac{\partial\rho(x, y, t)}{\partial t} &= -\frac{\hbar^2}{2m}\left[\langle x|\psi(t)\rangle''\langle\psi(t)|y\rangle - \langle x|\psi(t)\rangle\langle\psi(t)|y\rangle''\right] \\ &\quad \times\rho(x, y, t) + [V(x) - V(y)]\,\rho(x, y, t). \end{aligned} \tag{11.48}$$

(For simplicity, "prime" notation is used for partial derivatives with respect to position.)

We will now start to convert equation 11.47 to the hydrodynamic form. In 1989, Skodje et al., in the appendix to their insightful paper, were apparently the first to derive equations of motion for the amplitude and phase functions that appear in the polar form for the density matrix (see equations A6 and A7 in [11.36] and Section 4 in [11.37]). They considered both pure states and mixtures, but the direct solution of these equations using ensembles of trajectories was not pursued. Years later, in 2001–2002, Bittner and Maddox [11.32, 11.33] also expressed the complex-valued density matrix elements in the Madelung–Bohm polar form

$$\rho(x, y, t) = \exp\left(\chi(x, y, t) + i\delta(x, y, t)/\hbar\right), \tag{11.49}$$

where $\chi(x, y, t)$ and $\delta(x, y, t)$ are real-valued amplitude and action functions. These two functions can be expressed in terms of the C-amplitude and action function associated with the wave function and its complex conjugate ($\psi(x, t) = \exp[C(x, t) + iS(x, t)/\hbar]$),

$$\chi(x, y, t) = C(x, t) + C(y, t), \quad \delta(x, y, t) = S(x, t) - S(y, t). \tag{11.50}$$

Continuing with the general development, we first substitute the polar form, equation 11.49, into the equation of motion, equation 11.47, and then separate into real and imaginary parts. This gives the two coupled equations of motion

$$\begin{aligned} \frac{\partial\chi}{\partial t} &= -\frac{1}{2m}\left[\frac{\partial^2\delta}{\partial x^2} - \frac{\partial^2\delta}{\partial y^2}\right] - \frac{1}{m}\left[\frac{\partial\chi}{\partial x}\frac{\partial\delta}{\partial x} - \frac{\partial\chi}{\partial y}\frac{\partial\delta}{\partial y}\right], \\ \frac{\partial\delta}{\partial t} &= -\frac{1}{2m}\left[\left(\frac{\partial\delta}{\partial x}\right)^2 - \left(\frac{\partial\delta}{\partial y}\right)^2\right] - V(x, y) - Q(x, y). \end{aligned} \tag{11.51}$$

In the second equation, the classical and quantum "difference potentials" are given by

$$V(x, y) = V(x) - V(y), \quad Q(x, y) = Q(x) - Q(y). \tag{11.52}$$

These equations will now be modified for dissipative coupling between an open quantum subsystem and a thermal bath at the temperature T.

For an *open quantum system*, several additional terms must be added to the Liouville–von Neumann equation to provide for coupling to the thermal bath. In this case, the focus will be on the *reduced density matrix* for the system, obtained from the full density matrix by tracing over the bath modes. The specific terms added to equation 11.47 depend on the details of the model, especially regarding the system-bath coupling and the spectral properties of the bath. For the Caldeira–Leggett (CL) model (described earlier, in Section 11.4), the two additional coupling terms are given by the CL dissipation operator action on the density matrix,

$$L_{CL}\rho = -i\hbar\gamma(x - y)\left(\frac{\partial}{\partial x} - \frac{\partial}{\partial y}\right)\rho - i\gamma(2mk_BT/\hbar)(x - y)^2\rho, \tag{11.53}$$

where γ is again the friction coefficient that links the system and the bath. The first term on the right side of this equation provides dissipative system–bath coupling, and the second term, involving $(x - y)^2$, leads to *decoherence*, a symptom of which is the dynamical reduction in the magnitude of the off-diagonal elements of the density matrix. The decoherence term is frequently expressed in terms of the thermal de Broglie wave length, defined by $\lambda = \hbar/\sqrt{2mk_BT}$. In terms of this quantity, equation 11.53 becomes

$$L_{CL}\rho = -i\hbar\gamma(x - y)\left(\frac{\partial}{\partial x} - \frac{\partial}{\partial y}\right)\rho - i\hbar\frac{\gamma}{\lambda^2}(x - y)^2\rho. \tag{11.54}$$

If we substitute the polar form for the density matrix into this equation, and add the real and imaginary parts to the right sides of equations 11.47, we then obtain

$$\begin{aligned}
\frac{\partial\chi}{\partial t} &= \frac{1}{2m}\left[\frac{\partial^2\delta}{\partial x^2} - \frac{\partial^2\delta}{\partial y^2}\right] - \frac{1}{m}\left[\frac{\partial\chi}{\partial x}\frac{\partial\delta}{\partial x} - \frac{\partial\chi}{\partial y}\frac{\partial\delta}{\partial y}\right] \\
&\quad -\gamma(x - y)\left[\frac{\partial\chi}{\partial x} - \frac{\partial\chi}{\partial y}\right] - \frac{\gamma}{\lambda^2}(x - y)^2, \\
\frac{\partial\delta}{\partial t} &= -\frac{1}{2m}\left[\left(\frac{\partial\delta}{\partial x}\right)^2 - \left(\frac{\partial\delta}{\partial y}\right)^2\right] - V(x, y) - Q(x, y) \\
&\quad -\gamma(x - y)\left[\frac{\partial\delta}{\partial x} - \frac{\partial\delta}{\partial y}\right].
\end{aligned} \tag{11.55}$$

Finally, in terms of derivatives of the action, the momenta with respect to x and y may be expressed in terms of derivatives of $\delta(x, y)$:

$$p_x = \frac{\partial\delta(x, y)}{\partial x}, \quad p_y = \frac{\partial\delta(x, y)}{\partial y}. \tag{11.56}$$

For a pure state, the latter equation can also be expressed in terms of $p_y = -\partial S(y)/\partial y$, the gradient of $S(y)$, where the unusual minus sign arises because this action function occurs in the complex conjugate of a wave function.

In order to express the diagonal and off-diagonal character of the density matrix, it is convenient to introduce a new set of *rotated coordinates*. These coordinates are

$$\xi = (x + y)/\sqrt{2}, \quad \eta = (y - x)/\sqrt{2}. \tag{11.57}$$

The density matrix in the new coordinate system is labeled $\rho(\xi, \eta, t)$, with $\rho(\xi, 0, t)$ referring to a diagonal population. In the rotated coordinates, the CL relaxation and decoherence terms are given by

$$L_{CL}\rho = -i\hbar(2\gamma\eta)\frac{\partial \rho}{\partial \eta} - i\hbar(2\gamma/\lambda^2)\eta^2\rho. \tag{11.58}$$

With the addition of these terms to the equations of motion, we expect to obtain partial *dynamical diagonalization of the density matrix*: a narrowing of the density in the η direction along with broadening along the ξ axis due to relaxation.

Before continuing, we note that the density matrix may be converted into a *phase space distribution* by performing the partial Fourier transform with respect to the off-diagonal variable η:

$$W(x, p, t) = \frac{1}{h}\int_{-\infty}^{\infty} \rho(\xi, \eta, t)e^{ip\eta}d\eta. \tag{11.59}$$

Following standard convention, on the left side of this expression, ξ has been replaced by the position x.

In order to convert the previous equations of motion into the new coordinates, it is useful to use the following derivatives:

$$\frac{\partial}{\partial x} = \frac{1}{\sqrt{2}}\left[\frac{\partial}{\partial \xi} - \frac{\partial}{\partial \eta}\right], \quad \frac{\partial}{\partial y} = \frac{1}{\sqrt{2}}\left[\frac{\partial}{\partial \xi} + \frac{\partial}{\partial \eta}\right]. \tag{11.60}$$

Using these relations, the equations of motion for χ and δ then become

$$\frac{\partial \chi}{\partial t} = \frac{1}{m}\frac{\partial^2 \delta}{\partial \xi\, \partial \eta} + \frac{1}{m}\left[\frac{\partial \chi}{\partial \eta}\frac{\partial \delta}{\partial \xi} + \frac{\partial \chi}{\partial \xi}\frac{\partial \delta}{\partial \eta}\right] - 2\gamma\eta\frac{\partial \chi}{\partial \eta} - \frac{2\gamma}{\lambda^2}\eta^2, \tag{11.61}$$

$$\frac{\partial \delta}{\partial t} = \frac{1}{m}\frac{\partial \delta}{\partial \xi}\frac{\partial \delta}{\partial \eta} - V(\xi, \eta) - Q(\xi, \eta) - 2\gamma\eta\frac{\partial \delta}{\partial \eta}. \tag{11.62}$$

In addition, the quantum (difference) potential in the new coordinates becomes

$$Q(\xi, \eta, t) = \frac{\hbar^2}{m}\left[\frac{\partial^2 C}{\partial \xi\, \partial \eta} + \frac{\partial C}{\partial \xi}\frac{\partial C}{\partial \eta}\right]. \tag{11.63}$$

The preceding equations of motion are expressed in the Eulerian (fixed-grid) frame. We will now convert these equations into the *Lagrangian moving frame*, wherein the convective term in equation 11.61 (the term in square brackets) disappears, and the right side of equation 11.62 becomes the Lagrangian. The equations derived by Maddox and Bittner are as follows:

$$\frac{d\chi}{dt} = -\frac{1}{2}\left[\frac{\partial v_\xi}{\partial \xi} + \frac{\partial v_\eta}{\partial \eta}\right] + \gamma - \frac{2\gamma}{\lambda^2}\eta^2,$$
$$\frac{d\delta}{dt} = -m v_\xi v_\eta + 2m\,\gamma\,\eta\, v_\xi - [V(\xi,\eta) + Q(\xi,\eta)]\,. \tag{11.64}$$

In these equations, d/dt is again used to express the time derivative of a function in the Lagrangian frame. The two velocities in these equations are given by

$$v_\xi = p_\xi/m, \quad v_\eta = p_\eta/m + 2\gamma\,\eta. \tag{11.65}$$

Because the η-component of the velocity depends on the friction coefficient, fluid elements far from the diagonal axis move very rapidly *away from the diagonal axis*. In the equation for $d\chi/dt$, the term involving η^2 damps off-diagonal elements of the density matrix that are far from the diagonal axis, which is a symptom of decoherence. For long-range coherences, the decay of these off-diagonal elements is very rapid. The second example in the next section will provide an illustration of the damping of these off-diagonal elements.

11.11 Examples of Density Matrix Evolution with Trajectories

In computational implementations of the trajectory formalism described in the previous section, the density matrix is discretized in terms of N fluid elements moving in a two-dimensional space bearing the coordinates (ξ, η). Each fluid element carries a *time-dependent descriptor* that stores the following dynamical quantities: the positions and momenta $(\xi, \eta, v_\xi, v_\eta)$, and the two fields $\chi(\xi, \eta, t)$ and $\delta(\xi, \eta, t)$, which define the amplitude and phase for the density matrix element along the trajectory. Each fluid element moves under the influence of forces from the two difference potentials $V(\xi, \eta)$ and $Q(\xi, \eta, t)$ along with frictional influences arising from coupling to the thermal bath.

In common with the quantum trajectories described in the previous chapters, problems are encountered due to *inflation and compression* of the moving grid points. Trajectories try to avoid nodes in the density, which in turn leads to undersampling and subsequent loss of accuracy in the derivative evaluation. On the other hand, trajectories tend to cluster together in regions of higher density, providing more information than is needed. In addition, a problem unique to dissipative systems is that trajectories are continuously forced toward large values of η (thus leading to decoherence), which causes a loss of accuracy in the derivatives. In order to counter this effect, the trajectories were *periodically remeshed back onto a uniform grid* whose size was not allowed to become too large. Initial values for the hydrodynamic fields at the new mesh points were found by least squares interpolation from the data carried by the old grid points.

The first application concerns a standard model that is frequently used to evaluate new methodologies for dissipative systems: the evolution of a *displaced coherent state in a harmonic potential* (frequency ω) coupled to a dissipative bath at temperature T [11.32]. The initial density matrix corresponding to the shifted oscillator has a C-amplitude given by

$$C(\xi, \eta, t = 0) = c_0 - \frac{1}{2\sigma_0^2}\left[(\xi - \xi_0)^2 + \eta^2\right], \tag{11.66}$$

where the width parameter is $\sigma_0 = \sqrt{\hbar/(m\omega)}$. Since the potential is harmonic, the density remains Gaussian, and the C-amplitude at later times is a quadratic form characterized by the widths in ξ and η

$$C(\xi, \eta, t) = c_0 - \frac{1}{2\sigma_\xi^2(t)}(\xi - \langle\xi(t)\rangle)^2 - \frac{1}{2\sigma_\eta^2(t)}\eta^2. \tag{11.67}$$

The two Gaussian widths $\sigma_\xi(t)$ and $\sigma_\eta(t)$ define the semiminor and semimajor axes of an ellipsoid centered at the average diagonal position $\langle\xi(t)\rangle$.

Figure 11.11 shows the time evolution of a set of trajectories corresponding to the diagonal elements of the density matrix ($\eta = 0$) in the underdamped ($\gamma/\omega = 0.25$) and critically damped ($\gamma/\omega = 1$) regimes. The dashed curve shows the time dependence of the center of the Gaussian distribution, and this is the same as the evolution of a *classical damped trajectory*. For the underdamped case (top plot), relaxation to a thermal stationary state occurs in about two vibrational periods, but for the critically damped case (lower plot), the stationary state is reached in about one period. Over the course of time, the distribution gradually expands as the center trajectory approaches the stationary-state diagonal position $\xi = 0$. These diagonal trajectories never cross, and at longer times each trajectory remains stationary. In contrast, the off-diagonal trajectories (representing coherences in the position representation) continuously evolve toward $\eta \to \pm\infty$. In the absence of coupling to the thermal bath, the initial coherent state would oscillate in the harmonic well without any change in shape. However, dissipative coupling to the bath causes the Gaussian to spread in ξ and contract in η, thus effectively diagonalizing the density matrix. An important dynamical balance is established at longer times: *as the density matrix becomes more concentrated about the diagonal, the quantum potential increases, and this forces trajectories away from this axis. At equilibrium, squeezing in the η direction is exactly counterbalanced by the outward force arising from the quantum potential.*

Arguments have been presented for decoherent systems to suggest that quantum trajectories should approach their classical analogues at long times [11.38]. For the damped harmonic oscillator, all *classical trajectories* converge to the bottom of the well, $\xi = 0$, at long times. This classical limit is completely different from the quantum trajectory results shown in figure 11.11, where the trajectories always maintain a *finite separation in the long-time limit*. It is especially noticeable that the quantum trajectories at the edges of the ensemble eventually settle in to equilibrium positions that are far removed from the central trajectory. Clearly, the *quantum*

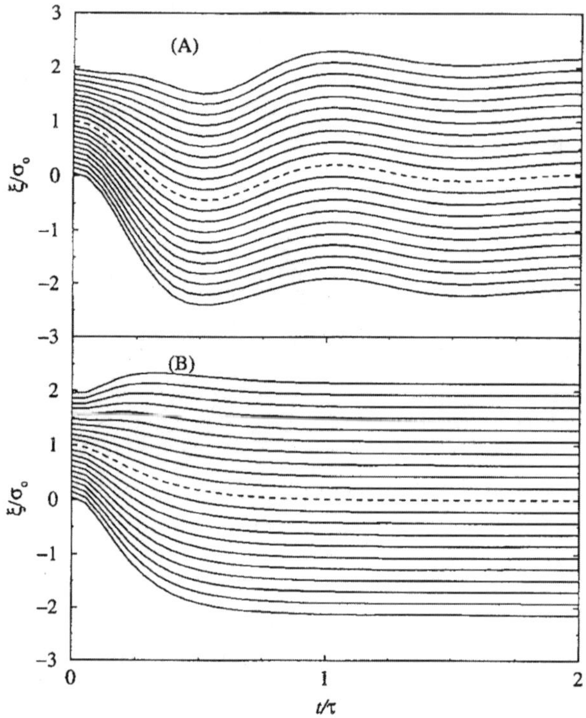

FIGURE 11.11. Diagonal trajectories ($\eta = 0$) representing the relaxation of a displaced Gaussian in a harmonic potential: (a) underdamped case ($\gamma/\omega < 1$); (b) critically damped case ($\gamma/\omega = 1$) [11.32]. The temperature is given by $k_B T/(\hbar\omega) = 2.5$, and the time in these plots is in units of the oscillator period, $\tau = 2\pi/\omega$. The dashed curve (a classical trajectory) follows the center of the Gaussian as it relaxes toward the stationary state centered at $\xi = 0$. The vertical axis shows the coordinate ξ divided by the initial width.

trajectories in the equilibrium ensemble do not bear any resemblance to classical trajectories [11.33].

The second example of the use of trajectories to evolve the density matrix concerns an initial minimum uncertainty Gaussian distribution centered at the barrier maximum of an Eckart potential [11.33], given by $V(x) = V_0 \sec h^2(\pi x/\omega)$. The parameters used in this study are given by $V_0 = 0.0248$ a.u., $\omega = 0.6613$ a.u., and the mass is $m = 2000$ a.u. In common with the preceding example, as time advances, the population width σ_ξ increases, while the coherence width σ_η decreases. However, in the absence of coupling to the thermal bath, the widths would evolve identically in time.

Figures 11.12 and 11.13 show the time evolution of the density matrix and the velocity field for the undamped and damped cases, respectively. In figure 11.12 for the undamped case, the initial density matrix gradually splits into four distinct lobes. As the density splits, it appears to be pulled toward four focal points located on the leading edge of each lobe. As time proceeds, the focal points move away

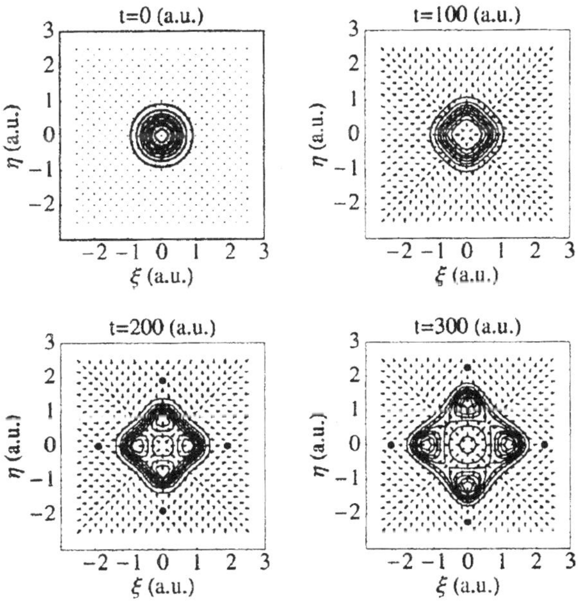

FIGURE 11.12. Density matrix evolution and velocity distribution for an *undamped* initial minimum uncertainty Gaussian distribution [11.33]. The initial distribution is centered at the maximum of an Eckart barrier. The filled circles are focal points that move away from the origin as time advances.

from the origin, seemingly pulling density as they go. At $t = 300$ a.u., the two peaks in the density near $\eta = 0$, $\xi = \pm 1.2$ represent the bifurcated initial state. The two off-diagonal peaks near $\eta = \pm 1.2$, $\xi = 0$ correspond to coherences between the peaks that are concentrated near the ξ axis.

The damped case shown in figure 11.13 displays a number of differences compared to the *undamped case*. Even at $t = 0$, the velocity field is directed away from the ξ axis. The off-diagonal coherent structures that appear to be forming at $t = 200$ a.u. disappear by about $t = 300$ a.u., which is a feature indicating decoherence. All the while, density in the η direction is depressed, and the density matrix evolves into two independent densities propagating away from the barrier region.

11.12 Summary

In the first part of this chapter, several classical and quantum-mechanical equations of motion for phase space distribution functions were described. For closed systems, the classical Liouville equation and the quantum-mechanical Wigner–Moyal or Husimi equations may be used to find the time evolution of the appropriate distribution functions. For open systems, including those coupled to a thermal bath,

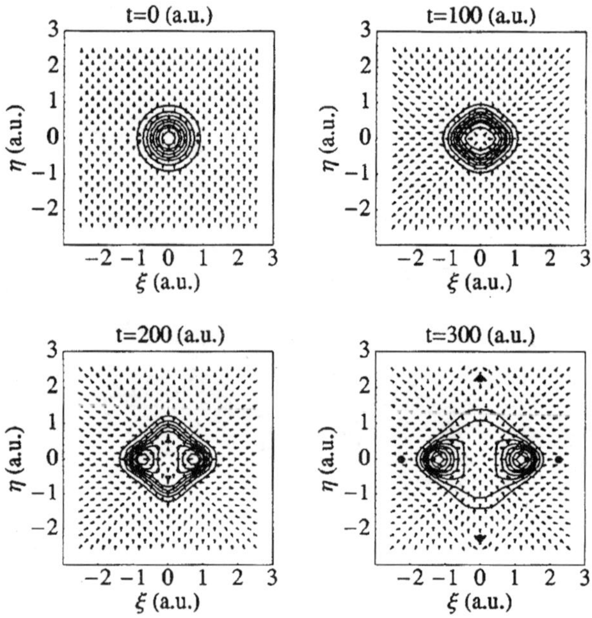

FIGURE 11.13. Density matrix evolution and velocity distribution for a *damped* initial minimum uncertainty Gaussian distribution [11.33]. The initial distribution is centered at the maximum of an Eckart barrier. The filled circles are focal points which move away from the origin as time advances. The bath temperature is $T = 1000\ K$, and the friction coefficient is $\gamma = 0.005$ a.u.$^{-1}$.

there are a number of equations of motion, but the classical Kramers equation, the quantum Caldeira–Leggett equation, and the modified Caldeira–Leggett equation have been emphasized in this chapter. Coupled trajectory ensembles may be used to study the time evolution of these distribution functions, but there are conceptual and computational problems connected with the formation of negative basins in the quantum distribution functions.

As an alternative to evolving a correlated ensemble in phase space, we also described use of the derivative propagation method, DPM. Rather than lockstep propagation of an ensemble of linked trajectories, analytic equations of motion for the x and p partial derivatives of the density were derived, and these derivatives were propagated along Lagrangian trajectories concurrently with the density itself. The various orders of partial derivatives are coupled together in an infinite hierarchy, but low-order truncations of this system yield useful approximations. The lowest-order spatial derivatives computed along the trajectory introduce *regional nonlocality* into the dynamics. In this chapter, the DPM was applied to the modified Caldeira–Leggett quantum-mechanical equation of motion. In this example, quantum trajectories were used to study the decay of a metastable state, that was coupled through friction terms to a thermal bath. The DPM may be readily extended to phase space dynamics in higher dimensionality and to a broad genre of transport equations.

In an extension of the moment equations presented earlier, in Chapter 3, in this chapter the momentum moment route to trajectory dynamics for dissipative quantum systems was described. Additional analysis and applications of these moment methods, especially for multidimensional systems, will be of interest.

In the last part of this chapter, the trajectory approach to density matrix evolution was described for open quantum systems. This method provides a novel way to approach the dynamics of dissipative quantum systems and opens up many opportunities to study tunneling processes and transition state dynamics in quantum subsystems coupled to a thermal environment. This trajectory approach provides new ways to compute and analyze the dynamics. Even though only a limited number of one-dimensional model problems have been studied, new insights concerning population evolution and the role played by relaxation and decoherence have been revealed by following the trajectory flow.

References

11.1. C. Trahan and R.E. Wyatt, *Evolution of classical and quantum phase space distributions. A new trajectory approach for phase-space dynamics*, J. Chem. Phys. **119**, 7017 (2003).

11.2. A. Donoso and C.C. Martens, *Quantum tunneling using entangled classical trajectories*, Phys. Rev. Lett. **87**, 223202 (2001).

11.3. A. Donoso and C.C. Martens, *Classical trajectory-based approaches to solving the quantum Liouville equation*, Int. J. Quantum Chem. **90**, 1348 (2002).

11.4. A. Donoso and C.C. Martens, *Solution of phase space diffusion equations using interacting trajectory ensembles*, J. Chem. Phys. **116**, 10598 (2002).

11.5. D.A. McQuarrie, *Statistical Mechanics* (Harper & Row, New York, 1973).

11.6. C.W. Gardiner, *Handbook of Stochastic Methods* (Springer, New York, 1985).

11.7. H.A. Kramers, *Brownian motion in a field of force and the diffusion model of chemical reactions*, Physica **VII**, 284 (1940).

11.8. G.D. Billing and K. V. Mikkelsen, *Molecular Dynamics and Chemical Kinetics* (Wiley Interscience, New York, 1996), Ch. 15.

11.9. E. Pollak, *Theory of activated rate processes: A new derivation of Kramers' expression*, J. Chem. Phys. **85**, 865 (1986).

11.10. E. Pollak, A. M. Berezhkovskii, and Z. Schuss, *Activated rate processes: A relation between Hamiltonian and stochastic theories*, J. Chem. Phys. **100**, 334 (1994).

11.11. P. Hanggi, P. Talkner, and M. Borkovec, *Reaction rate theory: Fifty years after Kramers*, Rev. Mod. Phys. **62**, 251 (1990).

11.12. M. Dresden, H. A. Kramers, *Between Tradition and Revolution* (Springer, New York, 1987).

11.13. D. Ter Haar, *Master of Modern Physics: The Scientific Contributions of H.A. Kramers* (Princeton Press, Princeton NJ, 1998).

11.14. M. Dresden, *Kramers's contributions to statistical mechanics*, Physics Today, Sept. 1988, p. 26.

11.15. E.P. Wigner, *On the quantum correction for thermodynamic equilibrium*, Phys. Rev. **40**, 749 (1932).

11.16. H.-W. Lee, *Theory and application of the quantum phase-space distribution functions*, Phys. Rept. **259**, 147 (1995).
11.17. H.-W. Lee and M. O. Scully, *A new approach to molecular collisions: Statistical quasiclassical method*, J. Chem. Phys. **73**, 2238 (1980).
11.18. H.-W. Lee and M. O. Scully, *The Wigner phase-space description of collision processes*, Found. of Phys. **13**, 61 (1983).
11.19. E.J. Heller, *Wigner phase space method: Analysis for semiclassical applications*, J. Chem. Phys. **65**, 1289 (1976).
11.20. E.J. Heller and R.C. Brown, *Errors in the Wigner approach to quantum dynamics*, J. Chem. Phys. **75**, 1048 (1981).
11.21. U. Weiss, *Quantum dissipative systems* (World Scientific, Singapore, 1993).
11.22. J.R. Chaudhuri, B. Bag, and D.S. Ray, *A semiclassical approach to the Kramers problem*, J. Chem. Phys. **111**, 10852 (1999).
11.23. P. Pechukas, *Reduced dynamics need not be completely positive*, Phys. Rev. Lett. **73**, 1060 (1994).
11.24. G. Lindblad, *On the generators of quantum dynamical semigroups*, Comm. Math. Phys. **48**, 119 (1976).
11.25. I. Percival, *Quantum State Diffusion* (Cambridge Press, Cambridge UK, 1998).
11.26. D. Kohen, C.C. Marsten, and D.J. Tannor, *Phase space approach to quantum dissipation*, J. Chem. Phys. **107**, 5236 (1997).
11.27. L. Diosi, *Caldeira–Leggett master equation and medium temperatures*, Physica A **199**, 517 (1993).
11.28. L. Diosi, *On high-temperature Markovian equation for quantum Brownian motion*, Europhys. Lett. **22**, 1 (1993).
11.29. I. Burghardt and K. B. Moller, *Quantum dynamics for dissipative systems: A hydrodynamic perspective*, J. Chem. Phys. **117**, 7409 (2003).
11.30. A.O. Caldeira and A.J. Leggett, *Path integral approach to quantum Brownian motion*, Physica A **121**, 587 (1983).
11.31. J. E. Moyal, *Quantum mechanics as a statistical theory*, Proc. Camb. Phil. Soc. **45**, 99 (1949).
11.32. J.B. Maddox and E.R. Bittner, *Quantum relaxation dynamics using Bohmian trajectories*, J. Chem. Phys. **115**, 6309 (2001).
11.33. J.B. Maddox and E.R. Bittner, *Quantum dissipation in unbounded systems*, Phys. Rev. E **65**, 026143 (2002).
11.34. J. Daligault, *Non-Hamiltonian dynamics and trajectory methods in quantum phase-spaces*, Phys. Rev. A **68**, 010501 (2003).
11.35. N.C. Dias and J.N. Prata, *Bohmian trajectories and quantum phase space distributions*, Phys. Lett. A **302**, 261 (2002).
11.36. R.T. Skodje, H.W. Rohrs, and J. Van Buskirk, *Flux analysis, the correspondence principle, and the structure of quantum phase space*, Phys. Rev. A **40**, 2894 (1989).
11.37. A. Spina and R.T. Skodje, *The phase-space hydrodynamic model for the quantum standard map*, Comp. Phys. Comm. **63**, 279 (1991).
11.38. D.M. Appleby, *Bohmian trajectories post-decoherence*, arXiv:quant-ph/9908029 (8 Aug. 1999).
11.39. F. McLafferty, *On classical paths and the Wigner path integral*, J. Chem. Phys. **78**, 3253 (1983).
11.40. K.H. Hughes and R.E. Wyatt, *Trajectory approach to dissipative phase space dynamics: Application to barrier scattering*, J. Chem. Phys. **120**, 4089 (2004).

11.41. M. Novaes, *Wigner and Huisimi functions in the double-well potential*, J. Optics B **5**, S342 (2003).
11.42. R.L. Hudson, *When is the Wigner quasi-probability density non-negative?* Repts. Math. Phys. **6**, 249 (1974).
11.43. F. Soto and P. Claverie, *When is the Wigner quasi-probability density of multi-dimensional systems non-negative?* J. Math. Phys. **24**, 97 (1983).
11.44. L. Shifren and D.K. Ferry, *Wigner function quantum Monte Carlo*, Physica B, **314**, 72 (2002).
11.45. C.-Y. Wong, *Explicit solution of the time evolution of the Wigner function*, arXiv:quantum-ph/0210112 (7 March 2003).

12
Mixed Quantum–Classical Dynamics

Trajectory methods for systems composed of a quantum subsystem interacting with a classical subsystem will be described, and several illustrative examples will be presented

12.1 Introduction

Many systems of interest involve a large number of degrees of freedom, such as molecule–surface collisional energy transfer including surface phonons, intramolecular proton transfer in an AH-B molecule that is embedded in a solvent, fragmentation and recombination of molecules in cryogenic lattices and clusters, and electron transfer between donor and acceptor sites in biological molecules. Because of the large number of degrees of freedom in these and related problems, it has been impossible to accurately develop the dynamics using conventional (fixed grid or basis set) quantum-mechanical techniques. For this reason, a number of mixed quantum–classical methods have been developed that typically employ quantum techniques for "light" particles such as electrons or protons and combine this with classical trajectory techniques for "heavy" particles such as nuclei. Conceptual problems arise at the interface between the quantum and classical subsystems: How do we permit each of the subsystems to both influence and be influenced by the other subsystem? There have been many formal and computational approaches, such as mean-field [12.1] and trajectory surface hopping [12.2], and it might appear that there is no unique solution to the coupling problem. In the frequently applied mean field method, the quantum subsystem is influenced by a time-dependent interaction potential, which arises from the dependence on the trajectories followed by the classical particles. The way that the quantum subsystem influences the classical one is through a mean force, the gradient of the average interaction potential which links the two subsystems. The latter method is one way of approaching what is called the *back reaction problem*. However, as we will see in this chapter, there are more consistent and accurate ways of handling both the forward and back-reaction problems. These methods use quantum trajectories and classical trajectories for the appropriate subsystems.

In the next section, the mean field method will be described in more detail. Then, in Section 12.3, the mixed hydrodynamical–Liouville phase space method developed by Burghardt and Parlant [12.3] will be described, and an application of this method will be presented in the following section. In Sections 12.5–12.7, the related trajectory methods that were developed independently by Gindensperger, Meier, and Beswick [12.4] and by Prezhdo and Brooksby [12.5] will be described. Finally, in Section 12.8, a discussion of these new trajectory approaches will be presented

12.2 The Ehrenfest Mean Field Approximation

The *mean field approximation* for mixed quantum–classical dynamics is straightforward to describe and implement. In order to simplify the notation, one-degree-of-freedom quantum and classical subsystems will be assumed to interact. However, the formalism can be readily extended to handle systems with additional degrees of freedom. The coordinates for the quantum and classical subsystems will be denoted by q and Q, P denotes the classical momentum, and m and M are the corresponding masses. The total potential energy will be partitioned into contributions for the quantum and classical subsystems along with an interaction term,

$$V(q, Q) = V_{qu}(q) + V_{cl}(Q) + V_{\text{int}}(q, Q). \tag{12.1}$$

The quantum and classical subsystems are then evolved as follows.

The quantum subsystem. When the interaction potential is evaluated along a specific classical trajectory $Q(t)$, this leads to a time-dependent driving potential for the quantum subsystem, $V_{\text{int}}(q, Q(t))$. The time-dependent wave function for the quantum subsystem is obtained by solving the Schrödinger equation with this parameterized time-dependent potential,

$$\left[-\frac{\hbar^2}{2m}\frac{\partial^2}{\partial q^2} + V_{qu}(q) + V_{\text{int}}(q, Q(t))\right]\psi(q,t) = i\hbar\frac{\partial\psi(q,t)}{\partial t}. \tag{12.2}$$

From the solution to this equation, the *mean interaction potential* is determined by averaging over this wave function (integration is over only the q coordinate),

$$\tilde{V}_{\text{int}}(Q;t) = <\psi|V_{\text{int}}(q, Q(t))|\psi> . \tag{12.3}$$

This mean potential will then enter the next stage of the calculation.

The classical subsystem. Using the mean interaction potential, the *average trajectory for the classical subsystem* is evolved using Hamilton's equations

$$\begin{aligned}\frac{dQ(t)}{dt} &= \frac{P(t)}{M},\\ \frac{dP(t)}{dt} &= -\frac{\partial}{\partial Q}\left[V_{cl}(Q) + \tilde{V}_{\text{int}}(Q;t)\right] = F_{cl}(Q) + \tilde{F}_{\text{int}}(Q,t),\end{aligned} \tag{12.4}$$

in which the last expression defines the *mean interaction force*, also called the *Ehrenfest force* [12.6]. It is through this force that the *back-reaction* of the quantum subsystem on the classical subsystem takes place. The classical trajectory thus never directly responds to the dynamical coordinate for the quantum subsystem; rather, it "feels" the quantum subsystem through an average force computed by averaging over the time-dependent wave function for the quantum subsystem. (In Section II of [12.16], several ways to evaluate the mean force are described.) In the following sections, when quantum trajectories are introduced to describe the dynamics of the quantum subsystem, the classical subsystem will be able to respond to the *instantaneous quantum coordinate*.

Historical comment. Paul Ehrenfest was born in Vienna in 1880 and obtained his doctorate from the University of Vienna in 1904, under Boltzmann's supervision. In the following years, he worked in both quantum mechanics and statistical mechanics. He also devised a commonly used classification for phase transitions that is based on the number of continuous derivatives of the chemical potential at the transition temperature. From 1914 until his death, he maintained a close friendship with Einstein. Einstein said of Ehrenfest that he was "the best teacher in our profession that I have ever known". At the age of 53, Ehrenfest committed suicide in Leiden, the Netherlands.

12.3 Hybrid Hydrodynamical–Liouville Phase Space Method

The novel *hybrid trajectory method* developed by Burghardt and Parlant [12.3] for *mixed quantum–classical dynamics* will be described in this section. Using notation introduced in the preceding section, we begin with the Wigner distribution function,

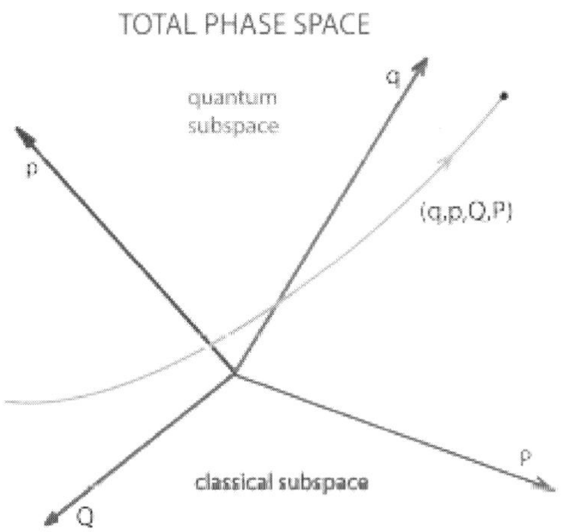

FIGURE 12.1. The four-dimensional full phase space with one trajectory.

$W(q, p, Q, P, t)$, in the four-dimensional *full phase space* shown in figure 12.1. This function, which was introduced in Section 3.3 and again made an appearance in Section 11.3, can be computed from the density matrix $\rho(q, Q, q', Q', t)$ for either a pure state or a statistical mixture. For the purposes of this chapter, it will be assumed that we are dealing with a *pure state* system described by the wave function $\psi(q, Q, t)$.

The four-dimensional total phase space will now be *contracted* to the three-dimensional *reduced phase space* in which the *independent* coordinates are (q, Q, P). As a result, the quantum momentum, now an *average value*, will become a function of the remaining three independent variables. The reduction will be performed and equations of motion will be derived, analogous to the moment methods introduced in Chapter 3, by computing *partial moments* $\bar{p}_n =< p^n W >$ of the distribution function $W(q, p, Q, P, t)$ with respect to *only* the quantum momentum p. If the system is in a pure state, then there are only two independent momentum moments: the 0-moment, which is the same as the reduced density, and the 1-moment, which is related to the average value of p. For these two moments, various notation may be used, including the following:

$$\rho(q, Q, P) = \bar{p}_0(q, Q, t) = < W > = \int W(q, p, Q, P) dp,$$

$$\bar{p}_1(q, Q, P) = < pW > = \int pW(q, p, Q, P) dp. \tag{12.5}$$

The 1-moment can in turn be expressed in terms of the average quantum momentum and the reduced density, $\bar{p}_1(q, Q, P) = p(q, Q, P)\rho(q, Q, P)$. The quantity $p(q, Q, P)$ is the *average quantum momentum* (also called the *hydrodynamic momentum*) at position q for the quantum subsystem, and it is parameterized by the coordinates (Q, P) in the classical subspace. It is important to note that the momentum moments $\bar{p}_n(q, Q, P)$ are actually *full phase space distributions with*

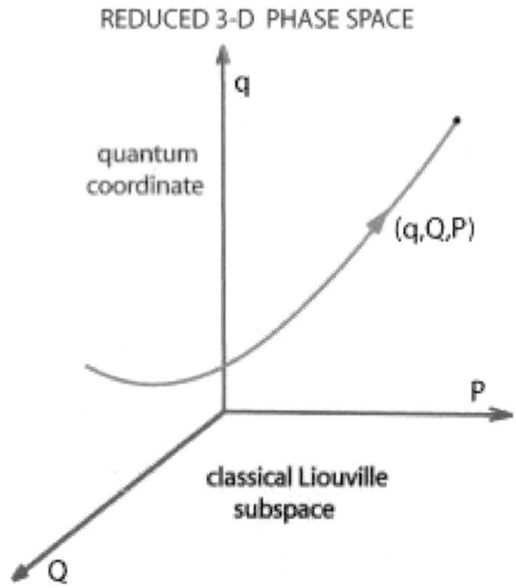

FIGURE 12.2. A trajectory evolving in the three-dimensional reduced phase space. Along this trajectory, the average quantum momentum $p(q, Q, P)$ is a function of the three independent variables (q, Q, P).

respect to the "classical" variables (Q, P) (i.e., *no reduction* has been performed with respect to these variables).

Before turning to equations of motion for these p-moments, consider a single trajectory in the *reduced phase space*, such as the one shown in figure 12.2. This trajectory is specified by the three time-dependent phase space coordinates $(q(t), Q(t), P(t))$, while the average quantum momentum $p(q, Q, P)$ is now a function of these coordinates. Although this momentum denotes an *average value*, it is important to emphasize that the classical momentum P is a *dynamical variable*, not an average value. Also, in this three-dimensional reduced phase space, there is no such thing as the "classical trajectory" or the "quantum trajectory"; rather, there may be many trajectories, *each of which is specified by a combination of classical and quantum coordinates*. Stated another way, each of these trajectories has a projection (or shadow) in both the quantum and classical subspaces. However, from the viewpoint of a monitor in the quantum subspace, the trajectory specified by $(q, p(q, Q, P))$ is parameterized by the classical dynamical variables (Q, P).

From the viewpoint of an observer in the three-dimensional reduced phase space, the distribution function that we are now dealing with is the hybrid form $W_{hybrid}(q, p, Q, P) = \rho(q, Q, P)\delta(p - p(q, Q, P))$, where the δ-function constrains the momentum p to take on values determined by the position in this reduced space.

An alternative way to view the dynamics is in the four-dimensional *partial hydrodynamic phase space*, which is spanned by the four coordinates $(q, p(q, Q, P), Q, P)$; one of these coordinates is the *average quantum momentum* (a dependent variable). The *hydrodynamic quantum subspace* is defined by the coordinate q and the average quantum momentum $p(q, Q, P)$, while the two remaining coordinates (Q, P) span the *classical Liouville subspace*.

Equations of motion for the partial moments can be derived using procedures analogous to those used earlier, in Chapter 3. We will begin with the Wigner–Moyal (quantum Liouville) equation of motion, which generates the evolution of the Wigner function $W(q, p, Q, P, t)$ *in the full four-dimensional phase space*. However, rather than writing it down all at once, we will begin by partitioning the right side of this equation into four contributions to the total rate of change

$$\frac{\partial W}{\partial t} = \boxed{Liouville_{q,p}} + \boxed{Quantum_{q,p}} + \boxed{Liouville_{Q,P}} + \boxed{Quantum_{Q,P}}.$$

The subscripts on each term indicate that it involves *derivatives* with respect to each member of this pair of independent variables. The two *Liouville terms* are given by

$$\boxed{Liouville_{q,p}} = -\frac{p}{m}\frac{\partial W}{\partial q} + \frac{\partial V}{\partial q}\frac{\partial W}{\partial p},$$

$$\boxed{Liouville_{Q,P}} = -\frac{P}{M}\frac{\partial W}{\partial Q} + \frac{\partial V}{\partial Q}\frac{\partial W}{\partial P}.$$

The two $\hbar$-dependent *quantum terms* are given by

$$\boxed{Quantum_{q,p}} = -\frac{\hbar^2}{24}\frac{\partial^3 V}{\partial q^3}\frac{\partial^3 W}{\partial p^3} + \text{higher-order terms},$$

$$\boxed{Quantum_{Q,P}} = -\frac{\hbar^2}{24}\frac{\partial^3 V}{\partial Q^3}\frac{\partial^3 W}{\partial P^3} + \text{higher-order terms}.$$

The quantum Liouville and quantum terms that involve the (q, p) partial derivatives also depend on the variables (Q, P), and vice versa for the other two terms. The total potential energy in these equations will again be partitioned into quantum, classical, and interaction components, as in equation 12.1. In addition, derivatives of the potential with respect to *either* of the coordinates (q, Q) bring in the derivative of the interaction term $V_{int}(q, Q)$.

In order to develop equations of motion for the partial moments (moments only with respect to the quantum momentum) for the quantum subsystem and to eliminate $\hbar$ dependence in the equations of motion for what will become the classical subsystem, we will make the following *approximation in the quantum Liouville equation*

$$\boxed{Quantum_{Q,P}} \to 0.$$

(Actually, this in *not an approximation* if the potential $V_{cl}(Q) + V_{int}(q, Q)$ is of polynomial form and contains terms no higher than Q^2, such as the potential $aQ^2 + Q(bq + cq^2)$.) The fact that we *do not* also throw out the term $\boxed{Quantum_{q,p}}$ is what gives rise to the *mixed aspect* of the quantum–classical dynamics.

From the *approximate Wigner–Moyal equation* described above, the 0 and 1 partial moments with respect to the *quantum momentum* may be derived. This leads to the *first two* equations shown below. The *second pair* of equations arise from the term $\boxed{Liouville_{Q,P}}$. The four equations of motion for a *single trajectory* evolving in phase space (the quantum hydrodynamic subspace plus the classical Liouville subspace) are given by

$$\boxed{Quantum:}\quad \frac{dq}{dt} = \frac{p}{m},$$

$$\frac{dp}{dt} = -\frac{\partial}{\partial q}\left[V_{qu}(q) + V_{\text{int}}(q, Q)\right] + F_{hydro}(q, Q, P), \tag{12.6}$$

$$\boxed{Classical:}\quad \frac{dQ}{dt} = \frac{P}{M},$$

$$\frac{dP}{dt} = -\frac{\partial}{\partial Q}\left[V_{cl}(Q) + V_{\text{int}}(q, Q)\right]. \tag{12.7}$$

These equations are augmented by the Lagrangian version of the continuity equation for the *reduced density,*

$$\frac{d\rho(q, Q, P)}{dt} = -\rho(q, Q, P)\frac{\partial v(q, Q, P)}{\partial q}, \tag{12.8}$$

in which the flow velocity is $v(q, Q, P) = p(q, Q, P)/m$. In addition, the hydrodynamic force in equation 12.6 is given by the gradient of the momentum variance,

$$F_{hydro}(q, Q, P) = -\frac{1}{m\rho(q, Q, P)}\frac{\partial\sigma(q, Q, P)}{\partial q}. \tag{12.9}$$

In this equation, the variance is computed with respect to the quantum momentum of the Wigner distribution at each point (q, Q, P) in the full phase space (the time dependence in not indicated explicitly),

$$\sigma(q, Q, P) = \bar{p}_2(q, Q, P) - p(q, Q, P)^2\rho(q, Q, P). \tag{12.10}$$

Equations 12.9 and 12.10 are identical to equations 3.31 and 3.19, respectively, except that the above versions carry additional labels, namely, the image of the trajectory in the classical subspace, specified by the dynamical coordinates Q and P. By setting $F_{hydro} = 0$ in equation 12.6, we obtain the classical equations of motion for each of the two subsystems. As a reminder, it is only through F_{hydro} that quantum effects influence the dynamics of the quantum subsystem.

The key to developing a viable implementation of this formalism lies in the evaluation of the momentum variance, from which the hydrodynamic force is obtained by taking the derivative with respect to q. For both pure and mixed-state systems, one possibility is to compute the moment $\bar{p}_2(q, Q, P)$ by propagating a truncated system of equations for the moments. Truncation schemes for these equations need further investigation. In addition, for pure states, this momentum moment can be computed from the coordinate space projections of the density and momenta [12.16], $\bar{\rho}(q, Q)$ and $\bar{p}(q, Q)$, where these quantities are defined by

$$\bar{\rho}(q, Q) = \int \rho(q, Q, P)dP, \tag{12.11}$$

$$\bar{p}(q, Q) = \int p(q, Q, P)\rho(q, Q, P)dP \Big/ \int \rho(q, Q, P)dP. \tag{12.12}$$

For the example given in the next section, the (Gaussian) form for the wave function and Wigner function are known at all times, and the hydrodynamic force can be evaluated in terms of shape parameters in the time-dependent wave function. Clearly, such a simple approach would not be possible for more general forms for the wave function.

Several additional remarks are in order regarding these equations of motion. The classical equations of motion may appear similar in form to those in the mean-field theory; see equations 12.4. However, the interaction potential in equation 12.7 depends on the *instantaneous quantum coordinate* $q(t)$ and is not equivalent to the mean-field expression appearing in equation 12.4. For the quantum subsystem, the first two force terms on the right side of equation 12.6 are not surprising, and appearance of the hydrodynamic force term is an expected consequence of the reduction from the full four-dimensional phase space to the three-dimensional reduced phase space. However, it is important to note that F_{hydro} and the (average) quantum momentum p are labeled by the classical dynamical variables Q and P, which define the shadow of the trajectory in the classical subspace. We also note that F_{hydro} in equation 12.9 takes on large values near nodes in the density, and near quasi-nodes where the density is small, $\rho(q, Q, P) \approx 0$. Because of the strong forces near these regions, the propagation of trajectories may be challenging.

A different method for mixed quantum–classical dynamics was developed by Kapral, Ciccotti, and coworkers [12.13, 12.17–12.19]. Starting from the operator form of the quantum Liouville equation, they performed a partial Wigner transform with respect to only the classical coordinate and then carried out an expansion in the parameter $\mu = (m/M)^{1/2}$, which is assumed to be small. Schmidt, Schütte, and coworkers [12.20, 12.21] have also developed an approach to quantum–classical dynamics in which the evolution of phase space densities is expressed in terms of propagating Gaussian phase-space-packets.

12.4 Example of Mixed Quantum–Classical Dynamics

A relatively simple model involving two coupled harmonic oscillators will be used to illustrate the method described in the preceding section [12.3]. This model was used in earlier comparisons of mean-field and surface-hopping approximations [12.7] and was also studied more recently by Gindensperger et al. [12.4] (the latter study will be described in Section 12.6). In this model, the potential energy is expressed in the form

$$V(q, Q) = \frac{1}{2}k(q - Q)^2 + \frac{1}{2}KQ^2, \tag{12.13}$$

from which the bilinear interaction term is obtained, $V_{\text{int}}(q, Q) = -kqQ$. The masses and force constants are given by (in atomic units) $m = 1$, $M = 10$, $k = 5$, and $K = 15$. For a fixed value of Q, this potential attains a minimum when $q = Q$. By transforming to normal coordinates, the eigenstates for this Hamiltonian are readily found (see the appendix in [12.7]). Because of the disparity in masses and force constants, effective sharing of energy between the zero-order vibrational modes may occur for some initial conditions.

In these studies, the initial wave function is taken to be the ground adiabatic (Born–Oppenheimer) state. The wave function for this state, $\psi_0(q, Q)$, is the product of the ground state function for the "fast" quantum oscillator, which depends parametrically on the coordinate for the heavy "slow" oscillator multiplied by the ground state function for the heavy oscillator,

$$\psi_0(q, Q) = \varphi_0(Q)\phi_0(q; Q). \tag{12.14}$$

The wave function for the quantum oscillator (which has been shifted by the amount Q to the minimum of the potential in the q direction), for example, is given by

$$\phi_0(q; Q) = (\beta^2/\pi)^{1/4} \exp(-(1/2)\beta^2(q - Q)^2), \tag{12.15}$$

where the width parameter for the Gaussian function is given by $\beta = (mk/\hbar)^{1/4}$.

At the initial time, the density was discretized onto the initial locations of a set of 27 quantum trajectories in the three-dimensional reduced phase space, and these were then propagated using equations 12.6 and 12.7. At each time, these trajectories $\{q_i(t), Q_i(t), P_i(t)\}_{i=1}^{27}$ form an unstructured moving grid in the three-dimensional reduced phase space. The hydrodynamic force and other spatial derivatives appearing in the equations of motion were evaluated from the analytical forms that were derived for a Gaussian pure-state density, which is the exact solution for this two-dimensional harmonic potential. Different computational procedures (such as the derivative evaluation techniques that have been used in the quantum trajectory method) would need to be implemented when the functional form for the evolving density is not known in advance.

In figure 12.3, the exact probability of surviving in the ground adiabatic state (continuous curve), $P_0^{ad}(t)$, is compared with results (open circles) from the hybrid quantum–classical method. It is clear that the highly nonadiabatic behavior is *captured exactly* by this method. These results are much more accurate than those obtained using either the surface-hopping or the mean-field methods (see figures 1 (e) and (c), respectively, in [12.7]), for which the results degrade considerably as time proceeds. In addition, the latter two methods are not able to capture the oscillation shown in the insert in figure 12.3. In contrast to these methods, for the coupled harmonic system considered in this section, the mixed quantum–classical method of Burghardt and Parlant provides the *exact* quantum-mechanical solution.

12.5 The Mixed Quantum–Classical Bohmian Method (MQCB)

In addition to the mixed quantum trajectory–classical trajectory method described in the two preceding sections, two other methods that employ Bohmian trajectories have been developed. These methods display a number of similarities, but differ

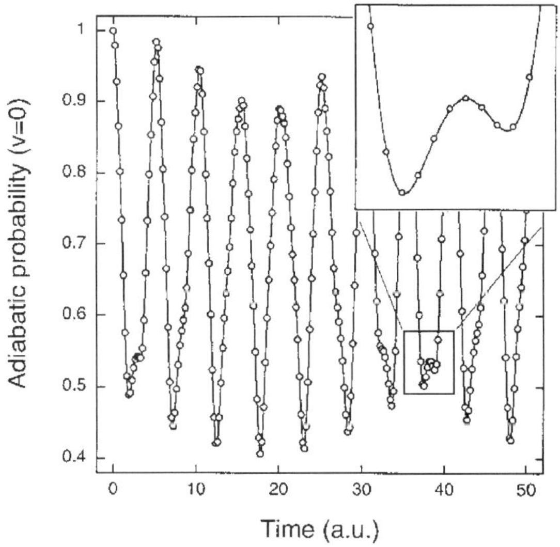

FIGURE 12.3. Time-dependent adiabatic ground state survival probabilities: the continuous curve gives the exact results and the open circles give results obtained using the mixed quantum–classical method [12.3]. The two sets of results are identical.

in both the ways they were derived and the details of implementation. The mixed quantum–classical Bohmian (MQCB) method developed by Gindensperger, Meier, and Beswick [12.4] will be described first, and then several applications will be presented. This will be followed by the method of Prezhdo and Brooksby [12.5] in Section 12.7.

In order to set the stage for these two methods, we will begin by applying Bohmian mechanics to a two-degree-of-freedom system. The system is in a pure state, and the exact quantum trajectory equations of motion are presented for *both* degrees of freedom. Then, approximations will be made in these equations so that coupled classical and quantum calculations can be performed. The same notation for coordinates, masses, etc. that was introduced in Section 12.2 will again be used in this section.

The time-dependent wave function for the system is written in the usual form

$$\psi(q, Q, t) = R(q, Q, t)e^{iS(q,Q,t)/\hbar}, \tag{12.16}$$

and this is substituted into the TDSE. After some algebra, we end up with the continuity equation and equations of motion for the quantum trajectory, which is specified by two coordinates $(q(t), Q(t))$:

$$\frac{d\rho(q, Q, t)}{dt} = -\rho(q, Q, t)\left[\frac{1}{m}\frac{\partial^2}{\partial q^2} + \frac{1}{M}\frac{\partial^2}{\partial Q^2}\right] S(q, Q, t), \tag{12.17}$$

$$\frac{dq}{dt} = \frac{p(q, Q, t)}{m}, \tag{12.18}$$

$$\frac{dQ}{dt} = \frac{P(q, Q, t)}{M}, \tag{12.19}$$

$$\frac{dp(q, Q, t)}{dt} = -\frac{\partial V(q, Q)}{\partial q} - \frac{\partial U(q, Q, t)}{\partial q}, \tag{12.20}$$

$$\frac{dP(q, Q, t)}{dt} = -\frac{\partial V(q, Q)}{\partial Q} - \frac{\partial U(q, Q, t)}{\partial Q}, \tag{12.21}$$

$$U(q, Q, t) = -\frac{\hbar^2}{2}\frac{1}{R(q, Q, t)}\left[\frac{1}{m}\frac{\partial^2}{\partial q^2} + \frac{1}{M}\frac{\partial^2}{\partial Q^2}\right]R(q, Q, t). \tag{12.22}$$

In the continuity equation, equation 12.17, the term on the right side multiplying the density is the divergence of the flow velocity, $\vec{\nabla} \cdot \vec{v}(q, Q, t)$. In equations 12.20 and 12.21, the second term on the right is the quantum force, and in equation 12.22, in order to avoid confusion with the coordinate Q, U is used to denote the quantum potential. Integration of these equations provides the density and two momenta $(p(t), P(t))$ along the quantum trajectory specified by the two coordinates $(q(t), Q(t))$. It is important to note that both $p(t)$ and $P(t)$ are *hydrodynamic* (average) *momenta,* and more complete notation would show the coordinate dependence of these dependent variables: $p(q, Q, t)$ and $P(q, Q, t)$. So far, *no approximations* have been made, and the *exact quantum dynamics* will be generated by quantum trajectory solutions to these equations.

Gindensperger et al. [12.4] make *several approximations* that lead to equations of motion for a quantum subsystem interacting with a classical subsystem. First, they assume that the mass for the quantum subsystem is much smaller than that for the classical subsystem, $M \gg m$. Next, they assume that the action $S(q, Q, t)$ and the amplitude $R(q, Q, t)$ have small curvatures *with respect to the "classical" coordinate* Q. With these assumptions, the two terms $\partial^2 S/\partial Q^2$ in equation 12.17 and $\partial^2 R/\partial Q^2$ in equation 12.22 are neglected. (Setting $\partial^2 R/\partial Q^2 = 0$ means that a source of curvature for the R-amplitude is neglected. Setting $\partial v_Q/\partial Q = \partial^2 S/\partial Q^2 = 0$ means that there is no change in the Q-component of the velocity. Taken together, these approximations prevent spreading (dispersion) of the wave packet in the Q-direction.)

As a result of these approximations, the *quantum subsystem*, with coordinate and momentum (q, p), evolves according to the approximate equations of motion

Quantum:

$$\frac{d\rho(q, Q(t), t)}{dt} = -\rho(q, Q(t), t)\frac{1}{m}\frac{\partial^2 S(q, Q(t), t)}{\partial q^2}, \tag{12.23}$$

$$\frac{dq}{dt} = \frac{p(q, Q(t), t)}{m}, \tag{12.24}$$

$$\frac{dp(q, Q(t), t)}{dt} = -\frac{\partial V(q, Q(t))}{\partial q} - \frac{\partial U(q, Q(t), t)}{\partial q}. \tag{12.25}$$

In addition, the quantum potential is approximated by the curvature of the amplitude with respect to (only) the q coordinate:

$$U(q, Q(t), t) = -\frac{\hbar^2}{2m}\frac{1}{R(q, Q(t), t)}\frac{\partial^2 R(q, Q(t), t)}{\partial q^2}. \tag{12.26}$$

It is interesting that the quantum force in equation 12.25 *does not involve the classical momentum*, whereas the hydrodynamic force in equation 12.6 does involve this dependence. In these four equations, the time dependence of the classical coordinate $Q(t)$ enters as a parameter. For consistency, the wave function for the composite system is also replaced by an approximate wave function for the quantum subsystem

$$\psi(q, Q(t), t) = R(q, Q(t), t)e^{iS(q,Q(t),t)/\hbar}. \quad (12.27)$$

This function also depends parametrically on the time-dependent classical coordinate.

The *classical subsystem*, with the dynamical variables (Q, P), evolves according to the two equations

$$\boxed{\textit{Classical:}}\quad \frac{dQ}{dt} = \frac{P(q, Q, t)}{M}, \quad (12.28)$$

$$\frac{dP(q, Q, t)}{dt} = -\frac{\partial V(q, Q)}{\partial Q} - \frac{\partial U(q, Q, t)}{\partial Q}. \quad (12.29)$$

Note that the Q-gradient of the quantum potential, the quantum force, enters the equation for $dP\ /\ dt$. A further approximation involves neglect of the quantum force in equation 12.29, thus leading to the simplified Newtonian equation

$$\frac{dP(q, Q, t)}{dt} = -\frac{\partial V(q, Q)}{\partial Q}. \quad (12.30)$$

Although not indicated explicitly, the three preceding equations depend parametrically on the time-dependent coordinate for the quantum trajectory, $q(t)$.

Rather than integrating the continuity equation, equation 12.23, along the quantum trajectory, as in the QTM, Gindensperger et al. elect to directly solve the TDSE, which is driven by the time dependent potential $V(q, Q(t))$:

$$\left[-\frac{\hbar^2}{2m}\frac{\partial^2}{\partial q^2} + V(q, Q(t))\right]\psi(q, Q(t), t) = i\hbar\frac{\partial\psi(q, Q(t), t)}{\partial t}. \quad (12.31)$$

This equation is formally equivalent to equation 12.2 in the mean-field approximation, although the notation has been changed slightly.

As a consequence of these approximations, the method of propagation for the mixed quantum–classical system is as follows. At the initial time, values are chosen for the classical coordinate and momentum, for one of the trajectories, $(Q_j(0), P_j(0))$. An initial wave function for the quantum subsystem is specified, and initial conditions are chosen for one quantum trajectory $(q_j(0), p_j(0))$. Through a sequence of time steps, the time-dependent wave function $\psi(q, Q_j(t), t)$ is evolved, equations 12.24–12.26 are used to update the quantum trajectory $(q_j(t), p_j(t))$, and equations 12.28 and 12.29 are used to update the classical trajectory $(Q_j(t), P_j(t))$. It is important to note that the *Schrödinger* equation is solved separately to find the wave function corresponding to *each trajectory* in the ensemble of N trajectories, $(q_j(t), p_j(t), Q_j(t), P_j(t))$, $j = 1, 2, \ldots, N$. *There is a unique time dependent wave function associated with each trajectory and these wave functions all branch from the same initial wave function.* Finally, we note that in contrast to the mean-field approximation, in the MQCB method there are no integrals to evaluate over the quantum wave function.

12.6 Examples of the MQCB Method

In this section, three applications of the MQCB method will be described. In the first study, Gindensperger et al. [12.4] computed the time evolution of the adiabatic populations for the same coupled harmonic oscillator model that was described earlier, in Section 12.4. Figure 12.4 shows the populations $P_i^{ad}(t)$ for $i = 0, 1, 2,$ assuming that the system is launched from the ground adiabatic state at $t = 0$. These populations are averages over the results obtained with 50 trajectories, this number being sufficient for convergence. In this figure, the exact quantum results are shown by the continuous curve and the dashed curve shows results that were obtained using the MQCB method when equations 12.28 and 12.29 were used to evolve the classical trajectory. The dotted line shows the result obtained

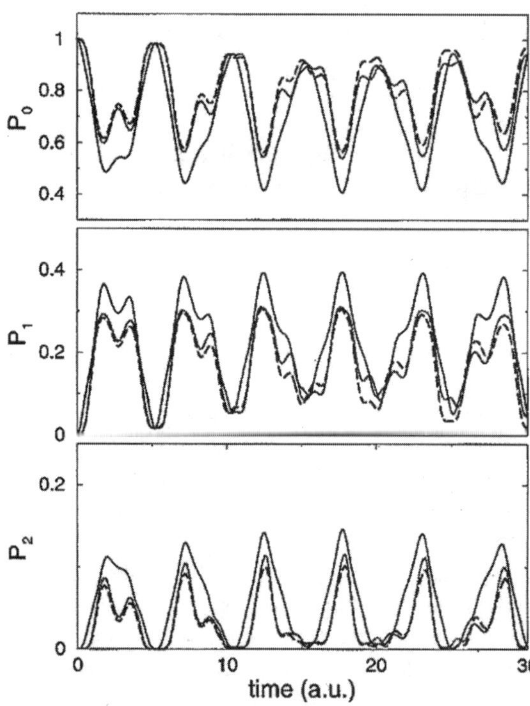

FIGURE 12.4. Adiabatic populations as a function of time [12.4]. The exact quantum results are shown by the continuous curve, the dashed curve shows the MQCB results, and the dotted curve shows the MQCB results that were obtained after dropping the quantum force back-reaction term from the classical equations of motion.

after dropping the quantum force in the classical equations (i.e., using equation 12.30 in place of equation 12.29). It is seen from this figure that good agreement was obtained between the MQCB results and the exact quantum results and that neglecting the quantum force in Eq. 12.29 produces only small changes in the mixed quantum–classical results. However, when this figure is compared with figure 12.3, we note that more accurate results (exact!) were obtained using the mixed quantum–classical method that was derived through the phase space route.

A second model problem studied with the MQCB method involves the collision of a light particle with a heavy particle that is adsorbed on an immobile surface [12.8]. This model was introduced by Sholl and Tully in their 1998 study [12.9]. In this example, there is energy exchange between the two degrees of freedom, and some of the incoming wave packet is temporarily trapped near the surface. The harmonic potential binding the adsorbed heavy particle to the surface is $V_{cl}(Q) = (1/2)M\Omega Q^2$, the interaction of the light quantum particle with the surface is represented by a Morse potential

$$V_{qu}(q) = a\left[e^{-2b(q-c)} - 2e^{-b(q-c)}\right], \tag{12.32}$$

and the interaction between the light and heavy particles is an exponential repulsive potential, $V_{\text{int}}(q, Q) = Ae^{-B(q-Q)}$. (A list of values for the parameters is given in Table 1 in [12.8].) At the initial time, the total wave function is the product of a ground state harmonic oscillator function for the heavy particle and a Gaussian wave packet for the incoming light particle,

$$\psi(q, Q, t = 0) = \phi_0(Q) \cdot e^{-ik_0 q}\, Ne^{-(q-q_0)^2/\gamma^2}. \tag{12.33}$$

The quantity of interest is the time-dependent probability of finding the light particle scattered beyond the distance q_s above the surface. (The wave packet for the light particle is initially centered at $q_0 = 6$ a.u., and the detection distance is larger than this value, $q_s = 8$ a.u.)

Examples of quantum trajectories interacting with the surface and the adsorbed molecule are shown in figure 12.5. The trajectories proceed toward the surface and begin to interact with the heavy adsorbed particle around the time $t = 30$ fs. Some of the trajectories are quickly repelled from the surface (especially those on the "back side" of the incoming wave packet), but others undergo multiple small-amplitude oscillations near the surface, thus reflecting the formation of a temporarily trapped state. As we will describe shortly, this branching of the initial wave packet into fast and delayed reflected components is the reason that the mean-field method has poor accuracy for this system.

The time-dependent probabilities $P_s(t)$ for finding the light particle scattered beyond the distance q_s are shown in figure 12.6 for three initial wave packet translational energies. Starting from $P_s(t = 0) = 0$, these probabilities rapidly increase after a delay time that decreases with the collision energy. The MQC method yields probabilities that have the correct long-time limit $P_s(t \to \infty) \to 1$, a feature not shared with the mean-field results, especially at low collision energies. One defect of the MQCB results is that they rise too rapidly at short times. This is because zero-point energy in the bound heavy oscillator can be transferred

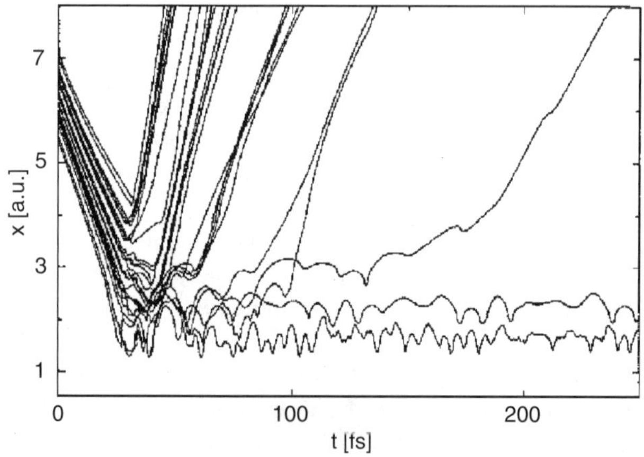

FIGURE 12.5. Quantum trajectories for the light particle as it encounters an adsorbed molecule trapped on an immobile surface [12.8]. The initial collision energy is $E_0 = 30$ kJ/mole.

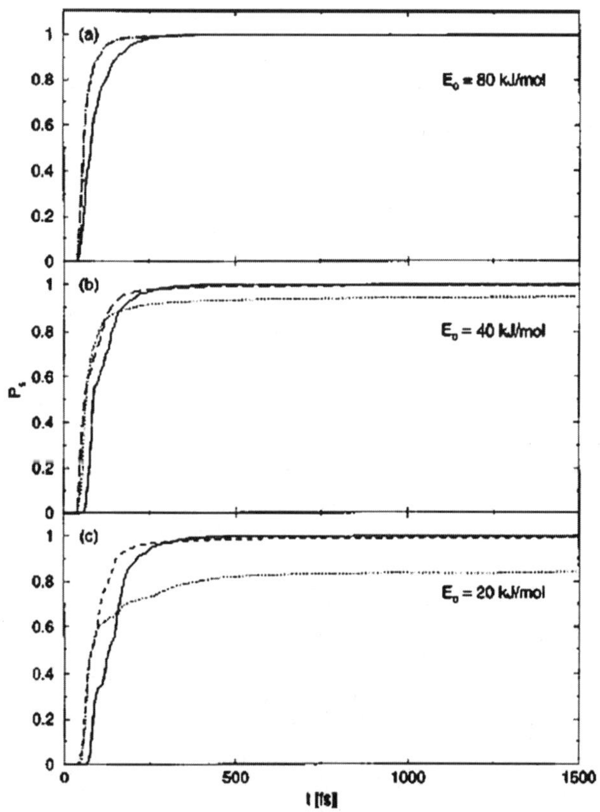

FIGURE 12.6. Time-dependent scattering probabilities at three initial collision energies [12.8]: full quantum results (continuous curve), mean-field results (dotted curve), and MQCB results (dashed curve).

to the quantum degree of freedom, leading to a rapid departure of the scattered trajectories. This feature is not shared by the exact quantum calculations.

In the third and final application of the MQCB method that will be described in this chapter, the diffractive scattering of diatomic molecules from a corrugated surface was studied [12.10]. The specific application focused on the rotational excitation of nitrogen molecules scattering from a corrugated lithium fluoride surface:

$$N_2(j = 0, m_j = 0) + \mathrm{LiF}(001) \rightarrow N_2(j, m_j) + \mathrm{LiF}(001).$$

Five coordinates were used to define the position and orientation of the rigid dumbbell molecule: Z denotes the perpendicular distance between the center of mass of the molecule and the surface, (X, Y) denotes displacements parallel to the surface, and the two angles (θ, ϕ) orient the molecular axis. Diffraction channels are specified by the box-normalized plane waves (a and b specify the size of the surface unit cell)

$$\Lambda_{m,n}(X, Y) = \frac{1}{\sqrt{ab}} e^{i(2\pi nX/a + 2\pi mY/b)}. \tag{12.34}$$

The rotation–diffraction channel is then expressed in terms of the product of a spherical harmonic times a diffraction function,

$$Y_{j,m_j}(\theta, \phi)\Lambda_{m,n}(X, Y). \tag{12.35}$$

Initially, the molecule is nonrotating and the incidence is normal to the surface, so that all four quantum numbers in equation 12.35 are zero. The initial translation is represented as a Gaussian wave packet centered above the surface at the position Z_0, and the direction of propagation is in toward the surface.

In this study, the *classical subsystem* comprises the three coordinates (Z, θ, ϕ), and the two diffraction coordinates (X, Y) define the *quantum subsystem*. With this partitioning, the three coordinates for a classical trajectory $(Z(t), \theta(t), \phi(t))$ enter the expression for the potential energy when the Schrödinger equation is solved for the time-dependent wave function $\psi(X, Y, Z(t), \theta(t), \phi(t))$. In these computations, for each of 20 values of the incident translational energy in the range 0.1–0.3 eV, 2000 trajectories were computed.

For the initial translational energy $E_0 = 0.150$ eV, figure 12.7 shows several time-dependent averages over the trajectories that were run at this energy. The top panel shows the average distance between the molecule and the surface, the middle panel shows the average rotational energy transfer (ΛRET), and the bottom panel shows the probabilities for excitation into two diffraction channels along with the probability for surviving in the entrance channel. Comparison of the top two panels shows that the average rotational energy grows significantly when the incoming molecules are moving slowly near the turning point in the Z direction. The bottom panel shows excitation of two of the Fourier components describing the (X, Y) motion. In this study, probabilities for excitation of seven diffraction channels were compared to full five-dimensional quantum-mechanical results obtained using the multiconfiguration time-dependent Hartree method (MCTDH). Figure 3 in [12.10]

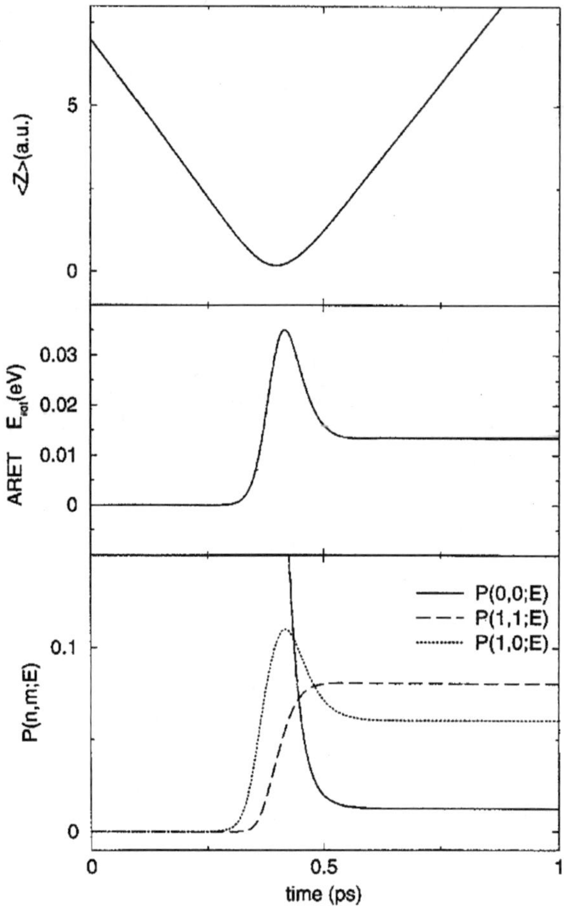

FIGURE 12.7. Time-dependent averages over 2000 trajectories at the collision energy 0.150 eV [12.10]. The top panel shows the average distance between the molecule and the surface, the middle panel shows the average rotational energy of the diffracted molecules, and the bottom panel shows the probabilities for excitation into two diffraction channels, (1, 0) and (1, 1), along with the survival probability in the incident (0,0) channel.

shows very good agreement between these two sets of results over the collision energy range 0.1–0.3 eV.

12.7 Backreaction Through the Bohmian Particle

Prezhdo and Brooksby described the method [12.5] *Quantum Backreaction Through the Bohmian Particle* as a solution to the long-standing problem, how do we permit a trajectory in the classical subsystem to respond to dynamics in the quantum subsystem? We will refer to the method introduced in this section as

QBBP. To develop this approximation, we begin with the mean quantum force in equation 12.4,

$$\begin{aligned}\tilde{F}_{\text{int}}(Q) &= \int |\psi(q,t)|^2 \left(-\partial V_{\text{int}}(q,Q)/\partial Q\right) dq \\ &= \int R^2(q,t) \left(-\partial V_{\text{int}}(q,Q)/\partial Q\right) dq. \end{aligned} \tag{12.36}$$

In the second version of this force, the density has been expressed in terms of the amplitude $R(q,t)$. In this form, the mean force can be viewed as the force at the position of each of a set of quantum trajectories $q(t)$ averaged over the density $R^2(q,t)$. The following approximation is then made: *the positions of the quantum trajectories are sampled from the initial density $R^2(q,t=0)$, but at each time step, no averaging of the quantum force is performed.* As a result, in the QBBP, the back-reaction problem is solved by propagating the classical trajectory using the equation of motion

$$\frac{dP(t)}{dt} = -\frac{\partial}{\partial Q}\left[V_{cl}(Q) + V_{\text{int}}(q,Q)\right] - f_{cl}(Q) + f_{\text{int}}(q,Q), \tag{12.37}$$

where q in the interaction component of the force is evaluated along the quantum trajectory $q(t)$. This equation is identical to the one used by Gindensperger et al., see equation 12.30. There is no quantum (hydrodynamical) force term in this equation.

A set of replicas of the coupled system, each involving one pairing of a quantum trajectory and a classical trajectory, $\{q_j(t), Q_j(t)\}$, is propagated, and at each time step the wave functions for each replica quantum subsystem, $\psi_j(q, Q_j(t), t)$, are updated. The wave function corresponding to each of these replicas is then used to update the quantum trajectory $q_j(t)$, which is guided by this wave function. Except for details of implementation, this method is equivalent to the MQCB method that was described in the two preceding sections. What differs is the route that culminates in the equations of motion. Additional comments about the QBBP will be made after we have presented an example; this is the only test case that was presented in the work by Prezhdo and Brooksby.

The QBBP method was applied to one of the model problems mentioned in the preceding section [12.5]: a light particle collides with a heavy particle that is harmonically bound to a surface. However, the parameters used in the potential (see Table I in [12.5]) are different from those used by Gindensperger et al. [12.4]. Using an average over 1500 trajectories, Prezhdo and Brooksby obtained the time-dependent scattering probabilities shown in figure 12.8. The probability obtained using the QBBP (continuous curve) has the correct long-time limit, and this lies above the mean field result (dashed curve). The results in this figure can be compared with the MQCB results that were presented earlier (see figure 12.6, lower panel). In a later study, the MQCB results were compared with results from three other approximate methods [12.14], including the *quantized mean field* (QMF) method, which was also developed by Prezhdo et al. [12.15].

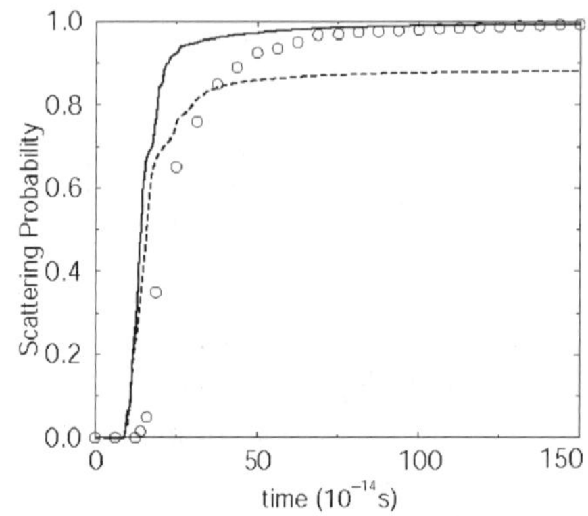

FIGURE 12.8. Time-dependent probabilities for the scattering of a light particle from a harmonically bound heavy particle at the collision energy $E_0 = 20$ kJ/mole [12.5]: quantum-mechanical results (open circles), QBBP (continuous curve), mean-field results (dashed curve).

12.8 Discussion

In this chapter, two different trajectory formalisms were presented for studying the dynamics of a quantum subsystem coupled to a classical subsystem. The first approach, developed by Burghardt and Parlant [12.3], starts with a quantum distribution function in the four-dimensional full *phase space* and ends up with equations of motion in a three-dimensional partial hydrodynamic phase space (with three independent variables). Because of the starting point in this derivation, this approach will be referred to as the *phase space method*. In the second approach (historically, the first of the two approaches to be published), which was developed independently by Prezhdo and Brooksby [12.5] and by Gindensperger, Meier, and Beswick [12.4], we start in the two-dimensional *configuration space*, derive the exact Bohmian equations of motion, and then make approximations in these equations. Because of the manner in which it was derived, this approach will be referred to as the *configuration space method*.

Although the final working equations may appear rather similar in these two methods, there are significant differences both in the derivations and in the interpretation and implementation of the final sets of equations. What are the differences between these two methods? We will attempt to answer this question by focusing on six aspects of the derivation and implementation.

(a) *The starting point for each derivation*

Phase space method. Equations of motion for the moments of the distribution function $W(q, p, Q, P, t)$ with respect to the quantum momentum can be derived from an *approximate* Wigner–Moyal equation in the four-dimensional phase space: the quantum terms involving derivatives with respect to Q and P are dropped *before* proceeding with the steps that lead to the partial moments. (It is also possible to derive the same final equations by starting from the *exact* Wigner–Moyal equation followed by evaluating the exact partial moment equations, and then making the appropriate approximations.) The 0-moment and the 1-moment are (nonreduced) densities with respect to the classical variables (Q, P).

Configuration space method. Quantum trajectory equations of motion are derived in the two-dimensional configuration space spanned by the coordinates (q, Q). The derivation is initiated by substituting the polar form for the wave function $\psi(q, Q, t)$ into the TDSE. When making the partition into quantum and classical subsystems, the second derivative with respect to Q of the amplitude and action, $R(q, Q, t)$ and $S(q, Q, t)$, are set to zero.

(b) *Independent and dependent coordinates*

Phase space method. There are *three* independent variables, q, Q, and P. The average quantum momentum $p(q, Q, P, t)$ is a function of time along the trajectory $(q(t), Q(t), P(t))$ in the three-dimensional partial hydrodynamic phase space.

Configuration space method. There are *two* independent variables, q and Q. The two average momenta $p(q, Q, t)$ and $P(q, Q, t)$ are functions of time along the configuration space trajectory $(q(t), Q(t))$.

(c) *Form of the equations of motion*

Phase space method. Five equations are derived for the time evolution of the following quantities along each trajectory: $\rho(t)$, $q(t)$, $p(t)$, $Q(t)$, and $P(t)$. The three quantities $q(t)$, $Q(t)$, and $P(t)$ locate the trajectory in the three-dimensional partial hydrodynamic phase space, while $\rho(t)$ and $p(t)$ are evaluated along the trajectory. The hydrodynamic force enters the equation of motion for the average quantum momentum $p(q, Q, P, t)$, but *not* for the classical momentum.

Configuration space method. Five equations are derived for the time evolution of the following quantities along each trajectory: $\rho(t)$, $q(t)$, $p(t)$, $Q(t)$, and $P(t)$. The two quantities $q(t)$ and $Q(t)$ locate the trajectory in the two-dimensional configuration space, while $\rho(t)$, $p(t)$, and $P(t)$ are functions of time along the trajectory. Approximate equations of motion are developed for the two average momenta, $p(q, Q, t)$ and $P(q, Q, t)$. Components of the approximate quantum force enter the equations of motion for *both* of the *average momenta* $p(q, Q, t)$ and $P(q, Q, t)$.

(d) *Initial conditions on the trajectories*

Phase space method. Given the initial value for the phase space distribution function $W(q, p, Q, P, t = 0)$, the average quantum momentum is given by $p(q, Q, P) = < pW > / < W >$ (where integration is *only* over p). The values of q, Q, and P may be chosen at random, on a grid, or sampled from the reduced density $\rho(q, Q, P) = < W >$.

Configuration space method. Given the initial value for the wave function $\psi(q, Q, t = 0)$, the two average momenta are given by gradients of the action function: $p(q, Q) = \partial S(q, Q)/\partial q$ and $P(q, Q) = \partial S(q, Q)/\partial Q$. The coordinates q and Q may be chosen in various ways: at random, on a grid, or sampled from the probability density $|\psi(q, Q, t = 0)|^2$.

(e) *The density riding along each trajectory*

Phase space method. The density riding along a trajectory in the partial hydrodynamic phase space is given by $W_{ps}(q, p, Q, P, t) = \rho(q, Q, P, t)\delta(p - p(q, Q, P, t))$.

Configuration space method. The density in the hydrodynamic phase space is given by $W_{cs}(q, p, Q, P, t) = \rho(q, Q, t)\delta(p - p(q, Q, t))\delta(P - P(q, Q, t))$.

(f) *Average versus instantaneous values*

Phase space method At each point in the partial hydrodynamic phase space, (q, Q, P), only the quantum momentum p is an average value.

Configuration space method Both of the quantum and classical momenta, p and P, are average values at each configuration space point (q, Q).

For both of these methods, there are additional properties that arise from the mathematical structure of the equations. There properties include the following: (1) The dynamics are reversible in time, and time slicing does not alter the dynamics. The latter property means that within expected numerical errors (truncation, discretization, etc.), the system propagated in two steps from t_1 to $t_1 + \Delta t$ yields the same results as a single propagation between these two times. (2) Although the total energy is not rigorously conserved, good conservation (within a few %) can be attained in practice. Further discussion is presented in Salcedo's comments [12.11] on the MQCB method and in the reply by Prezhdo and Brooksby [12.12].

References

12.1. G.D. Billing, *Classical path method in inelastic and reactive scattering*, Int. Rev. Phys. Chem. **13**, 309 (1994).

12.2. J.C. Tully, *Nonadiabatic dynamics*, in D.L. Thompson (ed.), *Modern Methods for Multidimensional Dynamics Computations in Chemistry* (World Scientific, Singapore, 1998).

12.3. I. Burghardt and G. Parlant, *On the dynamics of coupled Bohmian and phase-space variables: A new hybrid quantum-classical approach*, J. Chem. Phys. **120**, 3055 (2004).

12.4. E. Gindensperger, C. Meier, and J.A. Beswick, *Mixing quantum and classical dynamics using Bohmian trajectories*, J. Chem. Phys. **113**, 9369 (2000).

12.5. C. Brooksby and O.V. Prezhdo, *Quantum back-reaction through the Bohmian particle*, Phys. Rev. Lett. **86**, 3215 (2001).

12.6. P. Ehrenfest, *Bemerkung überdie angenähertegültigkeitder klassischen mechanik innerhalb der quantenmechanik*, Z. Physik. **45**, 455 (1927).

12.7. D. Kohen, F.H. Stillinger and J.C. Tully, *Model studies of nonadiabatic dynamics*, J. Chem. Phys. **109**, 4713 (1998).

12.8. E. Gindensperger, C. Meier, and J.A. Beswick, *Quantum–classical dynamics including continuum states using quantum trajectories*, J. Chem. Phys. **116**, 8 (2002).

12.9. D.A. Sholl and J.C. Tully, *A generalized surface hopping method*, J. Chem. Phys, **109**, 7702 (1998).

12.10. E. Gindensperger, C. Meier, and J.A. Beswick, *Quantum–classical description of rotational diffractive scattering using Bohmian trajectories: Comparison with full quantum wave packet results*, J. Chem. Phys. **116**, 10051 (2002).

12.11. L.L. Salcedo, *Comment on "Quantum backreaction through the Bohmian particle"*, Phys. Rev. Lett. **90**, 118901 (2003).
12.12. O.V. Prezhdo and C. Brooksby, *Reply to comment*, Phys. Rev. Lett. **90**, 118902 (2003).
12.13. R. Kapral and G. Ciccotti, *Mixed quantum–classical dynamics*, J. Chem. Phys. **110**, 8919 (1999).
12.14. O.V. Prezhdo and C. Brooksby, *Non-adiabatic molecular dynamics with quantum solvent effects*, J. Mol. Structure (Theochem) **630,** 45 (2003).
12.15. C. Brooksby and O.V. Prezhdo, *Quantized mean-field approximation*, Chem. Phys. Lett. **346**, 463 (2001).
12.16. F.A. Bornemann, P. Nettesheim, and C. Schutte, *Quantum–classical molecular dynamics as an approximation to full quantum dynamics*, J. Chem. Phys. **105**, 1074 (1996).
12.16. I. Burghardt, K. Moller, and G. Parlant, to be published.
12.17. S. Nielsen, R. Kapral, and G. Ciccotti, *Statistical mechanics of quantum classical systems*, J. Chem. Phys. **115**, 5805 (2001).
12.18. A. Sergi, D. MacKernan, G. Ciccotti, and R. Kapral, *Simulating quantum dynamics in classical environments*, Theor. Chem. Acta, **110**, 49 (2003).
12.19. A. Sergi and R. Kapral, *Nonadiabatic reaction rates for dissipative quantum classical systems*, J. Chem. Phys. **119**, 12776 (2003).
12.20. I. Horenko, C. Salzmann, Bukhard Schmidt, and Christof Schütte, *Quantum–classical Liouville approach to molecular dynamics: Surface hopping Gaussian phase-space packets*, J. Chem. Phys. **117**, 11075 (2002).
12.21. I. Horenko, M. Weisner, Bukhard Schmidt, and Christof Schütte, *Fully adaptive propagation of the quantum–classical Liouville equation*, J. Chem. Phys. **120**, 8913 (2002).

13
Topics in Quantum Hydrodynamics: The Stress Tensor and Vorticity

Two concepts from classical fluid dynamics are extended into the quantum domain: The quantum stress tensor appears in an extension of the Navier–Stokes equations, and quantized vortices develop around wave function nodes.

13.1 Introduction

Vortex formation and the stress tensor are two topics that play an important role in classical fluid dynamics [13.1]. Vortices (from the Latin *vertere*, to turn) are commonly observed in flowing fluids, and their formation is closely connected with the onset of turbulent flow. Vortices such as eddies in rivers, the red spot of Jupiter, tornadoes, and smoke rings all have the characteristic feature of matter flowing around a central core. In quantum hydrodynamics, both vortices and the stress tensor again play a significant role. In the quantum transport equations, the stress tensor has terms that are formally equivalent to those appearing in the classical dynamical equations plus additional $\hbar$-dependent contributions [13.2–13.18]. A remarkable feature of quantum vortices, which form around nodes in the wave function, is that they carry only specific amounts of "circulation" and angular momentum [13.19]. These vortices have been detected in calculations on electron transport in wave guides and molecular wires, atomic and molecular collision processes, and scattering from the surfaces of solids [13.20–13.30]. In this chapter, the origin and interpretation of quantized vortices and quantum contributions to the stress tensor will be described. A number of examples will be given, especially for quantum vortex dynamics.

The five fundamental equations of motion that govern classical fluid flow can be developed from conservation relations: conservation of mass (the continuity equation), conservation of the three components of linear momentum (the Navier–Stokes equations), and conservation of energy [13.1]. The momentum conservation equations express the rate of change in the *momentum density* (the product of linear momentum and fluid density) as the sum of two contributions: the force density arising from external fields (for example, electromagnetic or gravitational) and a contribution from "internal" forces. The latter depends on spatial derivatives of the stress tensor, which arises from forces acting on the surface of a small element of fluid. These forces may in turn be decomposed into components normal and tangential to the surfaces bounding a fluid element. These components are the compressive and shear contributions, respectively. In the quantum-mechanical extension of the classical Navier–Stokes equations, there are $\hbar$-dependent quantum contributions to both the compressive and shear components of the stress tensor.

In quantum mechanics, streamlines surrounding the vortex core form approximately circular loops, the wave function phase undergoes a 2π (or integer multiple of 2π) winding, and the circulation integral and angular momentum are always quantized. In time-dependent studies, vortices may be created and destroyed, and probability density may be temporarily trapped in the vicinity of each vortex. Quantized vortices covering vastly different length scales have been found in both computational studies and in experiments. They play a significant role in a remarkably diverse range of phenomena, including ultracold atomic Bose–Einstein condensates [13.31–13.33], superfluid liquid ^{3}He [13.34], type-II low-temperature superconductors [13.35], and pulsars (rotating neutron stars) [13.36, 13.37].

In Section 13.2, the quantum stress tensor is introduced in the simplest possible setting, the one-dimensional quantum fluid (where the tensor is actually a scalar). Then, in Section 13.3, the full stress tensor is presented in the context of the quantum extension of the Navier–Stokes equations. A computational model showing the various contributions to the stress tensor is given in Section 13.4. Quantized vortices are introduced in Section 13.6, and a number of examples from the literature are given in Sections 13.5–13.6. The interconnection between dynamical tunneling, vortices, and quantum trajectories is described in Sections 13.7–13.8. To end this chapter, a summary is provided in Section 13.9.

13.2 Stress in the One-Dimensional Quantum Fluid

The route by which the stress tensor makes its appearance in one version of the quantum-mechanical equations of motion can be illustrated for the one-dimensional case. To be sure, all features of the stress tensor cannot be brought out with this example, but the method of derivation is analogous to that used in higher dimensionality. First, relevant features of the stress tensor in the context of classical fluid dynamics are described in Box 13.1.

Box 13.1. The stress tensor in classical fluid mechanics

For a fluid flowing in three-dimensional space, the stress $\Pi(x, y, z, t)$, a rank-3 *tensor field*, has nine components, three of which, the diagonal elements Π_{xx}, Π_{yy}, and Π_{zz}, refer to *normal (compressive) stress*, while the six off-diagonal elements refer to *shear stress*. The stress tensor, with each component labeled by a pair of coordinates, has the following form:

$$\Pi(x, y, z, t) = \begin{bmatrix} \Pi_{xx} & \Pi_{xy} & \Pi_{xz} \\ \Pi_{yx} & \Pi_{yy} & \Pi_{yz} \\ \Pi_{zx} & \Pi_{zy} & \Pi_{zz} \end{bmatrix}. \quad (1)$$

If we focus on a tiny cube of fluid, the normal components of the stress either compress or dilate the cube, while the shear components produce angle deformations that squash the cube. The stress has units of force/area, pressure units, or what is equivalent, the units of momentum flux: momentum/(area · time).

The stress components are labeled with two indices. As an example, for the shear component denoted by Π_{xy}, the second label indicates the direction of the force, along the y axis in this case. The first label shows that this force lies in a plane perpendicular to the x axis. This tangential force creates angular strain on the fluid element shown in the first figure in this box. A positive value for one of the diagonal components signifies "pulling" or tension, whereas a negative value is compressive. For both normal and shearing components, if the normal to the plane is in the positive (negative) coordinate direction, then a positive stress component acts in the positive (negative) coordinate direction. The following figure shows two normal and two shear stress components acting on a small chunk of fluid.

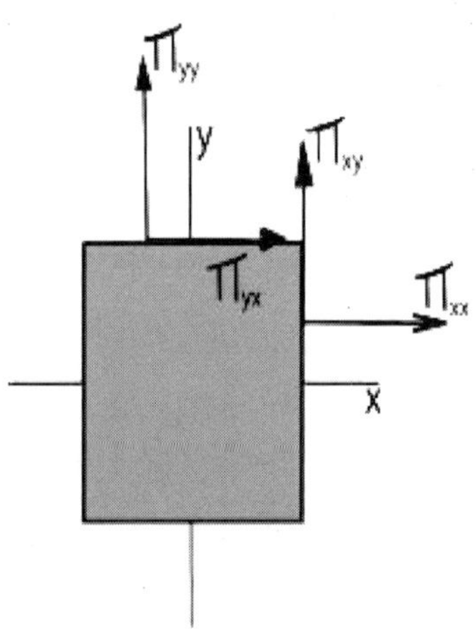

For small deformations in an elastic solid material, the deformation (strain, denoted by ε) is related to the normal component of the stress Π_{xx} through the modulus of elasticity (E) : $\Pi_{xx} = E\varepsilon$. This constitutive relation is *Hooke's law*. For a fluid, the rate of strain $(\dot{S})$ depends on the stress through the equation $\Pi = a\dot{S} + b1$, where a is the *dynamic viscosity*, and 1 is the unit tensor. This *constitutive relation* was suggested by Stokes in 1845.

The relationship between shear stress and the rate of angle deformation can be illustrated by consideration of *Newton's law of viscosity*. We imagine sheets of fluid in laminar flow in the x−direction. The fluid in the top plane (with the area $\Delta x \Delta z$) of the box shown below has the velocity $v_x(\Delta y)$ relative to the fluid in the bottom plane, which is taken to be at rest.

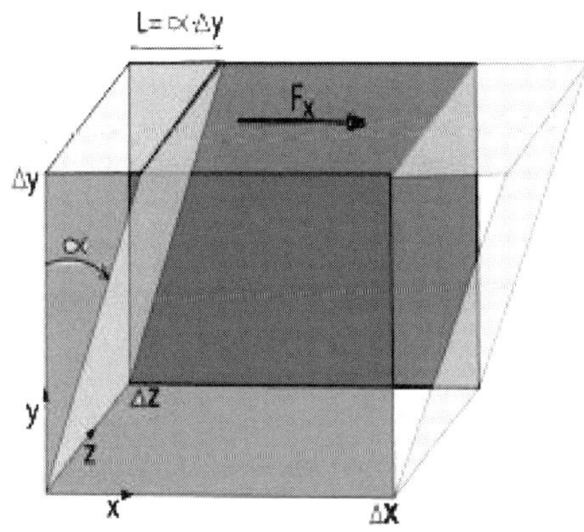

The fluid in the top plane is in motion because the shear force (F_x) applied to this plane is in the positive x-direction. The response to this force is the establishment of a gradient in the x-component of the velocity $\Delta v_x/\Delta y$. Newton's law then relates this velocity gradient and the applied force

$$F_x = \mu \cdot \Delta x \Delta z \cdot \frac{\Delta v_x}{\Delta y}, \tag{2}$$

in which μ is the dynamic viscosity. After the shear stress is defined as the force per unit area $\Pi_x = F_x/(\Delta x \Delta z)$, the preceding equation then becomes

$$\Pi_x = \mu \frac{\Delta v_x}{\Delta y}. \tag{3}$$

During the time interval Δt, the top plane moves the distance $L = v_x \Delta t = \alpha \Delta y$, where α is the angle measured from the y axis. The velocity gradient is then

$$\Delta v_x/\Delta y = \alpha/\Delta t = \dot{\alpha}. \tag{4}$$

As a result, the stress is related to the rate of angle deformation by

$$\Pi_x = \mu \dot{\alpha}. \tag{5}$$

This connection between the stress and the strain rate is *Newton's law of viscosity*. More general relations between off-diagonal elements of the stress tensor and the angle deformation are derived in texts on classical fluid mechanics [13.1].

Proceeding with the one-dimensional quantum-mechanical example, the two familiar hydrodynamic equations of motion in the Eulerian frame are

$$\frac{\partial \rho}{\partial t} = -\frac{\partial(\rho v)}{\partial x}, \tag{13.1}$$

$$m\left(\frac{\partial}{\partial t} + v\frac{\partial}{\partial x}\right)v = -\frac{\partial}{\partial x}(V + Q). \tag{13.2}$$

Before using these equations in the following derivation, a few definitions are in order: the *momentum density* is $\rho m v$, the *classical force density* is $\rho\,(-\partial V/\partial x)$, and the *quantum force density* is $\rho\,(-\partial Q/\partial x)$. We will now start the derivation of an *equation of motion for the momentum density*:

$$\frac{\partial(\rho m v)}{\partial t} = \rho m\frac{\partial v}{\partial t} + m v\frac{\partial \rho}{\partial t}. \tag{13.3}$$

If we substitute equations 13.1 and 13.2 into the right side of equation 13.3 and then evaluate $\partial Q/\partial x$ in terms of derivatives of the density, we finally obtain

$$\begin{aligned}\frac{\partial(\rho m v)}{\partial t} &= -m v^2\frac{\partial \rho}{\partial x} - 2\rho m v\frac{\partial v}{\partial x} + \frac{\hbar^2}{4m}\left(\frac{\partial^3 \rho}{\partial x^3}\right)\\ &\quad -\frac{\hbar^2}{2m}\frac{1}{\rho}\frac{\partial \rho}{\partial x}\left(\frac{\partial^2 \rho}{\partial x^2}\right) + \frac{\hbar^2}{4m}\frac{1}{\rho^2}\left(\frac{\partial \rho}{\partial x}\right)^3 - \rho\frac{\partial V}{\partial x}.\end{aligned} \tag{13.4}$$

The right side of this rather complicated expression can be written in terms of the gradient of the *stress* (a scalar for this one-dimensional example) plus the classical force density,

$$\frac{\partial(\rho m v)}{\partial t} = -\frac{\partial \Pi}{\partial x} - \rho\frac{\partial V}{\partial x}. \tag{13.5}$$

This equation is *formally the same* as one of the *Navier–Stokes equations* in classical fluid dynamics.

The first term on the right side of equation 13.5, the gradient of the stress, has an informative interpretation. At some fixed time, a plot of $\Pi(x)$ might reveal peaks and valleys representing regions of high and low stress, respectively. The first term on the right side of this equation has the following effect: during a short time interval, the momentum density changes so as to smooth out the stress. The moral of the story is, Fluids abhor stress! However, the rate at which the momentum density changes is also influenced by the last term in this equation, namely, the classical force acting on the fluid element.

The stress term in equation 13.5 is given by

$$\Pi = \rho m v^2 + \rho m\left\{\frac{\hbar^2}{4m^2}\frac{1}{\rho^2}\left(\frac{\partial \rho}{\partial x}\right)^2\right\} - \frac{\hbar^2}{4m}\frac{\partial^2 \rho}{\partial x^2}. \tag{13.6}$$

The first term on the right, lacking dependence on $\hbar$, is the *classical stress*, and the two remaining terms are both of quantum origin (they both come from derivatives of the quantum potential). The second term on the right is the *quantum stress*, and the last term, which will be denoted by P, is the *quantum pressure*.

The quantum stress can be compactly expressed in terms of a quantity introduced by Einstein when he developed a classical theory of diffusion. The *osmotic velocity* u will be defined by $u = -(D/\rho)(\partial\rho/\partial x)$, where the *quantum diffusion coefficient* (units: length^2/time) is $D = \hbar/(2m)$. The stress can conveniently be expressed in terms of the pressure P and the two velocities u and v,

$$\Pi = P + \rho m(v^2 + u^2), \tag{13.7}$$

where the first and third quantum terms depend on $\hbar$. The flow velocity and the osmotic velocity are components of a *complex-valued velocity*, which is defined in Box 13.2. Another, less compact, expression for the stress is

$$\Pi = \rho m v^2 - \frac{\hbar^2}{4m}\left[\frac{\partial^2\rho}{\partial x^2} - \frac{1}{\rho}\left(\frac{\partial\rho}{\partial x}\right)^2\right]. \tag{13.8}$$

Also, note from equations 13.7 and 13.8 that there are two kinds of stress: *flow stress* depending on the flow velocity v, and *shape stress* arising from spatial derivatives of the density.

In order to calibrate our analysis of the stress tensor, consider a one-dimensional Gaussian density, centered at $x = 0: \rho(x) = A\exp(-2\beta x^2)$. The single component of the stress tensor is given by $\Pi(x) = (\hbar^2\beta/m)\rho(x)$, so that the stress is highest at the peak of the density. We also note that the rate of change in the momentum density in the quantum Navier–Stokes equation is given by $-\partial\Pi/\partial x = (4\hbar^2\beta^2 x/m)\rho(x)$; this quantity is equal to $\rho(x)f_q$, where f_q is the quantum force. Since $x\rho(x) > 0$ for positive x and $x\rho(x) < 0$ for negative x, this equation requires, as time proceeds, that fluid elements carry momentum density away from the region of high stress near $x = 0$. Consequently, the density spreads symmetrically away from the origin as the stress is relieved.

Box 13.2. The complex-valued velocity

The complex-valued velocity $\vec{w}$ is defined through the gradient of the wave function,

$$\vec{w} = \frac{1}{m}\frac{\hbar}{i}\frac{\vec{\nabla}\psi}{\psi} = \vec{v} + i\vec{u}. \tag{1}$$

If we substitute the polar form of the wave function ($\psi = \rho^{1/2}e^{iS/\hbar}$) into this expression, we obtain

$$\vec{w} = \frac{1}{m}\vec{\nabla}S + i\left[-\frac{\hbar}{2m}\frac{1}{\rho}\vec{\nabla}\rho\right]. \tag{2}$$

The real and imaginary parts of $\vec{w}$ are recognized as the familiar flow velocity and the osmotic velocity, respectively, where $\hbar/(2m) = D$ plays the role of a diffusion coefficient. The complex velocity appears in the Hirschfelder et al. 1974 study [13.25] of quantum vortices and in papers related to Nelson's stochastic quantum mechanics [13.38–13.41].

Some ingredients of the quantum Navier–Stokes equation for the momentum density were exposed in this example, but the multidimensional version in the next section will bring additional terms into the equations.

13.3 Quantum Navier–Stokes Equation and the Stress Tensor

In a space with dimensionality D, the generalization of equation 13.5 is

$$\frac{\partial(\rho m v_i)}{\partial t} = -\sum_{j=1}^{D} \nabla_j \Pi_{ji} - \rho \nabla_i V, \quad i = 1, 2, \ldots, D. \tag{13.9}$$

This equation is *formally* the same as the *Navier–Stokes equation* in classical fluid dynamics. As an example, in two dimensions, there are two equations, one for each component of the momentum density, and each right side involves the sum of three terms:

$$\begin{aligned} \frac{\partial(\rho m v_x)}{\partial t} &= -\nabla_x \Pi_{xx} - \nabla_y \Pi_{yx} - \rho \nabla_x V, \\ \frac{\partial(\rho m v_y)}{\partial t} &= -\nabla_x \Pi_{xy} - \nabla_y \Pi_{yy} - \rho \nabla_y V. \end{aligned} \tag{13.10}$$

Equation 13.9 gives the rate of change in the i-th component of the momentum density in terms of derivatives of components of the $D \times D$ real-symmetric *stress tensor* Π, plus the "classical force density" (∇_j in this equation is the j-th component of the gradient operator). As described earlier, in Box 13.1, elements of the stress tensor carry two subscripts. For the element Π_{ij}, the force per unit area acts in the j-th direction but is in a plane perpendicular to the i-th coordinate axis. The diagonal elements of the stress tensor produce compressive deformation, and the off-diagonal shear terms lead to squashing (angular) deformation of a small fluid element.

The stress tensor has both *classical and quantum components*. These elements are given by

$$\Pi_{ij} = P\delta_{ij} + \rho m \left(v_i v_j + u_i u_j\right) = \rho m v_i v_j + \Pi_{ij}^{quantum}, \tag{13.11}$$

in which the (quantum) diagonal pressure term is $P = -(\hbar^2/(2m))\nabla^2 \rho$ and where the flow and diffusive velocity components are given by

$$v_i = \frac{1}{m} \nabla_i S, \quad \text{and} \quad u_i = -\frac{D}{\rho} \nabla_i \rho. \tag{13.12}$$

The (quantum) diffusive velocity is determined by the gradient of the density, $\vec{u} = -(D/\rho)\vec{\nabla}\rho$. In common with the one-dimensional case described in Section 13.2, the quantum components of the stress tensor depend on derivatives of the density, rather than on the components of the flow velocity. In contrast, the "classical" component of the stress tensor depends only on components of the flow velocity.

13.4 A Stress Tensor Example

There are very few quantum systems for which the stress tensor has been computed and only one case in which the computations and analysis were carried out using quantum trajectories. In Section 8.3, the quantum trajectory method was used to analyze the mechanism for decoherence in a model two-mode system [13.18]. The stress tensor for this system has also been computed, but before turning to these results, the coordinates, Hamiltonian, and initial wave function will be briefly reviewed. Initially, a coherent superposition of two well-separated Gaussian wave packets was prepared in a composite system involving a subsystem (coordinate x) coupled to a harmonic "bath" mode (coordinate y). The centers of the two Gaussians are located at $x = \pm a$, where the separation parameter $a = 1$ is chosen to be much larger than the width of either Gaussian. The Hamiltonian for this two-degree-of-freedom system can be decomposed into subsystem, bath, and coupling contributions, $H = H_s + H_b + H_c$. The coupling potential is bilinear in the system and bath coordinates, $H_c = cxy$. When the coupling coefficient c is not zero, this Hamiltonian is separable in the rotated coordinates (the normal coordinates) y_0 and y_1, which are obtained from the original coordinates (x, y) by rotation through the angle given by $\vartheta = -\tan^{-1}(c/k)$, where k is the force constant for the harmonic oscillator. For the parameter values used in this study, $\vartheta = -19^\circ$.

In order to evolve the wave packet that originates from this initial state, the quantum hydrodynamic equations of motion were integrated using an ensemble of 200 quantum trajectories. For the uncoupled case, $c = 0$, as time advances, the trajectories associated with each component Gaussian evolve on the potential surface, and by the time $t = 450$ a.u., a buildup of density takes place near the origin of the coordinates. However, when the coupling in nonzero, by the same time, decoherence is indicated by a significant *suppression of density* in the same region where there was buildup in the uncoupled case.

We will now consider features of the stress tensor at one time ($t = 400$ a.u.) for this two-mode system. In this case, there are three independent components of the stress tensor Π, denoted $\Pi_{0,0}$, $\Pi_{0,1}$, and $\Pi_{1,1}$, but we will focus only on the diagonal "bath–bath" element $\Pi_{1,1}$ (the subscripts (0,1) refer to the pairs of coordinates (x, y) or (y_0, y_1) for the uncoupled or coupled cases, respectively). For the uncoupled case, figures 13.1–13.4 display contour maps of all three terms that contribute to $\Pi_{1,1}$ along with the sum of these contributions. Then, figures 13.5–13.7 show contour maps of two of the terms that contribute to $\Pi_{1,1}$ and the sum of these contributions for the coupled case. The three terms that contribute to each element of the stress tensor are as follows: the *classical term* ($m\rho\, v_1^2$), the *quantum pressure term* (which depends on $\nabla^2\rho$), and the *quantum stress term* ($m\rho\, u_1^2$).

The quantum pressure term makes only a small contribution, at most 6% for the uncoupled case and less than 1% for the coupled case. Figure 13.1 shows a contour map of this term for the uncoupled case. The pressure is highest near the two peaks in the density, at $x = \pm 1$, since this is where the curvature of the density is largest. Because the pressure term makes such a small contribution for the coupled case, the corresponding plot will not be shown.

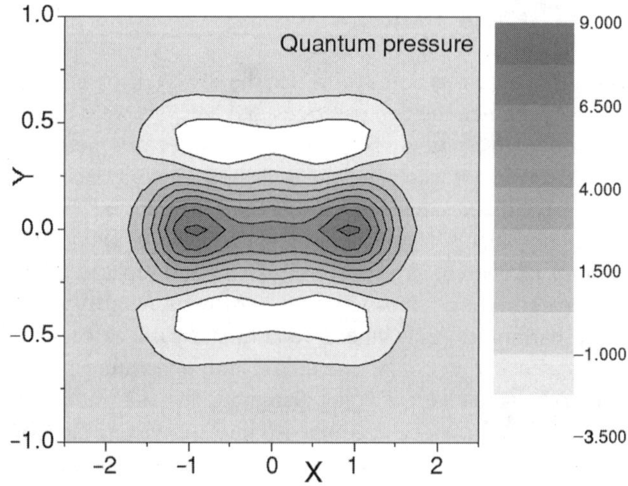

FIGURE 13.1. Contour map of the quantum pressure contribution to the stress tensor for the uncoupled case ($c = 0$). The pressure is highest close to the two peaks in the density, near $x = \pm 1$. Atomic units are used in this and figures 13.2–13.7.

For the uncoupled case, figures 13.2 and 13.3 show contour maps of the classical and quantum contributions to the stress tensor. The classical stress makes at most a 2% contribution to the total stress tensor. The reason is that the y-component of the velocity is very small: $v_1^2 \ll u_1^2$. Consequently, the quantum stress totally dominates the other two contributions to $\Pi_{1,1}$ for this case. As mentioned earlier,

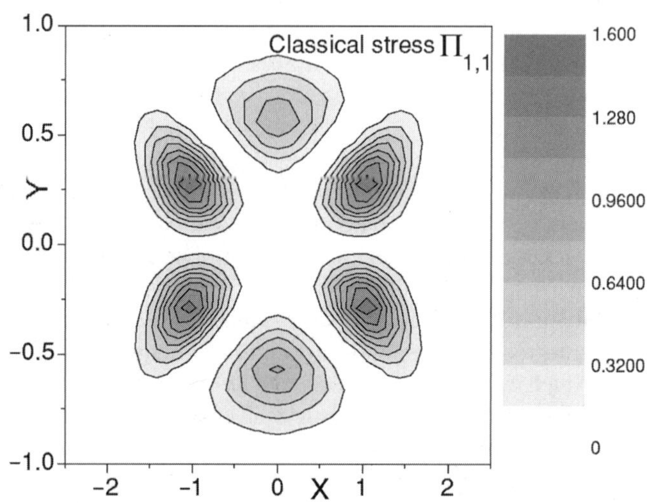

FIGURE 13.2. Contribution of the classical term ($m\rho v_1^2$) to the $\Pi_{1,1}$ stress tensor component for the uncoupled case. For this case, the classical stress is very small.

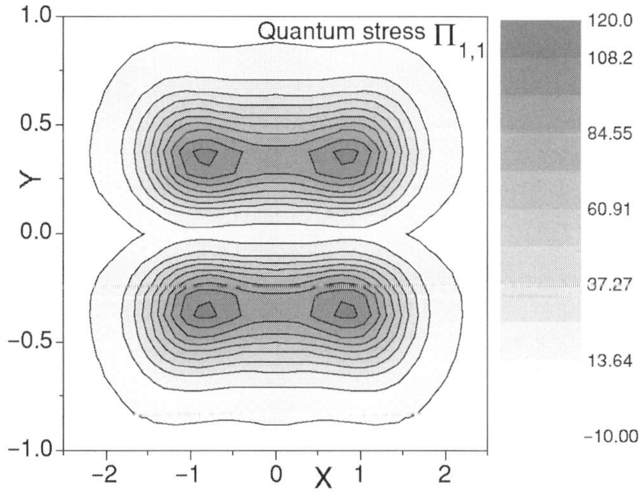

FIGURE 13.3. Contribution of the quantum term $(m\rho u_1^2)$ to the $\Pi_{1,1}$ stress tensor component for the uncoupled case.

at this time step the density is concentrated close to the y axis between the points $x = -1$ and $x = +1$. The osmotic velocity component is approximately linear in y, and the quantum stress is proportional to $y^2\rho$. Consequently, this leads to a bimodal distribution with symmetrical peaks above and below the y axis, as shown in figure 13.3. For this reason, the total stress shown in figure 13.4 is very similar to the quantum component shown in figure 13.3.

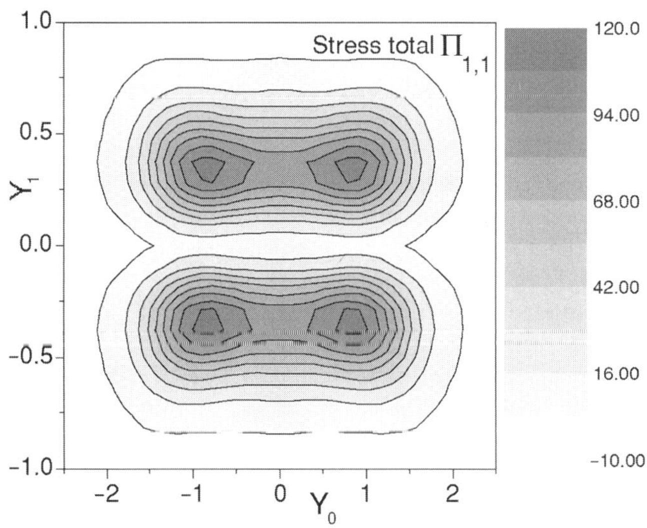

FIGURE 13.4. The stress tensor component $\Pi_{1,1}$ for the uncoupled case. The total stress is the sum of the three contributions displayed previously in figures 13.1–13.3.

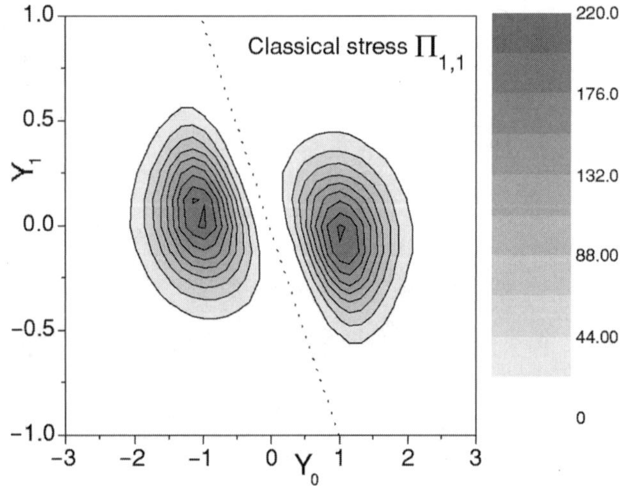

FIGURE 13.5. Contribution of the classical term ($m\rho v_1^2$) to the $\Pi_{1,1}$ stress tensor component for the coupled case.

For the uncoupled case, the flux map (see figure 8.4) shows vectors pointing away from the high-density–high-stress regions near $x = -1$ and $x = +1$. Vectors located between vertical lines drawn at $x = -1$ and $x = +1$ point inwards toward the vertical y axis, which acts as an *attractor* for these flux vectors. The result is a large buildup of density near the origin, between the peaks of the two Gaussians in the initial wave packet. The stress contour maps thus provide assistance in understanding the flux vector maps, and vice versa.

Figures 13.5 and 13.6 illustrate the classical and quantum contributions to the stress tensor for the coupled case. The situation here is qualitatively different from the uncoupled case: the maximum value of the quantum stress is only about 50% of the maximum value of the classical stress. The large value for the y-component of the velocity leads to the inequality $v_1^2 \gg u_1^2$. Consequently, the classical stress dominates the other contributions to $\Pi_{1,1}$. The stress is very large just to the left and right of the dotted line shown in figures 13.5 and 13.7, and the maximum value for the stress is about twice as large as the maximum value for the uncoupled case.

For the coupled case, the flux map (again, see figure 8.4) shows vectors pointing toward the upper left and lower right regions of figure 13.7, as though they were attracted to the upper and lower segments of the dotted line shown in this figure. These vectors point away from the vertical y axis, which acts a *repeller* for these flux vectors. The result is movement of density away from the central region between the peaks of the two Gaussians in the initial wave packet. Flux directed away from this repeller prevents buildup of the interference component of the density and thus contributes to decoherence. The type of analysis presented for $\Pi_{1,1}$ is readily extended to the other two components of the stress tensor, $\Pi_{0,0}$ and $\Pi_{0,1}$.

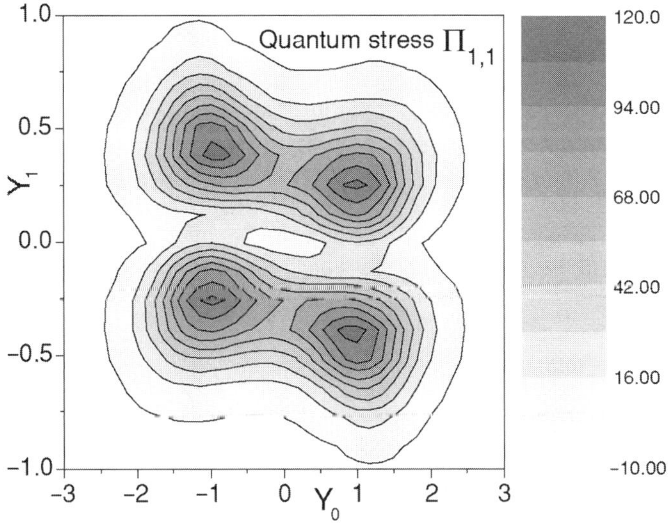

FIGURE 13.6. Contribution of the quantum term $(m\rho u_1^2)$ to the $\Pi_{1,1}$ stress tensor component for the coupled case.

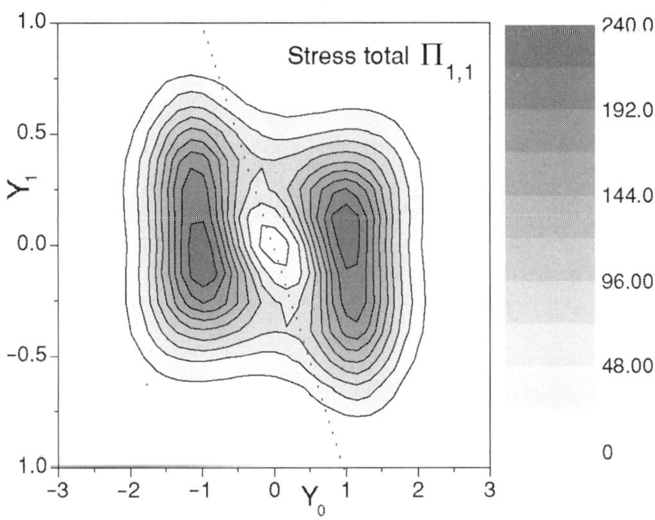

FIGURE 13.7. The stress tensor component $\Pi_{1,1}$ for the coupled case. The total stress is the sum of the two contributions displayed previously in figures 13.5 and 13.6 along with a very small pressure contribution (not displayed).

13.5 Vortices in Quantum Dynamics

The formation of quantized vortices around wave function nodes was predicted by Dirac in 1931 [13.19]. The proof that they are quantized depends only on the requirement that the wave function be single-valued and continuous. The wave function is not required to satisfy any particular dynamical equation.

In vortex dynamics, a fundamental role is played by the *circulation integral*, a measure of the *strength* of the vortex (for example, see [13.20]). This quantity is closely related to the net action accumulated around a closed path. For a problem with two degrees of freedom, let C denote a small closed curve encircling the point (x_0, y_0). The tangent vector to a point on the curve is $\vec{l}(x, y)$. The line integral of the tangential component of the flow velocity is the *circulation integral*

$$\Gamma = \oint_C \vec{v} \cdot d\vec{l} = \frac{1}{m} \oint_C p \cdot d\vec{l} = \frac{1}{m} \oint_C \vec{\nabla} S \cdot d\vec{l} = \frac{1}{m} \Delta S. \tag{13.13}$$

In the last expression, $\Delta S = S_2 - S_1$ is the change in action for one transit around the loop. If the loop *does not enclose a wave function node*, the action is continuous and the net change around the loop is zero, $\Delta S = 0$. This is the case because the loop can be contracted to a small circular path on which the phase does not change. In regions such as this, the curl of the velocity field (which itself is a measure of circulation, as elaborated in Box 13.3) is given by $\vec{\omega} = \vec{\nabla} \times \vec{v} = (1/m)\vec{\nabla} \times \vec{\nabla} S = 0$. The vector field $\vec{\omega}(x, y)$ is defined as the *vorticity*, and this field is *irrotational* when the curl of the velocity vanishes, $\vec{\nabla} \times \vec{v} = 0$. This situation is also

Box 13.3. Stokes's theorem and circulation

Stokes's theorem relates the line integral of a vector field around a closed curve to the integral over the enclosing surface of the curl of the field. Consider a surface Σ with the boundary curve $\partial\Sigma$ and the vector field $\vec{F}$. Also, let the vector surface element be $d\vec{A} = (dA)\vec{n}$, where $\vec{n}$ is a unit normal to the surface, and let $d\vec{l}$ be a unit tangent vector to the bounding curve $\partial\Sigma$. Then the theorem states that

$$\int_\Sigma \vec{\nabla} \times \vec{F} \cdot d\vec{A} = \oint_{\partial\Sigma} \vec{F} \cdot d\vec{l}. \tag{1}$$

In the limit in which the bounding curve shrinks around an enclosed point, the normal component of the curl becomes the circulation of the vector field per unit area:

$$\vec{\nabla} \times \vec{F} \cdot \vec{n} = \lim_{\Sigma \to 0} \left[\frac{1}{A} \oint_{\partial\Sigma} \vec{F} \cdot d\vec{l} \right]. \tag{2}$$

This relation shows that the curl of a vector field is a measure of circulation around a point.

referred to as *potential flow*, with $\Omega = S/m$ acting as a potential for determination of the velocity, $\vec{v} = \vec{\nabla}\Omega$ (note that the curl of a gradient is always zero).

Now assume that the loop C *encloses a node* in the wave function at the point (x_0, y_0). The phase of the wave function is no longer a continuous function around the entire loop. In this case, the net phase change around the loop is $\Delta S/\hbar = n \cdot 2\pi$, and from equation 13.13, *the circulation integral is quantized* because the phase changes by an integer multiple of 2π

$$\Gamma - \frac{1}{m}\Delta S = \frac{n}{m}(2\pi)\hbar. \tag{13.14}$$

The integer n can be positive or negative, the sign determining the *vortex chirality* [13.21], while the magnitude determines the *state of excitation* (and angular momentum) of the vortex. The ground state corresponds to $|n| = 1$, the first excited state to $|n| = 2$, etc. The remainder of the argument is well expressed by Hirschfelder et al. [13.20], whom we quote:

If one makes a small change in the loop, the phase change of Ψ at the points on the loop changes by only a small amount, and so the value of $\Delta(S/\hbar)$ can change by only a small amount. However, since $\Delta(S/\hbar)$ is quantized, the value of $\Delta(S/\hbar)$ cannot change by an amount smaller than 2π. Thus the value of $\Delta(S/\hbar)$ must remain unchanged. Continuing to make small changes in the loop, one concludes that the circulation is the same around all loops which can be changed into one another.... Thus, we conclude that the circulation is the same around all loops which encircle (in the same sense) the same nodal regions.

In order to be more specific about the form of the wave function close to a node [13.21], we will assume a p-th order node, so that

$$\Psi(r, \varphi) = r^p e^{in\varphi}, \tag{13.15}$$

where $r = \left[(x - x_0)^2 + (y - y_0)^2\right]^{1/2}$ is the distance to the node and where φ is the twist angle. The action is then given by $S = n\varphi\hbar$. The simplest case corresponds to a first-order node in the ground state, $n = 1$ ($\Psi \approx re^{i\varphi}$), for which the density increases quadratically, $\rho \approx r^2$. Near the node, assuming the form of the wave function given in equation 13.15, the quantum potential is given by

$$Q = -\frac{\hbar^2}{2m}\frac{\nabla^2 R}{R} = -\frac{\hbar^2}{2m}\frac{n^2}{r^2}. \tag{13.16}$$

Thus the quantum potential acts as an attractive $-1/r^2$ potential, and the quantum number n determines the strength of the potential. This potential supports bound states, which are the quantized vortices, and the streamlines around this attractive sink are circular (at least for small values of r). The latter feature is shown nicely in studies by Wu and Sprung [13.21] on the transmission of an electron beam in a right-angled wave guide. Figure 13.8, from this study, shows a streamline enclosing three vortices. The streamlines around the node within the small dotted box are circular, as shown in the inset.

In a space of dimensionality D, the condition $\mathrm{Re}(\Psi) = 0$ is a (*D-1*)-dimensional surface, and the condition $\mathrm{Im}(\Psi) = 0$ gives another (*D-1*)-dimensional surface.

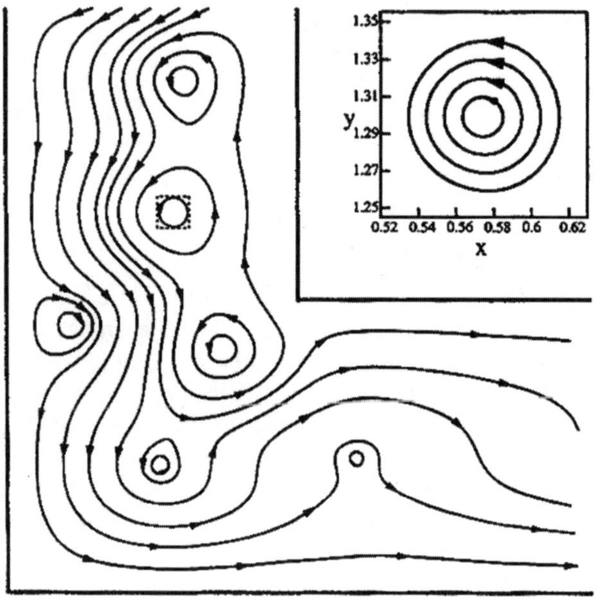

FIGURE 13.8. Streamlines for an electron beam in a right-angled wave guide [13.21]. The beam enters from the top left. The streamlines within the small dotted box (upper left) are shown magnified in the insert. The streamlines form circular paths around the vortex core.

The intersection of these two surfaces usually gives the (D-2)-dimensional nodal surface. This nodal surface has nonvanishing circulation only if it "extends over a region which is capable of trapping the loop; it must be able to prevent the loop from being shrunk away" [13.20]. In three-dimensional space, $D = 3$, a point or a finite line segment will not work, but an endless line, a *thread*, will be capable of supporting nonvanishing circulation. In this case, quantized vortices can form only along infinitely long (possibly curved) lines and around closed curves. In planes perpendicular to the axis of a thread vortex, the streamlines form concentric tubes, and around circular "smoke ring" vortices, the streamlines form concentric tori. An example illustrating the formation of ring vortices will be given in the next section.

13.6 Examples of Vortices in Quantum Dynamics

Vortices in quantum scattering were first identified by McCullough and Wyatt in quantum wave packet studies of the collinear $H + H_2 \rightarrow H_2 + H$ exchange reaction [13.22, 13.23]. As the wave packet starts to turn the corner on the potential surface and move toward the product valley, a node develops temporarily in the wave function on the inside of the bend in the potential surface. This feature forms near the barrier maximum and temporarily traps some of the probability fluid that

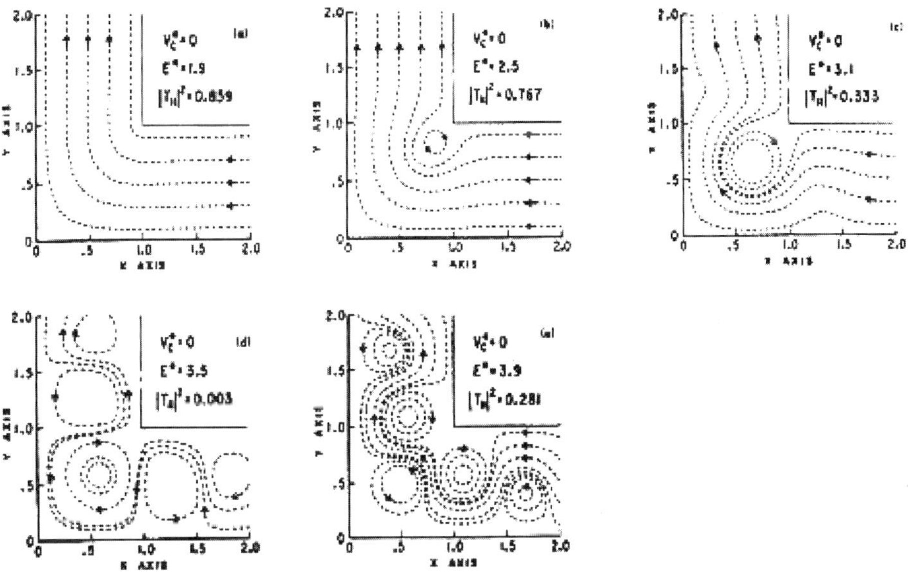

FIGURE 13.9. Streamlines at five collision energies for scattering in the right-angled wave guide [13.26].

is turning the corner on the potential surface. This whirlpool motion produces multiple recrossings of the H_3 symmetric stretch dividing line between the reactant and product valleys. Using time-independent scattering wave functions, Kuppermann and coworkers have calculated and plotted streamlines for this reaction [13.24].

In 1974, Hirschfelder, Christoph, and Palke [13.25] noted vortices and *stagnation points* (a point where the flux vanishes but the density is not zero) in the scattering from a two dimensional square barrier. A few years later, Hirschfelder and Tang [13.26] made extensive studies of the wave function and flux for a model atom–diatomic molecule reactive scattering problem. The potential surface was a right-angled duct with a flat but adjustable potential (V^*) in the corner region. Figure 13.9 shows the streamlines at five scattering energies for the case in which the corner potential is $V^* = 0$. Clearly, flux enters through the reactant channel in the lower right of the figure. In part (a), for $E^* = 1.9$, smooth streamlines make the turn from the reactant channel on the lower right side to the product channel in the upper left. For energies above $E^* - 2.0$, one or more vortices develop around nodes in the scattering wave function. In this case, some of the streamlines form closed loops around the vortex cores, while others meander around the vortices while attempting to link the exit and entrance regions of the wave guide. In part (b), for the energy $E^* = 2.5$, a clockwise rotating vortex appears near the corner at $x = y = 1.0$, and as the energy increases, the vortex moves down toward the origin as it increases in size. In part (c), for the energy $E^* = 3.1$, the streamlines that make it into the product channel are squeezed into the lower left corner. In part (d), the energy $E^* = 3.5$ is just below the energy ($E^* = 3.535$) of a reflection

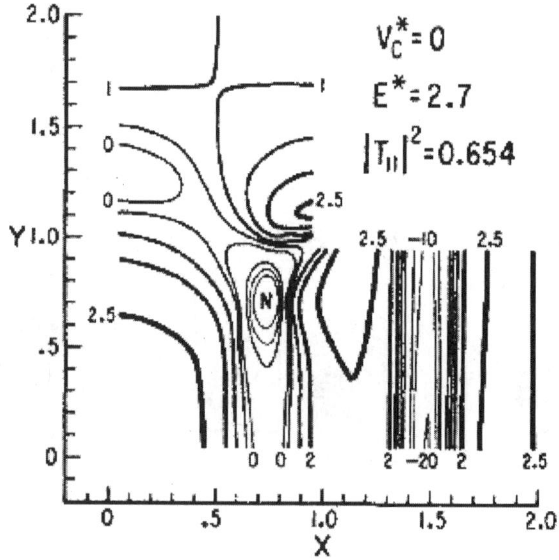

FIGURE 13.10. Contour map of the quantum potential at the energy $E^* = 2.7$, with the position of a node marked N [13.26].

resonance. At the precise resonance energy, *all of the streamlines disappear, and at a slightly higher energy, the streamlines reappear, but the vortices all rotate in the opposite direction*! A result of this switch in chirality is that the streamlines are no longer crowded into the lower left corner, but cross near the middle of the $x = y$ dividing line, as shown in part (e) for $E^* = 3.9$. For the energy $E^* = 3.5$, the contour map of the probability density shows that although the streamlines in part (c) are symmetric about the diagonal, the probability density lacks this symmetry. Very little change occurs in the density map as the energy is increased through the resonance, although the flux maps change dramatically.

The quantum potential at an energy intermediate between those in figures 13.9 (b) and (c) is shown in figure 13.10. The quantum potential becomes increasingly negative near the node, marked N, near the center of the figure. Finally, the phase of the scattering wave function is shown in figure 13.11 at the energy $E^* = 3.9$, corresponding to the flux map in figure 13.9 (e). Note that the phase increases in magnitude by 2π when a circular loop is made around each vortex core, thus showing each to be in its ground state, $|n| = 1$. For example, for the vortex shown in the lower right of this figure, the phase increases by 360° when a counterclockwise loop is traversed around the core. Because of the *phase dislocations* that occur when each vortex is encircled, Nugent et al. [13.51] refer to these as *phase vortices*. At the center of the core, the phase is undefined. In addition, the streamlines shown earlier, in figure 13.9 (e), cut the curves of constant phase at right angles, emphasizing again that surfaces of constant phase and the streamlines are analogous to wave fronts and rays in optics.

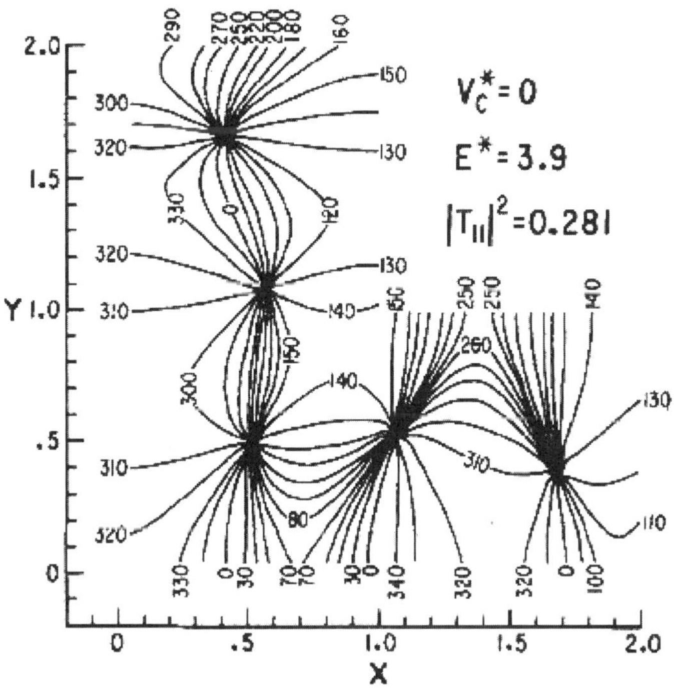

FIGURE 13.11. Curves of constant wave function phase (given in degrees) at the energy $E^* = 3.9$ [13.26]. These curves merge at the positions of five nodes. Around each node, the magnitude of the phase increases by 2π, but the chirality (sense of rotation) alternates from one node to the next.

Vortex structure that develops during intramolecular electron transfer has been studied by Stuchebrukhov and coworkers [13.27, 13.28]. In one example [13.28], the tunneling current was computed in the model charge transfer system shown in figure 13.12. This system represents a 35-angstrom stretch of polypeptide chain (Gly_5) sandwiched between a Cu^+ center and a Ru^{3+} site that serve as the source and sink for electron transfer along the "molecular wire". The goal is to determine the current distribution through the wire connecting the donor and acceptor sites.

The distribution of flux in the plane of the polypeptide backbone is shown in figure 13.13. A remarkable feature is the convoluted path followed by the current between the donor and acceptor sites. The dominant feature is the presence of multiple vortices around which the flux circulates. Nodes in the wave function occur at the intersections of the continuous and dashed curves, and it is around these points that the vortices form. The current that is not trapped around the vortices follows a very complicated path from one end of the chain to the other.

An outstanding example of smoke ring tori from the work by Wu and Sprung [13.21] is shown in figure 13.14. Several ring tori are formed on the back side of a thin impenetrable disk when a beam incident from the left side impinges on it.

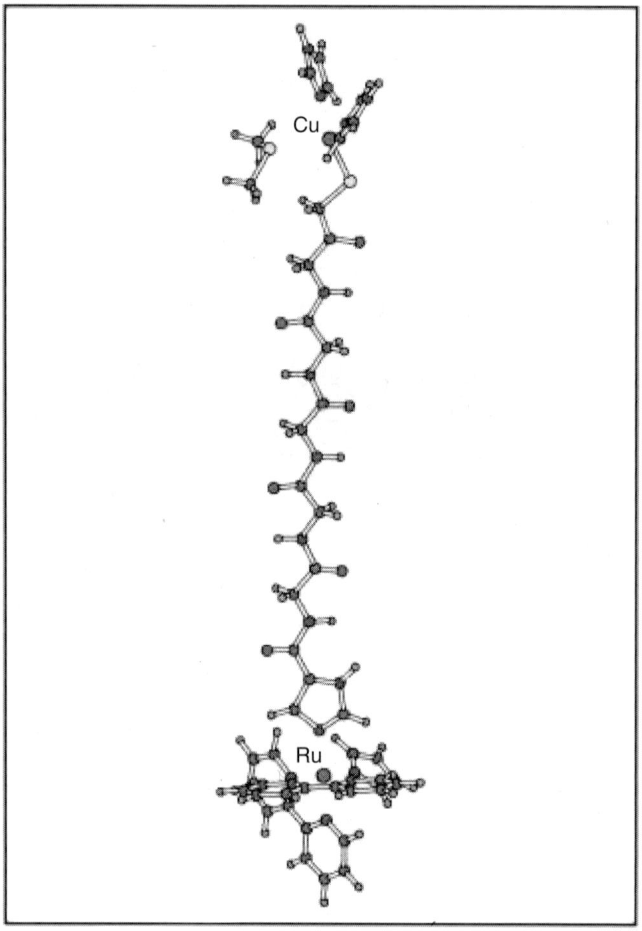

FIGURE 13.12. The model electron transfer system. The geometry of the polypeptide chain is planar except for the two H atoms nearest the donor and acceptor sites [13.28].

Quantum trajectories have been calculated for the scattering of a beam of atoms from an absorbate on a metal surface. In the two-dimensional model studied by Sanz et al. [13.29], the interaction potential between the He atom and the CO adsorbate on the Pt(111) surface was expressed as a sum of two terms:

$$V(r) = V_{He-Pt}(z) + V_{He-CO}(r - r_{CO}),$$

in which z is the perpendicular distance to the surface, r_{CO} locates the center of mass of the adsorbate, and $r = (x, z)$ locates the He atom relative to the adsorbed molecule (x is the coordinate parallel to the surface). The He–Pt interaction was represented as a Morse potential, while a Lennard-Jones function described the He–CO interaction. The incoming wave packet was a superposition of Gaussians,

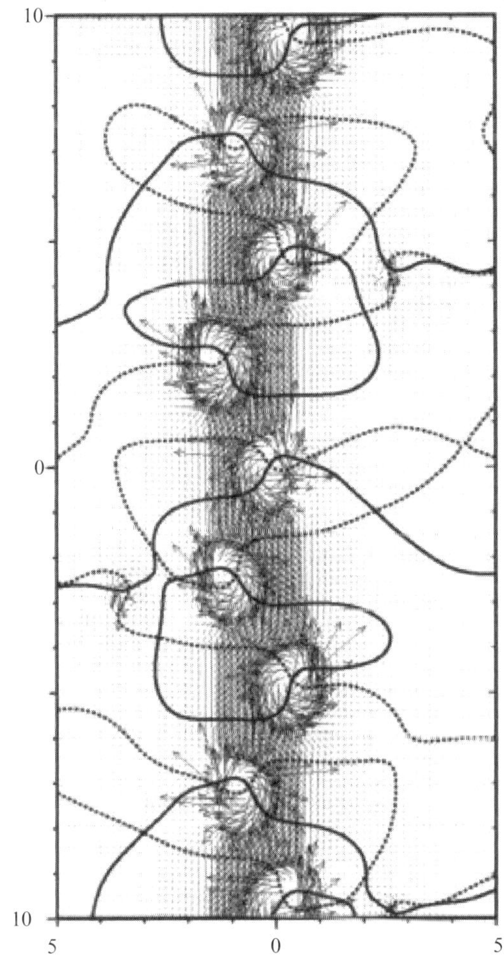

FIGURE 13.13. Spatial distribution of the electron flux in the x-y plane of the charge transfer system shown in figure 13.12 [13.28]. The fragment shown here is the middle part of the peptide system. The wave function nodes form at the intersections of the dashed and solid curves along which the real and imaginary parts of the wave function vanish.

with the centers of the Gaussians located a distance z_0 from the surface and parallel to it. Figure 13.15 shows probability densities for the incoming beam (left) and the outgoing scattered fragments (right) at 5.5 ps. Part of the incoming amplitude is trapped near the surface on both sides of the single CO adsorbed molecule that is located at the origin. From the time-dependent scattering wave function, quantum trajectories were computed. Figure 13.16 shows the fate of a number of quantum trajectories that were launched from a range of x values between $x = 0$ and $x = 50$ a.u. at the initial distance $z = 13.4$ a.u. above the surface. The trajectories follow

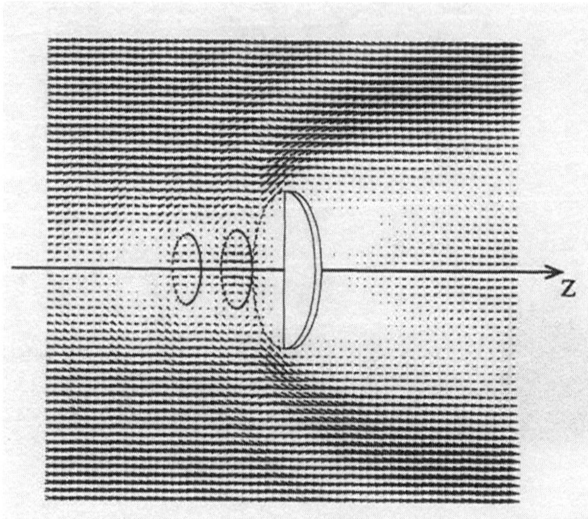

FIGURE 13.14. Three-dimensional flux distribution that develops when a beam incident from the left is scattered from a thin disk [13.21]. The incoming beam propagates to the right and parallel to the Z axis. Several ring vortices form on the back side of the target.

the maximum density in the wave packet and can be divided into sets that contribute to each peak in the scattered density. In addition, some of the trajectories launched downward onto the CO bump near the origin scatter sideways and nearly parallel to the surface. As a result, a quantum vortex is formed near $x = 65$ a.u. by these swirling trajectories. In a more detailed analysis, it was shown that nodes formed in the scattering wave function near the metallic surface act as organizing centers for swirling trajectories [13.49, 13.50].

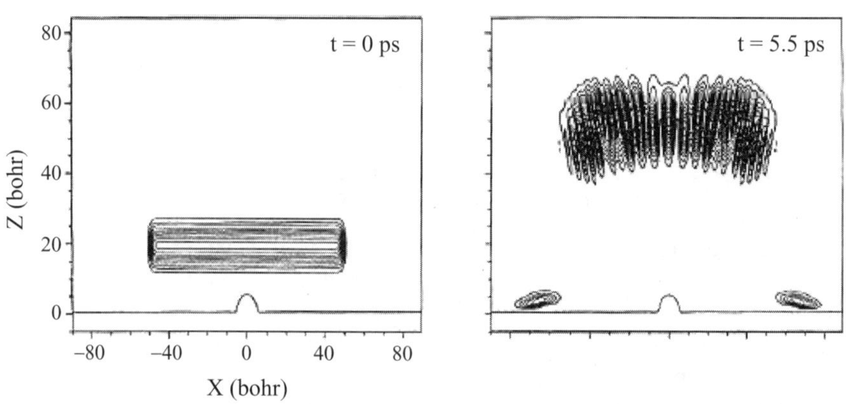

FIGURE 13.15. Probability density for the incoming wave packet (left) and outgoing scattered wave (right) at 5.5 ps [13.29]. Part of the density is trapped near the Pt surface on both sides of the adsorbed CO molecule located at the origin.

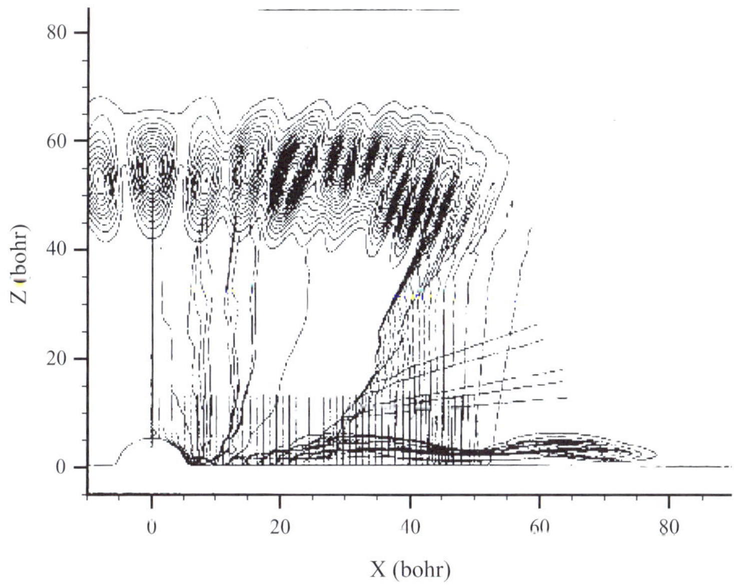

FIGURE 13.16. Contour map of the scattered probability density (at 5.5 ps) and the quantum trajectories calculated from the time-dependent wave function [13.29]. In this example, the trajectories were launched from the distance $Z = 13.4$ a.u. above the surface.

Ring and line vortices move in time and may be created, sometimes singly and other times in groups. Figures 1 through 4 in the study by Bialynicki-Birula and coworkers illustrate the creation of a ring vortex at a point, the creation and destruction of a vortex pair with opposite circulation, the motion of a pair of entangled vortices, and the precession of a line vortex in a magnetic field [13.30].

13.7 Features of Dynamical Tunneling

Around 1980, the classical mechanical study of the vibrations of the water molecule revealed the existence of *local modes of vibration* [13.42]. Some of the classical trajectories execute asymmetric patterns even in the symmetric potential, reflecting the structure of local mode OH stretching vibrations. Later, a quantum-mechanical investigation revealed that the eigenvalues appear to be doublets in which the energy splitting decreases with an increase of the OH vibrational quantum number [13.43]. The superposition of a pair of delocalized almost degenerate eigenstates creates localized states that are concentrated along the OH bond directions. A quantum wave packet can make transitions from one localized state to the other, and the energy splitting is a measure of the transition rate. However, this type of transition between localized states is forbidden for classical trajectories. The system behaves as though it were under the influence of a double-well potential, though there is no identifiable potential barrier. This process has been described as *dynamical tunneling* [13.44] because of the possibility of a *transition in the quantum system between classically trapped regions of phase space in the absence of a potential barrier*. Unlike barrier tunneling, the mechanism for dynamical tunneling is not obvious, because the potential surface does not reveal any indication of possible classical trapping.

In a 2003 study, Babyuk and Wyatt [13.45] demonstrated that analysis of dynamical tunneling in terms of quantum trajectories can shed light on the nature of this process, especially the existence and location of potential energy barriers along some of the trajectories. In common with the preceding applications in this chapter, these quantum trajectories were guided by the predetermined time-dependent wave function. For this study of local mode states, the time dependent wave function was expressed in terms of the superposition of a pair of nearly degenerate states,

$$\Psi(x, y, t) = \frac{1}{\sqrt{2}}\left(\psi_1(x, y)e^{-iE_1t/\hbar} + \psi_2(x, y)e^{-iE_2t/\hbar}\right), \tag{13.17}$$

where ψ_1 and ψ_2 are the eigenfunctions with eigenvalues E_1 and E_2, correspondingly. In terms of the amplitudes R, R_1, and R_2 associated with the functions Ψ, ψ_1, and ψ_2, the following relation is obtained for the total energy along the trajectory (the total energy is the negative of the action derivative with respect to time):

$$E = -\frac{\partial S}{\partial t} = \frac{1}{2}(E_1 + E_2) + \frac{(E_2 - E_1)}{2} \cdot \frac{\left(R_2(x, y)^2 - R_1(x, y)^2\right)}{R(x, y, t)^2}. \tag{13.18}$$

Since the product of the two terms $(E_2 - E_1)$ and $R_2(x, y)^2 - R_1(x, y)^2$ is relatively small, the total energy along the trajectory is relatively constant away from the nodal regions. However, significant changes in the total energy may occur when the trajectory encounters nodal regions, where the R–amplitude becomes very small.

The model studied in the following section exhibits dense nodal structure around which vortices form. Some of the quantum trajectories are attracted by the vortices and circulate around these cores with high velocities. The fluid elements move very slowly when the density is relatively high, but singular properties appear in the velocity and the quantum potential when the trajectory approaches a node in the wave function. In addition, when quantum trajectories cross quasinodes (places where the density becomes very small, but not zero), the velocity also becomes large.

13.8 Vortices and Dynamical Tunneling in the Water Molecule

The close interplay between quantum trajectories, vortices, and dynamical tunneling will be illustrated by examining the vibrational dynamics of the water molecule in the ground electronic state, the system where dynamical tunneling was discovered. The bending motion of H_2O can be approximately separated so that the potential energy is a function of two variables, x and y, the OH bond displacements from the equilibrium positions. For the vibrational degrees of freedom, the potential may be approximated by the sum of two Morse potentials and a bilinear coupling term [13.46–13.48],

$$U(x, y) = D_1\left[1 - \exp(-\alpha x)\right]^2 + D_1\left[1 - \exp(-\alpha y)\right]^2 + fxy. \qquad (13.19)$$

This potential, with appropriate constants D_1, α, and f, is able to accurately reproduce experimental data for up to five quanta of stretch excitation. The kinetic energy operator in these coordinates is not diagonal, and the Hamiltonian becomes

$$\hat{H} = -\frac{1}{2}\frac{\partial^2}{\partial x^2} - \frac{1}{2}\frac{\partial^2}{\partial y^2} + \varepsilon\frac{\partial^2}{\partial x \partial y} + D\left[1 - \exp(-\alpha x)\right]^2 + D\left[1 - \exp(-\alpha y)\right]^2 + Fxy, \qquad (13.20)$$

where $D = \mu D_1$, $F = \mu f$, $\varepsilon = m_H/(m_H + m_O)\cos\Theta = 0.015$, $\mu = m_H m_O/(m_H + m_O)$, m_H and m_O are atomic masses, and $\Theta = 104.52°$ is the fixed bond angle.

In this study of dynamical tunneling, the first goal is the generation of nearly degenerate eigenstates rather than to match spectroscopic data for OH stretches in the water molecule. Therefore the constants D, α, and F in equation 13.20 were chosen such that a relatively small harmonic oscillator basis set would provide good convergence for the eigenvalues of the Hamiltonian matrix. (The parameter values, in atomic units, are given by $D = 11.86$, $\alpha = 0.205$, and $F = -0.013$.) Diagonalization of the Hamiltonian matrix was carried out in a harmonic oscillator direct product basis set with up to 900 terms. The energy splitting for the pair of states $(n, 0)$ and $(0, n)$ decreases with increasing excitation energy, and therefore the tunneling period is very long for the high-energy states.

We will consider as an example one pair of close eigenvalues, the sixth and seventh vibrational states (corresponding to the (3,0) and (0,3) local mode pair) with eigenvalues $E_1 = 3.7257$ and $E_2 = 3.7298$, respectively, for which the tunneling period is $\tau = 1550.98$. The time evolution of an initially localized state is shown in figure 13.17 for one-half of the tunneling period. Some of the quantum trajectories for the first half of the tunneling period are presented in figure 13.18. In order for their behavior to be better understood, these trajectories are superimposed on a flux map. During the first half-tunneling cycle, the flux angle is everywhere independent of time, but the magnitude changes with time. As time reaches one-half of the tunneling period, the flux vectors reverse direction and again preserve the new flux angle during the second half of the tunneling period.

The flux map in figure 13.18 has nine nodes at every time, except for $t = 0$ and $t = \tau/2$, and these play a major role in determining the trajectory evolution. If it is initially surrounded by nodes, a quantum trajectory cannot escape from them and subsequently circulates rapidly around the nodes for the entire half-tunneling period. For example, the trajectory with the initial condition ($x = 0$, $y = 0$) (trajectory 4 in figure 13.18), which is trapped by one vortex, makes more than

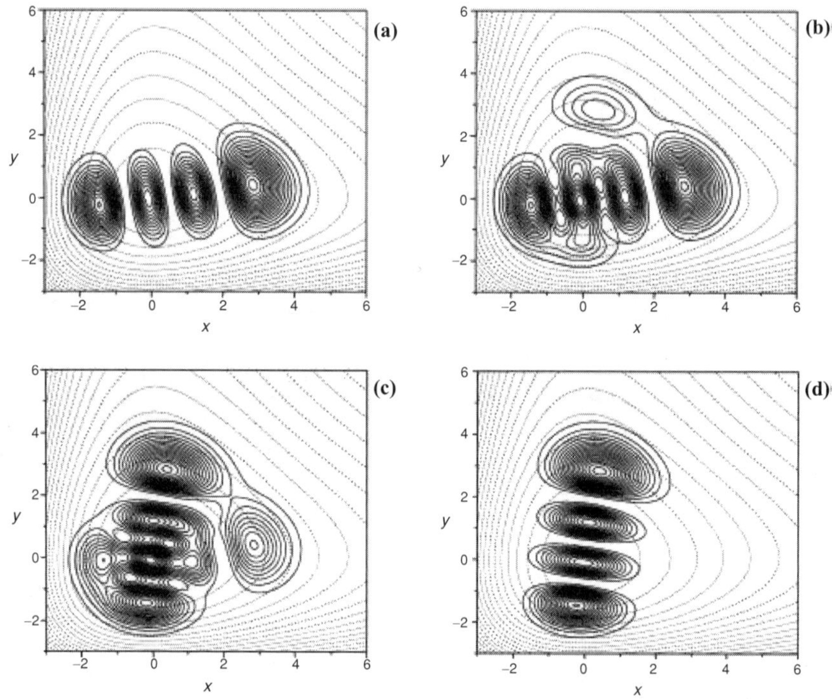

FIGURE 13.17. Transition between a pair of local-mode stretch vibrational states (shown in parts (a) and (d)) in the water molecule [13.45]. The probability density is shown at four times: (a) $t = 0$, (b) $t = 0.15\tau$; (c) $t = 0.3\tau$; and (d) $t = 0.5\tau$. The coordinates x and y are OH bond displacements from the equilibrium positions.

150 cycles. However, if the starting point of the trajectory is remote from the nodes, it moves very slowly. For example, trajectory 2 in this figure describes a fluid element that is initially almost motionless. Then it speeds up near the quasi-node at $x = 3,\ y = 3$ and later continues its motion around the vortex at $x = -1,\ y = -2$.

In order to reveal distinguishing features in the trajectory evolution, a quantum trajectory starting from $x = 3,\ y = -1$ will be analyzed in more detail. As mentioned above, if a trajectory encounters a nodal region, its energy components are expected to change rapidly. The trajectory shown in figure 13.19 is influenced by six nodes and three quasi-nodes. The slow motion of the trajectory outside of the nodal regions is reflected in the very low kinetic energy shown in figure 13.20 (a). As a result, the total energy shown in part (d) is essentially the sum of only the classical (part (b)) and quantum (part (c)) potentials, and they vary asynchronously, thus keeping the total energy almost constant. Acceleration of the trajectory on entering a nodal region leads to a sudden change of both the kinetic energy and the quantum potential. As a result, the total energy deviates from its previously constant value.

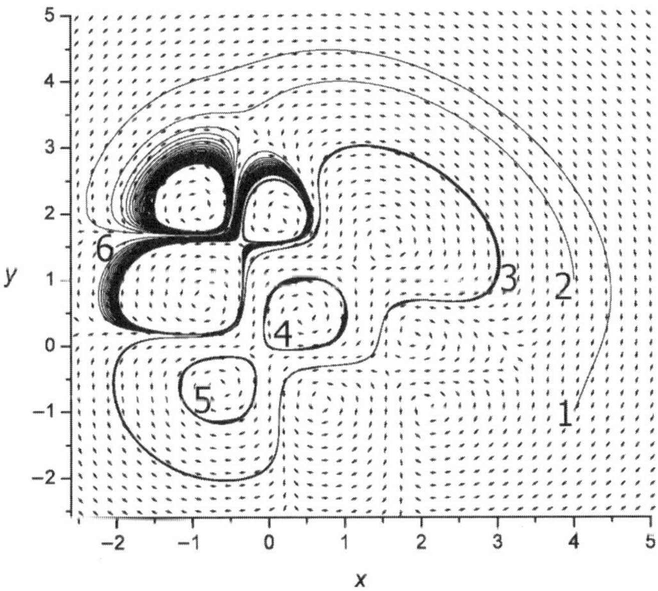

FIGURE 13.18. Quantum trajectories launched from six initial positions [13.45]: (1) $x = 4, y = -1$; (2) $x = 4, y = 1$; (3) $x = 3, y = 1$; (4) $x = 0, y = 0$; (5) $x = -1, y = -1$; (6) $x = -2, y = 1.5$. These trajectories are plotted on a flux map. Note that the quantum trajectories become trapped around vortices.

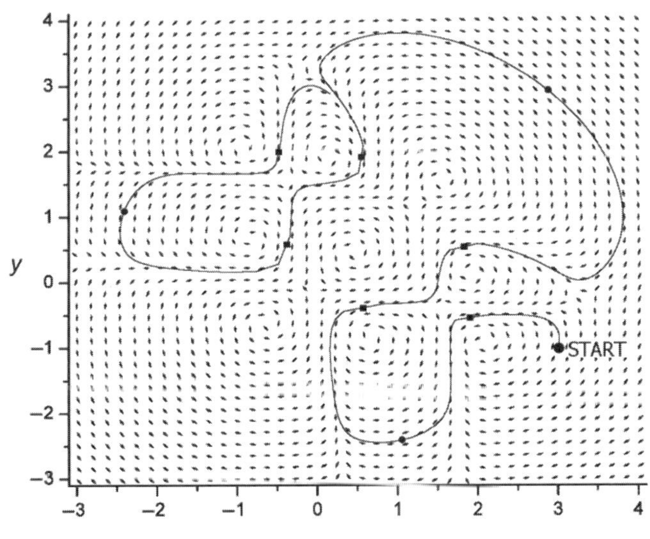

FIGURE 13.19. Trajectory starting at $x = 3,\ y = -1$ [13.45]. Markers show when it encounters nodes (squares) and quasi-nodes (circles). Times at nine markers along the trajectory are correspondingly 134.5, 138.3, 143.0, 149.0, 350.5, 475.4, 478.2, 486.1, 488.9.

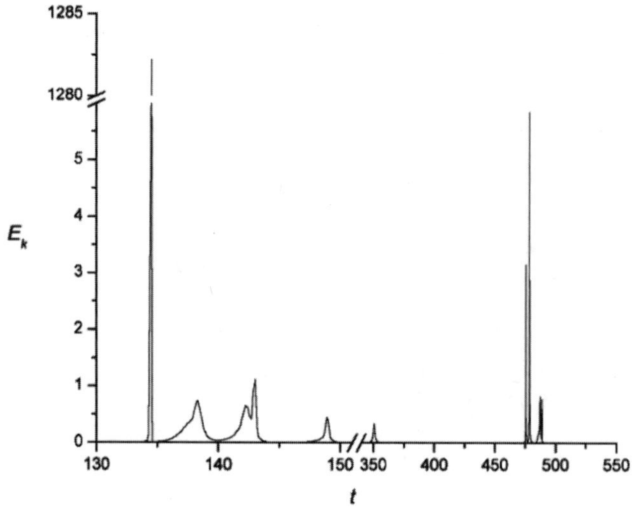

(a)

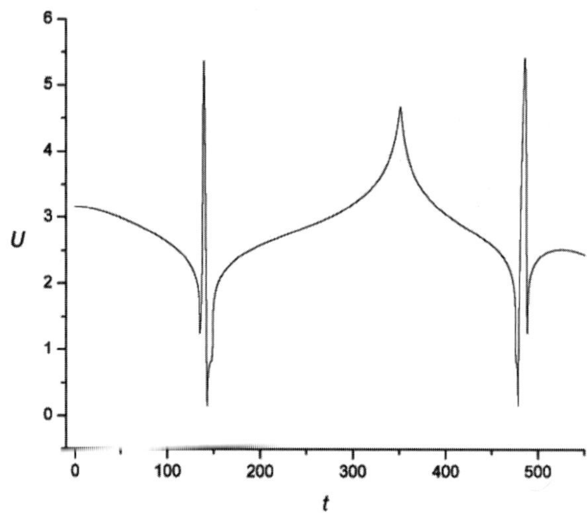

(b)

FIGURE 13.20. Energy components along the quantum trajectory shown in figure 13.19 [13.45]: (a) kinetic energy, (b) classical potential energy $U(x, y)$; (c) quantum potential $Q(x, y)$; (d) total energy E.

The potential energy of the fluid element reaches a maximum three times, and these are at the positions of the quasi-nodes (figure 13.20 (b)). Moreover, at these positions, the potential energy exceeds the total energy (figure 13.20 (d)), meaning that *the trajectory undergoes barrier tunneling*. Unlike conventional tunneling in a one-dimensional double-well potential, not all trajectories go through the barrier in

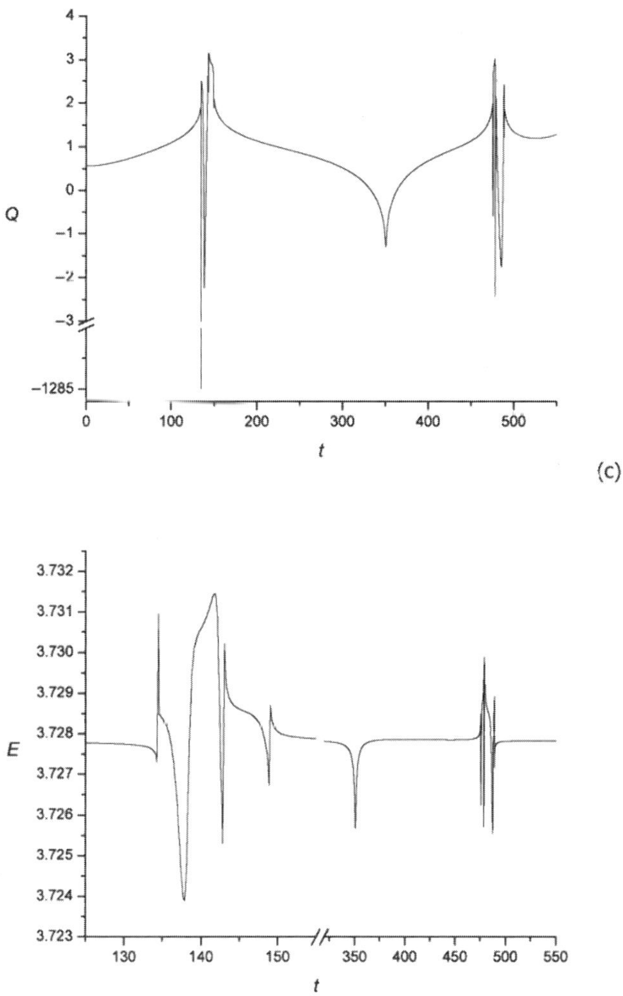

FIGURE 13.20. *Continued*

the dynamical tunneling process in two dimensions. Only those trajectories whose initial position is remote from the nodes go through the potential energy barriers. As we have already seen, other trajectories are trapped by vortices and will spend most of the period circulating around them. It is believed that for higher-energy states where the energy splitting is smaller, the trajectories will tunnel deeper; in other words, their potential energy exceeds the total energy to a greater extent than for the lower-energy tunneling states.

A summary of the dynamical tunneling process is as follows. For the two-dimensional coupled potential considered in this study, classical trajectories can be localized in two restricted regions of coordinate space. Dynamical restrictions prohibit these classical trajectories from making transitions to other regions even though there is no energy restriction, such as the presence of an easily identifiable potential energy barrier. In contrast, a quantum wave packet can make such a transition. Analysis of this process using quantum trajectories provides some new insights. Split nearly degenerate states have a coordinate space densely filled by permanent nodes around which the flux circulates. Because the transition between a pair of localized states is relatively slow, before this transition is complete, quantum trajectories in the nodal regions execute many circulation cycles around one or more of the vortex cores. In contrast, trajectories initiated far from the nodal region reflect more of the dynamics of the dynamical tunneling process because their time scale is comparable with that of the shift in density from one local mode to the other. They also have the important feature of passing through one or more regions where their *potential energy exceeds the total energy* (which includes the quantum potential). This feature is shared with that in barrier tunneling. Thus, *dynamical tunneling may be viewed as a veiled form of barrier tunneling.*

13.9 Summary

The stress tensor appearing in the quantum version of the Navier–Stokes equation includes both classical and quantum components. The classical component depends only on the flow velocity (proportional to the gradient of the action), while the quantum components depend on derivatives of the density and arise from spatial variations in the *shape* of the density. Spatial derivatives of the stress tensor, along with the classical force density, determine the rate at which the Navier–Stokes equation adjusts the local momentum density. Extremely few examples are available in the literature to demonstrate the manner in which components of the stress tensor contribute to the total stress. However, in this chapter, one example was presented for a two-mode system in which an interference feature was suppressed by turning on a mode coupling term in the Hamiltonian.

Near nodes in quantum-mechanical wave functions for nonstationary states, the quantum potential becomes very large, quantum trajectories follow nearly circular paths around the core, the phase of the wave function undergoes a 2π (or integer multiple of 2π) winding, and the circulation and angular momentum are always quantized. Examples of vortex formation were shown in this chapter for current passing through right-angled wave guides and along a model polypeptide chain. In addition, a ring vortex was shown for the three-dimensional scattering from a disk, vortices were shown for the scattering of a beam from adsorbed molecules on the surface of a solid, and vortex formation was described for the transition state region of the collinear $H + H_2$ exchange reaction. In the latter case, involving time dependent wave packet propagation, the formation and decay of the vortex could be studied. Nodes and quasi-nodes also play a significant role in dynamical tunneling. Using local mode vibrational states in the water molecule as an example, we saw that many quantum trajectories become trapped near vortices during the tunneling process that takes amplitude from one OH local mode to the other. For all of the examples presented in Sections 13.6 and 13.8, the *analytical approach to quantum hydrodynamics* was followed: the wave function was analyzed after we first solved the Schrödinger equation. Computing quantum trajectories near vortices by solving the hydrodynamic equations of motion on the fly poses a challenge for future studies. Because of the strong quantum potential near vortices, the use of adaptive grid techniques will probably be essential.

References

13.1. R.A. .Granger, *Fluid Dynamics* (Holt, Rinehart, Winston, New York, 1985).
13.2. T. Takabayasi, *On the formulation of quantum mechanics associated with classical pictures.* Prog. Theoret. Phys. **8**, 143 (1952).
13.3. R.J. Harvey, *Navier–Stokes analogue of quantum mechanics.* Phys. Rev. **152**, 1115 (1966).
13.4. N. Rosen, *A classical picture of quantum mechanics.* Il Nuovo Cimento **19 B**, 90 (1974).
13.5. S.T. Epstein, *Coordinate invariance, the differential force law, and the stress–energy tensor*, J. Chem. Phys. **63**, 3573 (1975).
13.6. C.Y. Wong, J.A. Murahn, and T.A. Welton, *Dynamics of Nuclear Fluid*, Nuclear Physics **A 253**, 469 (1975).
13.7. C.Y. Wong, *On the* Schrödinger *equation in fluid mechanical form*, J. Math. Phys. **17**, 1008 (1976).
13.8. M. Himi and K. Fukushima, *An application of quantum fluid mechanics*, Nuclear physics **A 431**, 161 (1984).
13.9. N. Rosen, *A semiclassical interpretation of wave mechanics*, Found. Phys. **14**, 579 (1984).
13.10. S. Sonego, *Interpretation of the hydrodynamical formalism of quantum mechanics*, Found. Phys. **21**, 1135 (1991).
13.11. B.M. Deb and A.S. Bamzai, *Internal stresses in molecules. I. One-electron systems*, Mol. Phys. **35**, 1349 (1978).
13.12. B.M. Deb and S.K. Ghosh, *On some local force densities and stress tensors in molecular quantum mechanics*, J. Phys. **B 12**, 3857 (1979).
13.13. B.M. Deb and A.S. Bamzai, *Internal stresses in molecules II. A local view of binding*, Mol. Phys. **38**, 2069 (1979).
13.14. A.S. Bamzai and B.M. Deb, *Internal stress and chemical binding*, Int. J. Quantum Chem. **20**, 1315 (1981).
13.15. J. Yvon, *Navier–Stokes et quanta*, J. de Physique-Lett. **39**, 363 (1978).
13.16. S.K. Ghosh and B.M. Deb, *Densities, density-functionals and electron fluids*, Phys. Repts. **92**, 1 (1982).
13.17. P. Holland, *Quantum Theory of Motion* (Cambridge University Press, Cambridge, 1993).
13.18. K. Na and R.E. Wyatt, *Quantum hydrodynamic analysis of decoherence: quantum trajectories and the stress tensor*, Phys. Lett. A **306**, 97 (2002).
13.19. P.A. Dirac, *Quantized singularities in the electromagnetic field*, Proc. R. Soc. A **133**, 60 (1931).
13.20. J.O. Hirschfelder, C.J. Goebel, and L.W. Bruch, *Quantized vortices around wave-function nodes*. II, J. Chem. Phys. **61**, 5456 (1974).

13.21. H. Wu and D.W.L. Sprung, *Inverse-square potential and the quantum vortex*, Phys. Rev. A **49**, 4305 (1994).
13.22. E.A. McCullough, Jr. and R.E. Wyatt, *Quantum dynamics of the collinear (H,H_2) reaction*, J. Chem. Phys. **51**, 1253 (1969).
13.23. E.A. McCullough and R.E. Wyatt, *Dynamics of the collinear H+H_2 reaction. I. Probability density and flux*, J. Chem. Phys. **54**, 3578 (1971).
13.24. A. Kuppermann, J.T. Adams, and D.T. Truhlar, Abstracts of Papers, VII International Conference of Physics of Electronic and Atomic Collisions, Belgrade, Yugoslavia, p. 149.
13.25. J.O. Hirschfelder, A.C. Christoph, and W.E. Palke, *Quantum mechanical streamlines. I. Square potential barrier*, J. Chem. Phys. **61**, 5435 (1974).
13.26. J.O. Hirschfelder and K.T. Tang, *Quantum mechanical streamlines. III. Idealized reactive atom-molecule collision*, J. Chem. Phys. **64**, 760 (1976).
13.27. I. Daizadeh, J. Guo, and A.A. Stuchebrukhov, *Vortex structure of the tunneling flow in long-range electron transfer reactions*, J. Chem. Phys. **110**, 8865 (1999).
13.28. A.A. Stuchebrukhov, *Toward ab initio theory of long-distance electron tunneling in proteins: Tunneling currents approach*, Adv. Chem. Phys. **118**, 1 (2001).
13.29. A.S. Sanz, F. Borondo, and S. Miret-Artes, *Particle diffraction studied using quantum trajectories*, J. Phys.: Cond. Matt. **14**, 6109 (2002).
13.30. I. Bialynicki-Birula, Z. Bialynicki-Birula, and C Sliwa, *Motion of vortex lines in quantum mechanics*, arXiv:quant-ph/9911007 (3 Nov. 1999).
13.31. M.P. Matthews, B.P. Anderson, P.C. Haljan, D.S. Hall, C.E. Wieman, and E.A. Cornell, *Vortices in a Bose–Einstein condensate*, Phys. Rev. Lett. **83**, 2498 (1999).
13.32. K.W. Madison, F. Chevy, W. Wohlleben, and J. Dalibard, *Vortex formation in a stirred Bose-Einstein condensate*, Phys. Rev. Lett. **84**, 806 (2000).
13.33. J.R. Abo-Shaeer, C. Raman, J.M. Vogels, and W. Ketterle, *Observation of vortex lattices in Bose–Einstein condensates*, Science, **292**, 476 (2001); http://science.nasa-.gov/headlines/y2002/03_apr.neutronstars.htm
13.34. R. Blaauwgeers, V.B. Eltsov, M. Krusius, J.J. Ruohio, R. Schanen, G.E. Volovik, *Double-quantum vortex in superfluid ^{3}He-A*, Nature, **404**, 471 (2000).
13.35. A. Bezryadin, Y.N. Ovchinnikov and B. Pannetier, *Nucleation of vortices inside open and blind microholes*, Phys. Rev. B **53**, 8553 (1996).
13.36. Y. Yu and A. Bulgac, *Spatial structure of a vortex in low density neutron matter*, Phys. Rev. Lett. **90**, 161101 (2003).
13.37. P. Donati and P.M. Pizzochero, *Is there nuclear pinning of vortices in superfluid pulsars?* Phys. Rev. Lett. **90**, 211101 (2003).
13.38. E. Nelson, *Derivation of the Schrödinger equation from Newtonian mechanics*, Phys. Rev. **150**, 1079 (1966).
13.39. E. Nelson, *Quantum Fluctuations* (Princeton University Press, Princeton, 1985).
13.40. M. Pavon, *Hamilton's principle in quantum mechanics*, J. Math. Phys. **36**, 6774 (1995).
13.41. M. Pavon, *Lagrangian dynamics for classical, Brownian, and quantum mechanical particles*, J. Math. Phys. **37**, 3375 (1996).
13.42. R.T. Lawton and M.S. Child, *Local mode vibrations of water*, Mol. Phys. **37**, 1799 (1979).
13.43. R.T. Lawton and M.S. Child, *Excited stretching vibrations of water: The quantum mechanical picture*, Mol. Phys. **40**, 773 (1980).
13.44. M.J. Davis and E.J. Heller, *Quantum dynamical tunneling in bound states*, J. Chem. Phys. **75**, 246 (1981).

13.45. D. Babyuk, R.E. Wyatt, and J.H. Frederick, *Hydrodynamic analysis of dynamical tunneling*, J. Chem. Phys. **119**, 6482 (2003).
13.46. J. Zhang and D.G. Imre, *Spectroscopy of photodissociation dynamics of water: Time-dependent view*, J. Chem. Phys. **90**, 1666 (1989).
13.47. D.F. Coker and R.O. Watts, *Structure and vibrational spectroscopy of the water dimer using quantum simulation*, J. Phys. Chem. **91**, 2513 (1987).
13.48. D.F. Coker, R.E. Miller, and R.O. Watts, *The infrared photodissociation spectra of water clusters*, J. Chem. Phys. **82**, 3554 (1985).
13.49. A.S. Sanz, F. Borondo, and S. Miret-Artes, *Role of quantum vortices in atomic scattering from single adsorbates*, Phys. Rev. B **69**, 115413 (2004).
13.50. A.S. Sanz, F. Borondo, and S. Miret-Artes, *Quantum trajectories in atom–surface scattering with single adsorbates: The role of quantum vortices*, J. Chem. Phys. **120**, 8794 (2004).
13.51. K.A. Nugent, D. Paganin, and T.E. Gureyev, *A phase odyssey*, Physics Today, August, 2001, p. 27.

14
Quantum Trajectories for Stationary States

For stationary states, two interrelated trajectory approaches are introduced, both of which exhibit major differences with Bohmian mechanics. Floydian trajectories and microstates are described first; then the quantum equivalence principle of Faraggi and Matone is presented.

14.1 Introduction

Starting around 1980, Edward Floyd began publishing a series of papers in which a novel trajectory approach for *bound stationary states* was developed [14.1–14.12]. This approach, referred to as the *trajectory representation* (TR), is very different from the de Broglie–Bohm interpretation and the hydrodynamic formulation that were developed in the preceding 13 chapters. Each Floydian trajectory evolves on an energy-dependent *modified potential* $U(x)$, which includes the classical potential plus a quantum correction, denoted by $Q(x)$. For a given bound eigenvalue, there are an *infinite number* of Floyd trajectories, *all of which map onto the same wave function.* These trajectories and their underlying modified potentials, referred to as *microstates,* are not "detected" by the Schrödinger equation and its wave function solutions.

Beginning in 1998, Faraggi and Matone proposed and then began exploring the consequences of a *quantum equivalence principle* [14.13–14.17]. This deep and far-reaching principle provides a route for developing quantum mechanics that is very different from previous approaches. Using the *equivalence of physical systems under coordinate transformations*, they were able to derive the form of the quantum potential for stationary states and relate it to a curvature that appears in projective geometry. Furthermore, they proposed that *quantum mechanics itself may arise from this principle.* A key equation in their work, as well as that of Floyd, is the *quantum stationary Hamilton–Jacobi equation*, which Faraggi and Matone *deduce* from the equivalence principle. This quantum mechanical equation, under certain geometric transformations of the coordinates, can be converted into the classical Hamilton–Jacobi equation. Since the coordinate transformation that accomplishes this depends on the quantum potential, quantum effects are subsumed into the structure of the underlying coordinate space. This formulation also has trajectory solutions, and these are analogous to those introduced years earlier by Floyd.

The aim of this chapter is to present an introduction to Floydian trajectories and to the equivalence principle of Faraggi and Matone. In order to set the stage for these developments, Bohmian trajectory analysis is applied to stationary states in Section 14.2, and the stationary state version of the Hamilton–Jacobi equation is presented in Section 14.3. Then, in Section 14.4, Floydian trajectories and microstates make their appearance. The Faraggi and Matone equivalence principle is introduced in Section 14.5, and some conclusions are presented in Section 14.6. (A short summary of the work by both Floyd and by Faraggi and Matone is presented in Sections I–III of the study by Brown [14.18].)

In Chapter 15, a related approach to both stationary and nonstationary states will be introduced. In the work by Poirier [15.1, 15.2], the wave function is decomposed into node-free counterpropagating components (see Sections 15.3 and 15.4 along with Box 15.1). Parameters that occur in the decomposition are chosen in accord with procedures from semiclasscial mechanics. When this is done, the trajectory results are in accord with what is expected from the correspondence principle.

14.2 Stationary Bound States and Bohmian Mechanics

For a stationary bound state at energy E, since the amplitude R and the density are now independent of time, the continuity equation (see equations 2.6 through 2.8) takes the simplified form

$$\frac{\partial}{\partial x}\left(R(x)^2\frac{\partial S(x)}{\partial x}\right) = 0. \tag{14.1}$$

In order to obtain a real-valued function for the space part of the time-dependent wave function, it is common in Bohmian mechanics to take the action to be constant (the constant might as well be zero), which does lead to a solution of the continuity equation. However, we then are led to the vanishing of the momentum $p = \partial S/\partial x = 0$, so that the velocity also vanishes. As a result, the Bohm trajectories are not very interesting: *they just sit glued to the initial position even though the clock continues to tick.* This is because the total force acting on the trajectory is zero: the quantum force is the negative of the classical force. This would also seem to cause difficulties in the classical limit, where the particles or fluid elements would need to start moving when $\hbar$ goes to zero. Some investigators have become distressed with these results and have sought other "more reasonable" trajectory approaches for bound stationary states. For these states, we will shortly introduce other approaches in which the *trajectories move along with time.*

(The type of state described above is a *static stationary state*, which includes, for example, the s states of the hydrogen atom. For *dynamic stationary states*, such as the $2p_{\pm 1}$ hydrogenic states, the moving Bohm trajectories possess angular momentum. A discussion is presented by Ghosh and Deb [14.23].)

14.3 The Quantum Stationary Hamilton–Jacobi Equation: QSHJE

For a stationary state at energy E, the action function can be separated into space and time parts, $S(x, t) = W(x) - Et$, where $W(x)$ is called the *reduced action* or Hamilton's *characteristic function*. The quantum Hamilton–Jacobi equation then becomes

$$\frac{1}{2m}\left(\frac{\partial W}{\partial x}\right)^2 + V + Q = E, \tag{14.2}$$

and this is referred to as the *quantum stationary Hamilton–Jacobi equation*, QSHJE. Both Floyd and Faraggi and Matone emphasize that this equation and its solutions provide *new dynamical information that is not contained in the Schrödinger equation*. The momentum obtained by solving this equation has upper and lower branches (the two branches of a *Lagrangian manifold*, as described later, in Box 15.1) given by

$$p(x) = \frac{\partial W}{\partial x} = \pm\sqrt{2m(E - V - Q)}, \tag{14.3}$$

which integrates to give a pair of values for the reduced action (expressed in terms of a *phase integral*)

$$W(x) = W(x_0) \pm \int_{x_0}^{x} \sqrt{2m(E - V - Q)}dx. \tag{14.4}$$

We will now approach the continuity equation in a slightly different way. Once again, since both the R-amplitude and the density are independent of time, the continuity equation takes the simplified form

$$\frac{\partial}{\partial x}\left(R^2\frac{\partial W}{\partial x}\right) = 0, \tag{14.5}$$

which integrates to $R^2\,(\partial W/\partial x) = k$, where k is a constant, which we will take to have the numerical value $k = 1$ (other choices are more appropriate in linking to semiclassical mechanics [14.22]).

The *flux* associated with this wave function is given by $J = R^2 \cdot (1/m)\,\partial W/\partial x$, so that the condition $k =$ constant also requires the flux to be constant [14.22]. (In higher dimensionality, equation 14.5 states that the divergence of $R^2\vec{\nabla} W$ is zero, $\vec{\nabla} \cdot (R^2\vec{\nabla} W) = 0$. Furthermore, this implies the existence of a "hidden" vector field such that $R^2\vec{\nabla} W = \vec{\nabla} \times \vec{B}$ [14.16].)

The R-amplitude for the one-dimensional case is then

$$R(x) = A\,(\partial W(x)/\partial x)^{-1/2} = A\,p(x)^{-1/2}, \tag{14.6}$$

where the constant A provides for normalization. This equation then leads to the following expression for the wave function:

$$\Psi(x,t) = A(\partial W/\partial x)^{-1/2} e^{iW/\hbar} e^{-iEt/\hbar}. \tag{14.7}$$

This form will be generalized in the next section.

(Floyd has pointed out that the continuity equations, equations 14.1 and 14.5, imply a probability definition for $\Psi^*\Psi$. However, these equations are equivalent to the Wronskian for the stationary Schrodinger equation being constant and for the Wronskian to be divergenceless. As a result, a probability interpretation is not really needed, although it may be satisfying to do so.)

14.4 Floydian Trajectories and Microstates

Floyd's investigations in quantum mechanics sprang from attempts to improve the WKB semiclassical approximation and from acoustical ray tracing algorithms. We will now explore some aspects of Floyd's approach. For *stationary bound states*, Floyd defined the *energy-dependent modified potential* $U = V + Q$ [14.1–14.2], in which Q is the quantum potential, which, as described later, *may be* the same quantity that appears in Bohmian mechanics. From the QSHJE, we previously obtained the *conjugate momentum*, the derivative of the reduced action, given by $p(x) = \partial\tilde{W}/\partial x = \pm\sqrt{2m(E-U(x))}$. (The symbol $\tilde{W}$ will now be used for the reduced action, to distinguish it from the function W that was used in the Madelung–Bohm polar form for the wave function.) As a result, the amplitude is given by $R(x) \sim [E-U(x)]^{-1/4} \sim p(x)^{-1/2}$ (the symbol $\sim$ means "goes like" or "has the form") and the following pair of *linearly independent traveling waves* can then be constructed:

$$\psi^{(\pm)}(x,t) = (E-U(x))^{-1/4} \exp\left[\pm\frac{i}{\hbar}\int\sqrt{2m(E-U(x))}dx\right]\exp\left[-iEt/\hbar\right]. \tag{14.8}$$

Recall that in Bohmian mechanics, the *unipolar ansatz* for the wave function is assumed, which for a stationary state is given by $\Psi(x,t) = R(x)\exp\left[iW(x)/\hbar\right]$ $\exp\left[-iEt/\hbar\right]$, where the amplitude satisfies the condition $R(x) \geq 0$. However, for these states, Floyd assumed the more general *bipolar decomposition of the wave function*

$$\psi(x,t) = A\psi^{(+)}(x,t) + B\psi^{(-)}(x,t), \tag{14.9}$$

where A and B *may be* complex. For bound states in one dimension, the real-valued combination of solutions in this equation leads to the *trigonometric ansatz* for the space part of the total wave function (α and β are real-valued coefficients)

$$\psi^{(r)}(x) = (E-U(x))^{-1/4}\left[\alpha\cos(\tilde{W}(x)/\hbar) + \beta\sin(\tilde{W}(x)/\hbar)\right]. \tag{14.10}$$

The reduced action is energy-dependent, but this is not indicated explicitly in the notation.

If we now substitute equation 14.8 into the time-dependent Schrödinger equation, we obtain a *nonlinear second-order differential equation for the modified potential:*

$$U + \frac{\hbar^2}{8m}\frac{\partial^2 U/\partial x^2}{E - U} + \frac{5\hbar^2}{32m}\left(\frac{\partial U/\partial x}{E - U}\right)^2 = V. \tag{14.11}$$

For real-valued wave functions, it is required that $U(x) < E$ for finite x. Otherwise, $R(x)$ in equation 14.10 will not remain real-valued. The limiting condition $U(x) \to E$ leads to a *critical point* in this equation, and these are the *turning points* at $x = \pm\infty$. It is worth pointing out that near a node in the wave function, *the modified potential remains finite,* and provided that $U(x) < E$, then $U = V + Q$, where Q is exactly the same function as the Bohm quantum potential. As an example, Floyd demonstrated that the modified potential remains finite near the node at $x = 0$ for the first excited state of the harmonic oscillator (see figure 1 in [14.1]). It will be shown later that *Floyd trajectories pass through nodes with finite momentum* ($p = \partial\tilde{W}/\partial x$) and nodes present no special problems in the dynamics.

Following analysis by Poirier, we can recast equation 14.11 into a different form. Starting from $p(x) = \pm\sqrt{2m(E - U(x))}$, this equation can be expressed in terms of $p(x)$ and its spatial derivatives (where each "prime" denotes a spatial derivative)

$$\frac{p^2}{2m} + V(x) + \frac{\hbar^2}{4m}\left[\frac{p''}{p} - \frac{3}{2}\left(\frac{p'}{p}\right)^2\right] = E, \tag{14.12}$$

in which the term in brackets, including the multiplicative prefactor, is the quantum potential. This term is related to the *Schwarzian derivative*, a quantity that appears in differential geometry (see equations 14.27–14.28). Because the equation given above is a form of the QSHJE, it might be said that equation 14.11 is a veiled form of the QSHJE.

We will now make some general remarks about this approach and then add details. Floyd pointed out that at a given eigenvalue, the solutions to equation 14.11 are *not unique*; *there is a series of solutions* $U_1, U_2, \ldots$, and associated with each of these is a trajectory $x_1(t), x_2(t), \ldots$. Each *Floydian trajectory* $x_j(t)$ is determined by (evolves on) the corresponding modified potential U_j. Each of these pairs $\{U_j(x), x_j(t)\}$ is referred to as a *microstate* of the Schrödinger equation. There are an infinite number of microstates at each energy eigenvalue, and these *different microstates all specify the same wave function.* No microstate is preferred over any other one, and each yields the same eigenvalue and wave function. In *Floyd's view*, the existence of microstates refutes "the assertion of the Copenhagen interpretation that the Schrödinger wave function be the exhaustive description of nonrelativistic quantum phenomena" [14.10]. This view is probably not shared universally in the quantum mechanics community!

There are several approaches that can be used to solve equation 14.11 for $U(x)$: numerical computations (including the use of finite difference approximations), perturbative calculations, power series expansions [14.5], and a closed-form

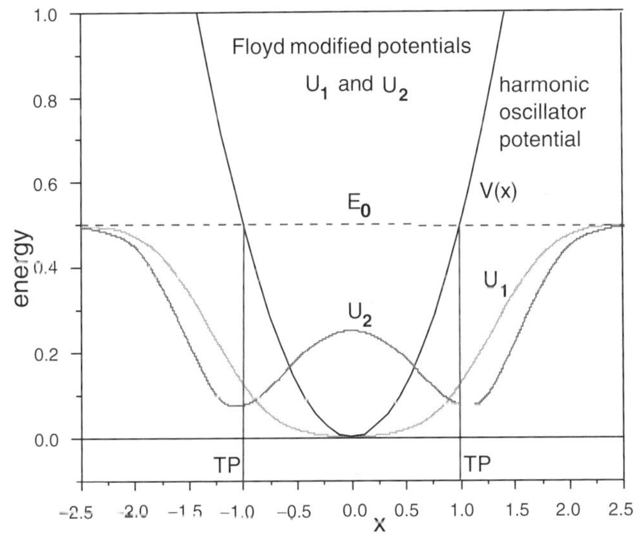

FIGURE 14.1. Two Floyd modified potentials $U_1(x)$ and $U_2(x)$ and the harmonic oscillator potential $V(x)$. The energies are in units of $\hbar\omega$, and x is in units of $(h/m\omega)^{1/2}$. The modified potentials are for the ground state energy $E_0 = (1/2)\hbar\omega$, marked by the horizontal dashed line. The modified potentials approach E_0 from below at large displacements from the origin. The classical turning points, marked TP, are $x = \pm 1$ (figure adapted from figure 2 in [14.1]).

solution in terms of solutions of the time-independent Schrödinger equation [14.4]. Before returning to the latter approach, we will first consider some general features of this differential equation. A particular solution can be developed once values are chosen for $U(x)$ and $\partial U/\partial x$ at some point x_0. Other choices for the values for these two quantities will lead to additional solutions for the modified potential.

As an example [14.1], we will consider solutions to equation 14.11 for the ground state energy ($E_0 = \hbar\omega/2$) of the harmonic oscillator potential, $V(x) = m\omega^2 x^2/2$. If we choose $x_0 = 0$, $U(0) = 0$, and $\partial U/\partial x|_{x_0=0} = 0$ and then solve for the modified potential, the curve labeled U_1 is obtained in figure 14.1. With another choice (at the same value of x_0), $U(0) = \hbar\omega/4$ and $\partial U/\partial x|_{x_0=0} = 0$, we obtain the new modified potential U_2. Both of these potentials approach the eigenvalue E_0 (the horizontal dashed line) *from below* as $x \to \pm\infty$. In this figure, the harmonic oscillator potential is also shown, and the classical turning points (TP) are indicated by the two vertical dashed lines. The modified potentials extend smoothly into the nonclassical regions.

The trajectories corresponding to these two modified potentials, of which x_1 is "eye-shaped" (a term suggested by Poirier [14.22]), are illustrated in phase space, shown in figure 14.2. These two trajectories are asymptotic to the x axis and form cusps at the two *turning points*, $x = \pm\infty$, where the momentum asymptotically

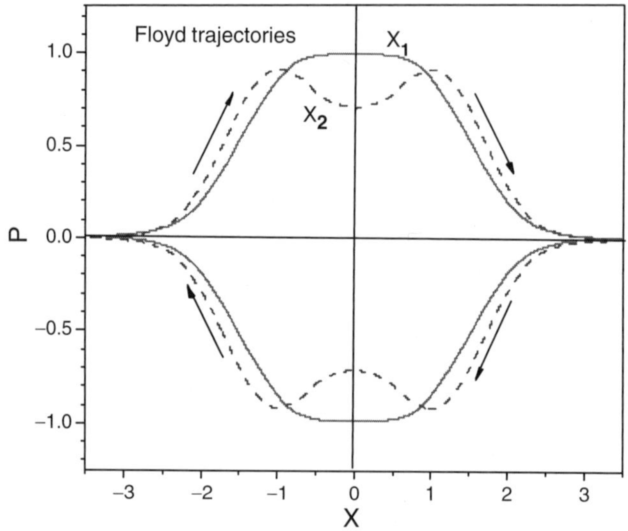

FIGURE 14.2. Phase space plots for Floyd trajectories, $x_1(t)$ and $x_2(t)$, evolving on the two modified potentials shown in figure 14.1. The coordinate x is in units of $(\hbar/m\omega)^{1/2}$, and the conjugate momentum $p = \partial\tilde{W}/\partial x$ is in units of $(m\hbar\omega)^{1/2}$. At large displacements from the origin, $p \to 0$ and a symmetric pair of cusps are created at the turning points. Both trajectories enclose the same area in phase space. These trajectories are real-valued and well behaved in the tunneling regions, $|x| > 1$. Note that the momentum plotted here is *not* the same as the mechanical momentum $(m\dot{x})$ (figure adapted from figure 3 in [14.1]).

approaches the horizontal axis. In the *classically forbidden regions*, $|x| > 1$, the trajectories have real- valued momenta and are well behaved at all points. In addition, the trajectories smoothly pass from the classically allowed to the classically forbidden regions.

As a consistency check, it is significant to note that during one cycle of motion, both of these trajectories *enclose the same area in phase space* and thus quantize to the same energy $(\hbar\omega/2)$ when the Bohr–Sommerfeld-type quantization scheme is used. In this respect, a single trajectory is sufficient to specify the eigenvalue and the wave function; there is no need for an ensemble of trajectories. The *integer quantization condition* is $\oint(\partial\tilde{W}/\partial x)dx = (n+1)2\pi\hbar$, for $n = 0, 1, 2\ldots$, in which the integral is over one cycle of motion in phase space. This integral yields the phase space area between the two branches (right-moving and left-moving) of the trajectory. In the classically forbidden regions, the contribution to the integrand is finite despite the infinite extent of the trajectory to the two turning points at $x = \pm\infty$. The integer quantization condition has also been verified for other potentials, including a quartic oscillator and the Morse potential [14.2]. It is important to note that this quantization rule is *not consistent* with the familiar half-integer *semiclassical* (WKB) rule, $\oint(\partial\tilde{W}/\partial x)_{sc}\,dx = (n+1/2)2\pi\hbar$, for $n = 0, 1, 2, \ldots$, where the integral in this case is evaluated between the two *classical* turning points. In the Floyd trajectory representation, there is no need for connection formulas and other mathematical techniques that are familiar in semiclassical analysis.

The solution to the QSHJE leads to a relationship between p and x, as illustrated for the two trajectories in figure 14.2. In order to obtain the time dependence of these trajectories, the following procedure can be used [14.1]. By a *theorem due to Jacobi*, the time t at which a particle passes point x is given by $t = t_0 + \partial S/\partial E$. Using the equation relating the action to the modified potential,

$$S(x,t) = \int_x \sqrt{2m(E - U(x))}dx - Et, \tag{14.13}$$

we obtain

$$t = t_0 + \sqrt{\frac{m}{2}} \int_x \frac{(1 - \partial U/\partial E)}{\sqrt{(E-U)}} dx. \tag{14.14}$$

The velocity is then given by differentiating this expression:

$$\frac{dx}{dt} = (dt/dx)^{-1} = (1 - \partial U/\partial E)^{-1} \cdot \frac{1}{m}\frac{\partial \tilde{W}}{\partial x} = \frac{p}{(1 - \partial U/\partial E)m}. \tag{14.15}$$

Through further differentiation, an equation can also be derived for the acceleration, d^2x/dt^2. Faraggi and Matone regard the quantity in the denominator of equation 14.15, $m_Q = (1 - \partial U/\partial E)m$, as the *effective quantum mass*, or the *quantum mass field* [14.15]. We also note from this equation that the *mechanical momentum* $m\dot{x}$ is not the same as the conjugate momentum $p = \partial \tilde{W}/\partial x$. The equations of motion given above are not those used for Bohmian trajectories. For example, a Bohmian trajectory always slices through a surface of constant action at right angles. In contrast to this, Floyd trajectories generally slice through at oblique angles.

For the trajectories shown earlier, in figure 14.2, Floyd plotted $\dot{x}$ against x in *mechanical phase space* (see figure 5 in [14.1]). Beyond the classical turning points at $x = \pm 1$, the trajectories reach *superluminal* velocities in this nonrelativistic theory. However, the *reflection time*, the amount of time spent in the tunneling regions, is finite. Because of this, the particle is able to traverse a complete cycle of motion in a finite period of time.

In *classical mechanics*, a unique trajectory can be launched from an initial point x_0 by specifying $\partial \tilde{W}/\partial x$ (which fixes the momentum) at that point. This is consistent with the classical stationary Hamilton–Jacobi equation being a first-order differential equation. However, for *Floyd trajectories*, the situation is more complicated because the QSHJE is a *third-order differential equation*. As a result, position and momentum form only a subset of the initial conditions that are necessary and sufficient to solve the QSHJE. At the point x_0, if we specify $p = \partial \tilde{W}/\partial x$, this fixes U and the initial velocity dx/dt. However, we also need to specify $p' = \partial^2 \tilde{W}/\partial x^2$ in order to have the value for $\partial U/\partial x$. Recall that *both* U and $\partial U/\partial x$ at point x_0 are needed in order to obtain a unique solution for $U(x)$. However, specification of $\partial^2 \tilde{W}/\partial x^2$ at point x_0 is equivalent to specification of the initial acceleration of the trajectory, denoted by $\ddot{x}_0$. Thus the three quantities $\{x_0, \dot{x}_0, \ddot{x}_0\}$, which specify the trajectory and the underlying modified potential, may be regarded as *hidden variables*; *they are hidden from the Schrodinger equation and the wave function.* If the energy eigenvalue is unknown, a fourth initial value is required to make the set of initial conditions $\{\tilde{W}, p = \partial\tilde{W}/\partial x, p' = \partial^2\tilde{W}/\partial x^2, p'' = \partial^3\tilde{W}/\partial x^3\}$, again specified at the point x_0, necessary and sufficient to determine the energy eigenvalue [14.3].

An instructive consistency check can be carried out on the harmonic oscillator modified potentials that were shown in figure 14.1 [14.1]. For small values of x, near the origin, the symmetric modified potential can be expanded in the power series

$$U(x) = c_0 + c_2x^2 + c_4x^4 + O(x^6). \tag{14.16}$$

The initial conditions that will be specified at $x = 0$ are $U = c_0 < E$ and $\partial U/\partial x = 0$. For the initial coefficient taking on either of the two values $c_0 = 0$ and $c_0 = \hbar\omega/4$, we obtain the potentials U_1 and U_2 in figure 14.1. If we then substitute this series for $U(x)$ into equation 14.11, the differential equation for the modified potential, and then equate the coefficients of x^0 and x^2, we obtain

$$c_2 = (4m/\hbar^2)c_0(c_0 - E), \tag{14.17}$$

and

$$c_4 = (2m/3\hbar^2)(E - c_0)\left[m\omega^2/2 + (4m/\hbar^2)c_0E - (18m/\hbar^2)c_0^2\right]. \tag{14.18}$$

The expression for $U(x)$, using these values for c_2 and c_4, can then be used to obtain the wave function in trigonometric form,

$$\psi(x) = (E - U(x))^{-1/4}\,\alpha\cos\Big(\int_x \sqrt{2m(E - U(x))}dx\Big). \tag{14.19}$$

(The coefficient β in equation 14.10 has been set to zero for this symmetric wave function.) After expanding the various terms in equation 14.19 in power series and then simplifying, this calculation gives for the ground state wave function (not normalized)

$$\psi(x) = 1 - (m\omega/\hbar)\frac{x^2}{2} + (m\omega/\hbar)^2\frac{x^4}{8} + \cdots, \tag{14.20}$$

and this is in fact the correct *series expansion of the Gaussian ground state wave function.* A very important point about this analysis is that *the coefficient c_0 that was used to specify the particular modified potential disappears when the wave function is computed. This is the reason why the different microstates all map to the same wave function.*

If a pair of linearly independent solutions χ_1 and χ_2 for the time-independent Schrödinger equation are known, Floyd has shown [14.4] that the modified potential is given in *closed form* by the expression

$$U(x) = E - \frac{1}{(a\chi_1^2 + b\chi_2^2 + c\chi_1\chi_2)^2}, \tag{14.21}$$

where a and b are positive definite and $ab > c^2/4$. In addition, the two solutions are scaled so that the Wronskian of the two solutions, given by $w = \chi_1(\partial\chi_2/\partial x) - \chi_2(\partial\chi_1/\partial x)$, has the value $w^2 = 2m/[\hbar^2(ab - c^2/4)]$. The three constants (a, b, c), which determine the particular microstate, play the role of *hidden variables*.

14.5 The Equivalence Principle and Quantum Geometry

The work by Faraggi and Matone [14.13–14.17], which began in the period 1998–1999, emphasizes the diverse consequences of a postulate called the *equivalence principle*: *all physical systems are equivalent under coordinate transformations*. The application of this principle leads to many general concepts, but we will consider only limited aspects of this theorem for one-dimensional systems. Consider a system with the coordinate x, potential $V(x)$, reduced action $\tilde{W}(x)$, and conjugate momentum $p = \partial\tilde{W}/\partial x$. For this system, the QSHJE is

$$\frac{1}{2m}\left(\frac{\partial\tilde{W}}{\partial x}\right)^2 + V + Q = E. \tag{14.22}$$

Now we will form a *new coordinate* $\hat{x}(x)$ that is obtained from the original one by stretching in some places or compressing in others as though the coordinate were a rubber band. The potential energy and reduced action in the new coordinate system are $\hat{V}(\hat{x})$ and $\hat{W}(\hat{x})$, respectively. For example, we might map the harmonic oscillator into the infinitely deep well, at least the part of the latter potential that lies between the turning points. The transformations being considered here are *point transformations*; they are not *canonical* in the sense used in classical mechanics because only the coordinate is mapped, while the momentum rides along as a dependent variable.

There are various ways to specify the coordinate mapping, including the *quantum* → *classical* transformation. In the latter case, we will demand that the stationary Hamilton–Jacobi equation in the new frame have the classical form

$$\frac{1}{2m}\left(\frac{\partial\hat{W}}{\partial\hat{x}}\right)^2 + \hat{V} = E. \tag{14.23}$$

In this new frame, all quantum effects have been subsumed in the elastic coordinate transformation $\hat{x}(x)$, so that particle trajectories *appear* to follow classical paths. *The equivalence principle is the requirement* $\hat{W}(\hat{x}) = \tilde{W}(x)$ and $\hat{V}(\hat{x}) = V(x)$, i.e., invariance of the reduced action and the potential under the coordinate deformation, or *physical systems are equivalent under coordinate transformations* [14.14]. If we accept these requirements, then we can write equation 14.23 as

$$\frac{1}{2m}\left(\frac{\partial W}{\partial x}\frac{\partial x}{\partial \hat{x}}\right)^2 + V = E, \tag{14.24}$$

which gives

$$\frac{1}{2m}\left(\frac{\partial W}{\partial x}\right)^2\left(\frac{\partial x}{\partial \hat{x}}\right)^2 = E - V. \tag{14.25}$$

Within a sign, we now have the relation between the new and old coordinates

$$d\hat{x} = \frac{(\partial W/\partial x)}{\sqrt{2m(E-V)}}dx = \frac{p}{\sqrt{p^2(1+2m\tilde{Q}/p^2)}}dx = \frac{1}{\sqrt{1+\tilde{Q}/(p^2/(2m))}}dx. \tag{14.26}$$

The last version of this equation shows how the quantum potential determines the Jacobian (the multiplier of dx) of the coordinate mapping $x \to \hat{x}$. In FM, the last version is also written $d\hat{x} = (1-\beta^2)^{-1/2}dx$, in which $\beta^2 = -2m\tilde{Q}/p^2$. These expressions show how $\tilde{Q}$ *deforms the geometry of the new space.*

An instructive example of the coordinate transformation mentioned previously concerns the trivializing map that takes the square well problem ($V(x) = 0$ for $|x| \leq L$ and $V(x) = V_0$ for $|x| > L$) into the "trivial" problem corresponding to $V - E = 0$. The details of this transformation are described in the FM review article (see pages 1882–1884 in [14.15]).

A more general application of the equivalence principle would take us from the original system with coordinate, reduced action, and quantum potential $\{x, \tilde{W}, \tilde{Q}\}$ into the new system $\{\hat{x}, \hat{W}, \hat{Q}\}$ by means of the coordinate deformation $x \to \hat{x}(x)$. (The old and new coordinates are independent variables in their own systems.) The equivalence principle again requires $\hat{W}(\hat{x}) = \tilde{W}(x)$ and $\hat{V}(\hat{x}) = V(x)$. By examining the group properties of sets of these transformations ($A \to B$, $B \to C$, $C \to A$), FM *deduced the form* of the QSHJE including an expression relating the quantum potential to the *Schwarzian derivative* of the reduced action. (Planck's constant multiplies the Schwarzian derivative and plays the role of a *covariantizing parameter*.) We will consider the latter relationship in the following paragraphs. It is important to appreciate that FM did not start from the Schrödinger equation, assume a form for the wave function, and then end up with an expression for the quantum potential. At this point in our presentation of FM's work, *there is no Schrödinger equation; it is later deduced from the QSHJE.*

Both Floyd and FM have derived an expression for the quantum potential in terms of derivatives of the reduced action. As mentioned above, FM derived this result as a consequence of the equivalence principle. Following Floyd [14.7, 14.10], from the form for $R(x)$ given in equation 14.6, the quantum potential can be expressed in terms of derivatives of the reduced action,

$$Q(x) = -\frac{\hbar^2}{4m}\left[\frac{3}{2}\left(\frac{\partial \tilde{W}}{\partial x}\right)^{-2}\left(\frac{\partial^2 \tilde{W}}{\partial x^2}\right)^2 - \left(\frac{\partial \tilde{W}}{\partial x}\right)^{-1}\left(\frac{\partial^3 \tilde{W}}{\partial x^3}\right)\right]. \tag{14.27}$$

In FM, though, repeating ourselves, this equation is *deduced from coordinate transformations that are guided by the equivalence principle*. This expression is closely related to a quantity used in differential geometry to measure curvature,

$$Q(x) = \frac{\hbar^2}{4m}\{\tilde{W}, x\}, \tag{14.28}$$

where $\{W, x\}$ is the *Schwarzian derivative*. Using this notation, the QSHJE derived by FM can be written

$$\frac{1}{2m}\left(\frac{\partial \tilde{W}}{\partial x}\right)^2 + V + \frac{\hbar^2}{4m}\{\tilde{W}, x\} = E. \tag{14.29}$$

(This equation, in different notation, appears in Messiah's *Quantum Mechanics* [14.21]). This complicated expression is a third order nonlinear partial differential equation for the reduced action. FM also make the important remark that this QSHJE implies the Schrodinger equation; mathematically, the Schrödinger equation arises when the wave function $\psi(x) = (\partial \tilde{W}/\partial x)^{-1/2} \exp(i\tilde{W}/\hbar)$ is introduced to convert the third-order QSHJE into an equation linear with respect to the new function $\psi(x)$.

Floyd showed [14.4] that the general solution to the QSHJE is given by

$$\frac{\partial \tilde{W}}{\partial x} = \pm\sqrt{2m}\left(a\chi_1^2 + b\chi_2^2 + c\chi_1\chi_2\right)^{-1}, \tag{14.30}$$

where χ_1 and χ_2 are linearly independent solutions to the time-independent Schrödinger equation and where (a, b, c) are real-valued coefficients with $a, b > 0$ and $ab - c^2/4 > 0$. For bound state eigenvalues, one of these two solutions is square integrable and the other is not. For example, for the potential well with finite potential walls, one of these solutions decays exponentially, and the other grows exponentially as $|x| \to \infty$. Within the potential well, the two solutions are $\sin(kx)$ and $\cos(kx)$. However, for energies falling within gaps between the eigenvalues, neither solution is square integrable.

We mentioned earlier that the QSHJE in the form given in equation 14.30 was first derived by Floyd and that this also plays a key role in the recent work of Faraggi and Matone. One further point is that in the work of both Floyd and FM, *the reduced action for a bound state is never a constant.*

Since the conjugate momentum associated with a trajectory is $p = \partial\tilde{W}/\partial x$, the trajectory is parameterized by the set of *three coefficients* (a, b, c), the *hidden variables*, and through this relation $\tilde{W}(x)$ is regarded as a generator of the motion. (Previously, we used $\{x_0, \dot{x}_0, \ddot{x}_0\}$, the initial values of the coordinate, velocity, and acceleration as the hidden variables used to specify the trajectory.) Because there are multiple solutions of the QSHJE for bound stationary states (the microstates) that do not arise directly from the Schrödinger wave equation, Floyd and FM regard the QSHJE as *more fundamental than the Schrödinger equation,* in the sense of providing additional information that is "lost" in going to wave function solutions of the Schrödinger equation. Stated differently, the QSHJE and the time-independent Schrödinger equation do not provide equivalent information.

Box 14.1. Is the moon held in place by the quantum potential?

In 2002, Matone published a paper [14.17] with the fascinating title *Equivalence postulate and the quantum origin of gravity*. In this paper he wrote, "We suggest that gravitation has a purely quantum mechanical origin" and "the aim ... is to investigate the possibility that such interaction *may be a consequence on the quantum potential*" (italics added for emphasis). We will now quote directly from the abstract of this paper.

We suggest that quantum mechanics and gravity are intimately related. In particular, we investigate the quantum Hamilton–Jacobi equation in the case of two free particles and show that the quantum potential is attractive and may generate the gravitational potential.... A consequence of this approach is that the quantum potential is always non-trivial even in the case of the free particle. We pursue this idea, by making a preliminary investigation of whether there exists a set of solutions for which the quantum potential can be expressed with a gravitational potential leading term which alone would remain in the limit $\hbar \to 0$.

In the analysis leading to this summary, Matone considers two seemingly "free" particles of masses m_1 and m_2 evolving (seemingly) in the absence of classical interaction, but under the influence of the quantum potential. The two-particle motion is expressed in terms of the motion of the center of mass and the relative motion, which is where the analysis is focused. The relative motion problem is then broken into radial and angular problems. The quantum potential, which is attractive, is then investigated to see whether it has a classical ($\hbar \to 0$) leading term given by $V_G(r) \approx -m_1 m_2/r$. The quantum potential is written $Q = V_G(r) + O(\hbar)$, where the higher-order quantum terms may also depend on r. This form for the quantum potential appears to be consistent with the quantum Hamilton–Jacobi equation and its classical limit. As emphasized by Matone, this preliminary study raises a number of questions for further study.

In this section, we have barely scratched the surface of the recent work by Faraggi and Matone. These authors have an ambitious program to extend their work to relativistic quantum mechanics, the derivation of expressions for the fundamental interactions, the synthesis of quantum mechanics and gravity, and string theory. Their work has been brought into this chapter because of its emphasis on the quantum stationary Hamilton–Jacobi equation and the origin and properties of the quantum potential. In addition, there are many fascinating issues raised by their work that should be investigated further. The 149-page exposition on these topics [14.15] should be consulted for further details.

14.6 Summary

Floyd's development of a novel trajectory approach for stationary bound states emphasizes the role played by the quantum stationary Hamilton–Jacobi equation. In addition, a third-order differential equation was derived for the modified potential. For each bound eigenvalue, there is an infinite number of solutions to this equation, each specified by a set of three parameters, the "hidden variables", and for each modified potential there is a corresponding trajectory. These trajectories and the associated modified potentials are referred to as *microstates*. All of the microstates at a specified energy yield the same wave function and are not detected by the Schrödinger equation.

In most of Floyd's papers, modified potentials and trajectories for only the one-dimensional case are considered, although in several papers solutions are briefly mentioned for separable models of higher dimensionality. Modified potentials and trajectories have apparently not been developed for nonseparable potentials in even two dimensions. It would be very interesting to develop the trajectory solutions for this case.

The development of quantum mechanics from the equivalence principle (EP) has far-reaching consequences. In this approach, no use is made of the usual Copenhagen axioms (such as the axiomatic interpretation of the wave function), and there is no "pilot wave" to guide trajectories. In this recent work of Faraggi and Matone, the quantum potential arises in the QSHJE from the equivalence principle itself. The Schrödinger equation is derived when the wave function is introduced to linearize the QSHJE. Energy quantization and tunneling come about directly from mathematical features of the QSHJE. Faraggi and Matone suggest that the EP, through the QSHJE and the quantum potential, may unify the fundamental interactions, possibly leading to the long-sought unification of quantum mechanics and general relativity [14.15–14.17]. This would be a most remarkable achievement!

Although Fataggi and Matone have made remarkable progress during the past few years in developing quantum mechanics from the EP, very few examples have been worked out, although the solutions for the free particle and the particle in the potential well provide valuable insights. Will this approach, especially the coordinate mapping, provide new computational algorithms for quantum-mechanical problems? It may be too soon to answer this question, but the hope is that other researchers, possibly including the authors of [14.18–14.20], will investigate this new approach to quantum mechanics.

References

14.1. E.R. Floyd, *Modified potential and Bohm's quantum mechanical potential*, Phys. Rev. D **26**, 1339 (1982).
14.2. E.R. Floyd, *Bohr–Sommerfield quantization with the effective action variable*, Phys. Rev. D **25**, 1547 (1982).
14.3. E.R. Floyd, *Arbitrary initial conditions of nonlocal hidden variables*, Phys. Rev. D **29**, 1842 (1984).
14.4. E.R. Floyd, *Closed-form solutions for the modified potential*, Phys. Rev. D **34**, 3246 (1986).
14.5. E.R. Floyd, *Higher order modified potentials for the effective phase integral approximation*, J. Math. Phys. **20**, 83 (1979).
14.6. E.R. Floyd, *Phase integral approximations for calculating energy bands*, J. Math. Phys. **17**, 881 (1976).

14.7. E.R. Floyd, *Where and why the generalized Hamilton–Jacobi representation describes microstates of the Schrödinger wave function*, Found. Phys. Lett. **9**, 489 (1996).
14.8. E.R. Floyd, *Which causality? Differences between trajectory and Copenhagen analyses of an impulsive perturbation*, Int. J. Mod. Phys. A **14**, 1111 (1999).
14.9. E.R. Floyd, *Classical limit of the trajectory representation of quantum mechanics, loss of information and residual indeterminacy*, Int. J. Mod. Phys. A **15**, 1363 (2000).
14.10. E.R. Floyd, *Extended version of The philosophy of the trajectory representation of quantum mechanics*, arXiv:quant-ph/0009070 (17 Sept. 2000).
14.11. E.R. Floyd, *Comments on Bouda and Djama's "Quantum Newton's law"*, Phys. Lett A **296**, 307 (2002).
14.12. E.R. Floyd, *Differences between the trajectory representation and Copenhagen regarding past and present in quantum theory*, arXiv:quant-ph/0307090 (12 Jul. 2003).
14.13. A.E. Faraggi and M. Matone, *Quantum transformations*, Phys. Lett. A **249**, 180 (1998).
14.14. A.E. Faraggi and M. Matone, *Quantum mechanics from an equivalence principle*, Phys. Lett. B **450**, 34 (1999).
14.15. A.E. Faraggi and M. Matone, *The equivalence postulate of quantum mechanics*, Int. J. Mod. Phys. **15**, 1869 (2000).
14.16. G. Bertoldi, A.E. Faraggi, and M. Matone, *Equivalence principle, higher-dimensional Möbius group and the hidden antisymmetric tensor of quantum mechanics*, Class. Quantum Grav. **17**, 3965 (2000).
14.17. M. Matone, *Equivalence postulate and quantum origin of gravitation*, Found. Phys. Lett. **15**, 311 (2002).
14.18. M.R. Brown, *Floydian trajectories for stationary systems: A modification for bound states*, arXiv:quant-ph/0102102 (20 Feb. 2001).
14.19. A. Bouda, *Probability current and trajectory representation*, Found. Phys. Lett. **14**, 17 (2001).
14.20. A. Bouda and T. Djama, *Quantum Newton's law*, Phys. Lett A **285**, 27 (2001).
14.21. A. Messiah, *Quantum Mechanics*, Vol. I (Wiley, New York, 1958), p. 232.
14.22. W. Poirier, notes and private communication, Dec. 2003–May 2004.
14.23. S.K. Ghosh and B.M. Deb, *Quantum fluid dynamics of many-electron systems in three-dimensional space*, Int. J. Quantum Chem. **22**, 871 (1982).

15
Challenges and Opportunities

Promising quantum trajectory methods are reviewed, and three approaches to the node problem are described. In two of these, the counterpropagating wave method and the covering function method, only node-free functions are propagated.

15.1 Introduction

In the preceding chapters, the equations of motion for quantum trajectories have been developed from several viewpoints. Methods have been introduced for evaluating spatial derivatives on the unstructured grids formed by moving ensembles of fluid elements. Problems that develop near nodal regions have been described, and adaptive methods have been presented for dealing with the trajectory instabilities that arise in these regions. Methods that approximate the quantum force have been applied to several problems. Although many of the examples of quantum trajectory evolution have focused on systems with one or two degrees of freedom, methods have been developed that are applicable to systems with higher dimensionality. Some of these promising methods and application areas will be briefly reviewed in this section.

Adaptive grids. Dynamic (moving) adaptive grids may be crafted according to various criteria. These *designer grids* overcome some of the problems that may develop with Bohmian trajectories by providing *complete user control* over the paths followed by the fluid elements. The simplest choice, expanding or contracting grids with internal points that are equally spaced, greatly simplifies the derivative evaluation problem. Other grid adaptation algorithms, such as the equidistribution principle, use monitor functions to guide grid points to *active regions* (for example, regions where the gradient or curvature of the solution becomes large). Adaptive grids will be considered further in Section 15.2.

Adaptive modification of the equations of motion. Approximate solutions to the hydrodynamic equations in nodal regions may be obtained by adding viscous forces to smooth over rapid variations in the quantum force. This greatly enhances the stability of quantum trajectories and permits propagation to long times. This method should be extensible to multidimensional systems.

Hybrid methods. These methods are designed to deal with the node problem by solving the hydrodynamic equations of motion exterior to the nodal region and the time-dependent Schrödinger equation within a "ψ-patch" covering the nodal region. Problems may arise at the boundary between these regions, where the solutions must be smoothly joined. Even so, accurate long-time propagation has been achieved for scattering in one-dimensional model systems. Extensions to higher dimensionality would be of interest.

Approximations to the quantum force. Approximations to the quantum force can be obtained by fitting the density or the log derivative of the density to a basis set expansion. Although the fitting can be done globally or locally, the latter usually provides better results in regions where there are ripples or nodes in the density. These promising methods have been used to compute energy-resolved reaction probabilities and the ground states for anharmonic systems.

Derivative propagation methods. By propagating spatial derivatives along individual quantum trajectories, it is possible to avoid the lockstep propagation of an ensemble of correlated trajectories. However, these approximate quantum trajectories are influenced only by "regional nonlocality" and do not properly bring in interference effects. In spite of this, the derivative propagation method and the trajectory stability method are able to accurately predict barrier tunneling. Is it possible to expand the horizon around each trajectory so that longer-range correlations can be built into the dynamics?

Mixed quantum–classical dynamics. New hydrodynamic methods for linking quantum and classical subsystems have been developed. In the phase space approach, the analysis begins by computing partial momentum moments of the phase space distribution function. In the configuration space approach, approximations are made in the hydrodynamic trajectory equations for the composite system. In the equations developed via the phase apace route, the quantum system may be in either a pure state or a statistical mixture. Although only a few model problems have been studied, these methods have great potential for application to interesting physical problems.

Electronic nonadiabatic dynamics. The hydrodynamic approach to electronic transitions involves ensembles of quantum trajectories propagating on each of the coupled potential energy surfaces. Unlike trajectory surface hopping, the amplitude and phase change smoothly and continuously on each of these potential surfaces. The trajectories are influenced by classical and quantum forces as well as transition forces, which depend on the coupling matrix elements linking the surfaces. Further analysis of the trajectory dynamics and additional applications will be significant.

Dissipative quantum systems. Phase space trajectories for these open systems evolve under the influence of both local and nonlocal (density-dependent) forces. The latter forces bring in quantum, thermal, and frictional effects. Adaptive grids and new methods for handling ripples and nodes (described in Section 15.3) will find applications in evolving ensembles of phase space trajectories. Further analysis is needed to devise and test methods for populating regions where the quantum distribution function has negative basins. Multidimensional applications of the methods described in Chapter 11 will be very informative.

We have seen in the preceding chapters that two main impediments to the application of quantum trajectory algorithms concern derivative evaluation on unstructured grids and trajectory instability near nodal regions. In Section 15.2, an effective and accurate procedure (the ALE method) for dealing with the derivative evaluation problem will be reviewed. Rather than reviewing approaches to the node problem from the preceding chapters (such as hybrid methods and use of an artificial viscosity), Section 15.3 introduces three methods for dealing with this problem. The first of these, *expansion of the wave function in counterpropagating waves* [15.1, 15.2], is illustrated with an example in Section 15.4. The second approach, the *covering function method* [15.3, 15.4], is illustrated in Section 15.5. The third approach, the *complex amplitude method* [2.32], is described near the end of Section 15.3. Brief remarks about future developments follow in Section 15.6.

15.2 Coping with the Spatial Derivative Problem

In order to propagate quantum trajectories, spatial derivatives of the hydrodynamic fields are needed at the positions of the moving fluid elements. Difficulties with derivative evaluation are exacerbated when there are regions of inflation and compression, which lead to undersampling and oversampling problems. When this happens, there is an undersupply of grid points in some regions and an oversupply in other regions. Chapter 5 dealt with the evaluation of spatial derivatives on unstructured grids. There are several ways of dealing with this problem, including use of the moving least squares and dynamic least squares algorithms (see Sections 5.2 and 5.3). Although these algorithms are very successful when the fields are relatively smooth, difficulties frequently arise when the density develops multiple ripples and nodes. To make matters worse, inflation and compression occur concurrently with the formation of ripples and nodes.

Fortunately, there is a way to avoid having to deal with unstructured grids. In Chapter 7, the ALE method for developing *adaptive moving grids* was described. Rather than using Lagrangian grids with points that move at the flow velocity of the probability fluid, the grid point velocities may be defined to produce *moving structured grids* with *uniform grid point spacings*. In order to do this, the

Lagrangian version of the quantum hydrodynamic equations of motion (QHEM) that was introduced in Chapter 2 is augmented with *additional dynamical terms,* and these *moving path transforms of the hydrodynamic equations of motion* are integrated. This technique completely eliminates problems connected with inflation and compression. In order to permit overall expansion or contraction of the grid, it is still convenient to allow the edge points to follow Lagrangian trajectories (other choices are also possible). However, at each time step, the internal grid points can be forced to be equally spaced. Spatial derivatives may then be accurately evaluated using any of the standard methods that have been developed for uniform grids, including high-order finite difference and pseuodospectral techniques. Use of the ALE method provides a straightforward solution to the unstructured grid and derivative evaluation problem, but this method alone will not solve problems connected with singularities in the quantum potential in nodal regions. Methods for dealing with this problem will be described in the next section.

15.3 Coping with the Node Problem

Is it possible to *completely circumvent* the node problem? It may be surprising to realize that the answer to this question is *always yes*, at least in principle! We will first describe three methods for handling the node problem for nonstationary states, and then a related procedure will be briefly described for dealing with stationary states. In order to introduce the first two methods, let us assume at time t that the time-dependent hydrodynamic fields $R(x, t)$ and $C(x, t)$ show evidence of node formation in some region, possibly on the reflected wave packet side of a barrier scattering problem. If we were to advance further in time, some of these prenodes would likely develop into full-fledged nodes. But at this time, a monitor function (such as the quantum potential) warns us that trouble is brewing, so now is the time to do something. *In order to avoid the nodes that form at later times, we will take advantage of the superposition principle.* The wave function at time t is decomposed into components that are node-free. Each component may be propagated without difficulty in the hydrodynamic formulation, at least over some time interval. Two methods for doing this will be described in the first part of this section, and examples of these decompositions will be demonstrated in the following two sections. The third approach to the node problem for nonstationary states will then be described in the last part of this section.

The first approach, which we will term the *counterpropagating wave method* (CPWM), is based on the analysis of Poirier [15.1, 15.2]. The nonstationary wave function is decomposed into a *carrier wave* multiplied by a real-valued prefactor that *unlike* the Madelung–Bohm amplitude $R(x, t)$, *can take on negative values.* The Poirier decomposition is then

$$\psi(x, t) = P(x, t)e^{iS_0(x,t)/\hbar}, \tag{15.1}$$

in which the carrier momentum is $p_0 = \partial S_0/\partial x$. It is required that the carrier action function be smooth and *not jump* by $\pm\pi\hbar$ when passing through a node.

The real-valued prefactor in turn is decomposed into *counterpropagating node-free components*

$$\begin{aligned} P(x,t) &= q(x,t)\cos\left[\delta(x,t)/\hbar\right], \\ &= \frac{1}{2}\left[q(x,t)e^{i\delta(x,t)/\hbar} + q(x,t)e^{-i\delta(x,t)/\hbar}\right] \\ &= \phi^{(+)}(x,t) + \phi^{(-)}(x,t), \end{aligned} \tag{15.2}$$

in which the amplitude function $q(x)$ is smooth, node-free, and nonnegative, $q(x) > 0$. It is clear that if any nodes develop in the amplitude $P(x,t)$, they must do so because of the oscillating cosine factor. Using this decomposition, the overall wave function may be written

$$\psi(x,t) = \frac{1}{2}\left[q(x,t)e^{iS_{+}(x,t)/\hbar} + q(x,t)e^{iS_{-}(x,t)/\hbar}\right] = \psi^{(+)}(x,t) + \psi^{(-)}(x,t). \tag{15.3}$$

The momenta associated with the two components $\psi^{(\pm)}$ are given by

$$p_{\pm}(x,t) = p_0(x,t) \pm p_{\delta}(x,t), \tag{15.4}$$

in which the momentum shift is given by $p_\delta = \partial\delta/\partial x$. Relative to an observer riding the carrier wave $\exp(iS_0/\hbar)$ with the "absolute" speed $v_0 = p_0/m$, the counterpropagating components move at speeds given by $v_+ = p_\delta/m$ and $v_- = -p_\delta/m$. The counterpropagating components are analogous to passengers walking in opposite directions in the aisle of an airplane that is heading toward Austin at the "carrier speed".

Each of the two components $\psi^{(+)}$ and $\psi^{(-)}$ is node-free. However, nodes can form in ψ at some points when these two components are added. For example, at some point x_0, if the phase takes on the value given by $\delta = \pm n\pi/2$ ($n = 1, 3, \ldots$), then the superposition, and the wave function itself, will have a node. The usefulness of this approach depends on developing procedures for finding the amplitude q as well as the two phase functions δ and S_0. Procedures for finding these functions have been reported [15.1]. For a model complex-valued nonstationary wave function, an example of this decomposition is given later in Section 15.4. For stationary states, where this type of decomposition has been performed for excited states with many nodes [15.2], additional detail is provided in Box 15.1.

The second approach for coping with the node problem, termed the *covering function method* (CFM), was developed by Babyuk and Wyatt [15.3, 15.4]. The superposition principle and two-term wave functions are again involved, but instead of decomposing $\psi(x,t)$, we choose to build a new function (called the "total" function) by *augmenting* the "actual" function with a covering function

$$\psi_T(x,t) = \psi(x,t) + \psi_C(x,t). \tag{15.5}$$

The purpose of the covering function is literally, to "cover up" the nodes in ψ. The total function may have ripples in the region where the actual function had nodes, but by design, it is guaranteed to be free of nodes (at least at time t).

Box 15.1. Counterpropagating wave method for stationary states

The CPWM can also be applied to real-valued stationary state wave functions. We will follow Poirier [15.2] and decompose the wave function (with notation that is slightly different from this reference) into a superposition of two linearly independent terms, which are complex conjugates of each other,

$$\psi(x) = e^{i\delta} q(x) e^{is(x)/\hbar} + e^{-i\delta} q(x) e^{-is(x)/\hbar} = \psi^{(+)}(x) + \psi^{(-)}(x), \tag{1}$$

in which δ is a constant (also see equations 14.9 and 14.10). In this *bipolar ansatz*, the standing wave on the left side is written as a superposition of two *counterpropagating traveling waves*. The amplitude $q(x)$ is smooth, node-free, and positive, no matter how many nodes are present in $\psi(x)$.

A standard calculation of the flux associated with each component $\psi^{(\pm)}(x)$ gives

$$J^{(\pm)} = \pm q^2(x) \cdot s'(x)/m. \tag{2}$$

Since each of the two components is stationary, the flux must be constant (independent of position), with equal and opposite values for the two components, $J^{(\pm)} = \pm F$. Poirier justifiably calls this the *invariant flux property*. Given a positive value for F, equation 2 may be solved for the amplitude $q(x) = \sqrt{mF/s'(x)}$. In order that $q(x)$ be real-valued and also satisfy the condition $q(x) > 0$ at all points, this equation requires that $s(x)$ be monotonically increasing. This in turn shows that $\psi^{(+)}$ propagates to the right. (Choosing appropriate values for δ and F is discussed in [15.2].)

The momentum associated with each propagating component is given by $p^{(\pm)}(x) = \pm s'(x)$. When $p^{(+)}(x)$ and $p^{(-)}(x)$ are plotted versus x, we obtain two sheets (or branches) of the *Lagrangian manifold* (LM) in the phase space having coordinates $\{x, p\}$. (An LM in this phase space is defined as the one-dimensional curve defined by $\{x, p(x)\}$.) These two sheets, each of which extends to $x = \pm\infty$, form the LM. The quantum momentum $p^{(\pm)}(x)$ satisfies the quantum stationary Hamilton–Jacobi equation given earlier, in equation 14.12.

By analogy with *semiclassical* LMs and trajectories, the following are required properties of *quantum* LMs and the associated *quantum trajectories* [15.2]: (1) the LM itself does not change in time; (2) trajectory flow on either sheet maintains invariant flux with respect to x; (3) the trajectory flow is *along* the LM and is equal and opposite on the two sheets. In addition (rather than the half-odd integer quantization rule used in semiclassical mechanics), energy quantization requires that the area enclosed between the two sheets be given by

$$\oint p(x)dx = 2\pi\hbar(n+1). \tag{3}$$

In this equation, the integral is evaluated for a complete circuit around the LM, and the quantum number n (the number of nodes) takes on the values, $n = 0, 1, 2, \ldots$. (This integer quantization rule was used in Section 14.4 with regard to Floydian trajectories.) In order to enclose a finite phase space area, the momentum must satisfy the property $s'(x) \to 0$ as $|x| \to \pm\infty$. However, this requires that $q(x) \to \infty$ as $|x| \to \pm\infty$. As a consequence, each propagating component $\psi^{(\pm)}(x)$ in equation 1 is non-L^2 and diverges at large $\pm x$. However, when added together, these two linearly independent components properly form an L^2 normalizable total wave function.

Poirier has given examples of "eye-shaped" quantum Lagrangian manifolds for the ground and several excited states of the harmonic oscillator as well as for an excited state of the Morse oscillator, and he has compared these with the usual semiclassical Lagrangian manifolds [15.2]. The quantum manifolds always lie outside of the semiclassical ones and enclose additional phase space area. (The semiclassical $p(x)$ is nonzero only within the classical turning points.) By choosing the value for F that is suggested by semiclassical mechanics (this value is the inverse of the period, $F = \omega/(2\pi)$, where ω is the angular frequency for the semiclassical trajectory), the quantum and semiclassical trajectories are practically identical in the classically allowed region. Differences arise near the classical turning points and in the tunneling regions, where the quantum trajectories have infinitely long tails extending toward $|x| = \infty$. These quantum LMs are well behaved everywhere, including near classical turning points and nodes in the wave function. In addition, the correspondence principle is satisfied because the quantum and semiclassical trajectories become increasingly similar as the energy increases, except for tunneling tails in the quantum manifolds.

The bipolar representation, similar to that given in equation 1, is commonly used in semiclassical mechanics, and this general form was also used by both Floyd and Faraggi and Matone in the studies described in Chapter 14. It is especially significant that Poirier has reconciled semiclassical mechanics with Bohmian mechanics, and has developed a unified approach that combines "the best of both worlds". There are many opportunities for further developments that combine concepts used in semiclassical and Bohmian mechanics. The work by Poirier represents a significant step in making these connections.

The covering function method proceeds as follows. When prenodes develop and the monitor function signals impending problems, this superposition is formed. Then, the functions ψ_T and ψ_C are propagated in the hydrodynamic representation for the additional time T, and the actual function is recovered,

$$\psi(x, t + T) = \psi_T(x, t + T) - \psi_C(x, t + T). \tag{15.6}$$

The function $\psi(x, t + T)$ is then examined for nodes and prenodes by testing the quantum potential to see whether it is above the threshold value. If there are no indications of problems, then ψ is propagated further in the hydrodynamic representation. However, if there are still problems with $\psi(x, t + T)$, then this function is covered again, "re-covered", using a new covering function. This sequence is then repeated as many times as necessary to advance the wave function to the target time.

In equation 15.6, we skipped over some details. If we express each component wave function in polar form, then the superposition can be written

$$R_T(x,t)e^{iS_T(x,t)/\hbar} = R(x,t)e^{iS(x,t)/\hbar} + R_C(x,t)e^{iS_C(x,t)/\hbar}. \tag{15.7}$$

It is clear from this expression that we must choose both the amplitude and phase of the covering function. Choosing the amplitude is relatively easy: it should extend over the region where the nodes or prenodes are forming.

How should we choose the phase of the covering function, $S_C(x,t)$? Some guidance is provided by writing the probability density of the total wave function in terms of the components,

$$R_T(x,t)^2 = R(x,t)^2 + R_C(x,t)^2 + 2R(x,t)R_C(x,t)\cos\left[S(x,t) - S_C(x,t)\right]. \tag{15.8}$$

In addition, the phase of the total function is given by

$$\tan(S_T) = \frac{R\sin(S) + R_c\sin(S_c)}{R\cos(S) + R_c\cos(S_c)}. \tag{15.9}$$

Equation 15.8 reveals that a sufficient, but not necessary, condition that the total density be node-free is that the phase of the covering function satisfy the condition

$$|S - S_c| \le \pi/2 + 2\pi N. \tag{15.10}$$

If this condition is satisfied, then any nodes in R will be "covered up". It appears that the simplest choice for the phase of the covering function is $S = S_c$, but in practice this may lead to some computational difficulties [15.3]. Another choice is to smooth the actual phase S so as to eliminate small undulations and oscillations. An example in which the latter choice was successful will be provided in Section 15.5.

Regarding the amplitude of the covering function, it is arbitrary as long as it is node-free. In multidimensional problems, it can even be constant ("flat covering") along one or more coordinates. However, there is no need to cover the actual wave function everywhere. A reasonable covering function should extend only over the region where node formation is expected (i.e., when the monitor function indicates impending trouble). This restricted covering function reduces the computational time compared with use of a global function. Usually, the covering function is chosen as a Gaussian localized in the nodal region. Outside of this covering zone, the total function then coincides with the actual one.

The ideal situation would be for a *single covering function* to permit propagation of both functions until nodes vanish in the actual wave function. The less than ideal solution to this problem is to repeat the covering process when necessary. When the total or covering functions start to form nodes (as detected by a monitor function), the actual function should be formed at this time, and a new covering function should be used to start the next cycle. This process may then be repeated as many times as necessary until the desired target time is reached. Another problem that may arise is that the total function, which does not have nodes, may develop "hard" density ripples, which yield stiff equations of motion. In this case, accurate time integration can be achieved if an implicit time integrator (such as the implicit Euler method) is used.

The third way of dealing with the node problem, which will be termed the *complex amplitude method* (CAM), was suggested by Garashchuk and Rassolov (GR) [2.32]. (The form of the wave function used in this method was briefly mentioned in Box 2.1.) In this approach, the following non-Madelung form is used for the wave function,

$$\psi(x,t) = r(x,t)\chi(x,t)e^{is(x,t)/\hbar}, \tag{15.11}$$

in which the *complex-valued function* $\chi(x,t)$ is chosen so that the real-valued amplitude $r(x,t)$ and the action $s(x,t)$ are *smooth functions*. The amplitude $r(x,t)$ describes the smooth envelope, and $\chi(x,t)$ builds in local features, including nodes. When this form for the wave function is substituted into the TDSE, the following Eulerian versions of the Hamilton–Jacobi and continuity equations are obtained:

$$-\frac{\partial s}{\partial t} = \frac{1}{2m}\left(\frac{\partial s}{\partial x}\right)^2 + V - \frac{\hbar^2}{2m}\frac{1}{r}\frac{\partial^2 r}{\partial x^2} - \hbar \mathrm{Im}\,[W/\chi], \tag{15.12}$$

$$\frac{\partial r}{\partial t} = -\frac{\partial}{\partial x}\left[r^2\frac{1}{m}\frac{\partial s}{\partial x}\right] - r\,Re\,[W/\chi], \tag{15.13}$$

in which $Re[..]$ and Im[..] denote the real and imaginary parts of the function W/χ. In these two equations, the terms dependent on spatial and temporal derivatives of $\chi(x,t)$ have been collected in the function $W(x,t)$, which is given by

$$W = \frac{\partial \chi}{\partial t} + \frac{1}{m}\frac{\partial r}{\partial x}\frac{\partial \chi}{\partial x} - \frac{i\hbar}{m}\left[\frac{1}{2}\frac{\partial^2 \chi}{\partial x^2} + \frac{\partial \ln(r)}{\partial x}\frac{\partial \chi}{\partial x}\right]. \tag{15.14}$$

By design, the *partial quantum potential* in equation 15.12 (the third term on the right) is smooth and non-singular, even in nodal regions.

How should the function $\chi(x,t)$ be chosen? GR suggest setting $W(x,t)$ to zero, $W(x,t) = 0$, which has the nice feature of eliminating the final term in equations 15.12 and 15.13. This condition then leads to the following differential equation for $\chi(x,t)$:

$$\frac{\partial \chi}{\partial t} = -\frac{1}{m}\frac{\partial r}{\partial x}\frac{\partial \chi}{\partial x} + \frac{i\hbar}{m}\left[\frac{1}{2}\frac{\partial^2 \chi}{\partial x^2} + \frac{\partial \ln(r)}{\partial x}\frac{\partial \chi}{\partial x}\right]. \tag{15.15}$$

In addition, the non-Lagrangian quantum trajectories are determined from the smooth action function $p(x,t) = \partial s(x,t)/\partial x$.

In applications to the propagation of various initial states for the harmonic oscillator and for a Gaussian initial wave packet scattering from an Eckart barrier, GR used a linear approximation for the complex-valued function $\chi(x,t)$, $\chi(x,t) = a_0(t) + a_1(t)x$ (the two coefficients are complex-valued), which leads to a simpler differential equation by eliminating the first term within the square brackets in equation 15.15.

In one application of this method [2.32], a superposition of the two lowest harmonic oscillator functions was used as the initial wave function. The smooth trajectories generated using the CAM described the overall expansion and contraction of the wave packet in the harmonic potential. The amplitude function $\chi(x, t)$ accurately described the local internal structure, including node formation. Additional applications of this method will be informative.

15.4 Decomposition of Wave Function into Counterpropagating Waves

In order to illustrate the decomposition of a complex-valued wave function into counterpropagating waves, we will consider a wave function that has a single node. This is an "artificial" wave function, constructed for the sole purpose of demonstrating the decomposition. By design, the node is located at $x = 0$. In the Madelung–Bohm unipolar form, this wave function is given, as usual, by the function $\psi(x) = R(x)\exp(iS(x)/\hbar)$, with $R(x) = |A(x)|$ (atomic units will be used for this example), where the amplitude and phase are given by

$$\begin{aligned}
A(x) &= \exp\left[-9(x+0.5)^2\right] - \exp\left[-9(x-0.5)^2\right],\\
S(x) &= S_1(x) = \frac{5}{4}\pi + c_1\left(\frac{\pi}{2} + a\tan\left[3(x+0.7)\right]\right), \quad x \le 0,\\
S(x) &= S_2(x) = S_1(x=0) - \pi + c_2\left(\frac{\pi}{2} + a\tan\left[3(x-0.7)\right]\right), \quad x > 0,\\
c_1 &= \frac{3}{4} - \frac{1}{9}, \qquad c_1 = \frac{3}{4} + \frac{1}{9}.
\end{aligned} \tag{15.16}$$

These amplitude and phase functions are shown in figure 15.1. The amplitude $R(x)$ plotted in part (a) has two lobes centered at $x = \pm 0.5$. The phase shown in part (b) decreases by π as x passes through the origin, although the phase within each of the two lobes $S_1(x)$ and $S_2(x)$ is monotonically increasing. When this wave function is plotted in the complex plane, as shown in figure 15.2, these two lobes are evident. The arrows in this figure point in the direction of increasing x, with lobe 1 ($x < 0$) mostly lying in quadrant 4 (lower right), while lobe 2 ($x > 0$) is concentrated in quadrant 2 (upper left). This input wave function will now be decomposed into counterpropagating components.

In order to decompose this wave function into counterpropagating components, it is necessary to choose three functions, the two phases $\delta(x)$ and $S_0(x)$, and the amplitude $q(x)$. The phase $\delta(x)$ is easily chosen because this wave function has a single node at $x = 0$, and this requires that $\delta(0) = \pi/2$, provided that $\delta(x)$ is constrained to fall in the interval $[0, \pi]$. An acceptable phase function is given by

$$\delta(x) = \pi/2 + a\tan(2x). \tag{15.17}$$

The monotonic carrier phase $S_0(x)$ is chosen to be equal to $S_1(x)$ when $x \le 0$ and to be equal to $S_2(x) + \pi$ when $x > 0$. Adding π to S_2 is necessary to eliminate the phase jump in $S(x)$ at $x = 0$. This carrier phase and the phases $S_\pm(x) = S_0(x) \pm \delta(x)$ are plotted in figure 15.3. The amplitude $q(x)$, a positive

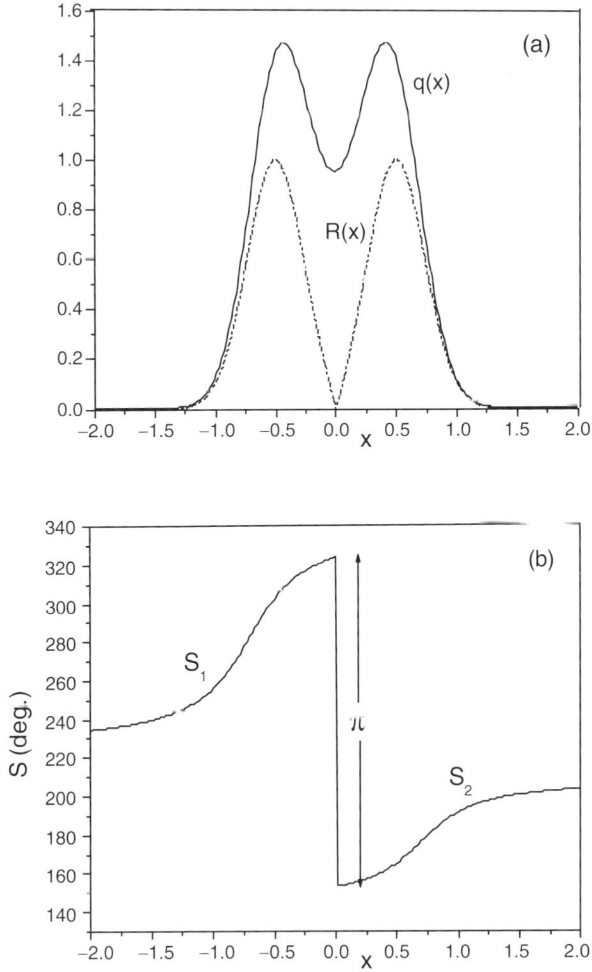

FIGURE 15.1. The amplitude $R(x)$ is plotted (dashed curve) in part (a), and the phase $S(x)$ (in degrees) is plotted in part (b), for the wave function given in equation 15.16. The phase has two branches, denoted by $S_1(x)$ for $x \leq 0$ and $S_2(x)$ for $x > 0$. The non-negative amplitude $q(x)$is also plotted in part (a).

function, is determined by the equation

$$q(x) = \frac{R(x)}{|\cos\delta(x)|}. \tag{15.18}$$

This function is also plotted in figure 15.1 (a). With this choice for $q(x)$, the function $P(x)$ is given by

$$P(x) = R(x)\frac{\cos\delta(x)}{|\cos\delta(x)|}. \tag{15.19}$$

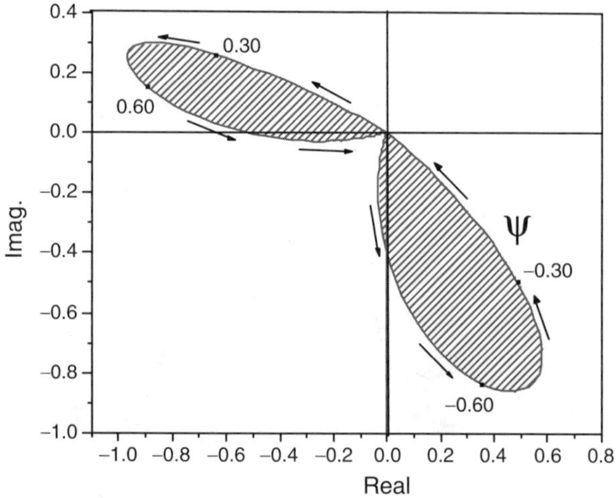

FIGURE 15.2. The wave function given in equation 15.16, plotted in the complex plane, $\mathrm{Im}[\psi(x)]$ versus $Re[\psi(x)]$. The arrows show the direction of increasing x (within each lobe, we move in the counter-clockwise sense as x increases). As $x \to \pm\infty$ and at the origin (where x also equals zero), $\psi = 0$ (at large $\pm x$, the wave function takes on values near the origin). Four x values are indicated by dots on the curve representing the wave function.

With these choices for the three functions appearing in equations 15.1–15.3, the decomposition of the wave function is completely defined. The two counterpropagating functions $\phi^{(+)}(x)$ and $\phi^{(-)}(x)$ are plotted in figure 15.4. The direction of increasing x is again shown by the arrows in this figure. Because these functions are complex conjugates, the two "heart-shaped" curves are symmetric with respect

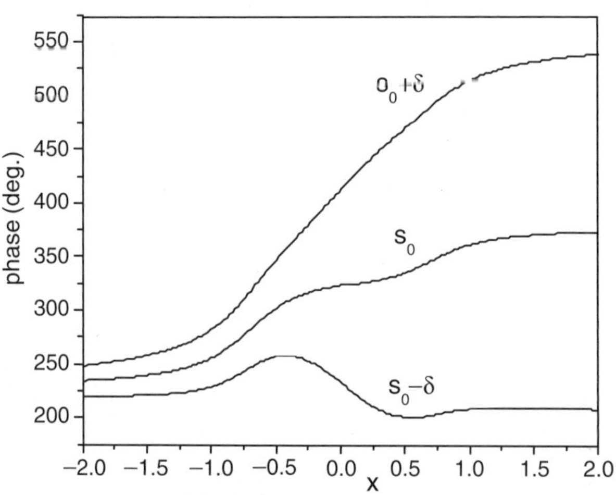

FIGURE 15.3. Carrier phase $S_0(x)$ and the two phases $S_0 \pm \delta$ (all phases are in degrees).

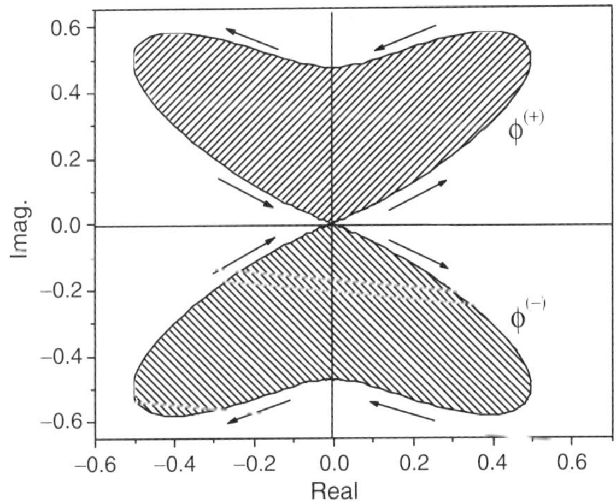

FIGURE 15.4. Counterpropagating functions $\phi^{(+)}(x)$ and $\phi^{(-)}(x)$, plotted in the complex plane, Im[$\phi(x)$] versus Re[$\phi(x)$]. For each lobe, the direction of increasing x is shown by the arrows.

to reflection in the real axis. When each of these functions is multiplied by the carrier wave $\exp(iS_0/\hbar)$, the counterpropagating waves $\psi^{(+)}(x)$ and $\psi^{(-)}(x)$ are formed. These two functions, plotted in figure 15.5, are not symmetric with respect to any axis. Several values of the x-coordinate are marked on each lobe of this

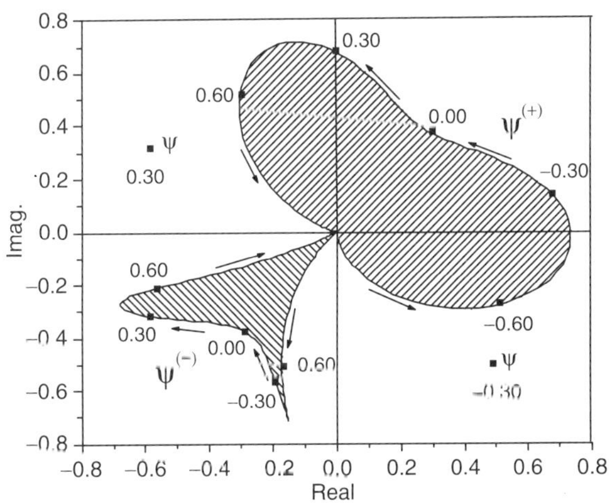

FIGURE 15.5. Counterpropagating wave functions $\psi^{(+)}(x)$ and $\psi^{(-)}(x)$, plotted in the complex plane, Im[$\psi(x)$] versus Re[$\psi(x)$]. For each function, the direction of increasing x is shown by the arrows. Several x values are shown by the dots on each lobe. When $x = \pm 0.3$, the two counterpropagating waves add to give the wave function values shown by the dots in the upper left and lower right of this figure. These two dots are at the positions of the corresponding dots shown earlier in figure 15.2.

figure. When the functions $\psi^{(+)}(x=-0.3)$ and $\psi^{(-)}(x=-0.3)$ are added, the point in the lower right of the figure is obtained. These two counterpropagating functions correctly add to give the wave function shown earlier, in figure 15.2.

The decomposition of the complex-valued wave function in figure 15.2 was carried out "geometrically". Clearly, it would be advantageous to follow a more formal approach, but this example shows that the decomposition can still be performed without resorting to such an approach.

15.5 Applications of the Covering Function Method

In this section, the CFM will be applied [15.3] to the scattering of an initial Gaussian wave packet from an Eckart barrier, $V = V_0 \sec h^2(\alpha\, x)$. The initial wave packet is given by

$$\psi(x,0) = (2\beta/\pi)^{1/4} \exp\left[ik(x-x_0)\right] \exp\left[-\beta\,(x-x_0)^2\right]. \tag{15.20}$$

The translational energy of the wave packet and the barrier height were both set to 3000 cm^{-1}, and the other parameters (in atomic units) have the values $m = 2000$, $\alpha = 1.2$, $\beta = 6.0$, and $x_0 = -6$. The initial wave packet, located far to the "left" of the barrier, was launched toward it. The initial packet was discretized with $N = 40$ grid points distributed in the interval $[-7.2, -5.0]$. As time progresses, the packet spreads and the spacing between the grid points increases. Every time that the grid point spacing exceeds a threshold value, chosen to be $\Delta x_0 = 0.075$, the number of points was increased by 50%.

At later times, multiple nodes and ripples form in the reflected wave packet. Node formation was indicated when a monitor function, the quantum potential Q, exceeded a threshold value, which was usually set in the range 0.010 to 0.025. However, the first stage of wave function propagation (without the covering function) was terminated at the time $t = 2000$, when the quantum potential was only 0.008. At time $t = 2000$, the first covering was applied through use of the Gaussian function

$$R_c = A \exp\left[-\gamma\,(x-x_c)^2\right], \tag{15.21}$$

in which the following parameter values were used: $A = 3$, $\gamma = 0.15$, $x_c = -2$. Regarding the phase of the covering function, the use of $S_c = S$ turned out to be suboptimal because the total and covering functions developed multiple ripples after a short propagation time. These ripples were completely resolved when the covering phase S_c was created by a slight smoothing of the actual phase S,

$$S_c(x_i) = \sum_j K(x_i - x_j) S(x_j) \Big/ \sum_j K(x_i - x_j), \tag{15.22}$$

where $K(x - x_i)$denotes a Gaussian kernel centered at the point x_j. The covering phase chosen in this manner still satisfies the condition given by equation 15.10.

Given the amplitude and phase for both the actual wave function and the covering function, these quantities for the total function were derived according to equations 15.8 and 15.9. However, the latter equation yields a total phase S_T that is a piecewise function with values in the interval $[-\pi, \pi]$. Since this piecewise continuous function cannot be used to initiate time propagation with the hydrodynamic equations, it was converted into a continuous function by adding or subtracting 2π whenever it experienced a sudden jump between a pair of neighboring points.

The C amplitudes and probability densities for the wave function ψ and the covering and total functions at $t = 2000$ are shown in figure 15.6. The covering function is a Gaussian centered at $x_c = -2$. Note the multiple ripples in the C-amplitude for the actual wave function. If propagated in the hydrodynamic picture *without* use of the covering function, node formation in ψ would quickly destroy the calculation.

Integration of the equations of motion for the total and covering functions was carried out simultaneously because both functions are needed at the same grid points in order to recover the actual wave function at later times. In order to do this, the amplitude and phase for the total function were first advanced for one time step. Then the velocities of the grid points for the total function were ascribed to the grid points of the covering function. As a result, new values of the total and covering functions may be defined at the same coordinates.

Following implementation of the CFM at $t = 2000$, the total and covering functions were propagated for an additional time interval, $\Delta t = 600$. Propagation with the covering function was terminated at this time because a node formed in the total function at later times. At time $t = 2600$, the actual wave function was formed by subtraction of the covering function from the total function. A piecewise phase was obtained for the actual function, but this was converted to a continuous one using the method described earlier. At this time step, figure 15.7 shows the C-amplitudes for both the total and actual functions. As seen from this figure, the actual function is not well bifurcated in the barrier region near $x = 0$ (this occurs at later times, near $t = 3600$), so that a new covering function was employed in order to continue propagation to later times. The propagation was split into two additional covering stages, with stage 2 going from $t = 2600$ to $t = 3100$ and with stage 3 proceeding from $t = 3100$ to $t = 3600$. (For these two stages, the parameters for the covering function were $A = 3$, $\gamma = 0.2$, and $x_c = -2$.) After the latter time, no covering function was required.

Figure 15.8 shows a comparison at $t = 3600$ of the density for the wave packet propagated using the CFM, with the exact density obtained from direct integration of the time-dependent Schrödinger equation. As seen, the transmitted and reflected packets are well bifurcated, and there is excellent agreement between the exact and hydrodynamic densities. Separate propagation of these two packets for much longer times, until $t = 1.2$ ps, still gives excellent results (see figure 4 in [15.3]). However, even this time is not an upper limit, because propagation in the hydrodynamic picture can be extended much further.

In summary, a new method based on the superposition principle was used to tackle the node problem that occurs in the hydrodynamic picture of quantum

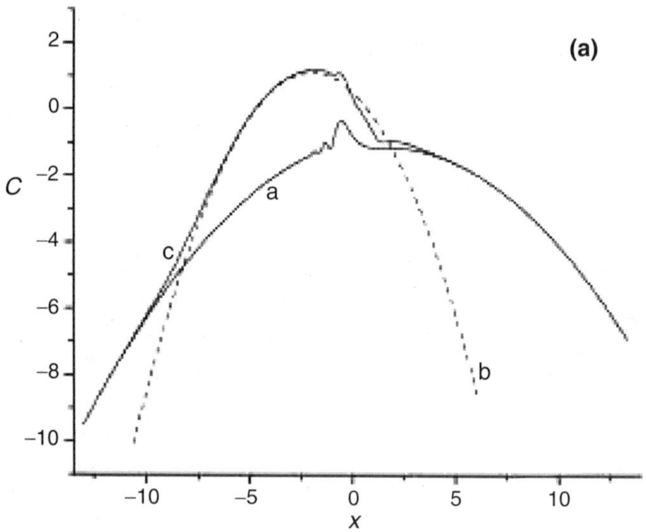

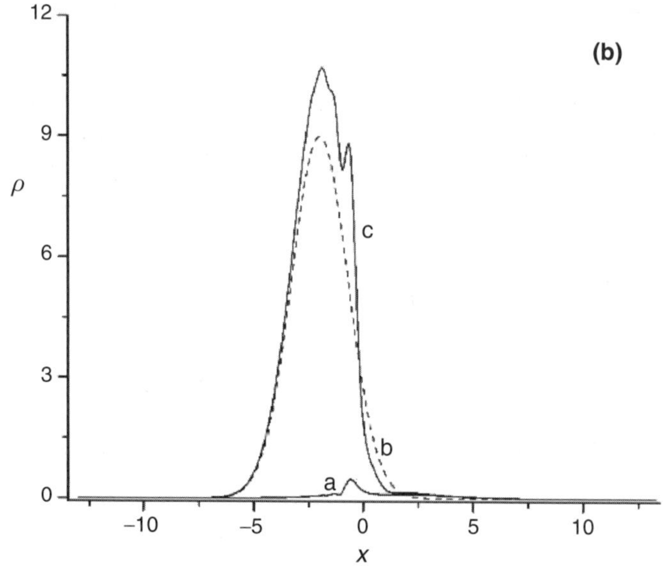

FIGURE 15.6. Curves (a), (b), and (c) denote the actual wave function, the covering function, and the total wave function at $t = 2000$ a.u., respectively. Part (a) shows the C-amplitudes at $t = 2000$ a.u. Part (b) shows the densities for these three functions. The covering function is a Gaussian centered at $x_c = -2$. Note the multiple ripples in the amplitude for the actual wave function.

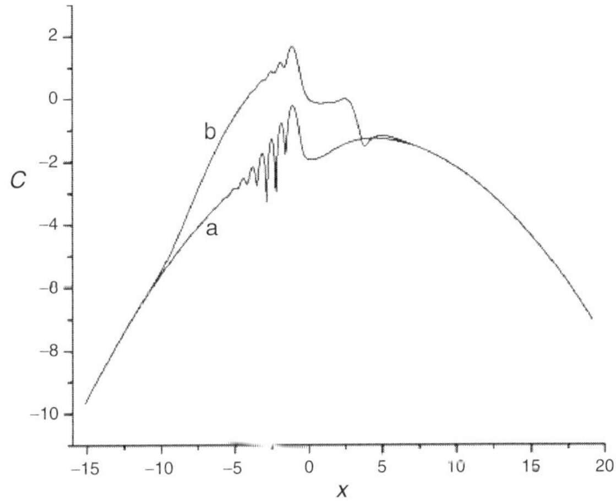

FIGURE 15.7. The C-amplitudes for the actual wave function (a) and the total function (b) [15.3] at time $t = 2600$ a.u. Multiple ripples and nodes are evident in the C-amplitude associated with the actual wave function.

mechanics. This problem has inhibited previous applications of hydrodynamic approaches and until now has been circumvented by means of approximate or hybrid methods. The covering function method introduced in this study is designed for application in the hydrodynamic picture even though it is based on the superposition principle in the wave function picture. This method provides a new tool for coping with the node problem, much as the use of ALE grids is a cure for the unstructured

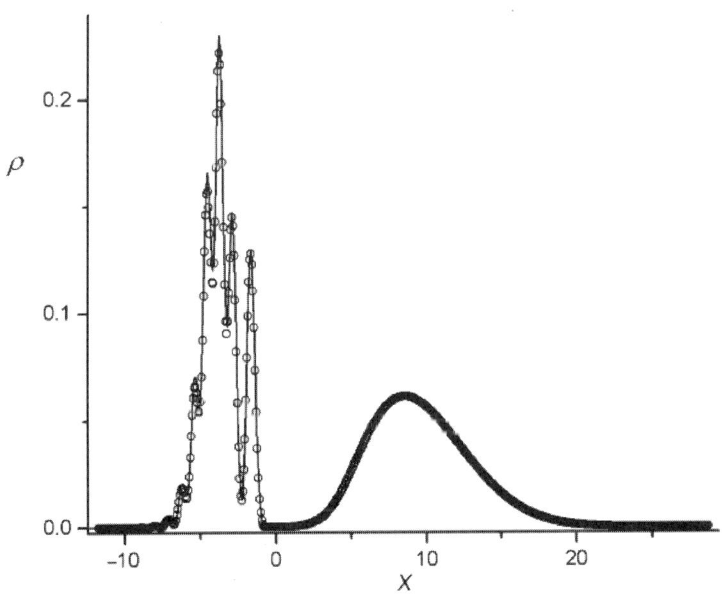

FIGURE 15.8. Densities (at time $t = 3600$ a.u.) computed by integrating the quantum hydrodynamic equations (circles) compared with the exact results (continuous curve) [15.3]. Multiple ripples are evident in the density for the reflected wave packet.

grid problem. Thus, combination of ALE grids and the CFM provides the potential for application of hydrodynamic approaches to multidimensional systems.

To demonstrate the capability of the CFM to deal with the node problem in more than one dimension, as a demonstration, we consider a two-dimensional model problem. The potential energy surface was constructed from an Eckart barrier along the x-coordinate coupled to a harmonic potential along the y-coordinate. The potential energy is given by

$$V(x, y) = V_0 \sec h^2(ax) + (1/2)k(x)y^2, \tag{15.23}$$

in which the variable force constant is given by

$$k(x) = 0.1909\left(1 - 0.15e^{-0.25x^2}\right) \tag{15.24}$$

The parameter values for the Eckart barrier are $V_0 = 8000\ \text{cm}^{-1}$ and $a = 0.4$.

Since the goal is to demonstrate the use of the CFM, this calculation was carried out entirely in the wave function representation, rather than in the hydrodynamic picture. The parameters for the initial wave packet and the potential surface were chosen in such a way that the first node in the reflected packet forms just after the bifurcation of the wave packet into reflected and transmitted components. The covering function was chosen to overlap the reflected packet, since this is where the nodes form. Following introduction of the covering function, the total and covering functions were separately propagated. The initial phase for the covering function was chosen to be constant along the y-coordinate. Along the x-coordinate, the phase was constructed by smoothing the phase of the actual wave function along the slice at $y = 0$ (following preliminary conversion of this phase into a continuous function).

In the top plot of figure 15.9, densities associated with the actual wave packet and the total function are shown in the reflected region to the left of the barrier at the time of application of the covering function, $t = 4464$ a.u. As seen from this figure, the total function is node-free, while the actual function has multiple nodes along the x-coordinate. In the lower plot in this figure, the densities for these two functions are shown after propagation for an additional 720 a.u. The covering function, which is not shown in this figure, remains very smooth for this integration time. If the total and covering functions are node-free, propagation can be easily performed in the hydrodynamic picture. Moreover, at the time corresponding to the lower panel of this figure, the actual function recovered by subtracting the covering function from the total function accurately coincides with the exact wave function. These results indicate that the CFM has the potential for dealing with multidimensional problems.

The capability of the CFM for multidimensional propagation in the hydrodynamic picture was further demonstrated in [15.3], where a vibrationally excited ($n = 5$) wave packet was scattered on a two-dimensional potential surface. The potential surface was identical to the one described above. Since the initial packet had five nodes in the y-direction, it was necessary to cover the initial function. The total and covering functions were then separately propagated (through direct solution of the TDSE), and at a later time the actual wave function was formed

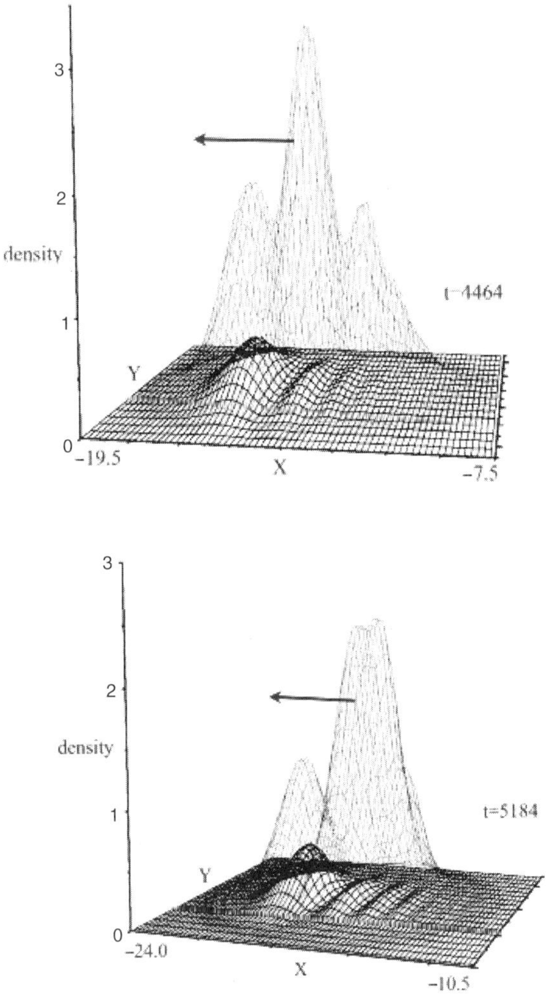

FIGURE 15.9. The densities for the actual wave function and the total function are shown in the top plot, just after application of the covering function (at $t = 4464$ a.u.), and in the lower plot after 720 a.u. of additional propagation time. In each part, the larger "outer" surface is that of the total function. Multiple nodes are evident at each time in the actual wave function. The x-coordinate is along the horizontal direction, and the vibrational coordinate runs perpendicular to this axis. In the top panel, the x and y axes cover the ranges $[-19.5, -7.5]$ and $[-0.5, 0.5]$, respectively. In the lower panel, the corresponding ranges are $[-24.0, -10.5]$ and $[-0.6, 0.6]$. In each figure, only the reflected region is shown, to the left of the barrier region. The barrier is centered at the origin in these coordinates.

by subtracting the covering function from the total function. The density obtained from this actual function was in excellent agreement with the exact result. In addition, through use of the CFM, propagation of wave packets for both ground and excited vibrational states has been accomplished *directly in the hydrodynamic representation* [15.4]. These examples demonstrate that the CFM might be widely applicable to quantum trajectory propagation for systems of greater complexity.

15.6 Quantum Trajectories and the Future

It has only been since 1999 that quantum trajectories have been used as a computational technique for solving the time-dependent Schrödinger equation. Since then, a number of promising methods have been introduced to greatly enhance and extend the basic methodology. In addition, quantum trajectories have been used to formulate new approaches to old problems, including mixed quantum–classical dynamics and dissipative phase space dynamics. A number of applications have been reported, many dealing with one-dimensional scattering problems, a few describing dynamics on model systems involving more than 10 degrees of freedom. From analysis of the trajectory dynamics, new insights have been revealed about fundamental dynamical processes such as barrier tunneling, decoherence, and electronic nonadiabaticity. Over the coming years, it appears that these and yet unforeseen hydrodynamic approaches may be extended to systems of increasing complexity and dimensionality.

References

15.1. B. Poirier, notes and private communication, Dec. 2003–May 2004.

15.2. B. Poirier, *Reconciling semiclassical and Bohmian mechanics: I. Stationary states,* J. Chem. Phys. **121**, 4501 (2004).

15.3. D. Babyuk and R. E. Wyatt, *Coping with the node problem in the hydrodynamic representation of quantum mechanics: The covering function method*, J. Chem. Phys. **121**, 9230 (2004).

15.4. D. Babyuk and R. E. Wyatt, *Application of the covering function method in quantum hydodynamics for two-dimensional scattering problems*, Chem. Phys. Lett., **400**, 145 (2004).

Appendix 1: Atomic Units

Quantity	Atomic Unit	SI Equivalent
mass	mass of electron	9.10938×10^{-31}kg
charge	charge of proton	1.602176×10^{-19}C
angular momentum	$\hbar$	1.05457×10^{-34}J · s
length (bohr)	Bohr orbit radius in H atom	$0.5291772 \times 10^{-10}$m
energy (hartree)	1/2 ionization energy of H atom	4.35974×10^{-18}J
time	period of ground state Bohr orbit in H	2.41888×10^{-17}s
velocity	velocity of electron in lowest Bohr orbit	2.18770×10^{6}m/s

1 hartree = 315,775 K = 219,475 cm^{-1} = 27.2114 eV.

Appendix 2: Example QTM Program

```
---------------------------------------------------------------------
PROGRAM QTM
Implicit none
!--------------------------------------------------------------------
!    This code integrates the POTENTIAL FORM of the quantum
!    hydrodynamic equations of motion (see equations 4.5 through 4.7
!    in Chapter 4). Note that the equation of motion for the action
!    function is integrated, but the quantum force is not used.
!
!    An Eckart potential is used as an example in this code.
!
!--------------------------------------------------------------------
!
!    For time integration, this code uses the first-order Euler
!    time-integrator.
!
!    For spatial derivatives, there are calls to a generic
!    subroutine deriv.
!    This can be one of many function approximation routines
!    (such as moving least squares)
!    Generally, these methods require the following specification
!    and arrays:
!
!                         deriv(np,x,f,d1x,d2x)
!    where...
!    np (input)         the total number of particles (grid points)
!    x   (input)        an array of length np containing the grid
!                       point locations
!    f   (input)        an array of length np containing the
!                       function values at the grid points
!    d1x (output)       an array of length np containing the first
!                       spatial derivative of function f
!    d2x (output)       an array of length np containing the second
!                       spatial derivative of function f
!--------------------------------------------------------------------
```

```
integer(kind=4), parameter :: np = 63, ntime = 150000
! ntime =  total time-steps.
integer(kind=4) i,j,k
real(kind=8) delv(np),c(np),d1x(np),d2x(np),quantum(np),pot(np),
             x(np),rho(np)
real(kind=8) vx(np),phase(np)
real(kind=8) dt,x0,beta,energy,conv,am,vb,xb,wx,xmin,xmax,h,pi,
             anorm,total_density
real(kind=8) kinetic_energy,lagrange

!----------------- Initial Wave Packet Parameters ------------------
!     Time-step in au
      dt = 0.5d0
!     Center of Gaussian function
      x0 = 0.d0
!     Width parameter for Gaussian wave packet
      beta = 9.d0
!     Initial translational energy in cm^-1
      energy = 8000.d0
!     Conversion factor from cm^-1 to a.u.
      conv = 219474.6d0
!     Translational energy in a.u.
      energy = energy/conv
!     Mass in a.u.
      am = 2000.d0

!----------------- Eckart potential parameters ---------------------
!     Barrier height in cm^-1
      vb=8000d.0
!     Barrier height in a.u.
      vb=vb/conv
!     Center of barrier in a.u.
      xb=6.d0
!     Width parameter in a.u.
      wx=5.d0

!----------------- Build a Uniform Initial Spatial Grid ------------
!     Minimum grid point
      xmin = -0.6d0
!     Maximum grid point
      xmax = 0.6d0
!     Particle spacings
      h = (xmax - xmin)/dble(np-1)
!     Particle positions
      do i = 0,np-1
      x(i+1) = xmin + i*h
      enddo
```

```
!--------- Initialize Wave Packet and Hydrodynamic Fields ----------
!     Normalization for an initial Gaussian wave function
      pi = 4.d0*atan(1.d0)
      anorm = (2.d0*beta/pi)**(1.d0/4.d0)

!     Build initial wave packet amplitude and density
      do i =1,np
        R(i) = anorm*exp(-beta*((x(i) - x0)**2))
        c(i) =log(R(i))
        rho(i) = R(i)**2
      enddo

!     Initial particle velocities (all particles are initially the
      same velocity and 0 divergence)
!     These velocities are obtained from the equation
      Etrans = 1/2mv^2.
      do i = 1,np
       vx(i) = sqrt(2.d0*energy/am)
       delv(i) = 0.d0
      enddo

!     Initial phase (action function) for each particle
      do i = 1,np
        phase(i)= sqrt(2.d0*am*energy)*x(i)
      enddo

!-------------------------------------------------------------------
!------------------- Time Propagation Loop ----------------------
do k = 1,ntime
  time = dble(k)*dt
 ! conversion from au to fs
 tfs = time*0.0242d0

!    Call subroutine for spatial derivatives of the C-amplitude
     call deriv(np,x,c,d1x,d2x)

!    Calculate quantum and classical potentials
     do i = 1,np
      ! Quantum Potential.
        quantum(i) = -1.d0/(2.d0*am)*(d2x(i) + d1x(i)**2)
      ! Classical Eckart Potential.
        pot(i) = vb*sech(wx*(x-x0))**2
     enddo

!    Calculate phase (S) using potentials (Quantum Lagrangian: T-(V+Q))
     do i = 1,np
       kinetic_energy = 0.5d0*am*vx(i)**2
       lagrange = kinetic-(pot(i)+quantum(i))
       phase(i) = phase(i) + lagrange*dt
     enddo
```

```
!   Update particle positions
    do i = 1,np
      x(i) = x(i) + vx(i)*dt
    enddo

!   Call subroutine for spatial derivatives of phase
    call deriv(np,x,phase,d1x,d2x)

!   Update velocities and probability density
    do i = 1,np
      vx(i) = (1.d0/am)*d1x(i)
      delv(i) = (1.d0/am)*d2x(i)
      rho(i) = rho(i)*exp(-delv(i)*dt)
    enddo

!----------------- End Time Propagation --------------------------
enddo

end
```

Index

Interdisciplinary Applied Mathematics

1. *Gutzwiller:* Chaos in Classical and Quantum Mechanics
2. *Wiggins:* Chaotic Transport in Dynamical Systems
3. *Joseph/Renardy:* Fundamentals of Two-Fluid Dynamics: Part I: Mathematical Theory and Applications
4. *Joseph/Renardy:* Fundamentals of Two-Fluid Dynamics: Part II: Lubricated Transport, Drops and Miscible Liquids
5. *Seydel:* Practical Bifurcation and Stability Analysis: From Equilibrium to Chaos
6. *Hornung:* Homogenization and Porous Media
7. *Simo/Hughes:* Computational Inelasticity
8. *Keener/Sneyd:* Mathematical Physiology
9. *Han/Reddy:* Plasticity: Mathematical Theory and Numerical Analysis
10. *Sastry:* Nonlinear Systems: Analysis, Stability, and Control
11. *McCarthy:* Geometric Design of Linkages
12. *Winfree:* The Geometry of Biological Time (Second Edition)
13. *Bleistein/Cohen/Stockwell:* Mathematics of Multidimensional Seismic Imaging, Migration, and Inversion
14. *Okubo/Levin:* Diffusion and Ecological Problems: Modern Perspectives (Second Edition)
15. *Logan:* Transport Modeling in Hydrogeochemical Systems
16. *Torquato:* Random Heterogeneous Materials: Microstructure and Macroscopic Properties
17. *Murray:* Mathematical Biology I: An Introduction (Third Edition)
18. *Murray:* Mathematical Biology II: Spatial Models and Biomedical Applications (Third Edition)
19. *Kimmel/Axelrod:* Branching Processes in Biology
20. *Fall/Marland/Wagner/Tyson (Editors):* Computational Cell Biology
21. *Schlick:* Molecular Modeling and Simulation: An Interdisciplinary Guide
22. *Sahimi:* Heterogeneous Materials I: Linear Transport and Optical Properties
23. *Sahimi:* Heterogeneous Materials II: Nonlinear and Breakdown Properties and Atomistic Modeling
24. *Bloch:* Nonholonomic Mechanics and Control
25. *Beuter/Glass/Mackey/Titcombe (Editors):* Nonlinear Dynamics in Physiology and Medicine
26. *Ma/Soatto/Košecká/Sastry:* An Invitation to 3-D Vision: From Images to Geometric Models
27. *Ewens:* Mathematical Population Genetics (Second Edition)
28. *Wyatt:* Quantum Dynamics with Trajectories: Introduction to Quantum Hydrodynamics